Parametric Modeling

with Pro/ENGINEER WildFire 3.0

Randy H. Shih
Oregon Institute of Technology

ISBN: 1-58503-309-X

SDC
PUBLICATIONS

Mission, Kansas

Schroff Development Corporation
P.O. Box 1334
Mission KS 66222
(913) 262-2664
www.schroff.com

Shih, Randy H.
Parametric modeling with Pro/ENGINEER Wildfire 3.0/
Randy H. Shih

ISBN 1-58503-309-X

Examination Copies:

Books received as an "Examination Copy" are for review purposes only and may not be made available for student use. Resale of examination copies is prohibited.

Electronic Files:

Any electronic files associated with this book are licensed to the original user only. These files may not be transferred to any other party.

Printed and bound in the United States of America.

Preface

The primary goal of ***Parametric Modeling with Pro/Engineer wildfire 3.0*** is to introduce the aspects of **Solid Modeling** and **Parametric Modeling**. This text is intended to be used as a training guide for students and professionals. This text covers Pro/Engineer and the lessons proceed in a pedagogical fashion to guide you from constructing basic shapes to building intelligent solid models and creating multi-view drawings. This text takes a hands-on, exercise-intensive approach to all the important *Parametric Modeling* techniques and concepts. This textbook contains a series of ten tutorial style lessons designed to introduce beginning CAD users to Pro/Engineer. This text is also helpful to Pro/Engineer users upgrading from a previous release of the software. The solid modeling techniques and concepts discussed in this text are also applicable to other parametric feature-based CAD packages. The basic premise of this book is that the more designs you create using Pro/Engineer, the better you learn the software. With this in mind, each lesson introduces a new set of commands and concepts, building on previous lessons. This book does not attempt to cover all of the Pro/Engineer's features, only to provide an introduction to the software. It is intended to help you establish a good basis for exploring and growing in the exciting field of **Computer Aided Engineering**.

Acknowledgments

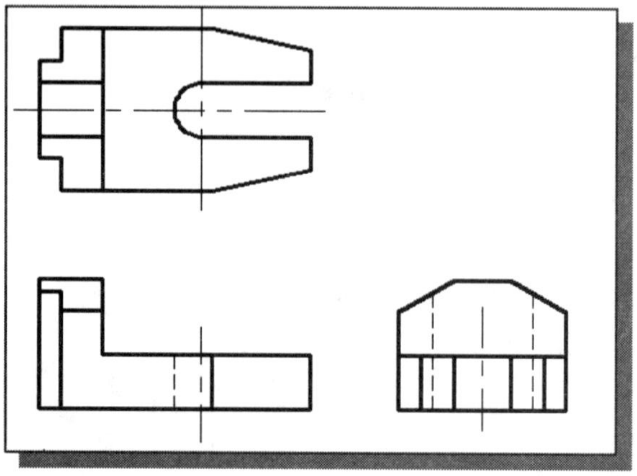

This book would not have been possible without a great deal of support. First, special thanks to two great teachers, Prof. George R. Schade of University of Nebraska-Lincoln and Mr. Denwu Lee, who taught me the fundamentals, the intrigue, and the sheer fun of Computer Aided Engineering.

The effort and support of the editorial and production staff of Schroff Development Corporation is gratefully acknowledged. I would especially like to thank Stephen Schroff and Mary Schmidt for their support and helpful suggestions during this project.

I am grateful that the Mechanical Engineering Technology Department of Oregon Institute of Technology has provided me with an excellent environment in which to pursue my interests in teaching and research.

Finally, truly unbounded thanks are due to my wife Hsiu-Ling and our daughter Casandra for their understanding and encouragement throughout this project.

Randy H. Shih
Klamath Falls, Oregon
Spring, 2006

Table of Contents

Chapter 2
Constructive Solid Geometry Concepts

Chapter 3
Model History Tree

Chapter 4
Parametric Relations and Constraints

Chapter 7
Symmetrical Features in Designs

Chapter 10
Assembly – Putting It All Together

INDEX

Notes:

Introduction

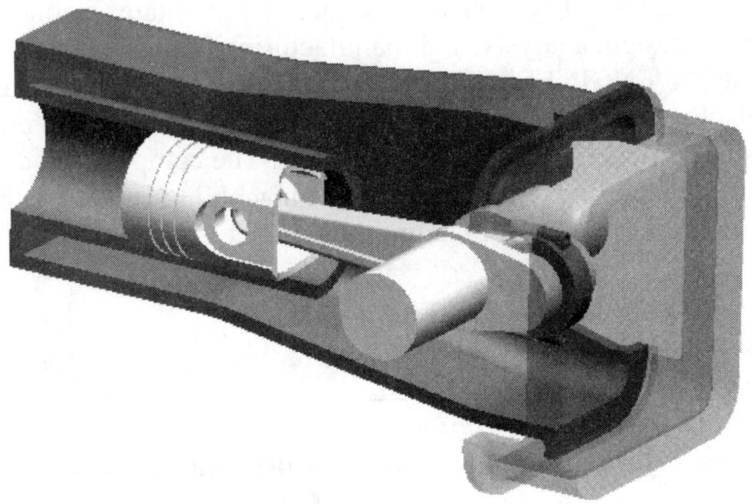

Learning Objectives

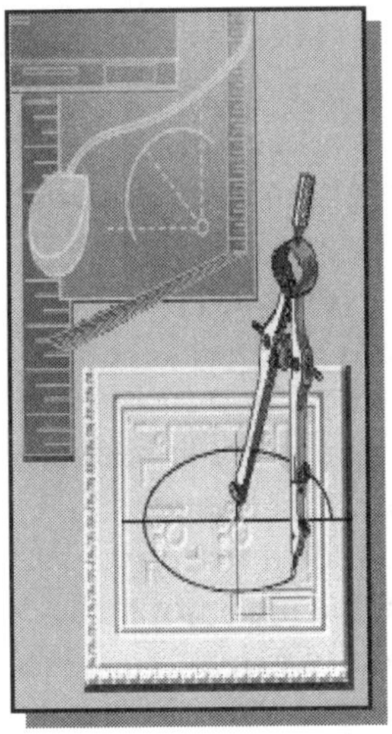

- **Development of Computer Geometric Modeling**
- **Feature-Based Parametric Modeling**
- **Getting Started with Pro/ENGINEER**
- **Startup Options and Units Setup**
- **Pro/ENGINEER Screen Layout**
- **Mouse Buttons**
- **Pro/ENGINEER On-Line Help**

Introduction

The rapid changes in the field of **Computer Aided Engineering** (CAE) have brought exciting advances in the engineering community. Recent advances have made the long-sought goal of **concurrent engineering** closer to a reality. CAE has become the core of concurrent engineering and is aimed at reducing design time, producing prototypes faster, and achieving higher product quality. Pro/ENGINEER is an integrated package of Mechanical Computer Aided Engineering software tools developed by *Parametric Technology Corporation* (**PTC**). Pro/ENGINEER is a tool that facilitates a concurrent engineering approach to the design, analysis, and manufacturing of mechanical engineering products. The Pro/ENGINEER software allows us to quickly create three-dimensional solid models. Real-life loads can be simulated on the computer to predict the behaviors of the designs under specific operating conditions. The computer models can also be used directly by manufacturing equipment such as machining centers, lathes, mills, or rapid prototyping machines to manufacture the product.

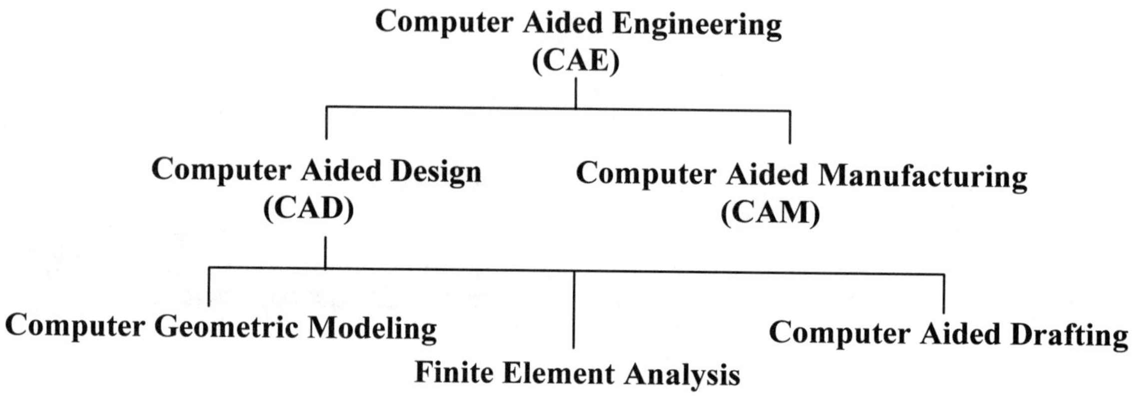

Development of Computer Geometric Modeling

Computer geometric modeling is a relatively new technology and its rapid expansion in the last fifty years is truly amazing. Computer-modeling technology advanced along with the development of computer hardware. The first generation CAD programs, developed in the 1950s, were mostly non-interactive; CAD users were required to create program-codes to generate the desired two-dimensional (2D) geometric shapes. Initially, the development of CAD technology occurred mostly in academic research facilities. The Massachusetts Institute of Technology, Carnegie-Mellon University, and Cambridge University were the lead pioneers at that time. The interest in CAD technology spread quickly and several major industry companies, such as General Motors, Lockheed, McDonnell, IBM, and Ford Motor Co., participated in the development of interactive CAD programs in the 1960s. Usage of CAD systems was primarily in the automotive industry, aerospace industry, and government agencies that developed their own programs for their specific needs. The 1960s also marked the beginning of the development of finite element analysis methods for computer stress analysis and computer aided manufacturing for generating machine toolpaths.

The 1970s are generally viewed as the years of the most significant progress in the development of computer hardware, namely the invention and development of **microprocessors**. With the improvement in computing power, new types of 3D CAD programs that were user-friendly and interactive became reality. CAD technology quickly expanded from very simple **computer aided drafting** to very complex **computer aided design**. The use of 2D and 3D wireframe modelers was accepted as the leading edge technology that could increase productivity in industry. The development of surface modeling and solid modeling technology were taking shape by the late 1970s; but the high cost of computer hardware and programming slowed the development of such technology. During this time period, the available CAD systems all required expensive room-sized mainframe computers that were extremely high in cost.

In the 1980s, improvements in computer hardware brought the power of mainframes to the desktop at less cost and with more accessibility to the general public. By the mid-1980s, CAD technology had become the main focus of a variety of manufacturing industries and was very competitive with traditional design/drafting methods. It was during this period of time that 3D solid modeling technology had major advancements, which boosted the usage of CAE technology in industry.

The introduction of the *feature-based parametric solid modeling* approach, at the end of the 1980s, elevated CAD/CAM/CAE technology to a new level. In the 1990s, CAD programs evolved into powerful design/manufacturing/management tools. CAD technology has come a long way, and during these years of development, modeling schemes progressed from two-dimensional (2D) wireframe to three-dimensional (3D) wireframe, to surface modeling, to solid modeling and, finally, to feature-based parametric solid modeling.

The first generation CAD packages were simply 2D **Computer Aided Drafting** programs, basically the electronic equivalents of the drafting board. For typical models, the use of this type of program would require that several to many views of the objects be created individually as they would be on the drafting board. The 3D designs remained in the designer's mind, not in the computer database. The mental translation of 3D objects to 2D views is required throughout the use of the packages. Although such systems have some advantages over traditional board drafting, they are still tedious and labor intensive. The need for the development of 3D modelers came quite naturally, given the limitations of the 2D drafting packages.

The development of three-dimensional modeling schemes started with three-dimensional (3D) wireframes. Wireframe models are models consisting of points and edges, which are straight lines connecting between appropriate points. The edges of wireframe models are used, similar to lines in 2D drawings, to represent transitions of surfaces and features. The use of lines and points is also a very economical way to represent 3D designs.

The development of the 3D wireframe modeler was a major leap in the area of computer geometric modeling. The computer database in the 3D wireframe modeler contains the locations of all the points in space coordinates and it is typically sufficient to create just one model rather than multiple views of the same model. This single 3D model can then be viewed from any direction as needed. Most 3D wireframe modelers allow the user to create projected lines/edges of 3D wireframe models. In comparison to other types of 3D modelers, 3D wireframe modelers require very little computing power and generally can be used to achieve reasonably good representations of 3D models. However, because surface definition is not part of a wireframe model, all wireframe images have the inherent problem of ambiguity. Two examples of such ambiguity are illustrated.

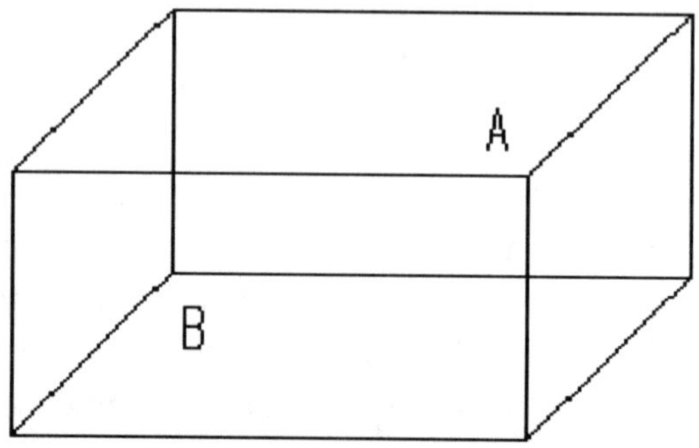

Wireframe Ambiguity: Which corner is in front, A or B?

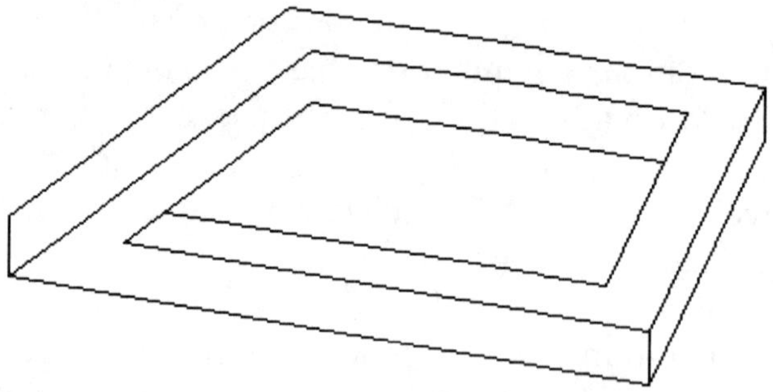

A non-realizable object: Wireframe models contain no surface definitions.

Surface modeling is the logical development in computer geometry modeling to follow the 3D wireframe modeling scheme by organizing and grouping edges that define polygonal surfaces. Surface modeling describes the part's surfaces but not its interiors. Designers are still required to interactively examine surface models to insure that the various surfaces on a model are contiguous throughout. Many of the concepts used in 3D wireframe and surface modelers are incorporated in the solid modeling scheme, but it is solid modeling that offers the most advantages as a design tool.

In the solid modeling presentation scheme, the solid definitions include nodes, edges, and surfaces, and it is a complete and unambiguous mathematical representation of a precisely enclosed and filled volume. Unlike the surface modeling method, solid modelers start with a solid or use topology rules to guarantee that all of the surfaces are stitched together properly. Two predominant methods for representing solid models are **constructive solid geometry** (CSG) representation and **boundary representation** (B-rep).

The CSG representation method can be defined as the combination of 3D solid primitives. What constitutes a "primitive" varies somewhat with the software but typically includes a rectangular prism, a cylinder, a cone, a wedge, and a sphere. Most solid modelers also allow the user to define additional primitives, which are shapes typically formed by the basic shapes. The underlying concept of the CSG representation method is very straightforward; we simply **add** or **subtract** one primitive from another. The CSG approach is also known as the machinist's approach as it can be used to simulate the manufacturing procedures for creating the 3D object.

In the B-rep representation method, objects are represented in terms of their spatial boundaries. This method defines the points, edges, and surfaces of a volume, and/or issues commands that sweep or rotate a defined face into a third dimension to form a solid. The object is then made up of the unions of these surfaces that completely and precisely enclose a volume.

By the 1980s, a new paradigm called *concurrent engineering* had emerged. With concurrent engineering, designers, design engineers, analysts, manufacturing engineers, and management engineers all work closely together right from the initial stages of the design. In this way, all aspects of the design can be evaluated and any potential problems can be identified right from the start and throughout the design process. Using the principles of concurrent engineering, a new type of computer modeling technique appeared. The technique is known as the *feature-based parametric modeling technique*. The key advantage of the feature-based parametric modeling technique is its capability to produce very flexible designs. Changes can be made easily and design alternatives can be evaluated with minimum effort. Various software packages offer different approaches to feature-based parametric modeling, yet the end result is a flexible design defined by its design variables and parametric features.

Feature-Based Parametric Modeling

One of the key-elements in the *Pro/ENGINEER* solid modeling software is its use of the **feature-based parametric modeling technique**. The feature-based parametric modeling approach has elevated solid modeling technology to the level of a very powerful design tool. Parametric modeling automates design and revision procedures by the use of parametric features. Parametric features control the model geometry by the use of design variables. The word ***parametric*** means that the geometric definitions of the design, such as dimensions, can be varied at any time in the design process. Features are predefined parts or construction tools in which users define the key parameters. A part is described as a sequence of engineering features, which can be modified/changed at any time. The concept of parametric features makes the modeling more closely match the actual design-manufacturing process than the mathematics of a solid modeling program. In parametric modeling, models and drawings are updated automatically when the design is refined.

Parametric modeling offers many benefits:

- **We begin with simple, conceptual models with minimal detail; this approach conforms to the design philosophy of "shape before size."**
- **Geometric constraints, dimensional constraints, and relational parametric equations can be used to capture design intent.**
- **The ability to update an entire system, including parts, assemblies and drawings after changing one parameter of complex designs.**
- **We can quickly explore and evaluate different design variations and alternatives to determine the best design.**
- **Existing design data can be reused to create new designs.**
- **Quick design turn-around.**

The feature-based parametric modeling technique enables the designer to incorporate the original **design intent** into the construction of the model, and the individual features control the geometry in the event of a design change. As features are modified, the system updates the entire part by re-linking the individual features of the model.

Pro/ENGINEER WILDFIRE is the twenty-fourth release, with many added features and enhancements, of the original *Pro/ENGINEER* software produced by Parametric Technology Corporation (PTC). *Pro/ENGINEER* is considered the industry leader in setting the standard for the feature-based modeling paradigm. *Pro/ENGINEER's* **Behavioral Modeling** and **Intent Referencing** allow users to concentrate on the design without depending on the associated parameters or constraints. Users can specify how features interact with each other and *Pro/ENGINEER* will automatically adjust sizes and positions as changes are made.

Getting Started with *Pro/ENGINEER*

- *Pro/ENGINEER* is composed of several application software modules (these modules are called *applications*), all sharing a common database. In this text, the main concentration is placed on the solid modeling modules used for part design. The general procedures required in creating solid models, engineering drawings, and assemblies are illustrated.

Starting *Pro/ENGINEER*

How to start *Pro/ENGINEER* depends on the type of workstation and the particular software configuration you are using. With most *Windows* and *UNIX* systems, you may select **Pro/ENGINEER** on the *Start* menu or select the **Pro/ENGINEER** icon on the desktop. Consult your instructor or technical support personnel if you have difficulty starting the software.

The program takes a while to load, so be patient. The tutorials in this text are based on the assumption that you are using *Pro/ENGINEER's* default settings. If your system has been customized for other uses, some of the settings may appear differently and not work with the step-by-step instructions in the tutorials. Contact your instructor and/or technical support personnel to restore the default software configuration.

Pro/ENGINEER Screen Layout

The default *Pro/ENGINEER drawing screen* contains the *pull-down menus*, the *Standard toolbar*, the *Navigator area*, the *Web Browser*, the *message area*, the *Datum toolbar* and the *Single Line Help*. A line of quick text appears next to the icon as you move the *mouse cursor* over different icons. You may resize the *Pro/ENGINEER* drawing window by click and drag at the edges of the window, or relocate the window by click and drag at the window title area.

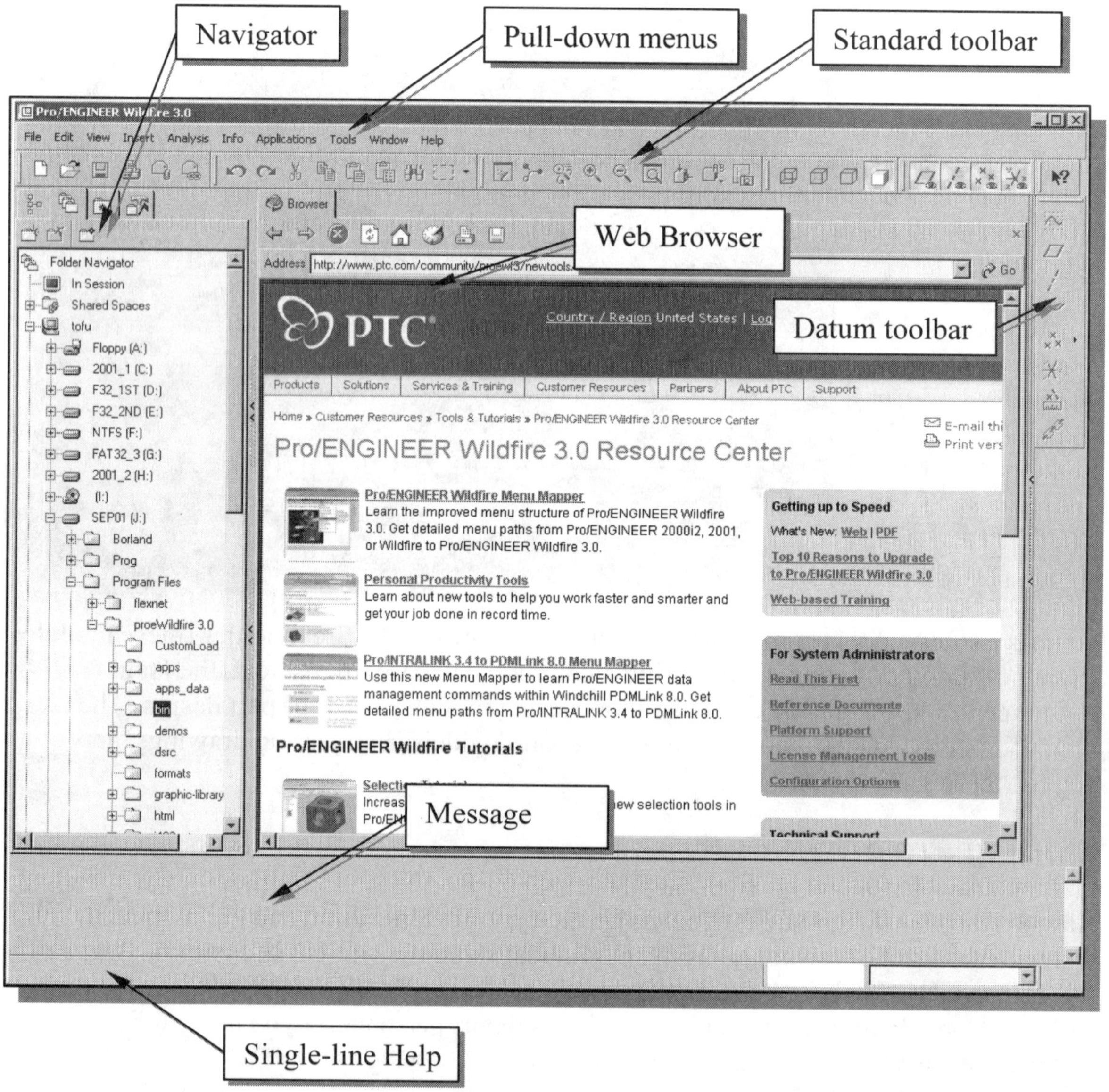

Pro/ENGINEER uses the **context sensitive menus** approach, which means additional selection menus and option windows will only be available when they are applicable to the current task. The appearance of the *Pro/ENGINEER* main window and toolbars, as shown in the figure above, remains the same on the screen throughout the different phases of modeling. Note that the menu items and icons of the non-applicable options are grayed out, which means they are temporarily disabled.

- **Pull-down menus**
  The pull-down menus at the top of the main window contain operations that you can use for all modes of the system. Note that the quick key combination to activate the same command is also shown in the displayed pull-down list.

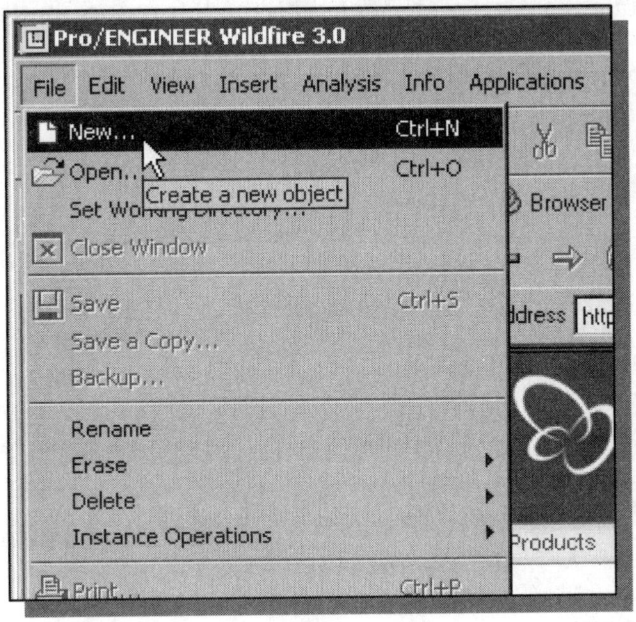

- **Standard toolbar**
  The *Standard* toolbar located at the top of the main window, below the pull-down menus, allows us quick access to frequently used commands. For example, the *view related* commands, such as Zoom, Pan, and Shaded Image, are tools to help manipulate the viewing of graphical objects. We can customize the toolbar by adding and removing sets of options or individual commands.

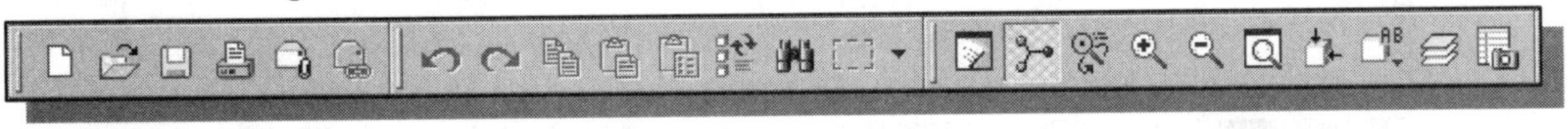

- **Message area**
  The section above the *Single Line Help* provides status information for an operation and it is also the area for data input.

- **Graphics display area**
  The graphics display area is the area where parts, assemblies, and drawings are displayed. The model's display is controlled by the *environment settings*.

- **Datum toolbar**
  The *Datum* toolbar, located toward the right section of the main window, provides us with quick access to create datum references such as datum planes and datum axes.

- **Navigator**
  The tabs across the top of the *Navigator* provide direct access to four groups of options. The initial content of the *Navigator* is the *Folder* information where *Pro/ENGINEER Wildfire* is launched. The *Navigator* window can be opened or closed by toggling the *Navigator Sash* with the left-mouse-button.

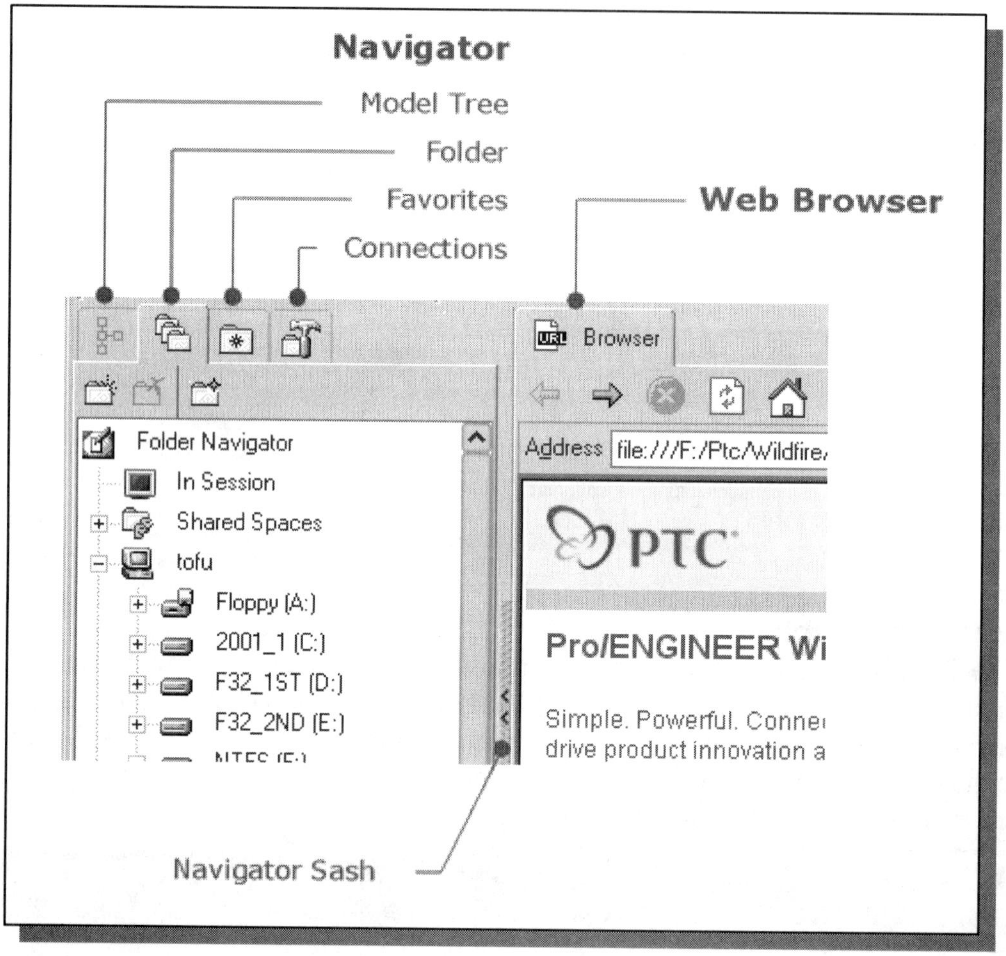

- **Web Browser**
  The *Pro/ENGINEER Web Browser* allows the user to quickly access **model information** and **on-line documentation**. It also provides general web browsing capabilities. The initial content of the *Pro/ENGINEER* web browser is the *Pro/ENGINEER Wildfire Web Tools*. The web browser will close automatically when you open a model; it can also be opened or closed by toggling the *Browser Sash* with the left-mouse-button.

- **Single Line Help**
  A line of quick text appears at the bottom of the *Pro/ENGINEER drawing screen* as the *mouse cursor* is moved over different icons.

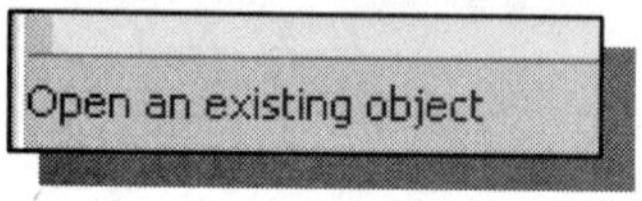

Basic Functions of Mouse Buttons

Pro/ENGINEER utilizes the mouse buttons extensively. In learning *Pro/ENGINEER's* interactive environment, it is important to understand the basic functions of the mouse buttons. A three-button mouse is highly recommended with *Pro/ENGINEER* since the package uses all three buttons for various functions. If a two-button mouse is used instead, some of the operations, such as the *Dynamic Viewing* functions, will not be available.

- **Left mouse button**
  The left mouse button is used for most operations, such as selecting menus and icons, or picking graphic entities. One click of the button is used to select icons, menus and form entries, and to pick graphic items.

- **Middle mouse button**
  The middle button is often used to accept the default setting of a prompt or to end a process. It is also used as the shortcut to perform the *Dynamic Viewing* functions: Rotate, Zoom, and Pan.

- **Right mouse button**
  The right mouse button is used to query a selection, and is also used in the *Pro/ENGINEER Sketcher* to create tangent arcs.

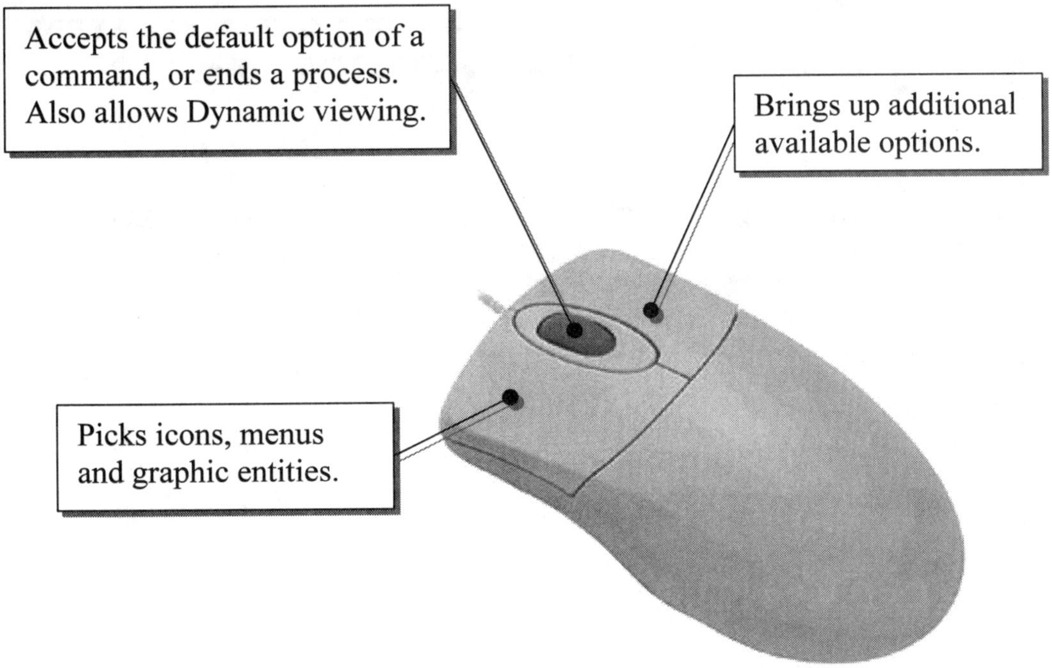

Model Tree window and Feature Toolbars

❖ The *Pro/ENGINEER* ***Model Tree*** window and ***Feature Toolbars*** are two of the most important *Pro/ENGINEER* interfaces; they will appear on the screen when we begin a modeling session.

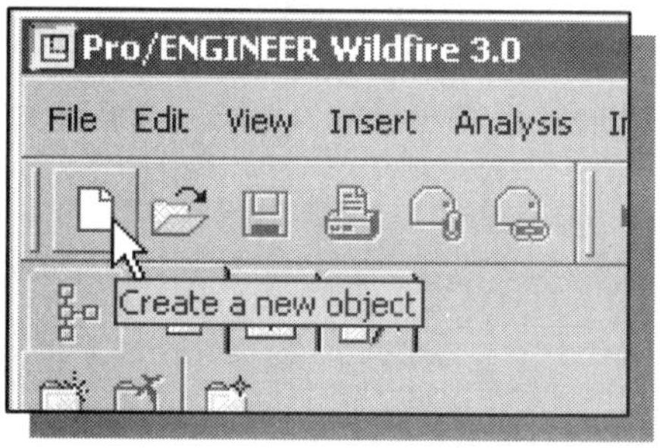

1. Click on the **New** icon, located in the *Standard* toolbar as shown.

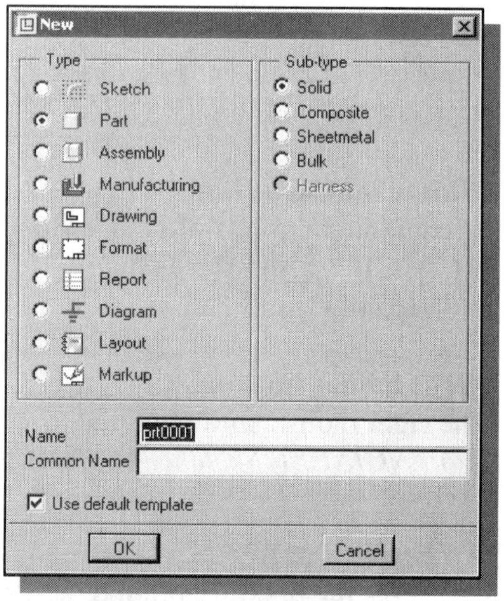

2. In the *New* dialog box, confirm the model's Type is set to Part (and the Sub-type is set to Solid).

3. Click on the **OK** button to accept the default settings and enter the *Pro/ENGINEER Part Modeling* mode.

❖ The ***Model Tree*** and the ***Feature Toolbars*** are on both sides of the *Pro/ENGINEER* main screen.

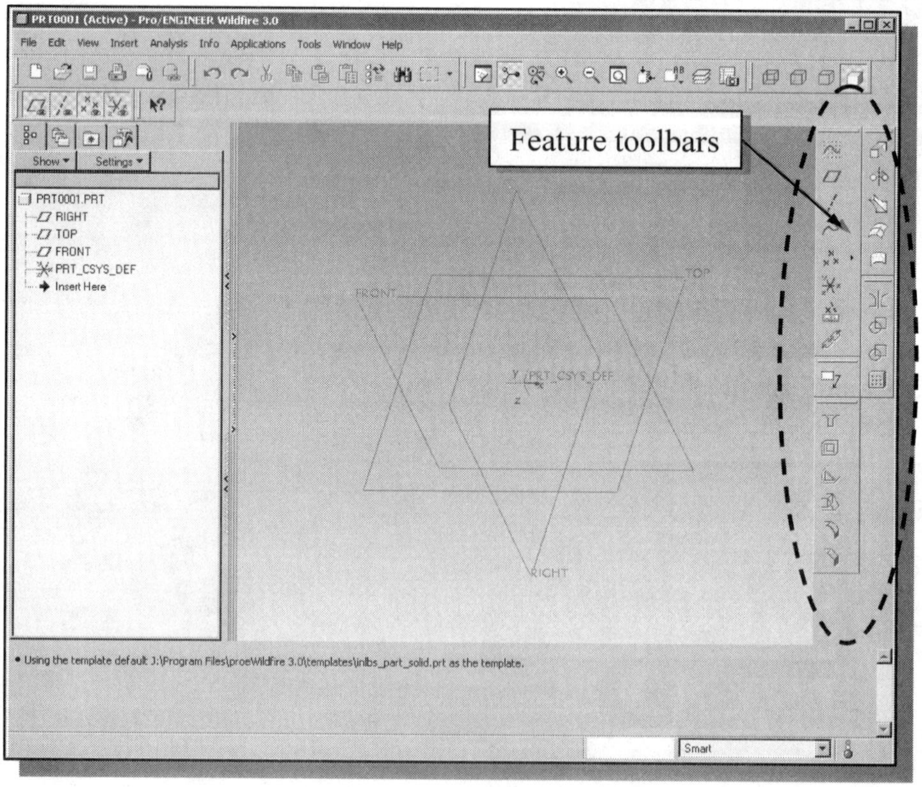

On-Line Help

❖ Several types of on-line *Help* are available at any time during a *Pro/ENGINEER* session. *Pro/ENGINEER* provides many help functions, such as:

- ***Pro/Engineer Web Browser***: The default contents of the *Pro/ENGINEER* web browser is the *Pro/ENGINEER Wildfire Web Tools*, which allow us to access the *Pro/ENGINEER* documentation installed in the local drives or through the internet.

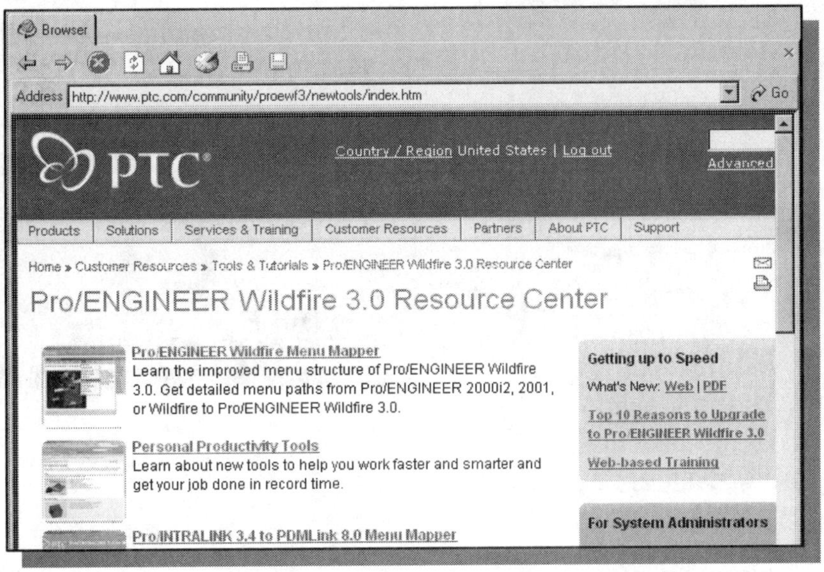

- ***Pro/ENGINEER Help Center***: Click on the **Help** option in the pull-down menus to access the *Pro/ENGINEER* ***Help Center***.

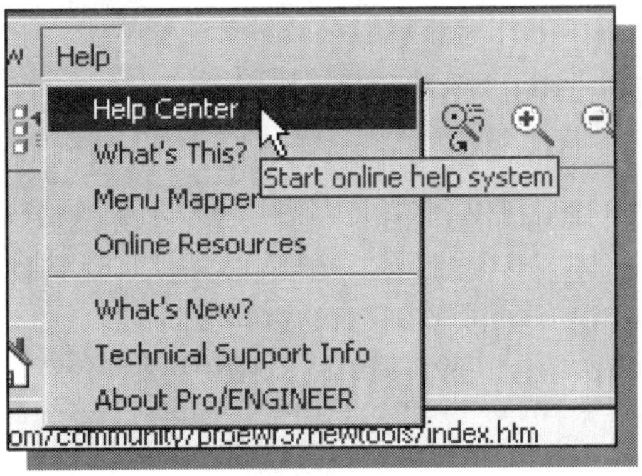

- ***Context Sensitive Help***: The context sensitive help can be used to quickly access the **Help** menu on specific tasks or icons displayed on the screen.

Leaving *Pro/ENGINEER*

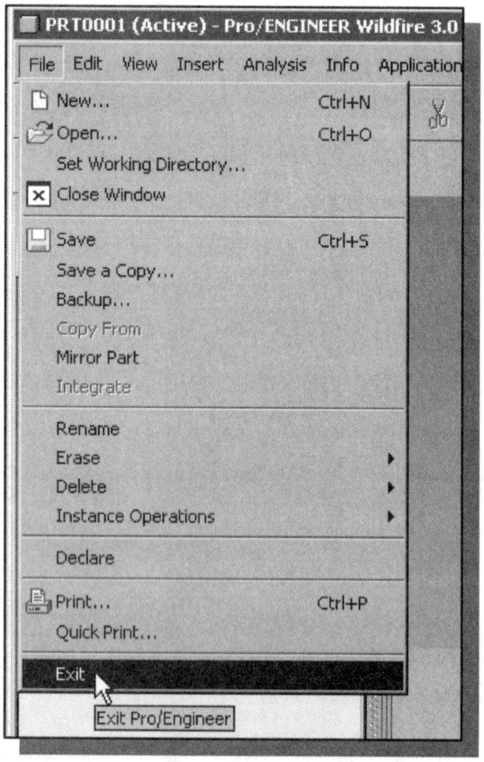

1. To leave *Pro/ENGINEER*, use the left-mouse-button and click on **File** at the top of the *Pro/ENGINEER* main window, then choose **Exit** from the pull-down menu.

2. In the confirmation dialog box, click on the **Yes** button to exit *Pro/ENGINEER.*

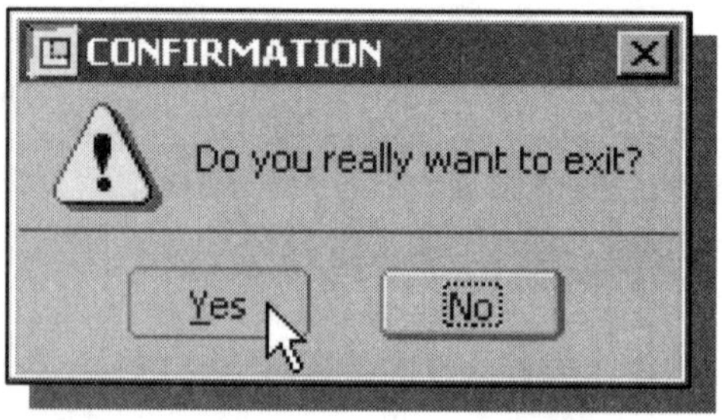

Creating a CAD files folder

It is a good practice to create a separate folder to store your CAD files. You should not save your CAD files in the same folder where the *Pro/ENGINEER* application is located. It is much easier to organize and back up your project files if they are in a separate folder. Making folders within this folder for different types of projects will help you organize your CAD files even further. When creating CAD files in *Pro/ENGINEER*, it is strongly recommended that you *save* your CAD files on the hard drive.

To create a new folder with most *Windows* systems:

1. In *My Computer*, or start the **Windows Explorer** under the *Start* menu, open the folder in which you want to create a new folder.

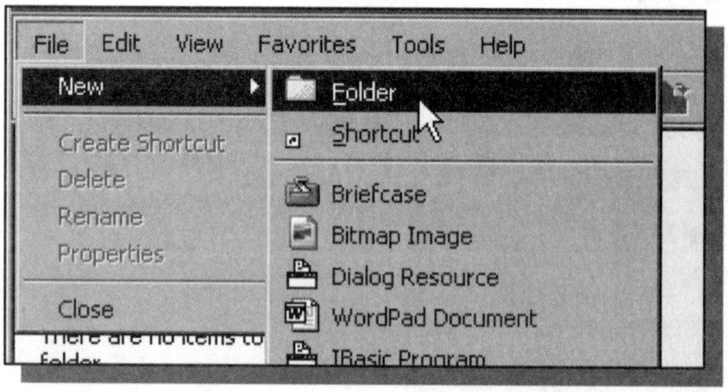

2. On the **File** menu, point to **New**, and then click **Folder**. The new folder appears with a temporary name.

3. Type a name for the new folder, and then press **ENTER**.

Chapter 1

Parametric Modeling Fundamentals

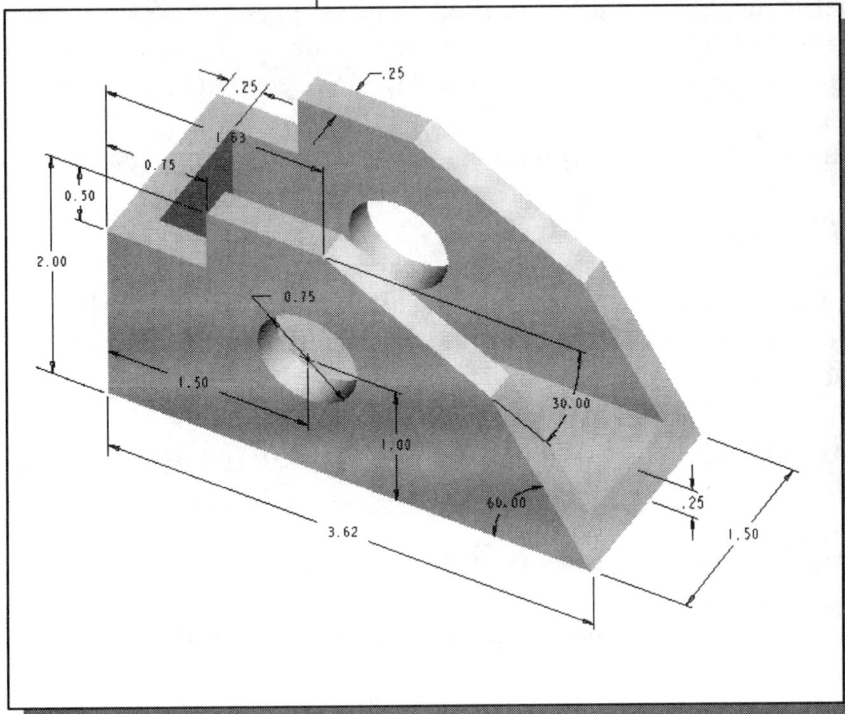

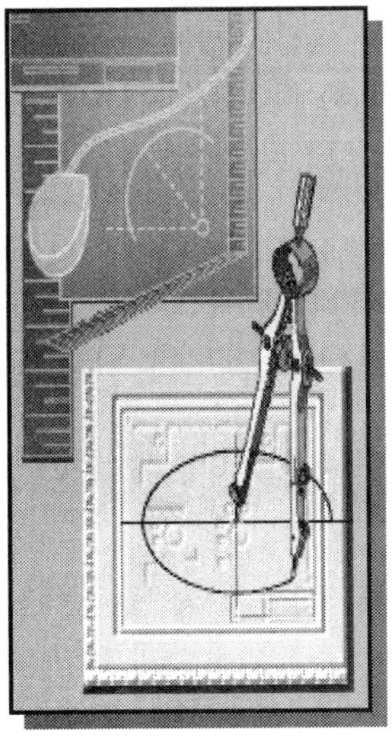

Learning Objectives

- **Create Simple Extruded Solid Models.**
- **Understand the Basic Parametric Modeling Process.**
- **Create 2-D Sketches.**
- **Understand the "Shape Before Size" approach.**
- **Use the Dynamic Viewing commands.**
- **Create and Modify Parametric Dimensions.**

Introduction

The **feature-based parametric modeling** technique enables the designer to incorporate the original **design intent** into the construction of the model. The word ***parametric*** means the geometric definitions of the design, such as dimensions, can be varied at any time in the design process. Parametric modeling is accomplished by identifying and creating the key features of the design with the aid of computer software. The design variables, described in the sketches and features, can be used to quickly modify/update the design.

In *Pro/ENGINEER*, the parametric part modeling process involves the following steps:

1. **Set up Units and Basic Datum Geometry.**

2. **Determine the type of the base feature, the first solid feature, of the design. Note that *Extrude*, *Revolve*, or *Sweep* operations are the most common types of base features.**

3. **Create a rough two-dimensional sketch of the basic shape of the base feature of the design.**

4. **Apply/modify constraints and dimensions to the two-dimensional sketch.**

5. **Transform the parametric two-dimensional sketch into a 3D solid.**

6. **Add additional parametric features by identifying feature relations and complete the design.**

7. **Perform analyses/simulations, such as finite element analysis (FEA) or cutter path generation (CNC), on the computer model and refine the design as needed.**

8. **Document the design by creating the desired 2D/3D drawings.**

The approach of creating three-dimensional features using two-dimensional sketches is an effective way to construct solid models. Many designs are in fact the same shape in one direction. Computer input and output devices we use today are largely two-dimensional in nature, which makes this modeling technique quite practical. This method also conforms to the design process that helps the designer with conceptual design along with the capability to capture the ***design intent***. Most engineers and designers can relate to the experience of making rough sketches on restaurant napkins to convey conceptual design ideas. Note that *Pro/ENGINEER* provides many powerful modeling and design tools, and there are many different approaches to accomplish modeling tasks. The basic principle of **feature-based modeling** is to build models by adding simple features one at a time. In this chapter, a very simple solid model with extruded features is used to introduce the general **feature-based parametric** modeling procedure.

The *Adjuster* design

Starting *Pro/ENGINEER*

How to start *Pro/ENGINEER* depends on the type of workstation and the particular software configuration you are using. With most *Windows* and *UNIX* systems, you may select **Pro/ENGINEER** on the *Start* menu or select the **Pro/ENGINEER** icon on the desktop. Consult your instructor or technical support personnel if you have difficulty starting the software.

1. Select the **Pro/ENGINEER** option on the *Start* menu or select the **Pro/ENGINEER** icon on the desktop to start *Pro/ENGINEER*. The *Pro/ENGINEER* main window will appear on the screen.

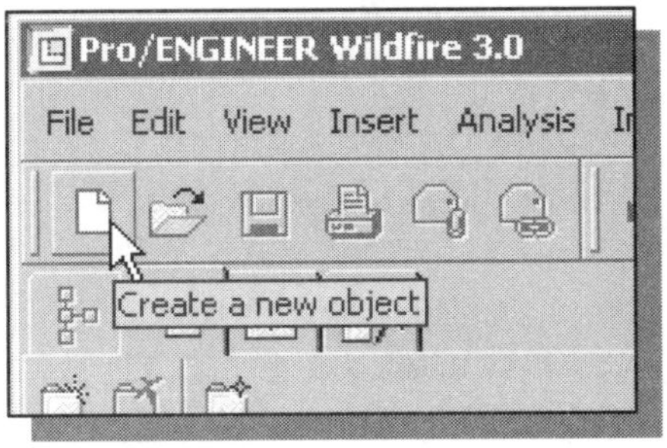

2. Click on the **New** icon, located in the *Standard* toolbar as shown.

3. In the *New* dialog box, confirm the model's **Type** is set to **Part** (**Solid** Sub-type).

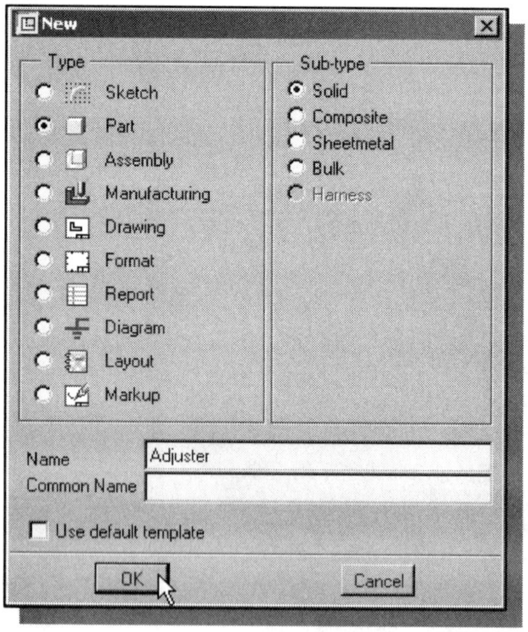

4. Enter ***Adjuster*** as the part **Name** as shown in the figure.

5. Turn *off* the **Use default template** option.

6. Click on the **OK** button to accept the settings.

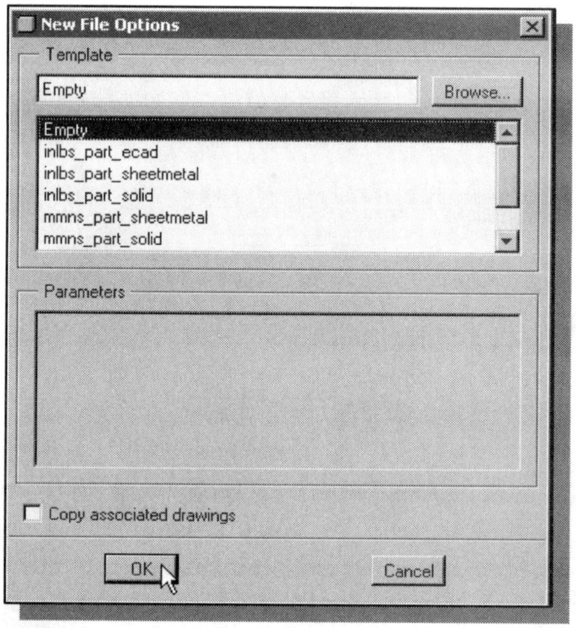

7. In the *New File Options* dialog box, select **EMPTY** in the option list to not use any template file.

8. Click on the **OK** button to accept the settings and enter the *Pro/ENGINEER Part Modeling* mode.

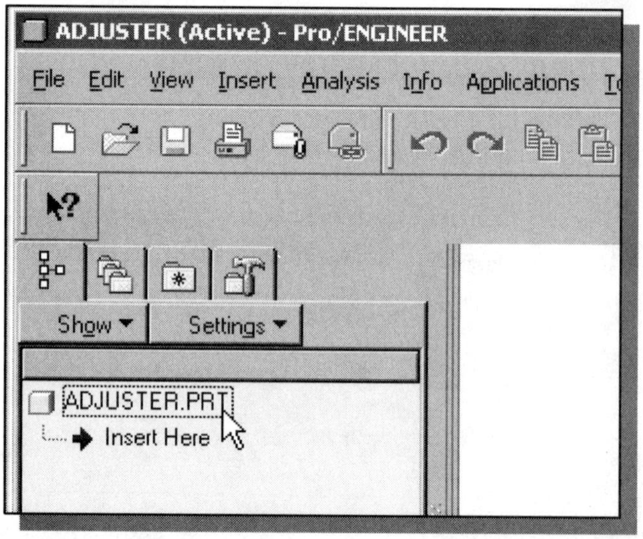

❖ Note that the part name, *Adjuster*, appears in the title area of the main window and in the *Navigator Model Tree* window.

Step 1: Units and Basic Datum Geometry Setups

♦ Units Setup and *Pro/ENGINEER* Menu Structure

When starting a new model, the first thing we should do is to choose the set of units we want to use.

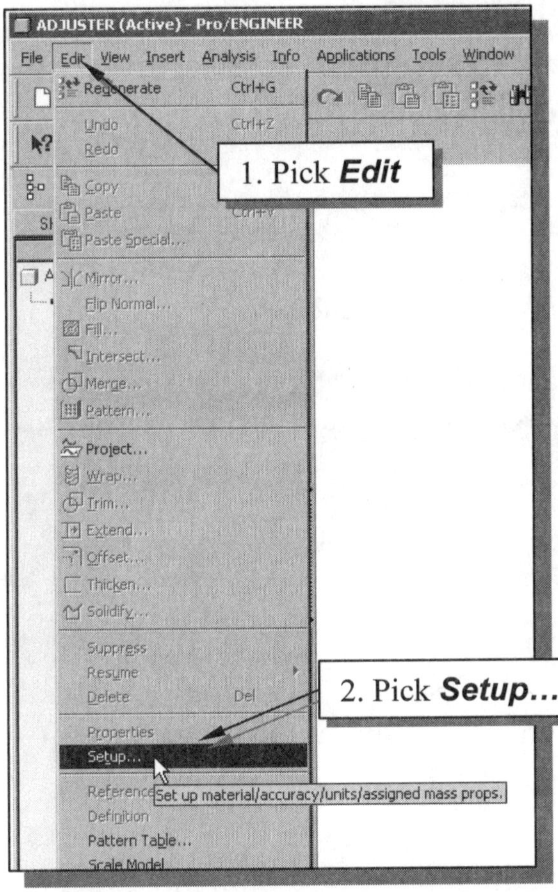

1. Use the left-mouse-button and select **Edit** in the pull-down menu area.

2. Use the left-mouse-button and select **Setup...** in the **pull-down list** as shown.

➢ Note that the *Pro/ENGINEER* menu system is context-sensitive, which means that the menu items and icons of the non-applicable options are grayed out (temporarily disabled).

3. Select the **Units** option in the **Menu Manager** window that appeared to the right of the *Pro/ENGINEER* main window.

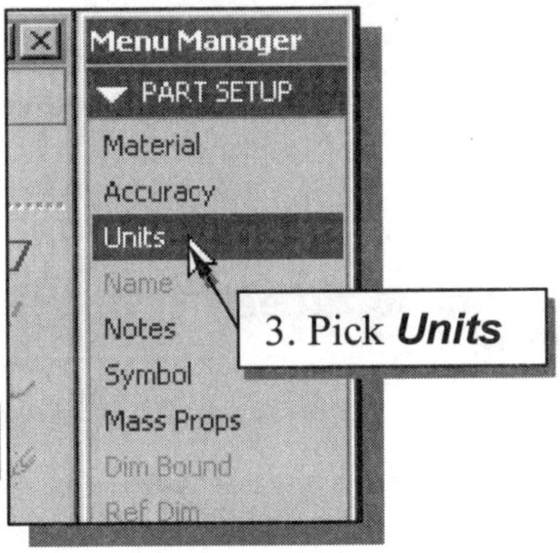

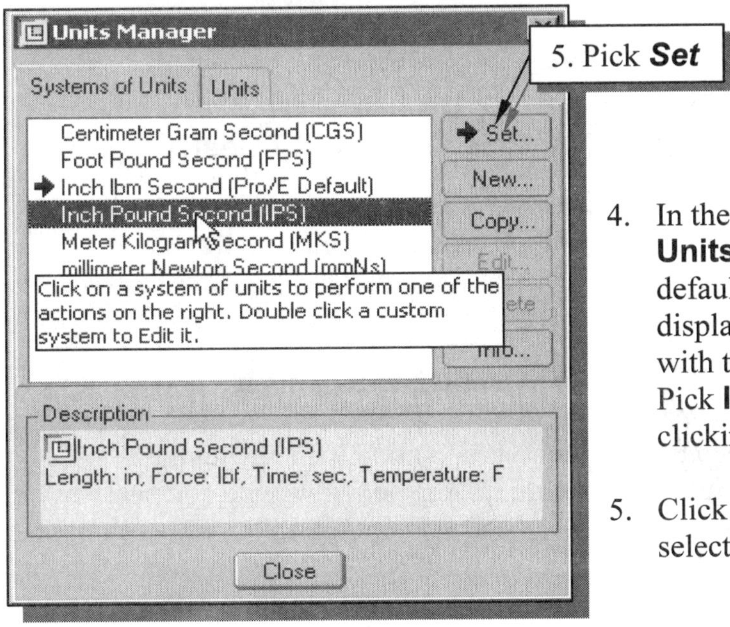

4. In the **Units Manager - System of Units** form, the *Pro/ENGINEER* default setting **Inch lbm Second** is displayed. The set of units is stored with the model file when you save. Pick **Inch Pound Second (IPS)** by clicking in the list window as shown.

5. Click on the **Set** button to accept the selection.

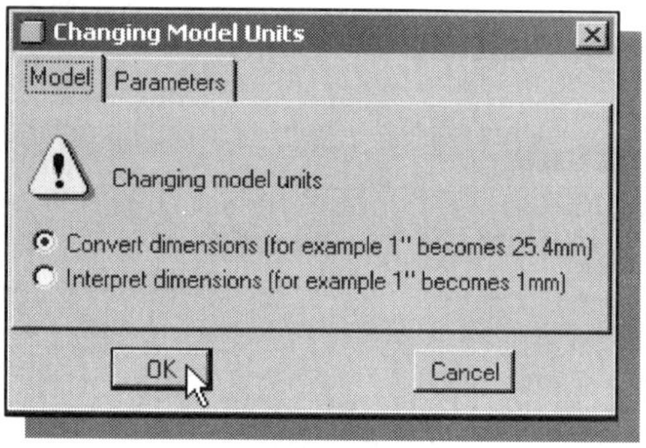

6. In the *Changing Model Units* dialog box, click on the **OK** button to accept the change of the units.

➢ Note that *Pro/ENGINEER* allows us to change model units even after the model has been constructed.

7. Click on the **Close** button to exit the *Units Manager* dialog box.

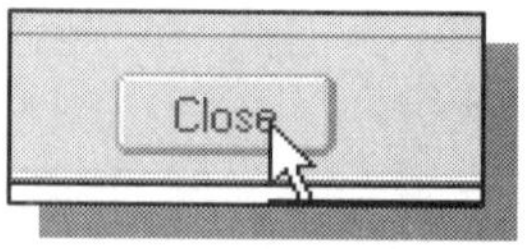

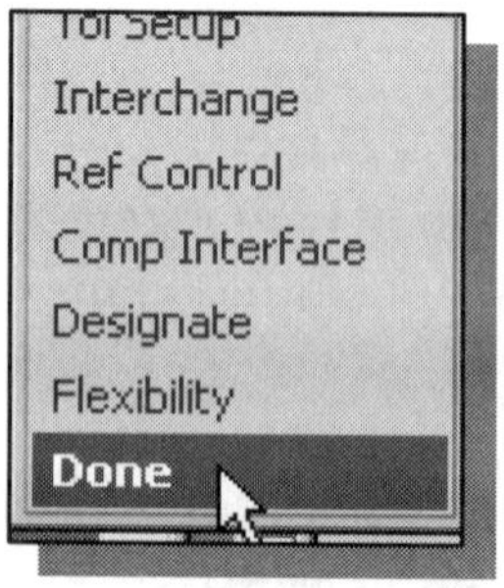

8. Pick **Done** to exit the **PART SETUP** submenu.

➢ Note that the submenu appeared and disappeared as different options were selected; this is known as the ***tree structure menu system***.

- **Tree Structure system**

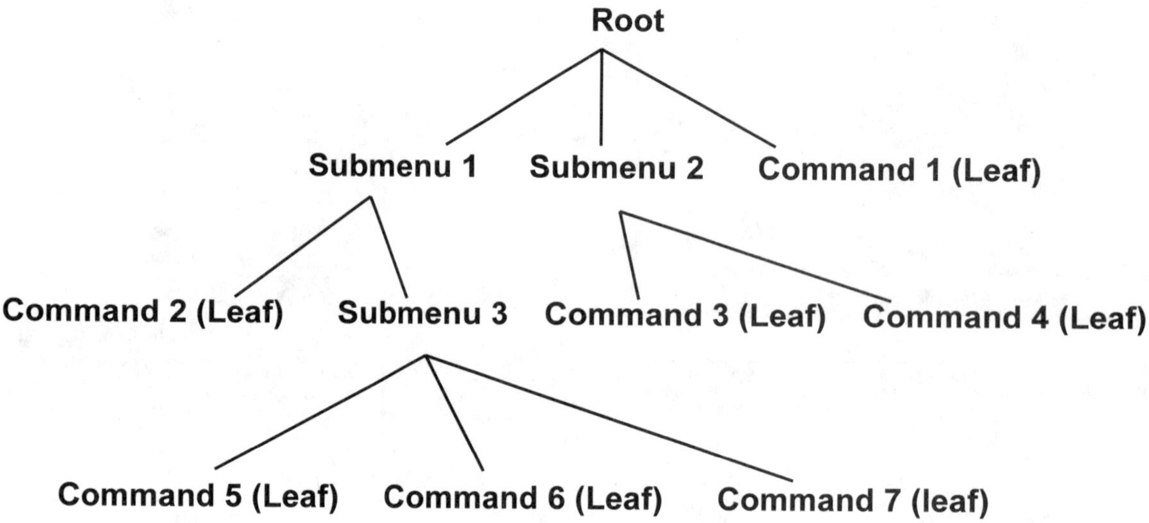

The tree structure is an effective way to organize menu items. Similar items are placed in a group that could belong to another subgroup based on the grouping method. The submenus represent different categories of items. The tree structure is used extensively in the majority of CAD software menu systems.

Using the tree structure shown, we will follow **Submenu 1** to **Submenu 3** and reach **Command 5**. If we then want to switch to **Command 4**, we will trace back to the root then branch off to **Submenu 2**. Keep this tree structure in mind while using the *Pro/ENGINEER* menu system. Think of the overall scheme and it will be quite easy to get to where you want to go. In *Pro/ENGINEER*, the **Done** option will usually return you to the previous level in the menu structure.

♦ Adding the First Part Features — Datum Planes

❖ *Pro/ENGINEER* provides many powerful tools for model creation. In doing feature-based parametric modeling, it is a good practice to establish three reference planes to locate the part in space. The reference planes can be used as location references in feature constructions.

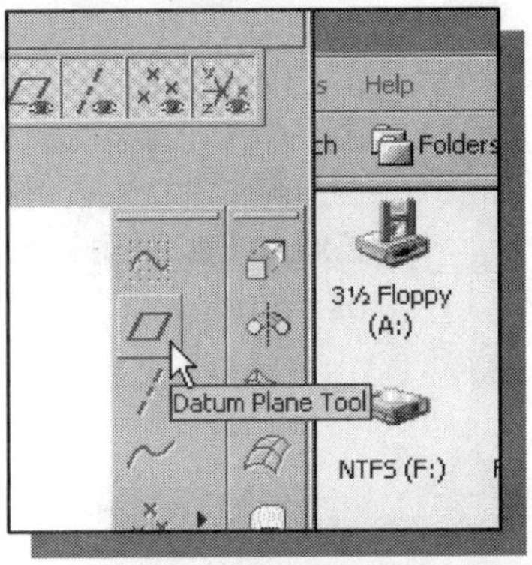

➢ Move the cursor toward the right side of the main window and click on the **Datum Plane Tool** icon as shown.

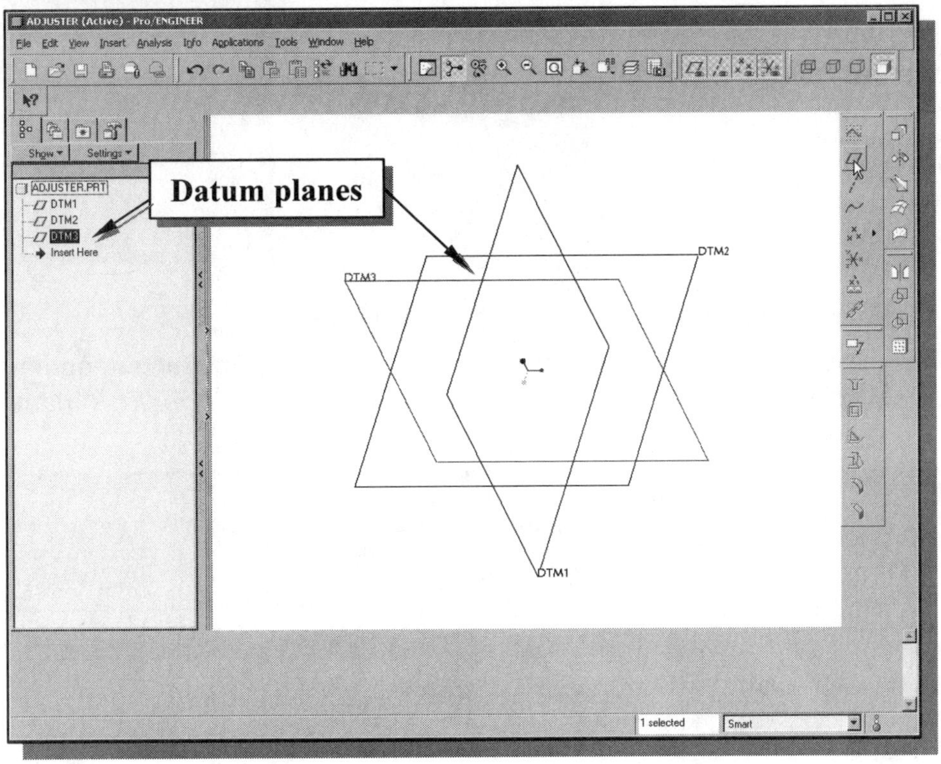

❖ In the *Navigator Model Tree* window and the display area, three datum planes represented by three rectangles are displayed. Datum planes are infinite planes and they are perpendicular to each other. We can consider these planes as XY, YZ, and ZX planes of a Cartesian coordinate system.

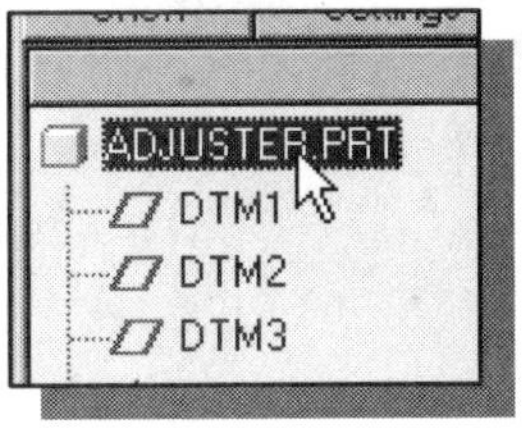

1. Note DTM3 is highlighted, click the Adjuster.PRT name in the navigator window to deselect DTM3.

Step 2: Determine/Set Up the Base Solid Feature

- For the *Adjuster* design, we will create an extruded solid as the base feature.

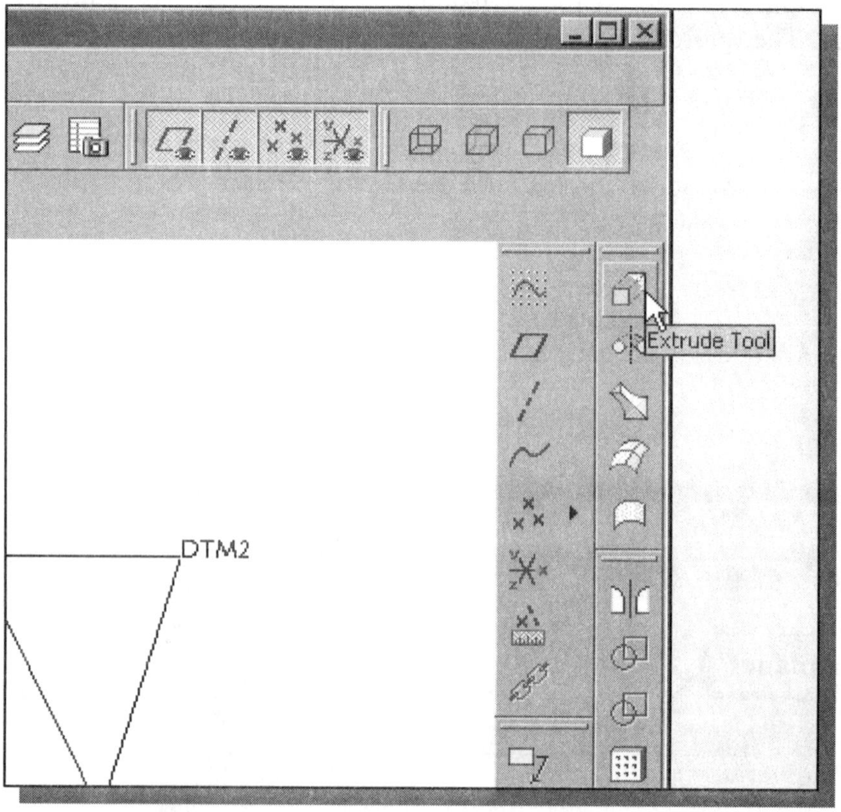

1. In the ***Feature* toolbars** (toolbars aligned to the right edge of the main window), select the **Extrude Tool** icon as shown.

- The ***Feature Option Dashboard***, which contains applicable construction options, is displayed above the message area near the bottom of the *Pro/ENGINEER* main window.

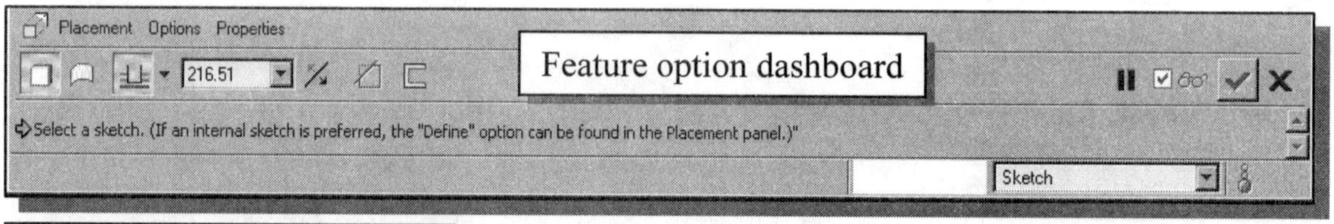

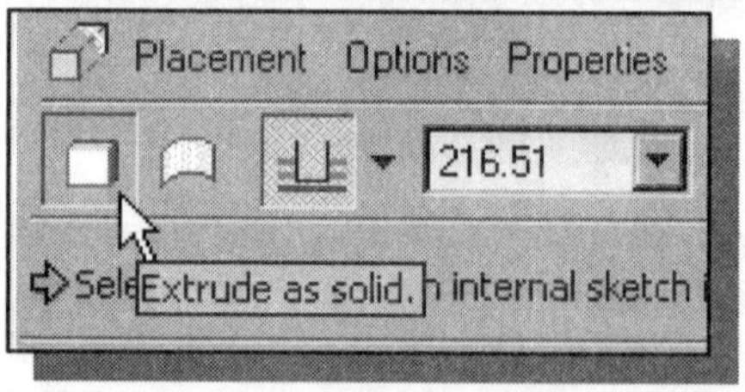

2. On your own, move the cursor over the icons and read the descriptions of the different options available. Note that the default extrude option is set to **Extrude as Solid**.

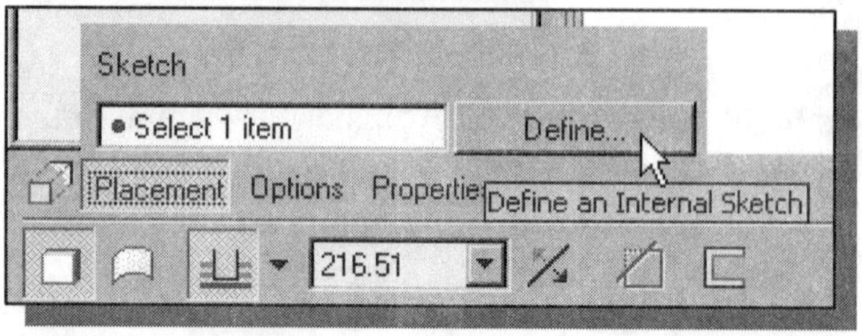

3. Click the *Placement Option* and choose **Define** to begin creating a new *Internal sketch*.

Sketching plane – It is an XY CRT, but an XYZ World

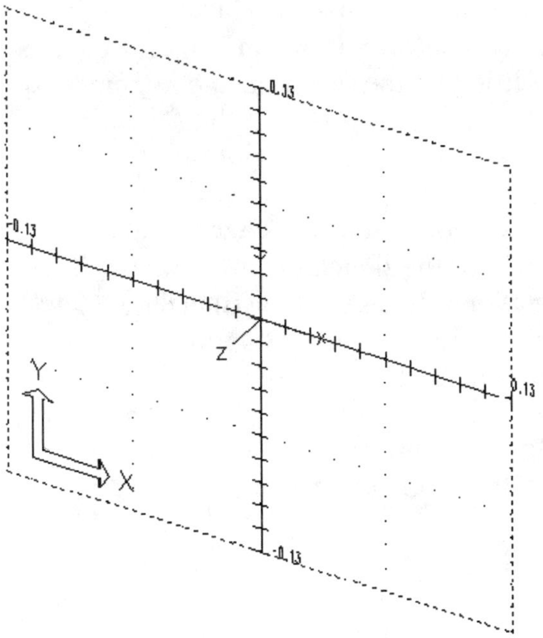

Design modeling software is becoming more powerful and user friendly, yet the system still does only what the user tells it to do. When using a geometric modeler, we therefore need to have a good understanding of what its inherent limitations are. We should also have a good understanding of what we want to do and what to expect, as the results are based on what is available.

In most 3D geometric modelers, 3D objects are located and defined in what is usually called **world space** or **global space**. Although a number of different coordinate systems can be used to create and manipulate objects in a 3D modeling system, the objects are typically defined and stored using the world space. The world space is usually a **3D Cartesian coordinate system** that the user cannot change or manipulate.

In most engineering designs, models can be very complex, and it would be tedious and confusing if only the world coordinate system were available. Practical 3D modeling systems allow the user to define **Local Coordinate Systems (LCS)** or **User Coordinate Systems (UCS)** relative to the world coordinate system. Once a local coordinate system is defined, we can then create geometry in terms of this more convenient system.

Although objects are created and stored in 3D space coordinates, most of the geometric entities can be referenced using 2D Cartesian coordinate systems. Typical input devices such as a mouse or digitizer are two-dimensional by nature; the movement of the input device is interpreted by the system in a planar sense. The same limitation is true of common output devices, such as CRT displays and plotters. The modeling software performs a series of three-dimensional to two-dimensional transformations to correctly project 3D objects onto the 2D display plane.

The *Pro/ENGINEER* ***sketching plane*** is a special construction approach that enables the planar nature of the 2D input devices to be directly mapped into the 3D coordinate system. The *sketching plane* is a local coordinate system that can be aligned to an existing face of a part, or a reference plane.

Think of the sketching plane as the surface on which we can sketch the 2D sections of the parts. It is similar to a piece of paper, a white board, or a chalkboard that can be attached to any planar surface. The first sketch we create is usually drawn on one of the established datum planes. Subsequent sketches/features can then be created on sketching planes that are aligned to existing **planar faces of the solid part** or **datum planes.**

Defining the Sketching Plane

- The *sketching plane* is a reference location where two-dimensional sketches are created. The *sketching plane* can be any planar part surface or datum plane. Note that *Pro/ENGINEER* uses a two-step approach in setting up the selection and alignment of the sketching plane.

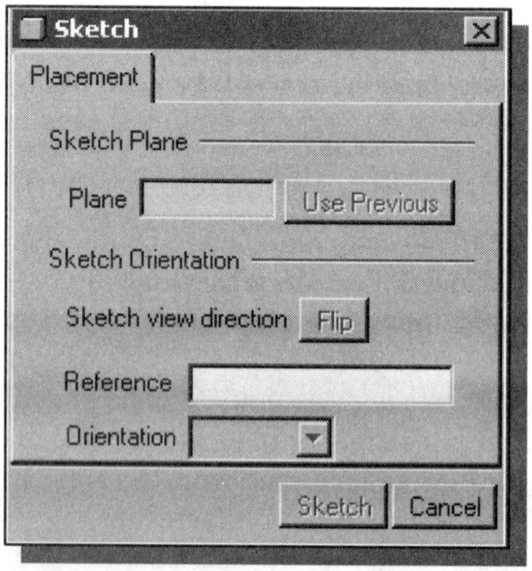

❖ In the Section Placement window, the selection of the sketch plane and the orientation of the sketching plane are organized into two groups as shown in the figure. The **Sketch Plane** can be set to any surfaces, including datum planes. The **Sketch Orientation** is set based on the selection of the Sketch plane.

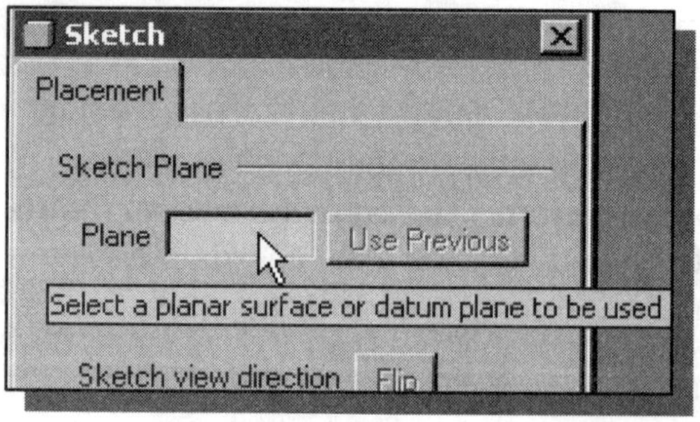

1. Move the cursor inside the **Plane** option box in *the Sketch* window as shown. The message "*Select a plane or surface to define sketch plane.*" is displayed in the message area and in the tool tip box.

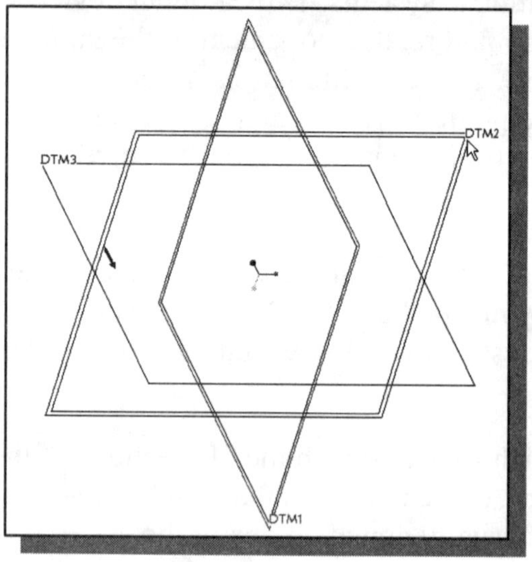

2. In the *graphic area*, select **DTM2** by clicking on the text DTM2 as shown.

❖ Notice an arrow appears on the edge of DTM2. The arrow direction indicates the viewing direction of the sketch plane. The viewing direction can be reversed by clicking on the Flip button in the *Sketch Orientation* section of the popup window.

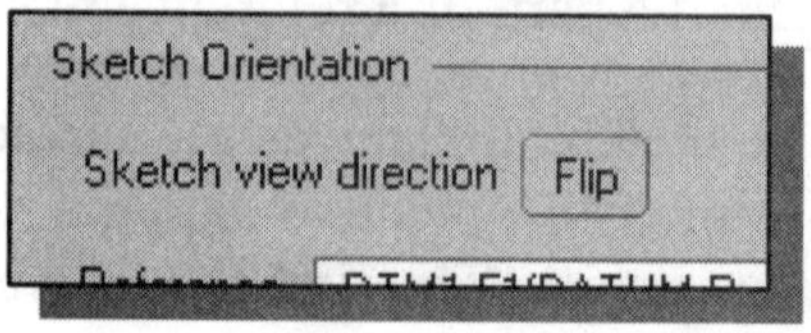

Defining the Orientation of the Sketching Plane

- Although we have selected the sketching plane, *Pro/ENGINEER* still needs additional information to define the orientation of the sketch plane. *Pro/ENGINEER* expects us to choose a reference plane (any plane that is perpendicular to the selected sketch plane) and the orientation of the reference plane is relative to the computer screen.

❑ **To define the orientation of the sketching plane, select the facing direction of the reference plane with respect to the computer screen.**

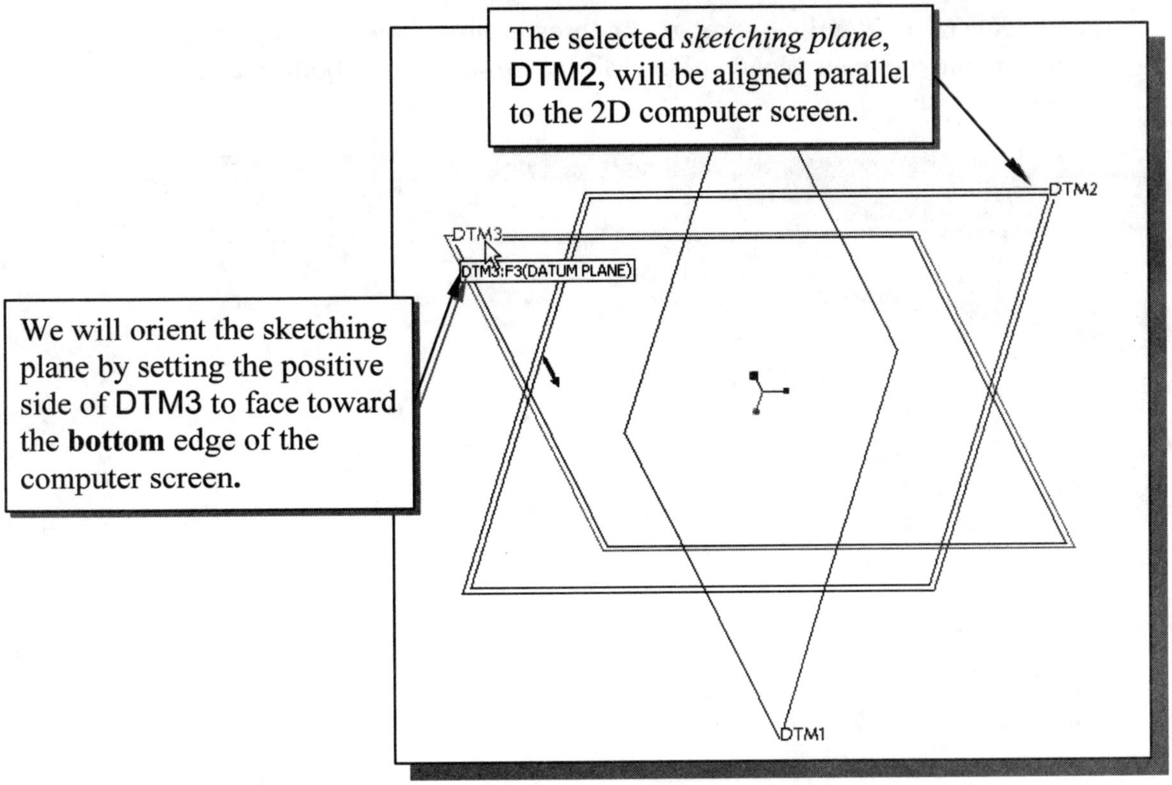

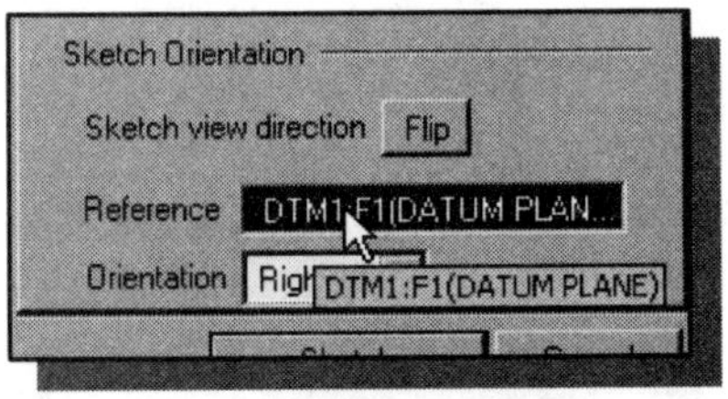

1. Click inside the **Reference** option box in *the Sketch-Orientation* window as shown. The message "*Select a reference, such as surface, plane or edge to define view orientation.*" is displayed in the message area.

2. In the graphic area, select **DTM3** by clicking on the text DTM3 as shown in the above figure.

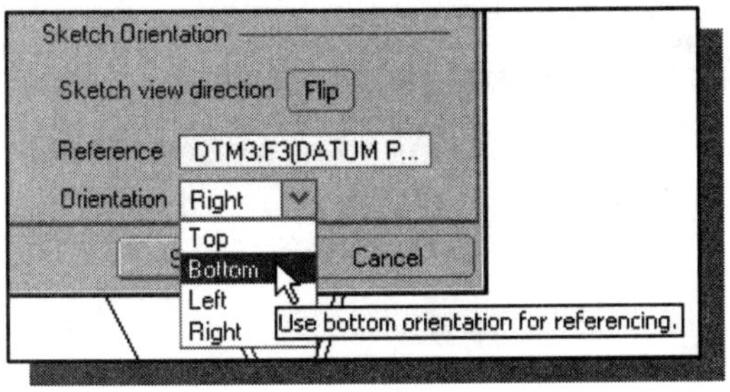

3. In the **Orientation** list, pick **Bottom** to set the orientation of the reference plane.

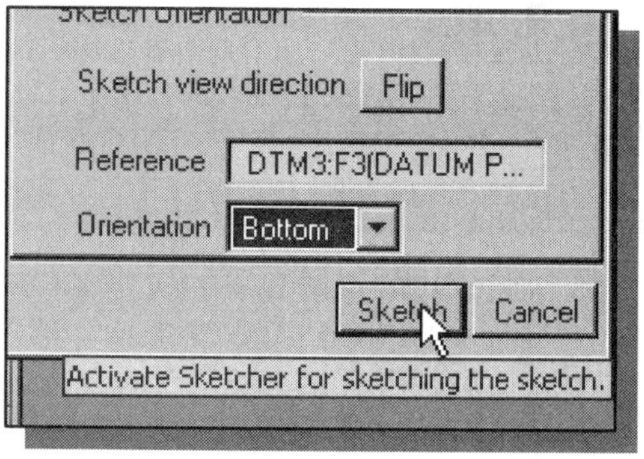

4. Pick **Sketch** to exit the *Section Placement* window and proceed to enter the *Pro/ENGINEER* sketcher mode.

- *Pro/ENGINEER* will now rotate the three *datum planes*: DTM2 aligned to the screen and the positive side of DTM3 facing toward the bottom edge of the computer screen.

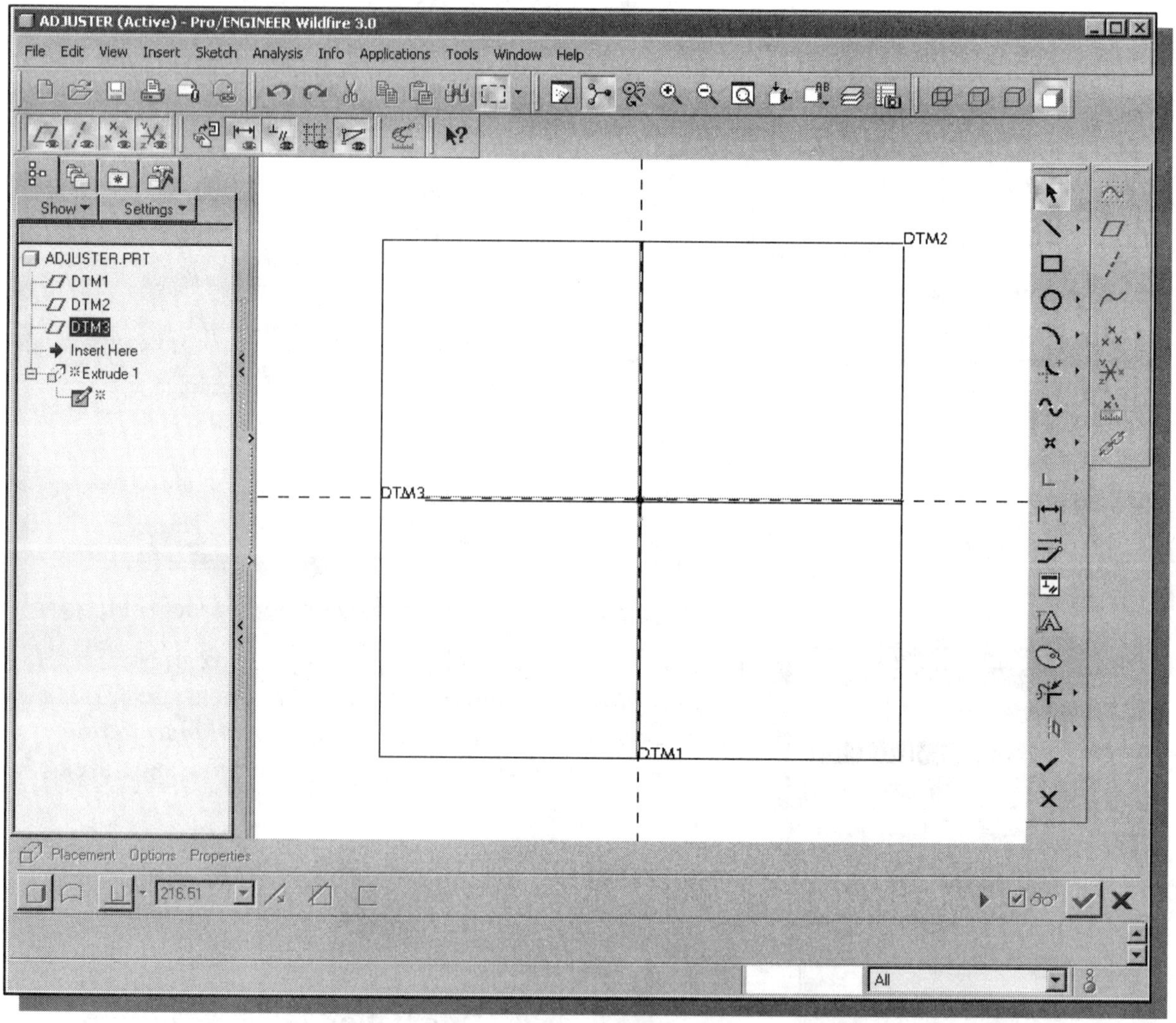

❖ The orientation of the *sketching plane* can be very confusing to new users. Read through this section again and make sure you understand the steps involved.

Step 3: Creating 2D Rough Sketches

♦ Shape Before Size – Creating Rough Sketches

Quite often during the early design stage, the shape of a design may not have any precise dimensions. Most conventional CAD systems require the user to input the precise lengths and location dimensions of all geometric entities defining the design, and some of the values may not be available during the early design stage. With *parametric modeling*, we can use the computer to elaborate and formulate the design idea further during the initial design stage. With *Pro/ENGINEER*, we can use the computer as an electronic sketchpad to help us concentrate on the formulation of forms and shapes for the design. This approach is the main advantage of *parametric modeling* over conventional solid-modeling techniques.

As the name implies, ***rough sketches*** are not precise at all. When sketching, we simply sketch the geometry so it closely resembles the desired shape. Precise scale or dimensions are not needed. *Pro/ENGINEER* provides us with many tools to assist in finalizing sketches, known as ***sections***. For example, geometric entities such as horizontal and vertical lines are set automatically. However, if the rough sketches are poor, much more work will be required to generate the desired parametric sketches. Here are some general guidelines for creating sketches in *Pro/ENGINEER*:

- **Create a sketch that is proportional to the desired shape.** Concentrate on the shapes and forms of the design.

- **Keep the sketches simple.** Leave out small geometry features such as fillets, rounds, and chamfers. They can easily be placed using the Fillet and Chamfer commands after the parametric sketches have been established.

- **Exaggerate the geometric features of the desired shape.** For example, if the desired angle is 85 degrees, create an angle that is 50 or 60 degrees. Otherwise, *Pro/ENGINEER* might assume the intended angle to be a 90-degree angle.

- **Draw the geometry so that it does not overlap.** The sketched geometry should eventually form a closed region. *Self-intersecting* geometric shapes are not allowed.

- **The sketched geometric entities should form a closed region.** To create a solid feature, such as an extruded solid, a closed region section is required so that the extruded solid forms a 3D volume.

➢ **Note:** The concepts and principles involved in *parametric modeling* are very different, and sometimes they are totally opposite, to those of the conventional computer aided drafting systems. In order to understand and fully utilize *Pro/ENGINEER's* functionality, it will be helpful to take a *Zen* approach to learning the topics presented in this text: **Temporarily forget your knowledge and experiences using conventional computer aided drafting systems.**

♦ The Pro/ENGINEER SKETCHER and INTENT MANAGER

In previous generation CAD programs, construction of models relies on exact dimensional values, and adjustments to dimensional values are quite difficult once the model is built. With *Pro/ENGINEER*, we can now treat the sketch as if it is being done on a napkin, and it is the general shape of the design that we are more interested in defining. The *Pro/ENGINEER* part model contains more than just the final geometry. It also contains the *design intent* that governs what will happen when geometry changes. The design philosophy of "**shape before size**" is implemented through the use of the *Pro/ENGINEER **Sketcher***. This allows the designer to construct solid models in a higher level and leave all the geometric details to *Pro/ENGINEER*.

One of the main improvements in *Pro/ENGINEER* since **Release 20** is the introduction and enhancements of the ***Intent Manager*** in the *Pro/ENGINEER Sketcher*.

The ***Intent Manager*** enables us to do:

- Dynamic dimensioning and constraints
- Add or delete constraints explicitly
- Undo any *Sketcher* operation

The first thing that *Pro/ENGINEER Sketcher* expects us to do, which is displayed in the *References* window, is to specify ***sketching references***. In the previous sections, we created the three datum planes to help orient the model in 3D space. Now we need to orient the 2D sketch with respect to the three datum planes. At least two references are required to orient the sketch in the horizontal direction and in the vertical direction. By default, the two planes (in our example, DTM1 and DTM3) that are perpendicular to the sketching plane (DTM2) are automatically selected.

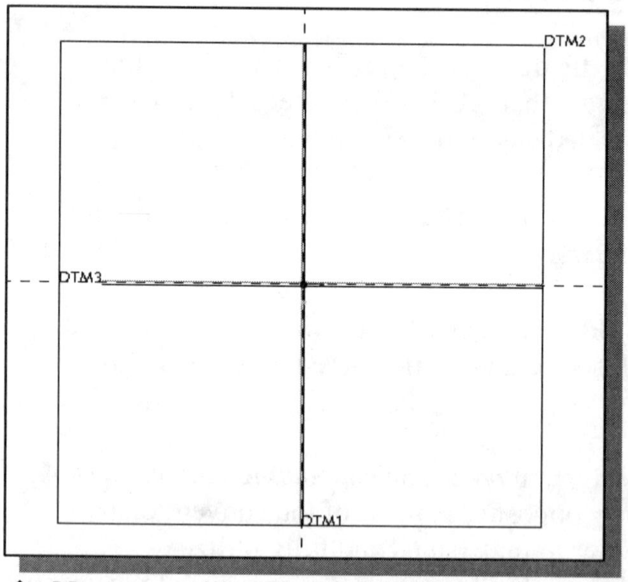

➢ Note that DTM1 and DTM3 are pre-selected as the sketching references. In the graphics area, the two references are displayed with two dashed lines.

- In Pro/E, a 2D sketch needs to be Fully Placed with respect to at least two references. In this case, DTM1 is used to control the horizontal placement of geometry, where DTM3 is used to control the vertical placements.

❖ Next, we will create a rough sketch by using some of the visual aids available, and then update the design through the associated control parameters.

- Move the graphics cursor to the **Line** icon in the *Sketcher* toolbar. A *help-tip* box appears next to the cursor and a brief description of the command options is displayed in the message area.

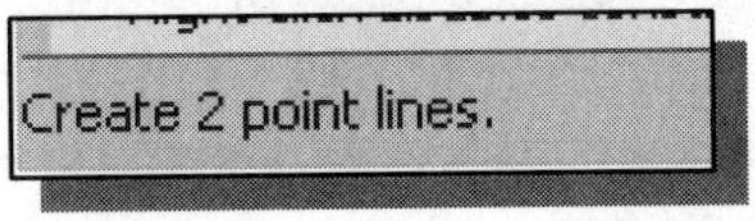

❖ The *Sketcher* toolbar, located on the right side of the main window, provides tools for creating the basic 2D geometry that can be used to create features and parts.

Graphics Cursors

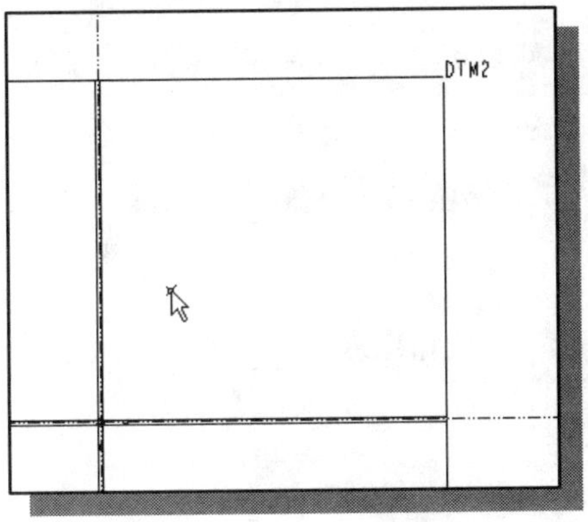

❖ Notice the cursor changes from an arrow to an arrow with a small crosshair when graphical input is expected.

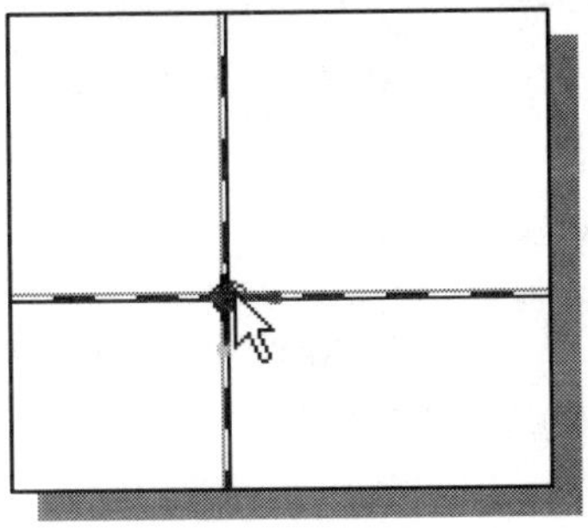

1. Move the cursor near the intersection of the two references, and notice that the small crosshair attached to the cursor will automatically snap to the intersection point. **Left-click** once to place the starting point as shown.

2. As you move the graphics cursor, you will see different symbols appear at different locations.

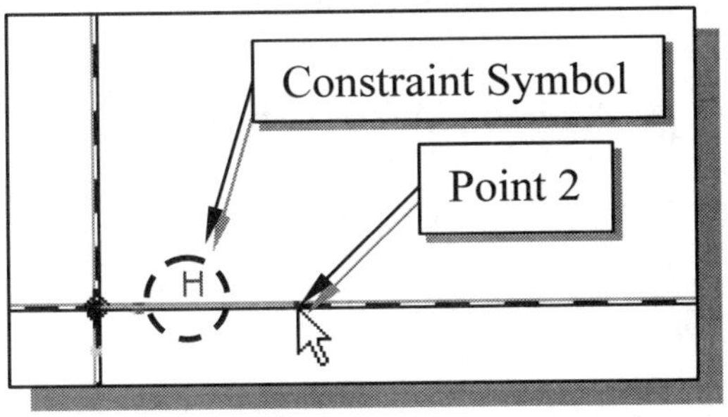

3. Move the cursor along the vertical reference and create a short horizontal line by clicking at a location to the right of the starting point (***Point 2***) as shown. Notice the geometric constraint symbol, **H**, indicating the created line is *horizontal*.

Geometric Constraint Symbols

❖ *Pro/ENGINEER* displays different visual clues, or symbols, to show you alignments, perpendicularities, tangencies, etc. These constraints are used to capture the *design intent* by creating constraints where they are recognized. *Pro/ENGINEER* displays the governing geometric rules as models are built.

V	Vertical	indicates a segment is vertical
H	Horizontal	indicates a segment is horizontal
L	Equal Length	indicates two segments are of equal length
R	Equal Radii	indicates two curves are of equal radii
T	Tangent	indicates two entities are tangent to each other
	Parallel	indicates a segment is parallel to other entities
	Perpendicular	indicates a segment is perpendicular to other entities
	Symmetry	indicates two points are symmetrical
O	Point on Entity	indicates the point is on another entity

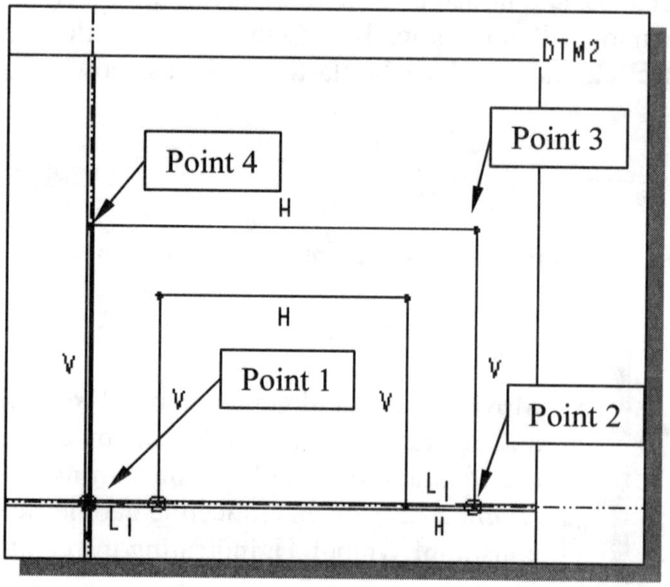

4. Complete the sketch as shown, a closed region ending at the starting point (***Point 1***). Watch the displayed constraint symbols while sketching, especially the applied ***Equal Length*** constraint, **L1**, to the two short horizontal edges.

➢ Note that all segments are either *vertical* or *horizontal*.

5. Inside the graphics area, click **twice** with the **middle-mouse-button** to end the current line sketch.

❖ *Pro/ENGINEER's* ***Intent Manager*** automatically places dimensions and constraints on the sketched geometry. This is known as the ***Dynamic Dimensioning and Constraints*** feature. Constraints and dimensions are added "on the fly." Do not be concerned with the size of the sketched geometry or the displayed dimensional values; we will modify the sketched geometry in the following sections.

Dynamic Viewing Functions

❖ *Pro/ENGINEER* provides a special user interface, ***Dynamic Viewing***, which enables convenient viewing of the entities in the display area at any time. The ***Dynamic Viewing*** functions are controlled with the combinations of the middle mouse button, the **[Ctrl]** key and the **[Shift]** key on the keyboard.

Zooming – [Ctrl] key and [middle-mouse-button]

Hold down the **[Ctrl]** key and press down the middle-mouse-button in the display area. Drag the mouse vertically on the screen to adjust the scale of the display. Moving upward will reduce the scale of the display, making the entities display smaller on the screen. Moving downward will magnify the scale of the display.

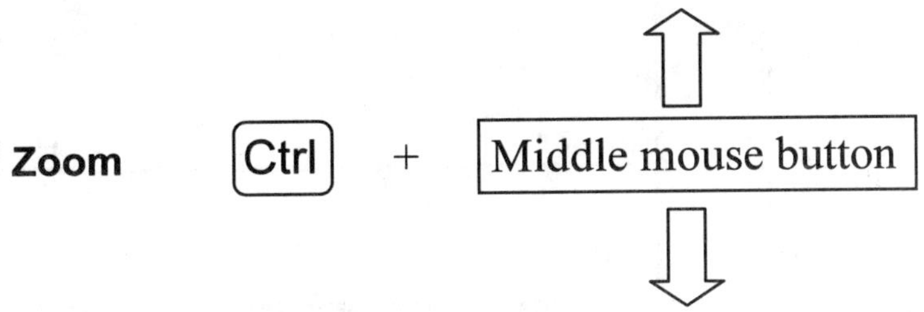

Panning – [Shift] key and [middle-mouse-button]

Hold down the **[Shift]** key and press down the middle-mouse-button in the display area. Drag the mouse to pan the display. This allows you to reposition the display while maintaining the same scale factor of the display. This function acts as if you are using a video camera. You control the display by moving the mouse.

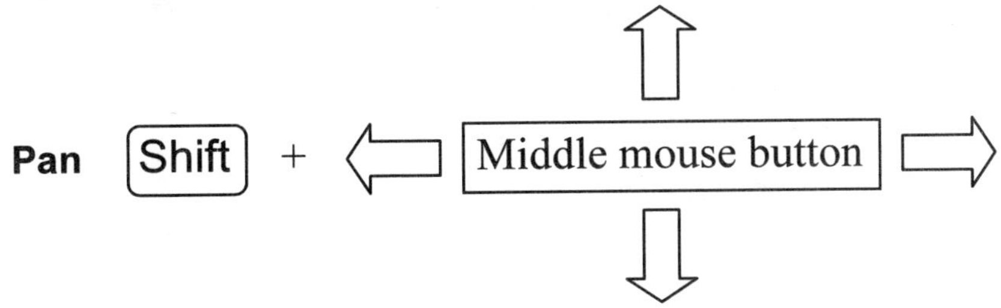

➢ On your own, use the *Dynamic Viewing* functions to reposition and magnify the scale of the 2D sketch to the center of the screen so that it is easier to work with.

Step 4: Apply/modify constraints and dimensions

➢ As the sketch is made, *Pro/ENGINEER* automatically applies geometric constraints (such as horizontal, vertical and equal length) and dimensions to the sketched geometry. We can continue to modify the geometry, apply additional constraints and/or dimensions, or define/modify the size and location of the existing geometry. It is more than likely that some of the automatically applied dimensions may not match with the design intent we have in mind. For example, we might want to have dimensions identifying the overall-height, overall-width, and the width of the inside-cut of the design, as shown in the figures below.

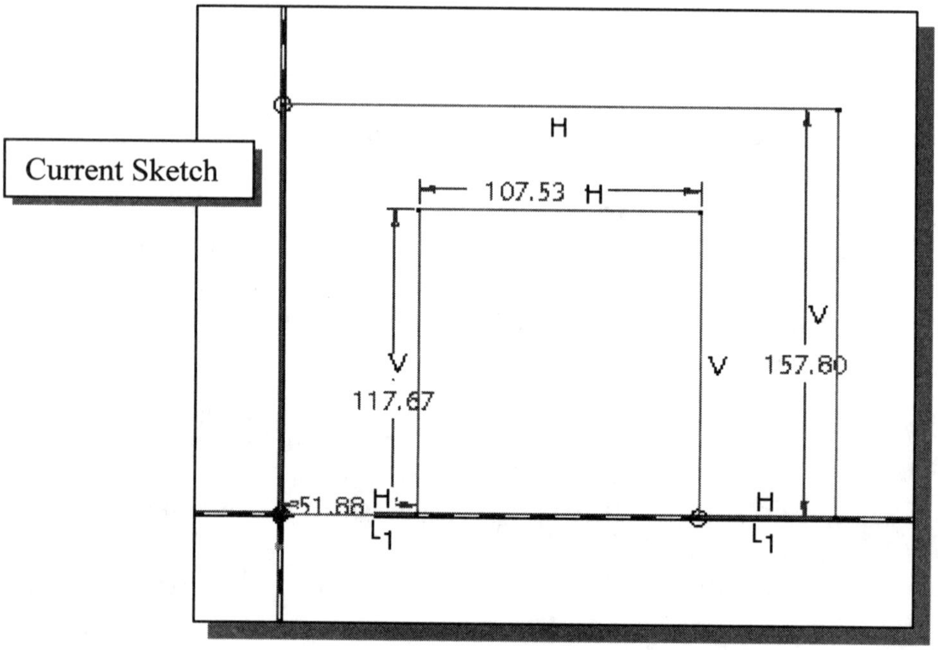

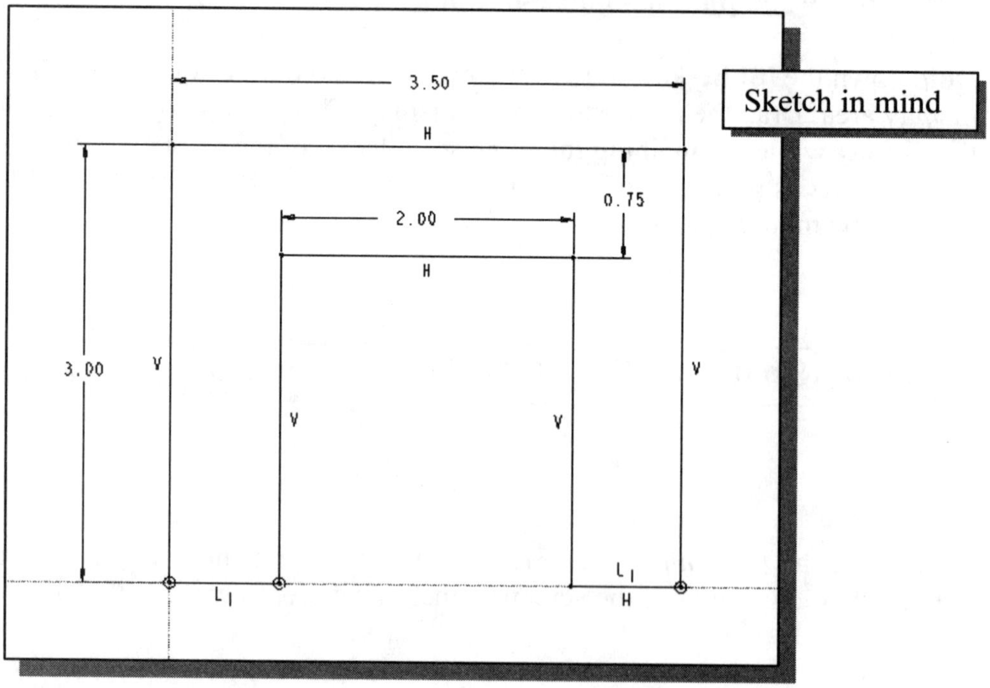

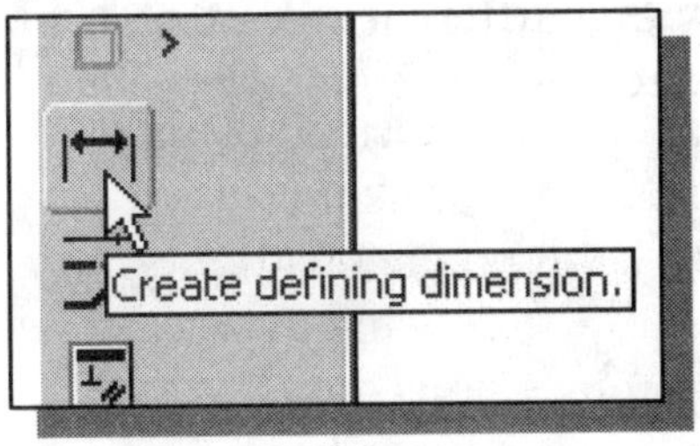

1. Click on the **Dimension** icon in the *Sketcher* toolbar as shown. This command allows us to create defining dimensions.

2. Select the inside horizontal line by left-clicking once on the line as shown.

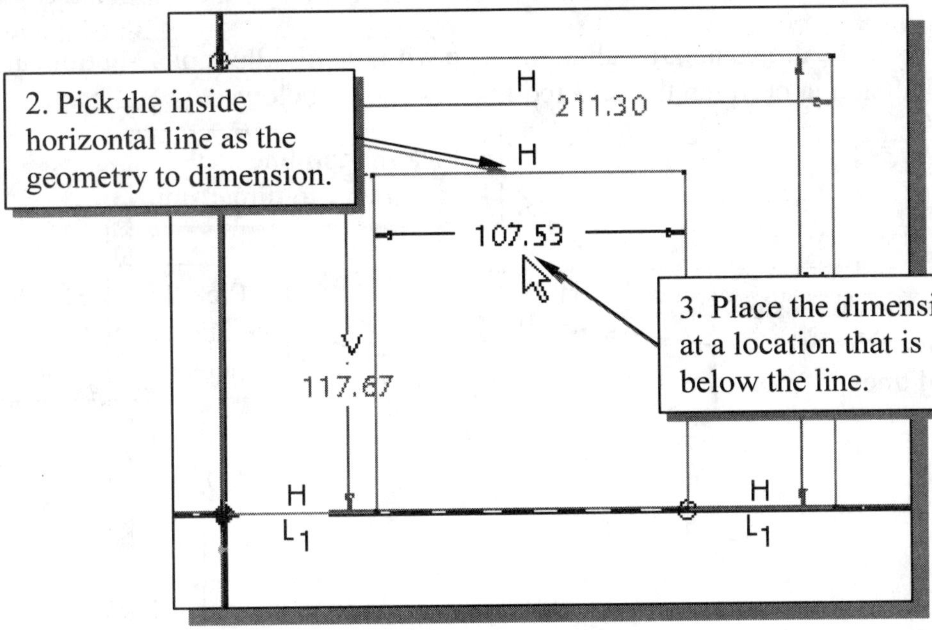

3. Move the graphics cursor below the selected line and click once with the **middle-mouse-button** to place the dimension. (Note that the value displayed on your screen might be different than what is shown in the above figure.)

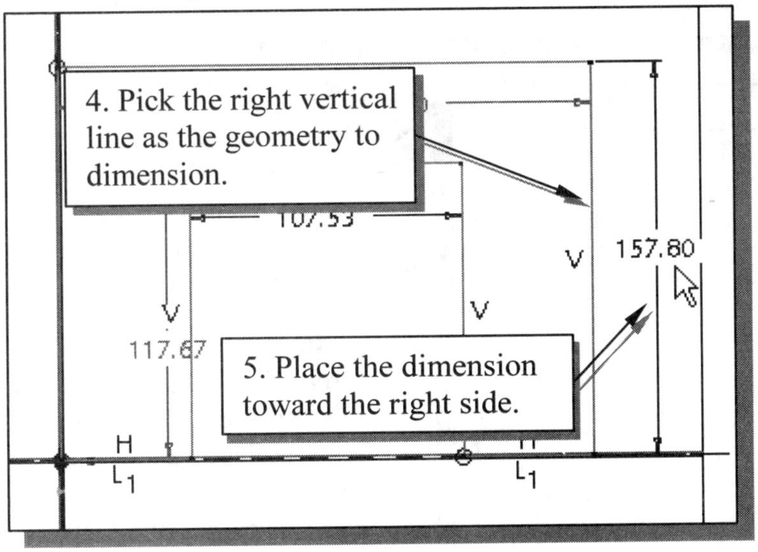

4. Select the right vertical line.

5. Place the dimension, by clicking once with the **middle-mouse-button** at a location toward the right of the sketch.

❖ The Dimension command will create a length dimension if a single line is selected.

- Notice the overall-height dimension applied automatically by the *Intent Manager* is removed as the new dimension is defined.

❖ Note that the dimensions we just created are displayed with a different color than those that are applied automatically. The dimensions created by the *Intent Manager* are called **weak dimensions**, which can be replaced/deleted as we create specific defining dimensions to satisfy our design intent.

6. Select the top horizontal line as shown below.

7. Select the inside horizontal line as shown below.

8. Place the dimension, by clicking once with the middle-mouse-button, at a location in between the selected lines as shown below.

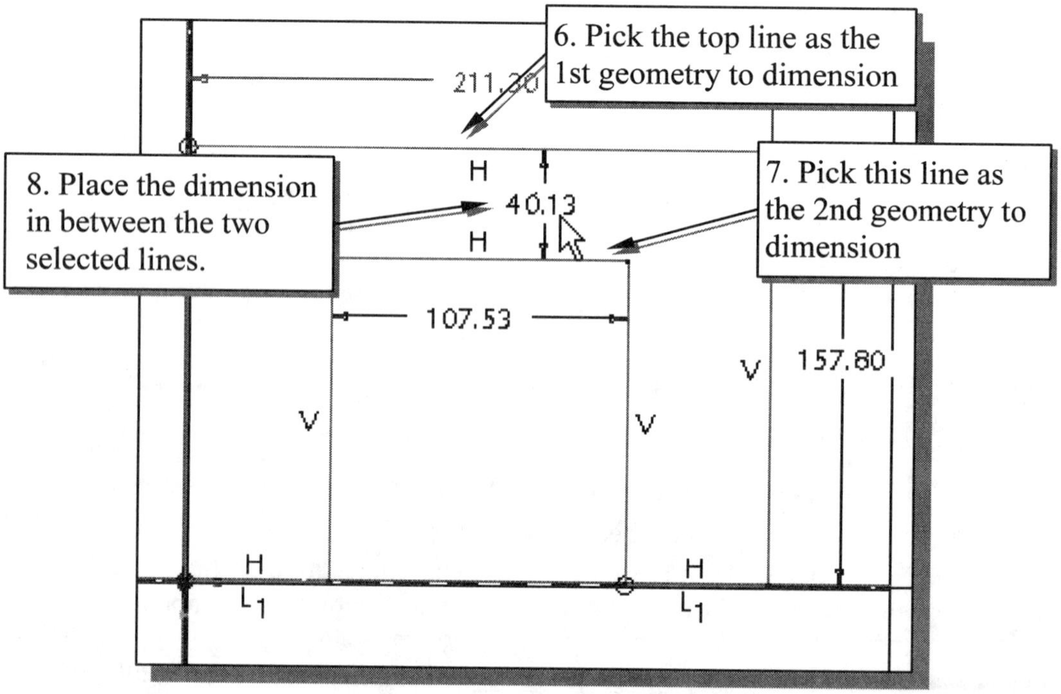

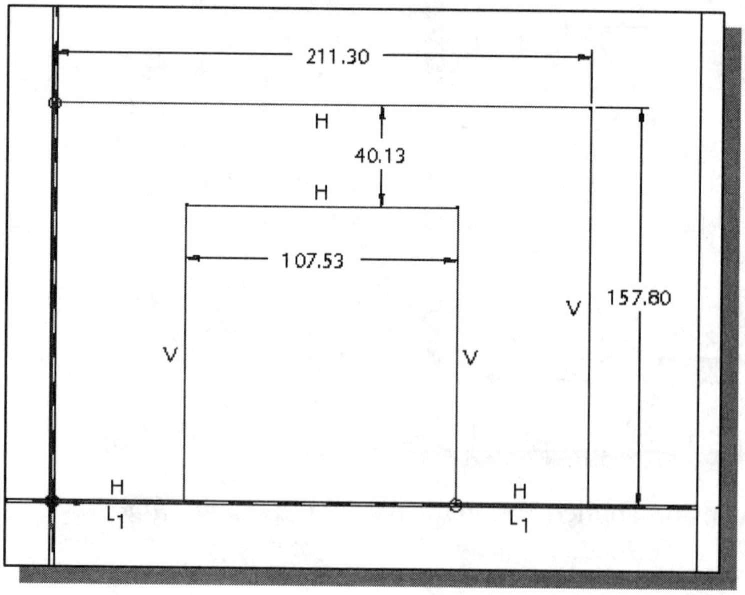

❖ When two parallel lines are selected, the Dimension command will create a dimension measuring the distance in between.

➤ Examine the established dimensions and constraints in the sketch that you have created, is the sketch fully described? Or should we add additional dimensions?

Modifying the dimensions of the sketch

1. Click on the **Select** icon in the *Sketcher* toolbar as shown. The Select command allows us to perform several modification operations on the sketched geometry and dimensions.

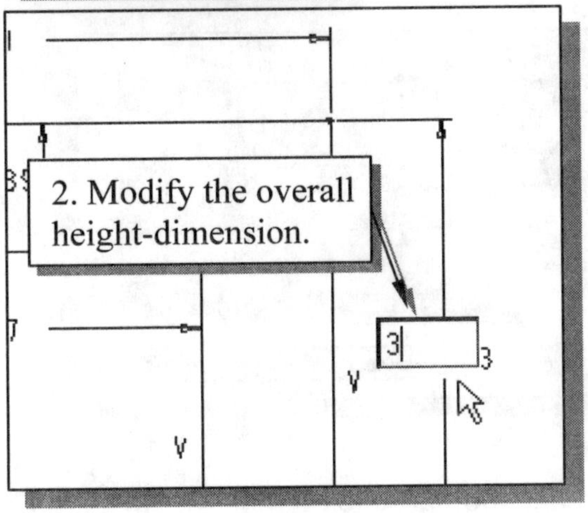

2. Select the overall height dimension of the sketch by **double-clicking** with the left-mouse-button on the dimension text.
3. In the *dimension value* box, the current length of the line is displayed. Enter **3** as the new value for the dimension.
4. Press the **ENTER** key once to accept the entered value.

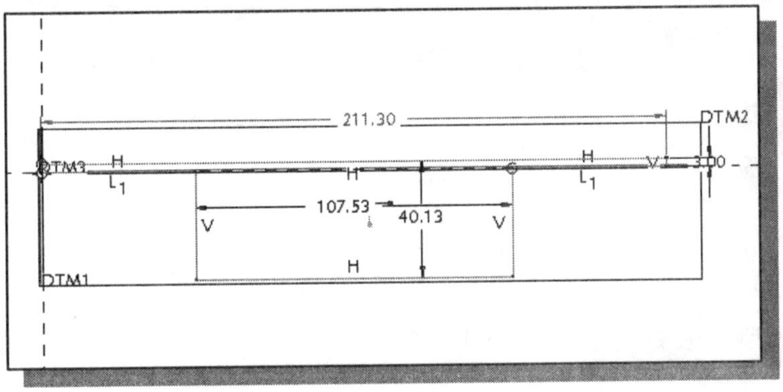

➢ *Pro/ENGINEER* will update the sketch using the entered dimension value. Since the other dimensions are much larger, the sketch becomes greatly distorted. We will take a different approach to modify the geometry.

5. Click on the **Undo** icon in the *Standard* toolbar to undo the Modify Dimension performed.

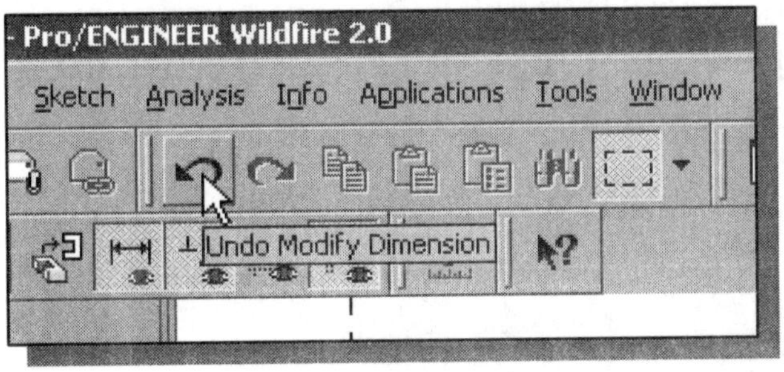

➢ Notice that the **Redo** icon is also available in the *Standard* toolbar.

6. In the pull-down menu area, click on **Edit** to display the option list and select the following option items:

Edit → Select → All (Note that Crtl+Alt+A can also activate this option.)

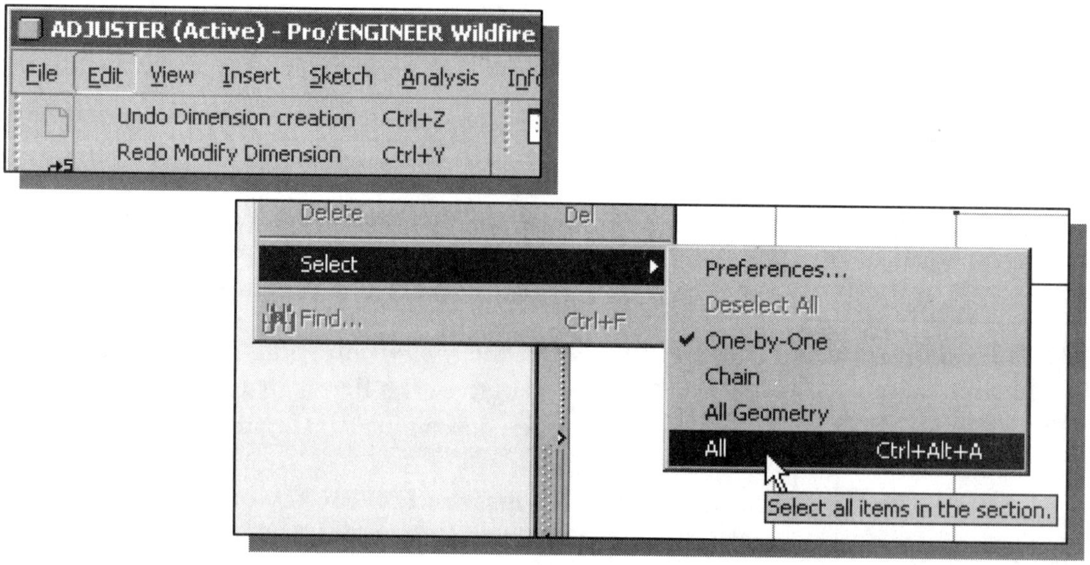

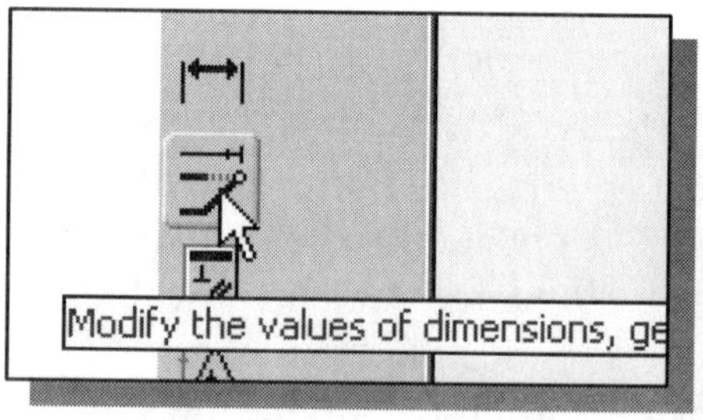

7. In the *Sketcher* toolbar, click on the **Modify** icon as shown.

- With the pre-selection option, all dimensions are selected and listed in the *Modify Dimensions* dialog box.

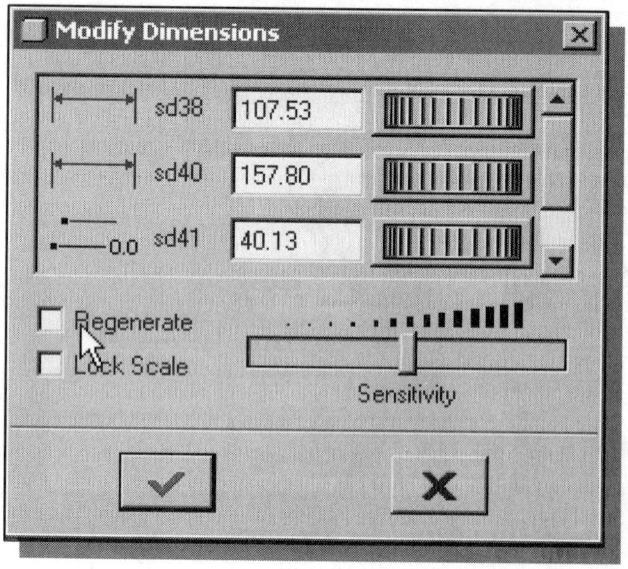

8. Turn **off** the ***Regenerate*** option by left-clicking once on the option as shown.

9. On you own, adjust the dimensions as shown below. Note that the dimension selected in the *Modify Dimensions* dialog box is identified with an enclosed box in the display area.

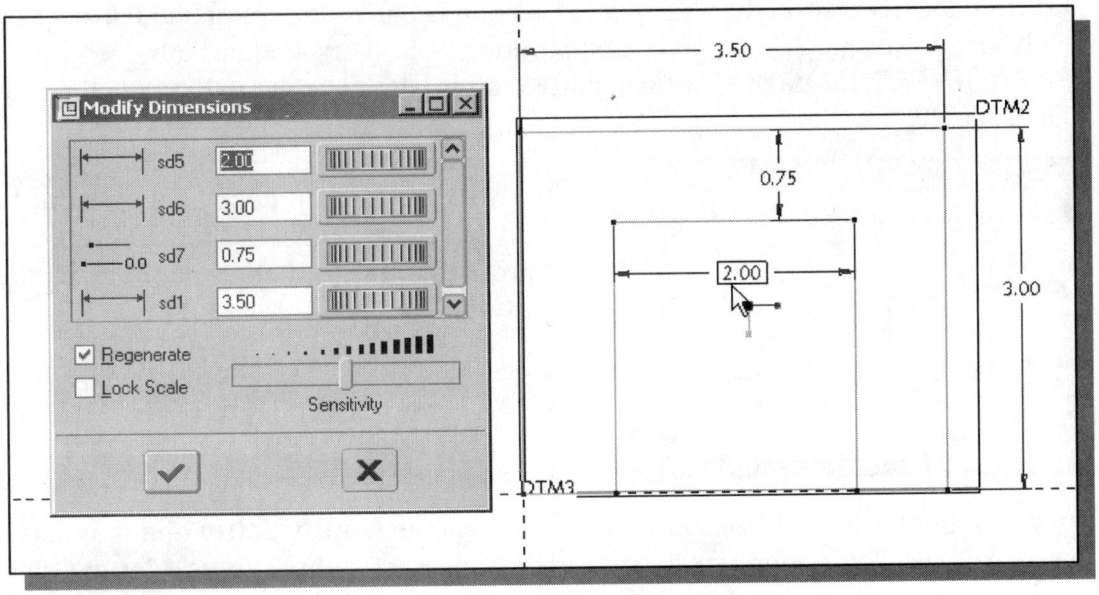

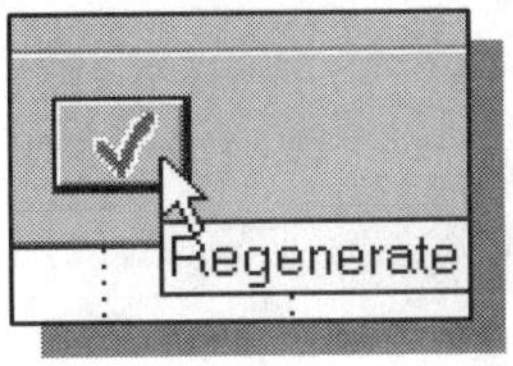

10. Inside the *Modify Dimensions* dialog box, click on the **Accept** button to regenerate the sketched geometry and exit the Modify Dimensions command.

Repositioning Dimensions

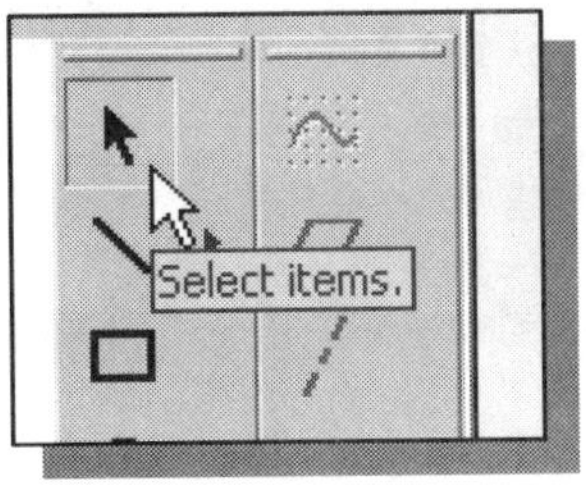

1. Confirm the **Select** icon, in the *Sketcher* toolbar, is activated as shown.

2. Press and hold down the left-mouse-button on any dimension text, then drag the dimension to a new location in the display area. (Note the cursor is changed to a moving arrow icon during this operation.)

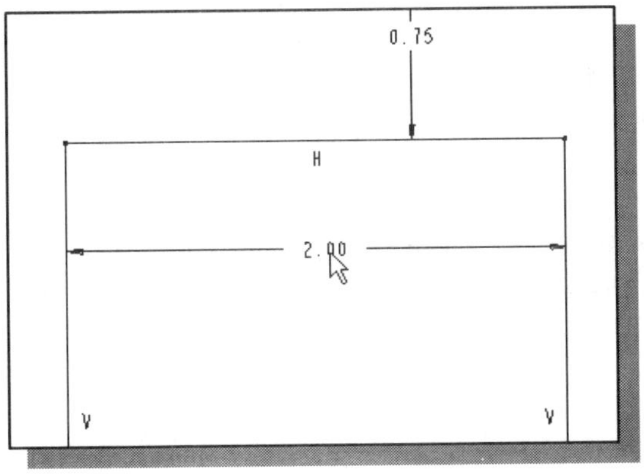

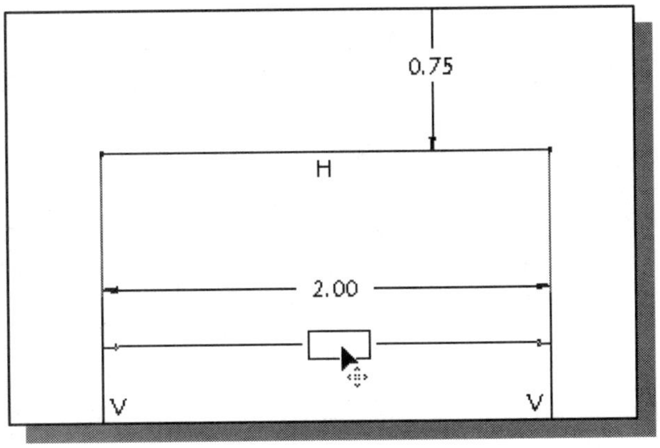

Step 5: Completing the Base Solid Feature

❖ Now that the 2D sketch is completed, we will proceed to the next step: creating a 3D part from the 2D section. Extruding a 2D section is one of the common methods that can be used to create 3D parts. We can extrude planar faces along a path. In *Pro/ENGINEER*, the default extrusion direction is perpendicular to the sketching plane, **DTM2**.

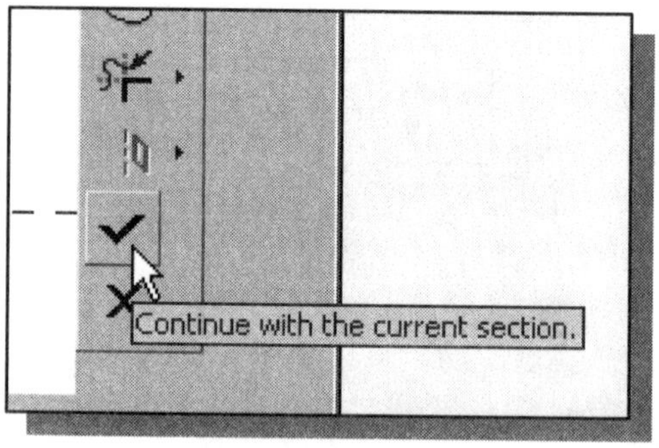

1. In the *Sketcher* toolbar, click on the **Accept** icon to end the *Pro/ENGINEER 2D Sketcher* and proceed to the next element of the feature definition.

2. In the *Feature Option Dashboard*, confirm the **Depth Value** option is set as shown. This option sets the extrusion of the section by **Extrude from sketch plane by a specific depth value**.

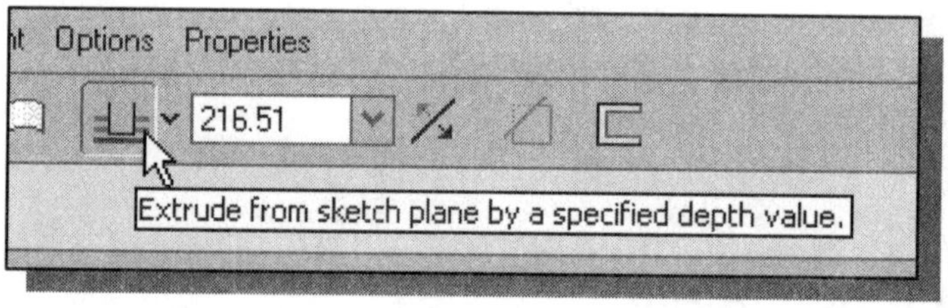

3. In the *depth value* box, enter **2.5** as the extrusion depth.

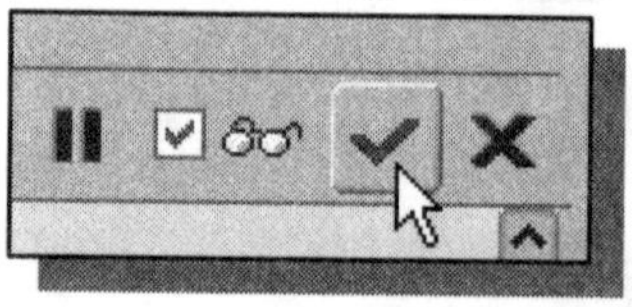

4. In the message area, click on the **Accept** button to proceed with the feature definition.

➢ Note that all dimensions disappeared from the screen. All parametric definitions are stored in the ***Pro/ENGINEER* database**, and any of the parametric definitions can be displayed and edited at any time.

The Third Dynamic Viewing Function

3D Dynamic Rotation – [middle mouse button]

Press down the middle-mouse-button in the display area. Drag the mouse on the screen to rotate the model about the screen.

3D Rotation Middle mouse button

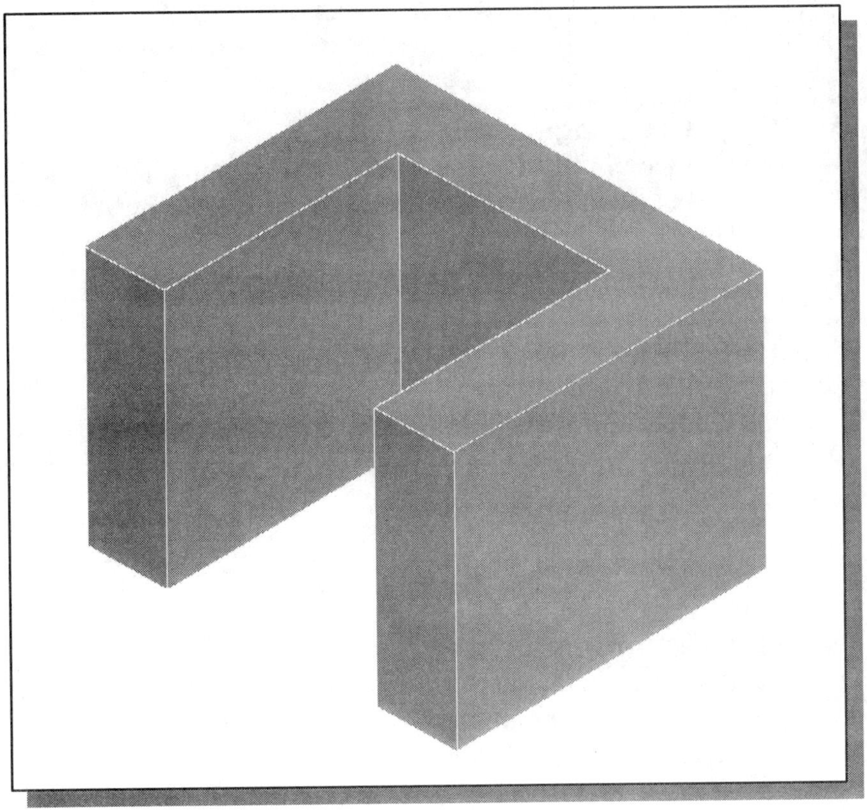

❖ On your own, practice the use of the ***Dynamic Viewing functions,*** which enables convenient viewing of the entities in the display area at any time.

Zoom Ctrl + Middle mouse button

Pan Shift + Middle mouse button

Display Modes: Wireframe, Shaded, Hidden Edge, No Hidden

- The display in the graphics window has four display-modes: wireframe, hidden edge displayed as hidden lines, no hidden lines, and shaded image. To change the display mode in the active window, click on one of the display mode buttons on the *Standard* toolbar, as shown in the figure below.

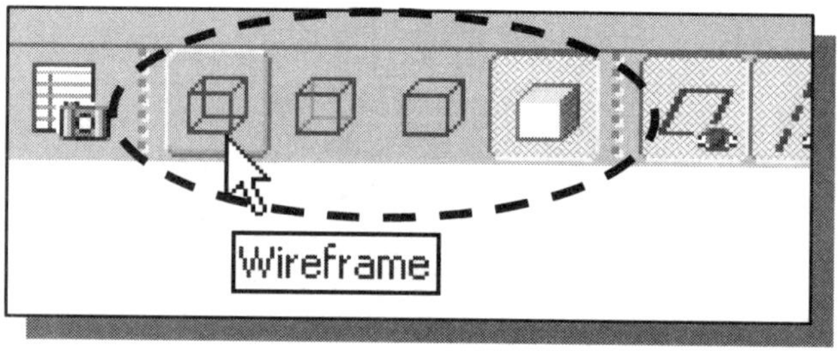

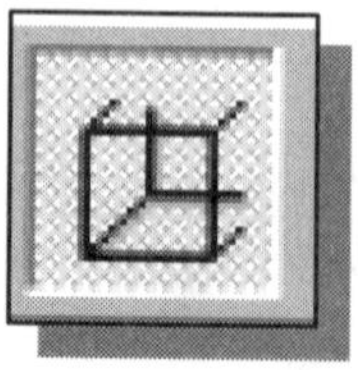

❖ **Wireframe Image:**
The first icon in the display mode button group allows the display of 3D objects using the basic wireframe representation scheme.

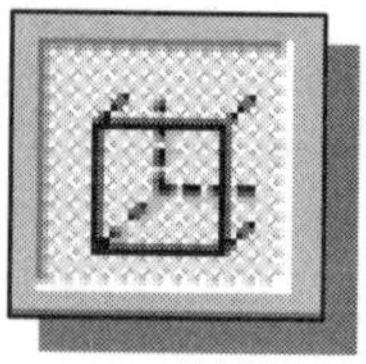

❖ **Hidden-Edge Display:**
The second icon in the display mode button group can be used to generate a wireframe image of the 3D object with all the back lines shown as hidden lines.

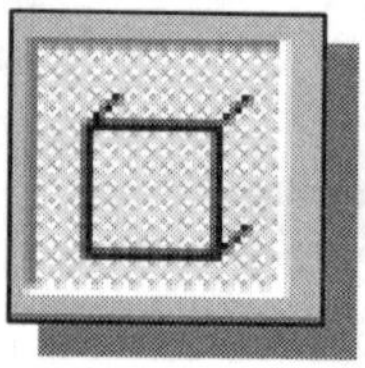

❖ **No Hidden-Edge Display:**
The third icon in the display mode button group can be used to generate a wireframe image of the 3D object with all the back lines removed.

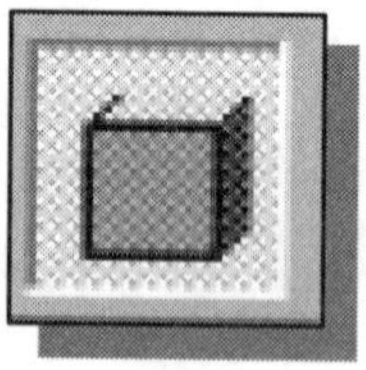

❖ **Shaded Solid:**
The fourth icon in the display mode button group generates a shaded image of the 3D object.

➢ On your own, use the different viewing options described in the above sections to familiarize yourself with the 3D viewing/display commands.

Step 6: Adding additional features

- Next, we will create another extrusion feature that will be added to the existing solid object.

1. In the *Feature Toolbars* (toolbars aligned to the right edge of the main window), select the **Extrude Tool** option as shown.

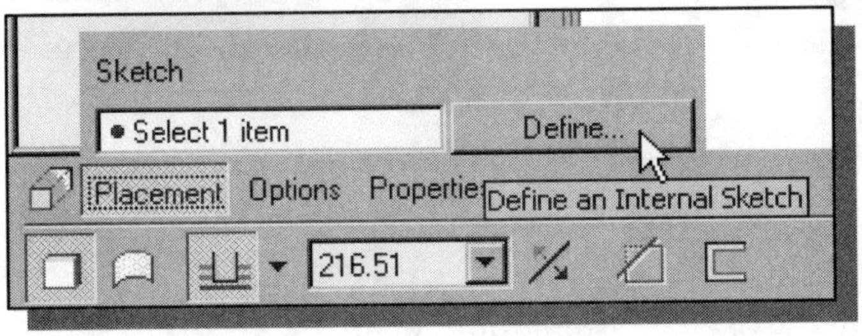

2. Click the **Placement** option and choose **Define** to begin creating a new *Internal sketch.*

3. Pick the right vertical face of the solid model as the sketching plane as shown in the below figure.

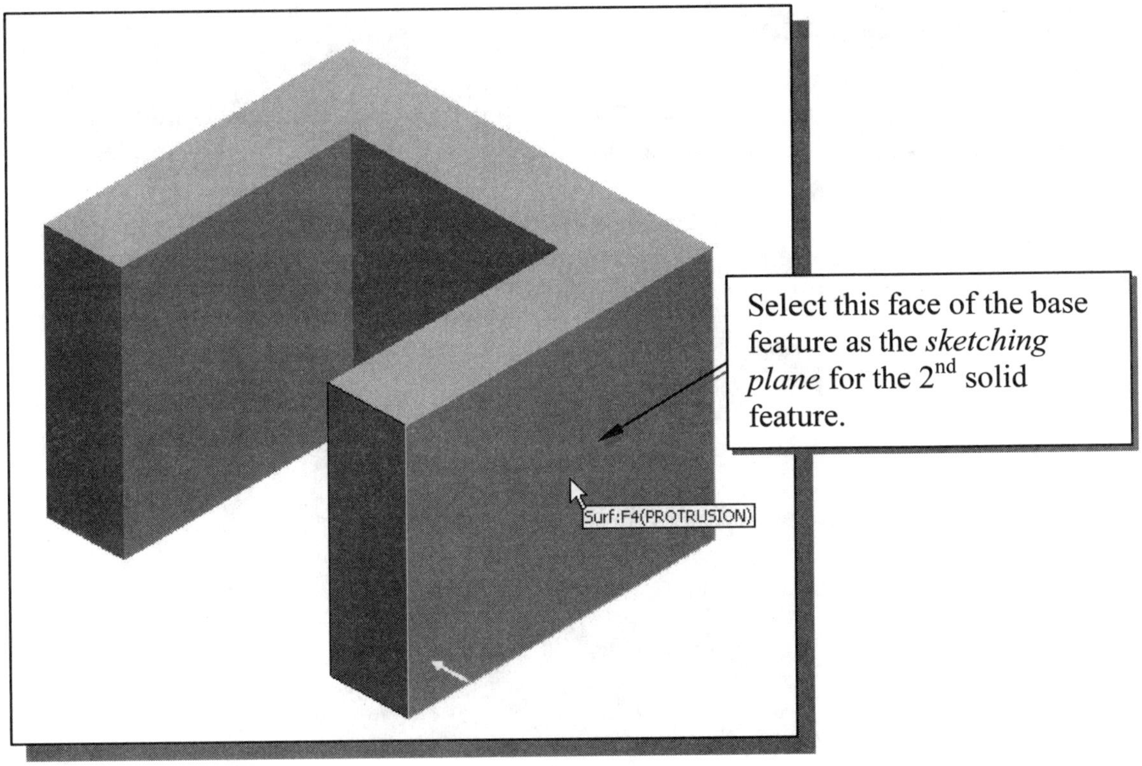

4. On your own, confirm the viewing direction is set as shown in the figure above.

5. In the display area, pick the top face of the base feature as shown.

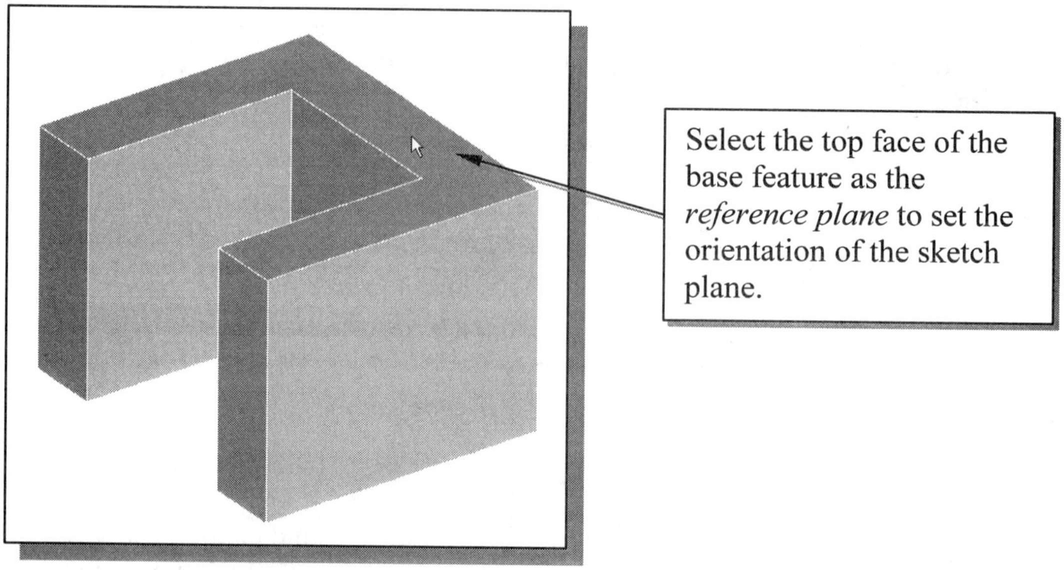

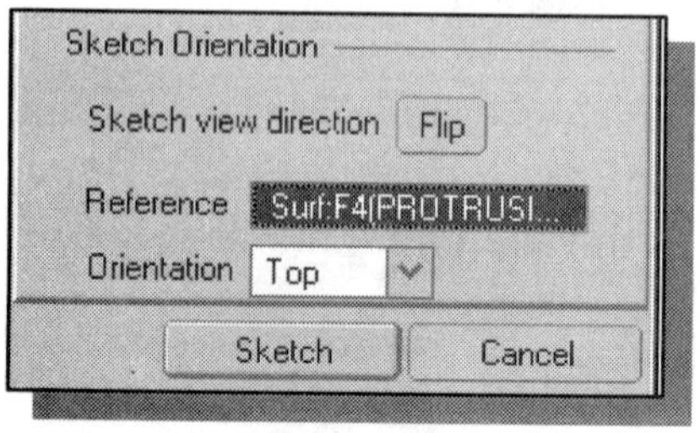

6. In the *Sketch Orientation* menu, pick **Top** to set the reference plane **Orientation**.

7. Pick **Sketch** to exit the *Section Placement* window and proceed to enter the *Pro/ENGINEER Sketcher* mode.

8. Note that the top surface of the solid model and DTM3 are pre-selected as the sketching references. In the graphics area, the two references are highlighted and displayed with two dashed lines.

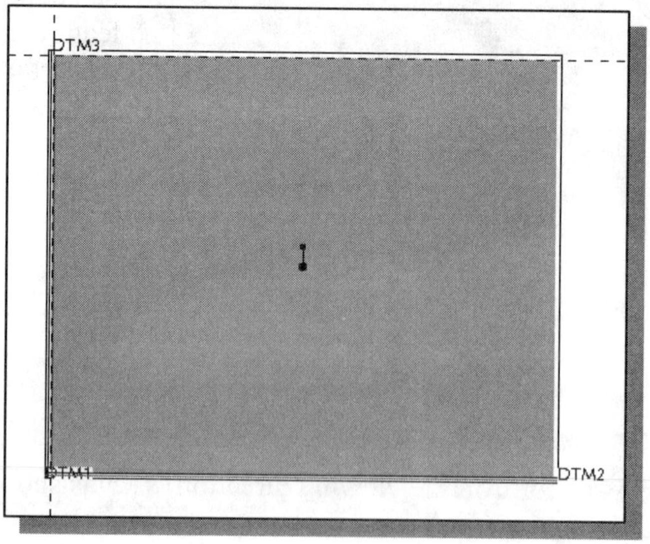

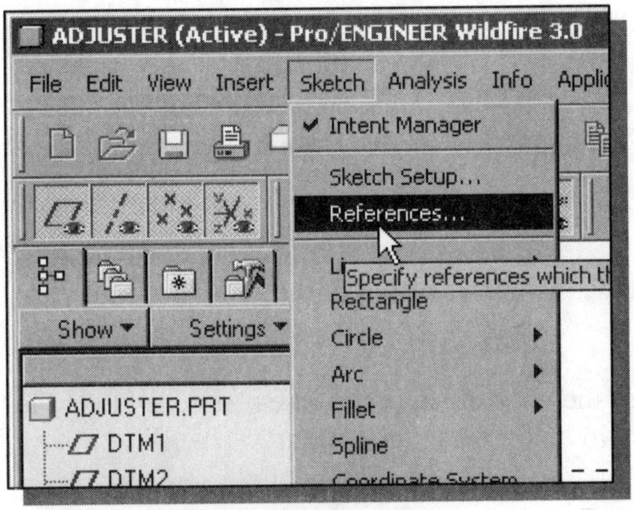

9. In the pull-down menu area, click on **Sketch** to display the option list and select the References option:

   Sketch → References...

➤ This will bring up the *References* dialog box.

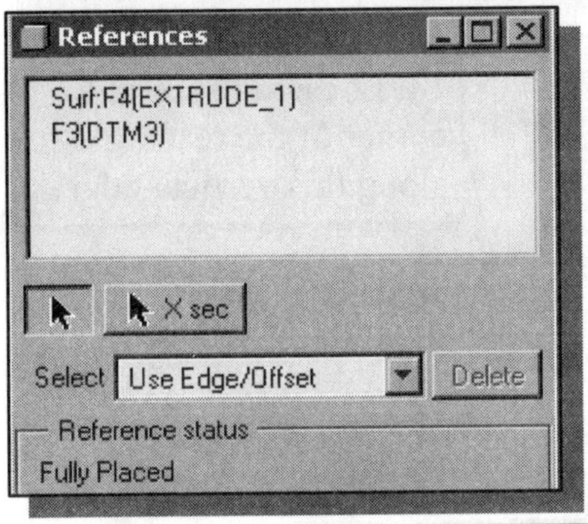

➤ Note that, in the *References* dialog box, the top surface of the solid model and DTM3 are pre-selected as the sketching references as shown.

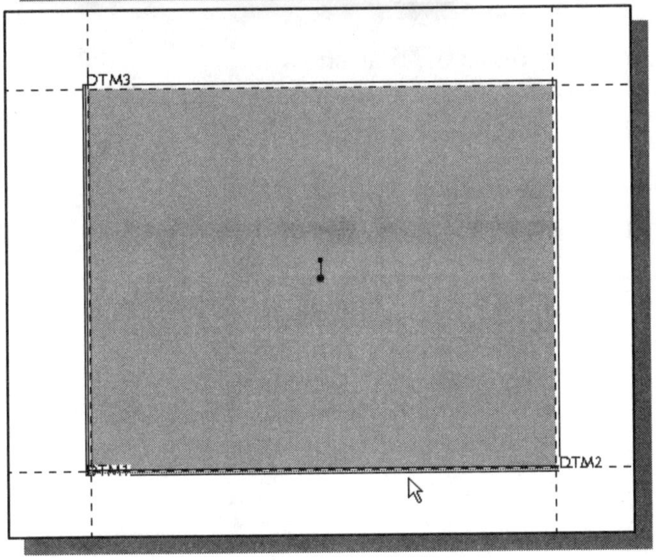

10. Select the right edge and the bottom edge of the base feature so that the four sides of the selected sketching plane, or corresponding datum planes, are used as references as shown.

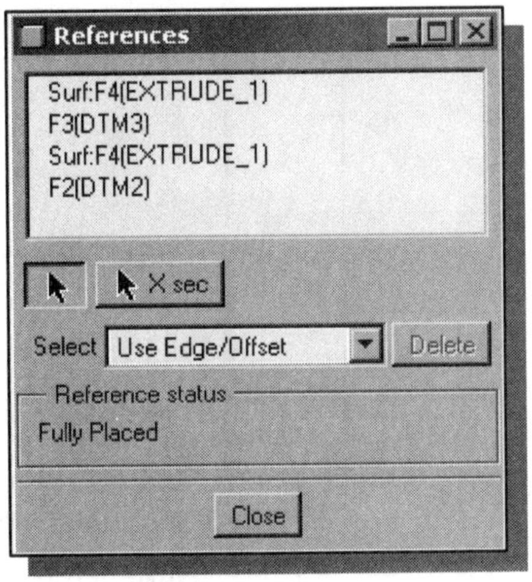

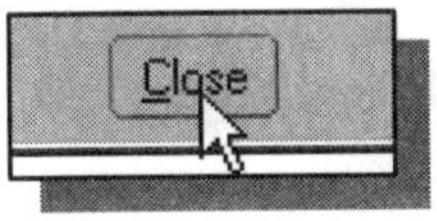

11. In the *References* dialog box, click on the **Close** button to accept the selections.

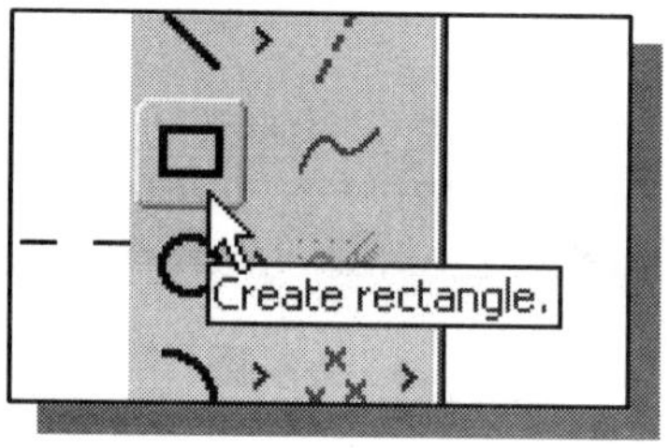

12. In the *Sketcher* toolbar, click on the **Rectangle** icon as shown to activate the **Create Rectangle** command.

13. Create a rectangle by clicking on the lower left corner of the solid model as shown below.

14. Move the cursor upward and place the opposite corner of the rectangle along the right edge of the base solid as shown below.

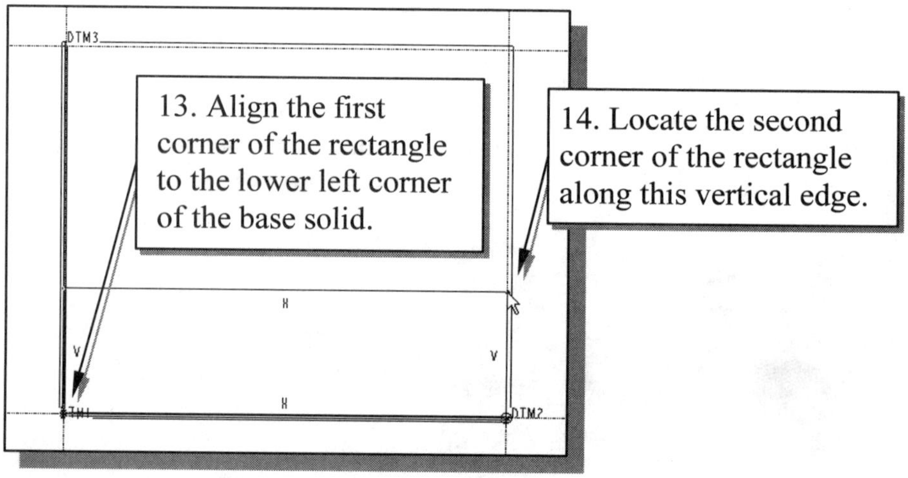

15. On your own, modify the height dimension to **0.75** as shown.

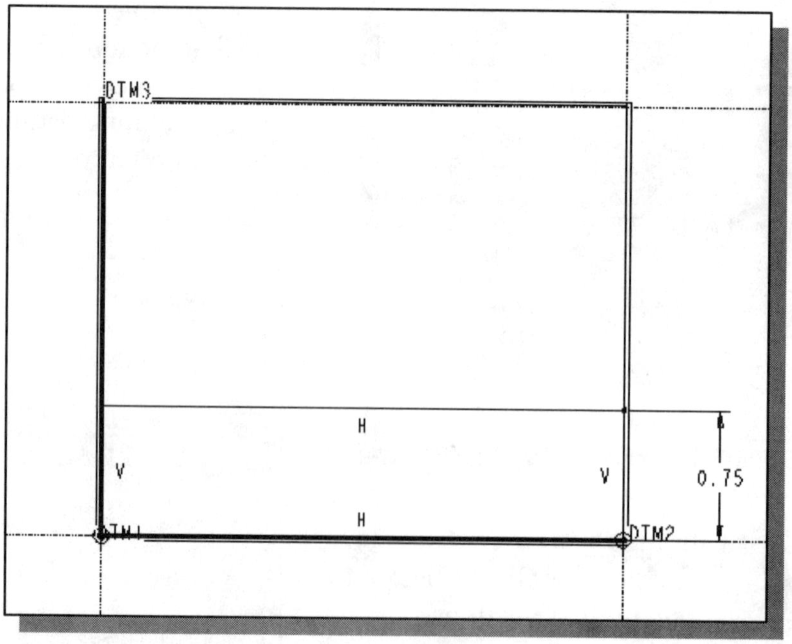

- Note that only one dimension, the height dimension, is applied to the 2D sketch; the width of the rectangle is defined by the references.

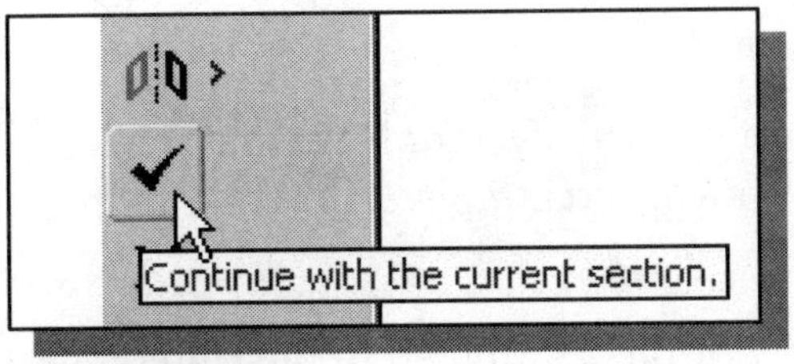

16. In the *Sketcher* toolbar, click on the **Accept** icon to end the *Pro/ENGINEER 2D Sketcher* and proceed to the next element of the feature definition.

17. Click **OK** to close the *Sketch* dialog box.

18. In the *Feature Option Dashboard*, confirm the **Depth Value** option is set and enter **2.5** as the extrusion depth as shown.

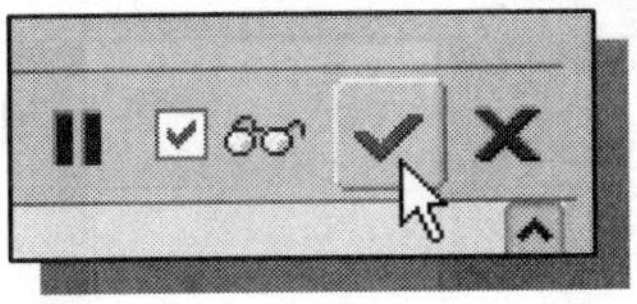

19. In the message area, click on the **Accept** button to proceed with the feature definition.

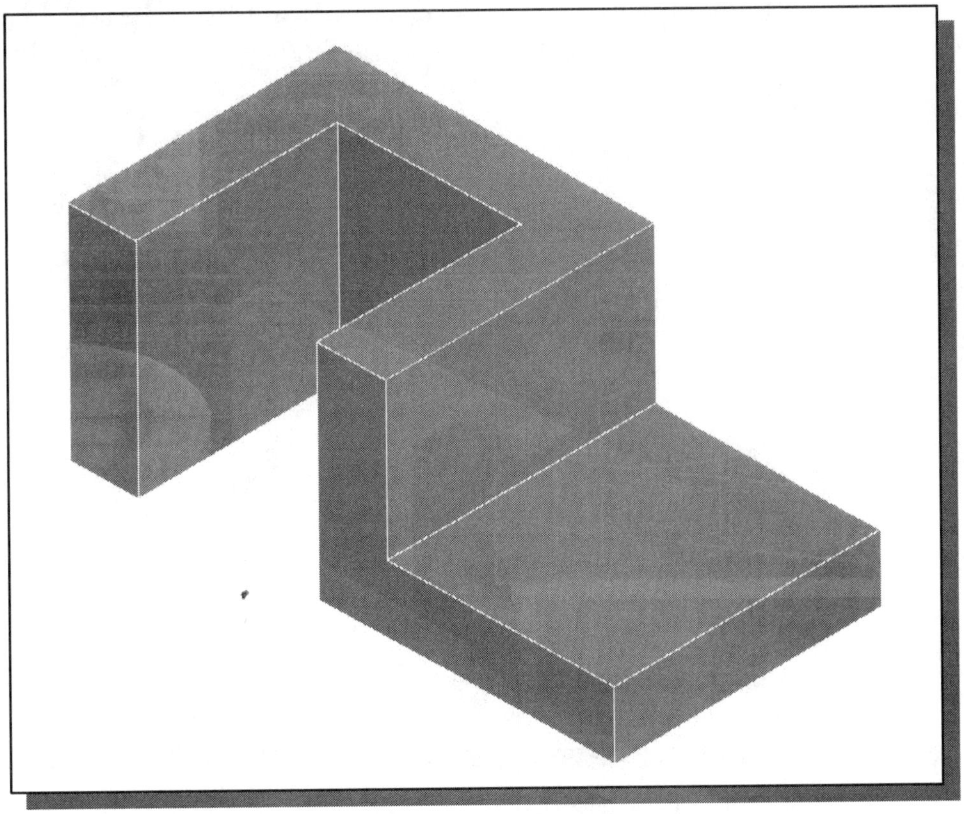

Creating a CUT Feature

❖ We will create a circular cut as the next solid feature of the design. Note that the procedure in creating a *cut feature* is almost the same as creating a *protrusion feature*.

1. In the *Feature Toolbars* (toolbars aligned to the right edge of the main window), select the **Extrude Tool** option as shown.

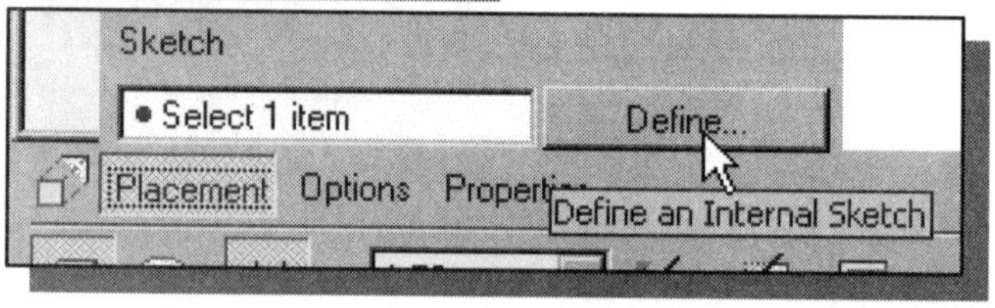

2. Click the **Placement** option and choose **Define** to begin creating a new *Internal sketch*.

3. We will use the top surface of the last feature as the sketching plane. Click once, with the **left-mouse-button**, inside the top surface of the rectangular solid feature as shown in the figure below.

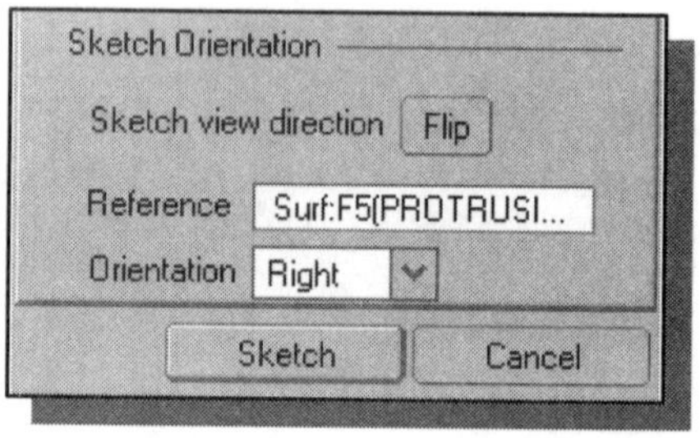

4. In the *Sketch Orientation* menu, confirm the reference plane **Orientation** is set to **Right**.

5. Pick the right vertical face of the second solid feature as the reference plane, which will be oriented toward the right edge of the computer screen.

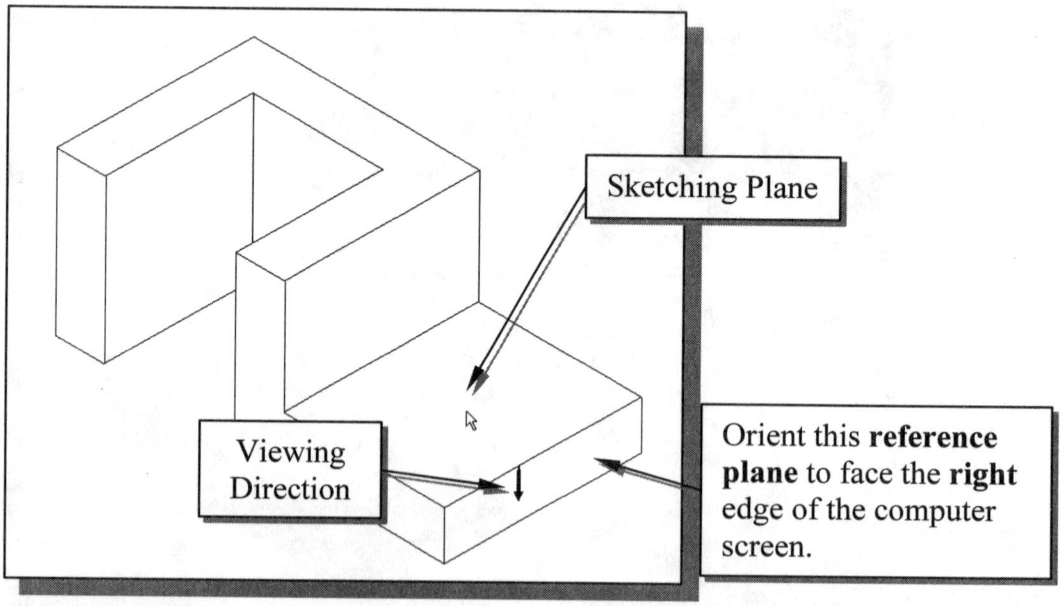

6. Pick **Sketch** to exit the *Sketch* dialog window and proceed to enter the *Pro/ENGINEER Sketcher* mode.

Creating the 2D Section of the CUT Feature

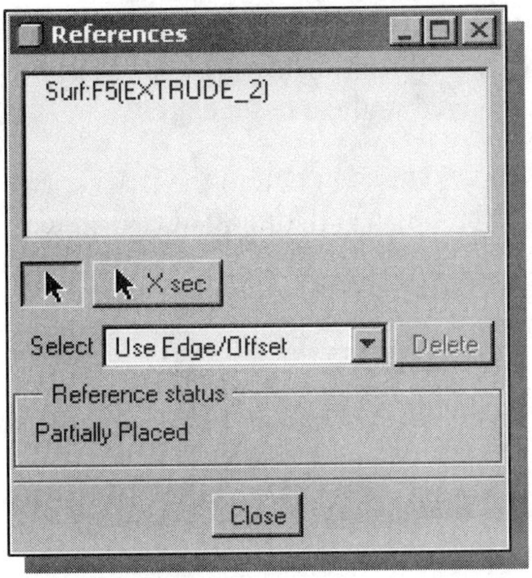

1. Note that the right vertical plane is pre-selected as a reference for the new sketch.

- Note that at least one horizontal reference and one vertical reference are required to position a 2D sketch. We will need at least one more vertical reference for this sketch.

2. Select **DTM3** as the vertical sketching references as shown. In the graphics area, the two references are highlighted and displayed with two dashed lines.

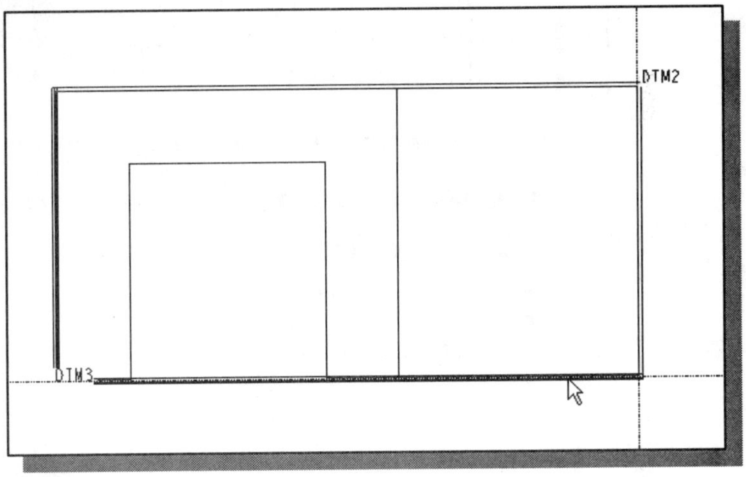

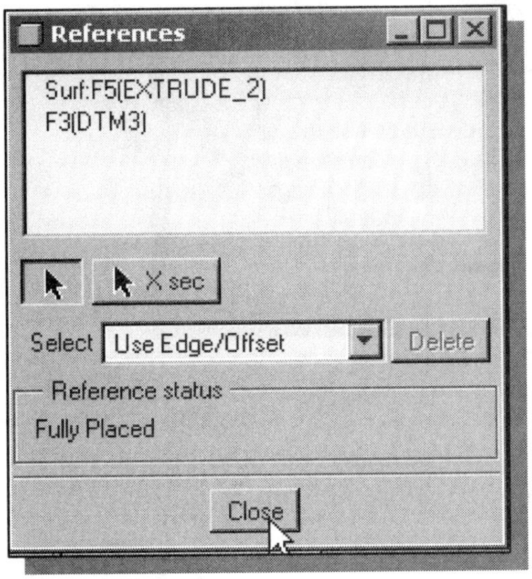

3. Click on the **Close** button to accept the selected references and proceed to entering the *Pro/ENGINEER Sketcher* module.

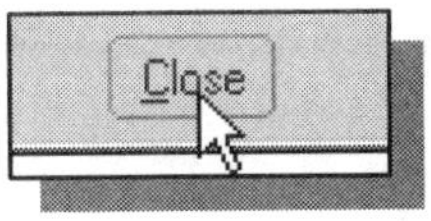

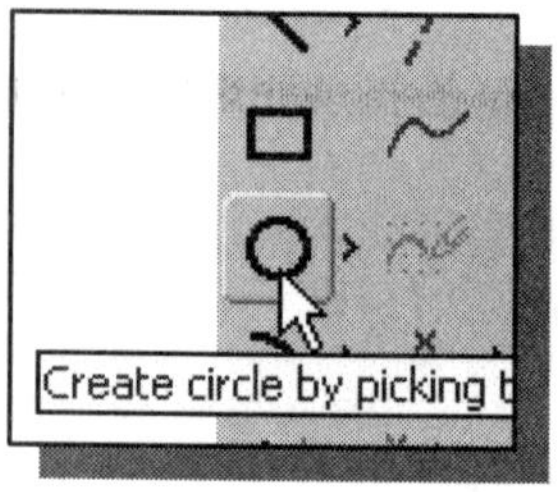

4. In the *Sketcher* toolbar, select **Circle** as shown. The default option is to create a circle by specifying the center point and a point through which the circle will pass. The message "*Select the center of a circle*" is displayed in the message area.

5. On your own, create a circle of arbitrary size on the sketching plane as shown.

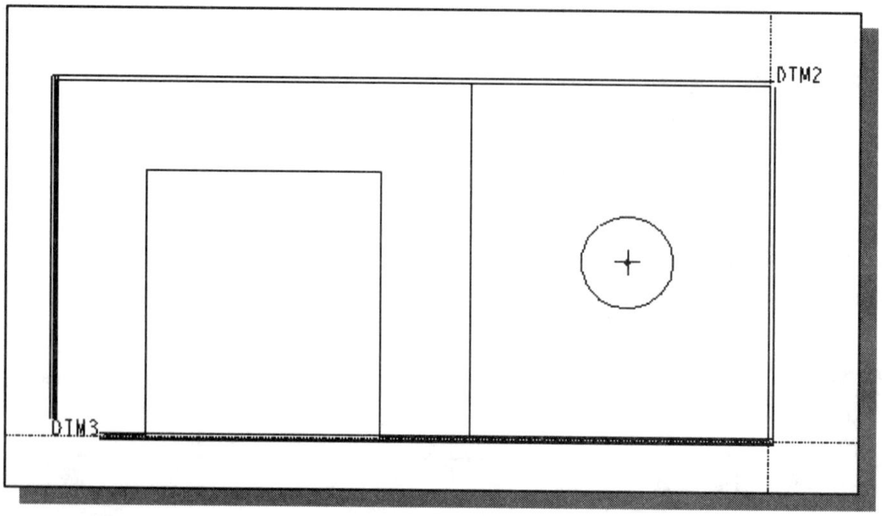

6. On your own, edit/modify the dimensions as shown.

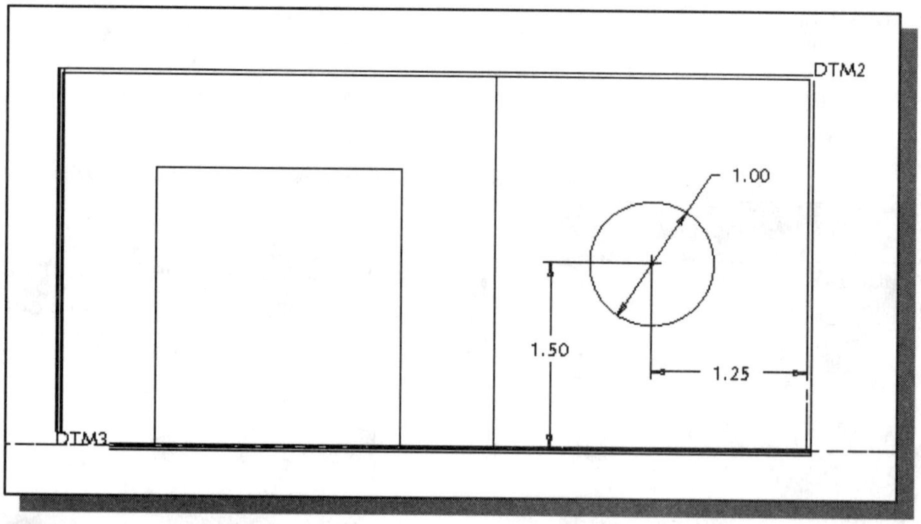

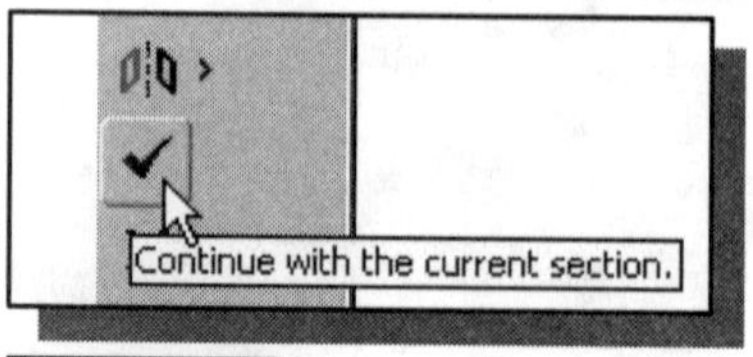

7. In the *Sketcher* toolbar, click on the **Accept** icon to exit the *Pro/ENGINEER 2D Sketcher* and proceed to the next element of the feature definition.

8. Click **OK** to close the *Sketch* dialog box..

9. Click on the **Remove Material** icon as shown in the below figure.

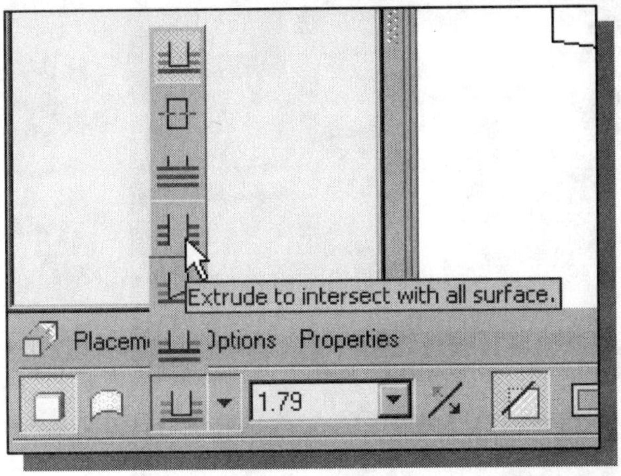

10. In the *Feature Option Dashboard*, select the **Extrude to intersect with all surface** option as shown.

- Note that this **Thru All** option does not require us to enter a value to define the depth of the extrusion; *Pro/ENGINEER* will calculate the required value to assure the extrusion is through the entire solid model.

11. On your own, use the *Dynamic Rotate* function to view the feature.

12. Click on the **Flip direction** icon as shown in the below figure to set the cut direction.

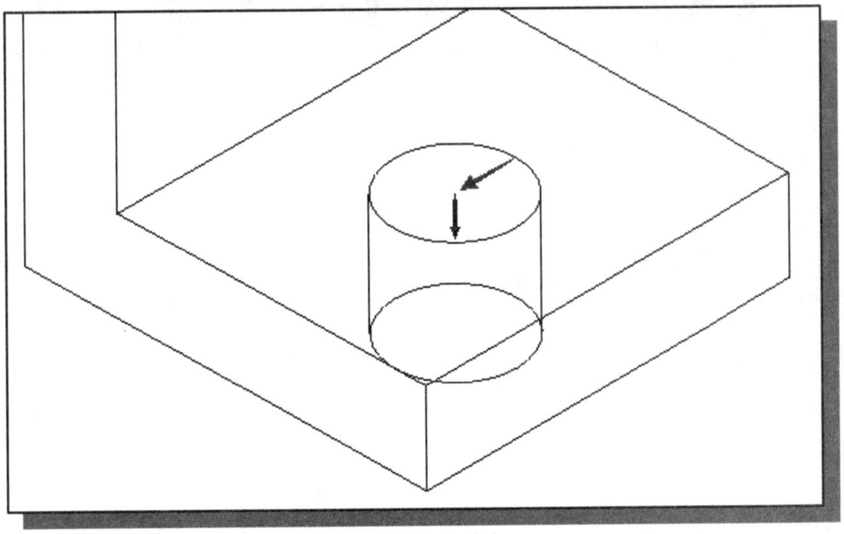

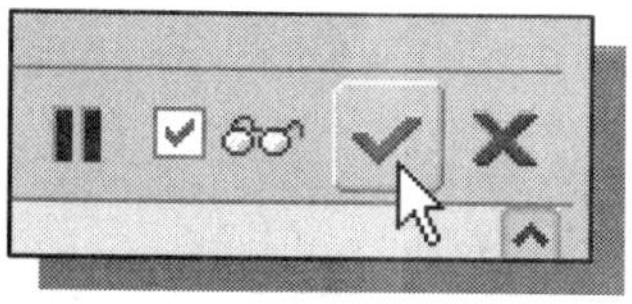

13. Click on **Accept** to proceed with the extrusion option.

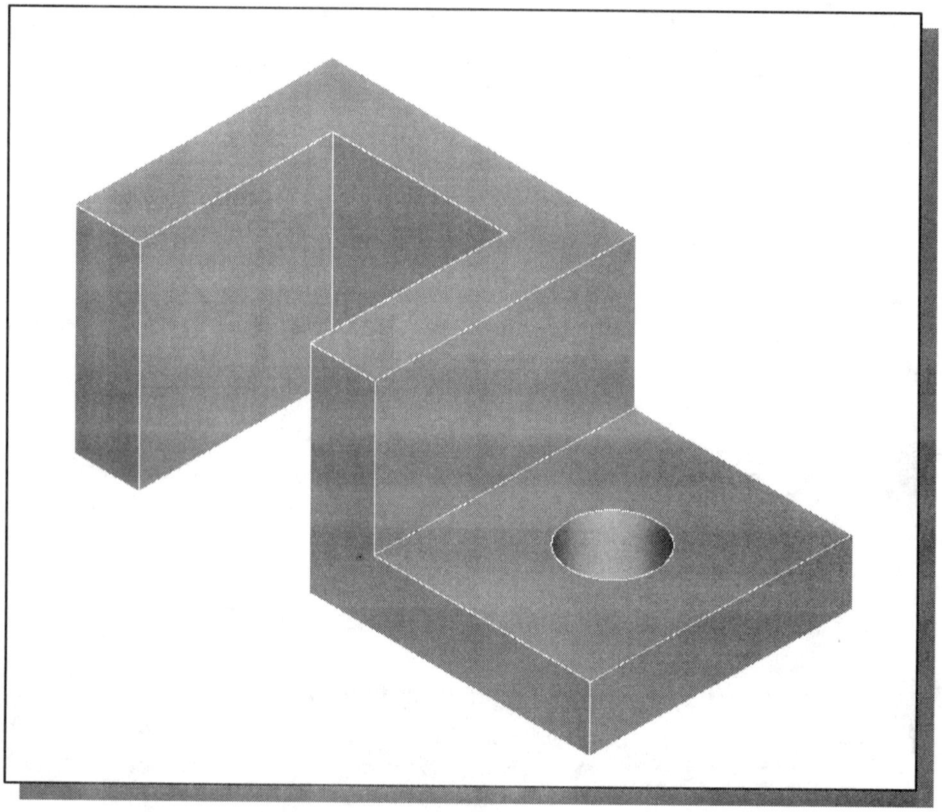

Save the Part and Exit

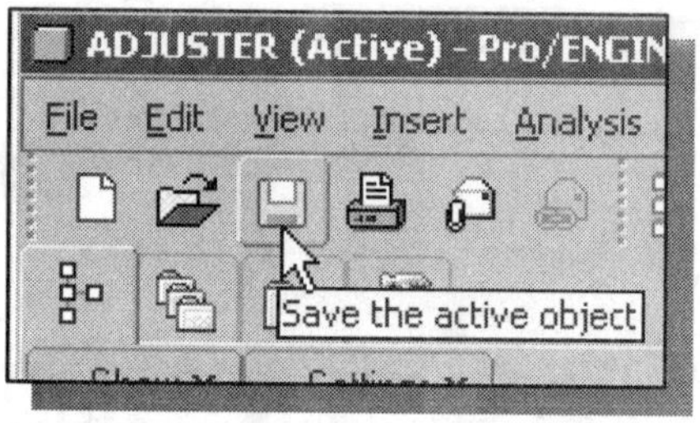

1. Select **Save** in the *Standard* toolbar, or you can also use the "**Ctrl-S**" combination (press down the **[Ctrl]** key and hit the **[S]** key once) to save the part.

2. In the message area, the part name is displayed. Click on the **Accept** button to save the file.

❖ It is a good habit to save your model periodically, just in case something might go wrong while you are working on it. In general, you should save your work onto the disk at an interval of every 15 to 20 minutes. You should also save before you make any major modifications to the model.

3. Use the left-mouse-button and click on **File** at the top of the *Pro/ENGINEER* main window, then choose **Exit** from the pull-down menu.

Questions:

1. What is the first thing we should set up in *Pro/ENGINEER* when creating a new model?

2. How do we modify more than one dimension in the *Sketcher*?

3. How do we reposition dimensions in the *Sketcher*?

4. List three of the geometric constraint symbols used by the *Pro/ENGINEER Sketcher*.

5. Describe two different ways to modify dimensions in the *Sketcher*.

6. Describe the steps required to define the orientation of the sketching plane?

7. Identify the following commands:

(a)

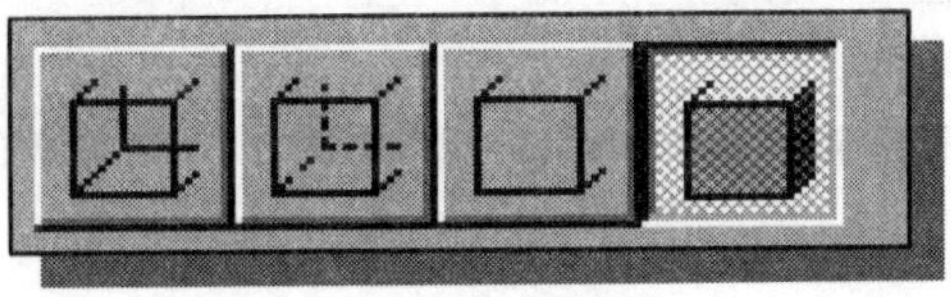

(b)

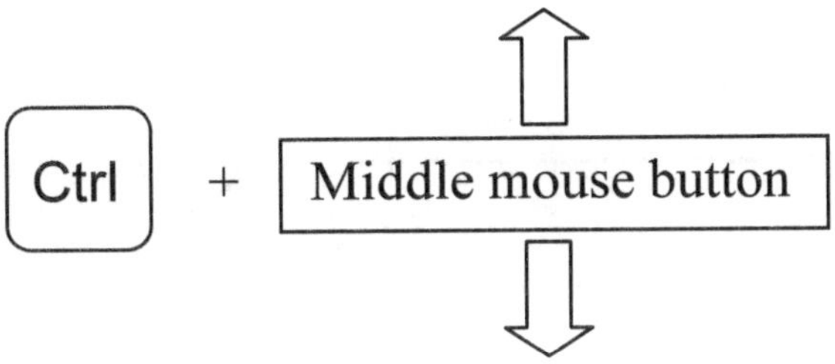

(c)

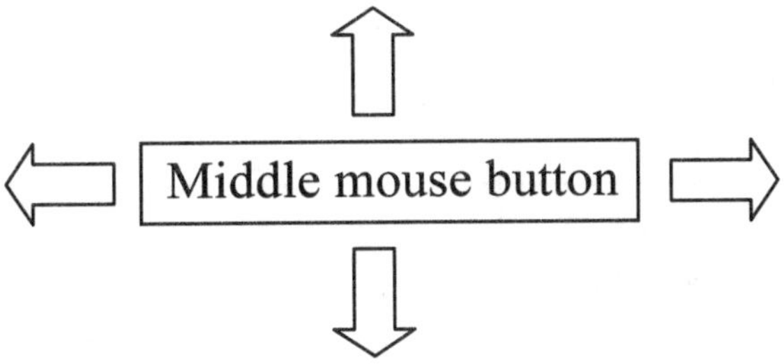

Exercises: (All dimensions are in inches.)

1. Plate Thickness: 0.25

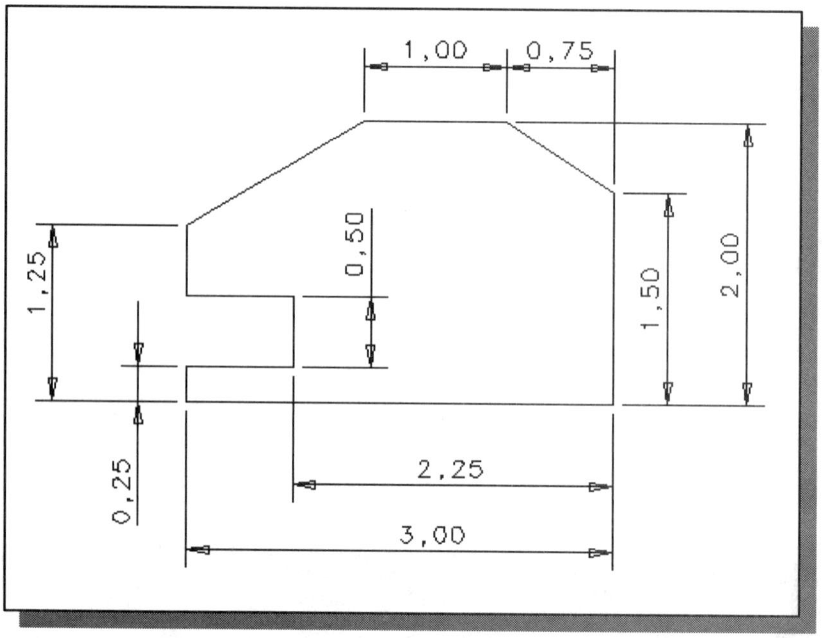

2. Plate Thickness: 0.5

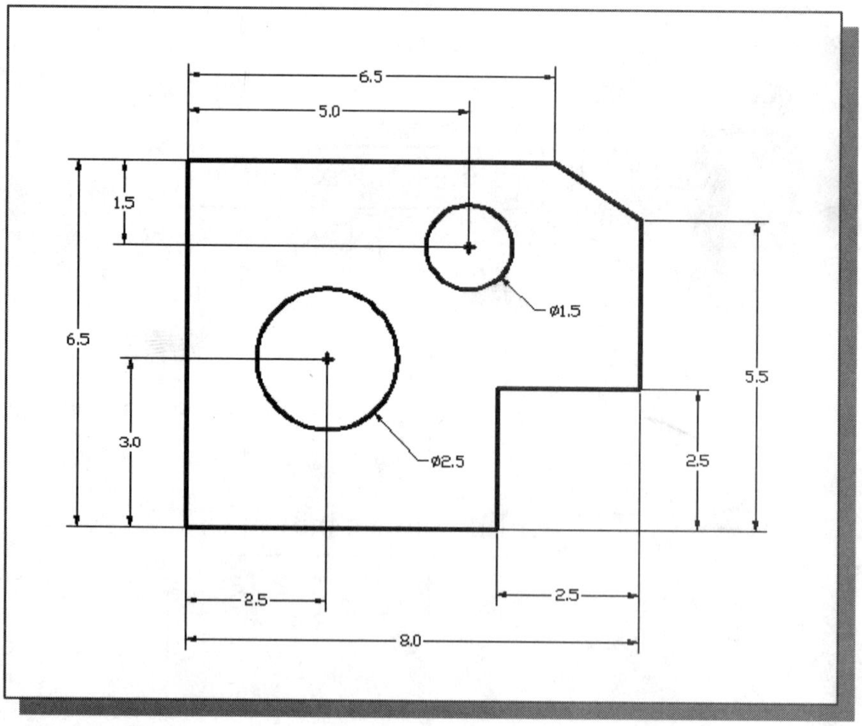

3.

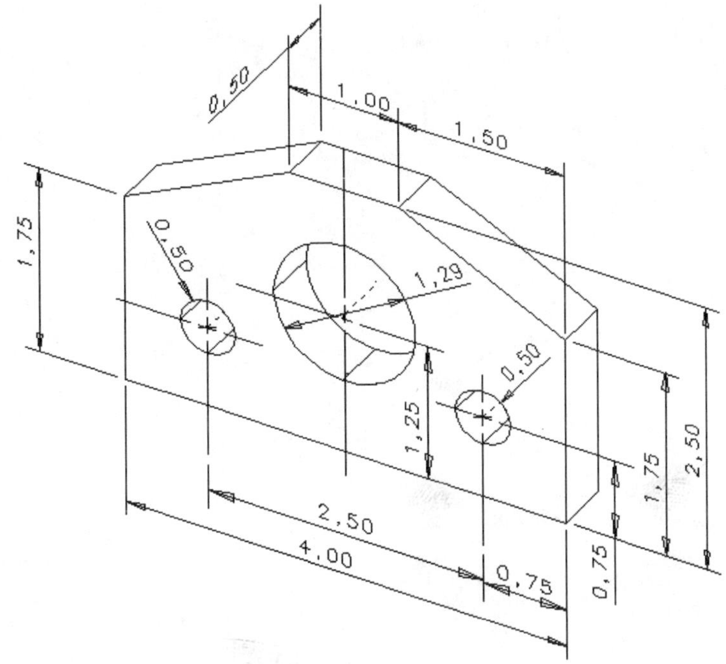
0,50
1,00
1,50
1,75
0,50
1,29
0,50
1,25
1,75
2,50
2,50
4,00
0,75
0,75

4.

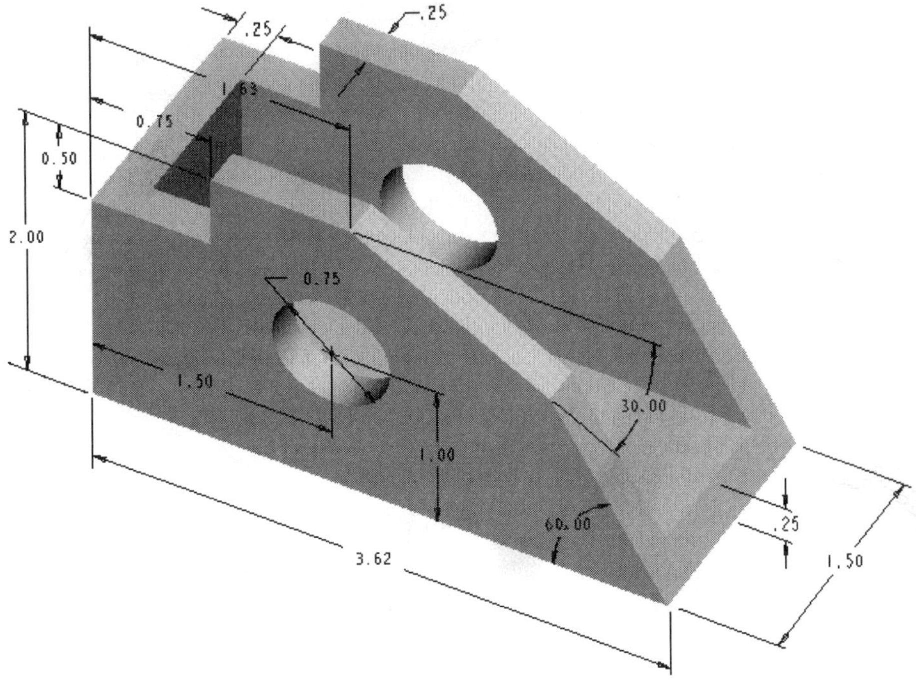
.25
.25
1.63
0.75
0.50
2.00
0.75
1.50
1.00
30.00
60.00
.25
1.50
3.62

Notes:

Chapter 2
Constructive Solid Geometry Concepts

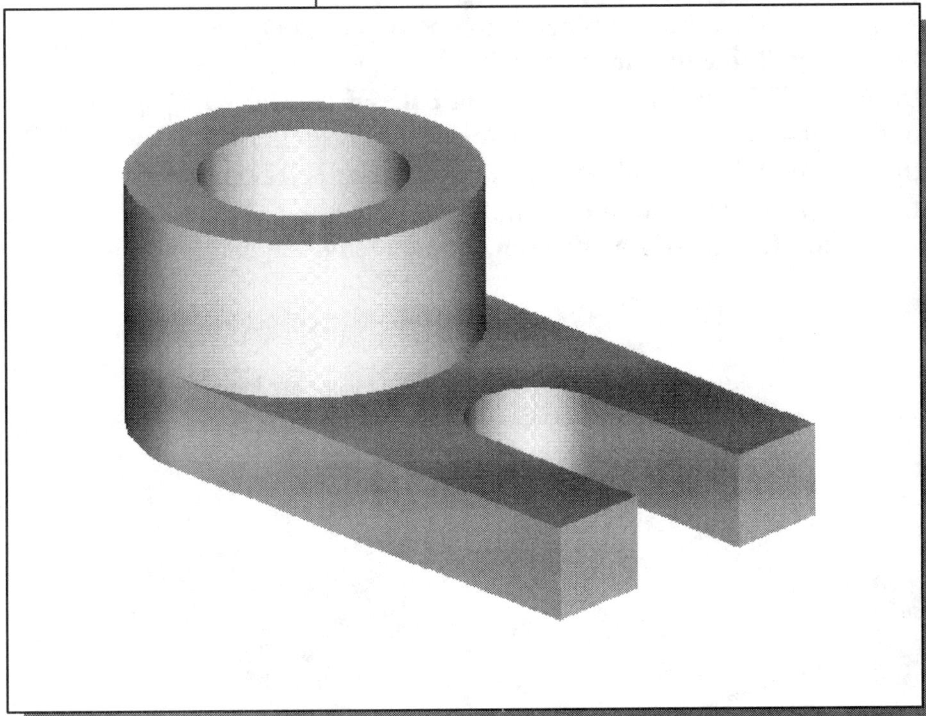

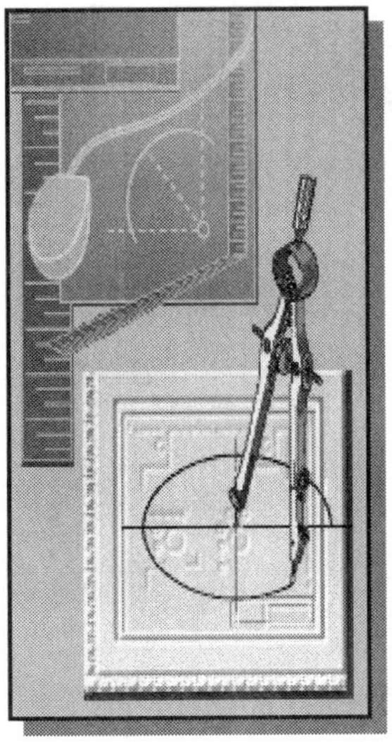

Learning Objectives

- **Understand the Constructive Solid Geometry Concepts.**
- **Create a Binary Tree.**
- **Understand the basic Boolean Operations.**
- **Understand the importance of Order of Features.**
- **Create Placed Features.**
- **Use the different Extrusion options.**

Introduction

In the 1980s, one of the main advancements in **solid modeling** was the development of the **Constructive Solid Geometry** (CSG) method. CSG describes the solid model as combinations of basic three-dimensional shapes (**primitive solids**). The basic primitive solid set typically includes: Rectangular-prism (Block), Cylinder, Cone, Sphere, and Torus (Tube). Two solid objects can be combined into one object in various ways, and these operations are known as **Boolean operations**. There are three basic Boolean operations: **JOIN (Union), CUT (Difference),** and **INTERSECT**. The JOIN operation combines the two volumes included in the different solids into a single solid. The CUT operation subtracts the volume of one solid object from the other solid object. The INTERSECT operation keeps only the volume common to both solid objects. The CSG method is also known as the **Machinist's Approach**, as the method is parallel to machine shop practices.

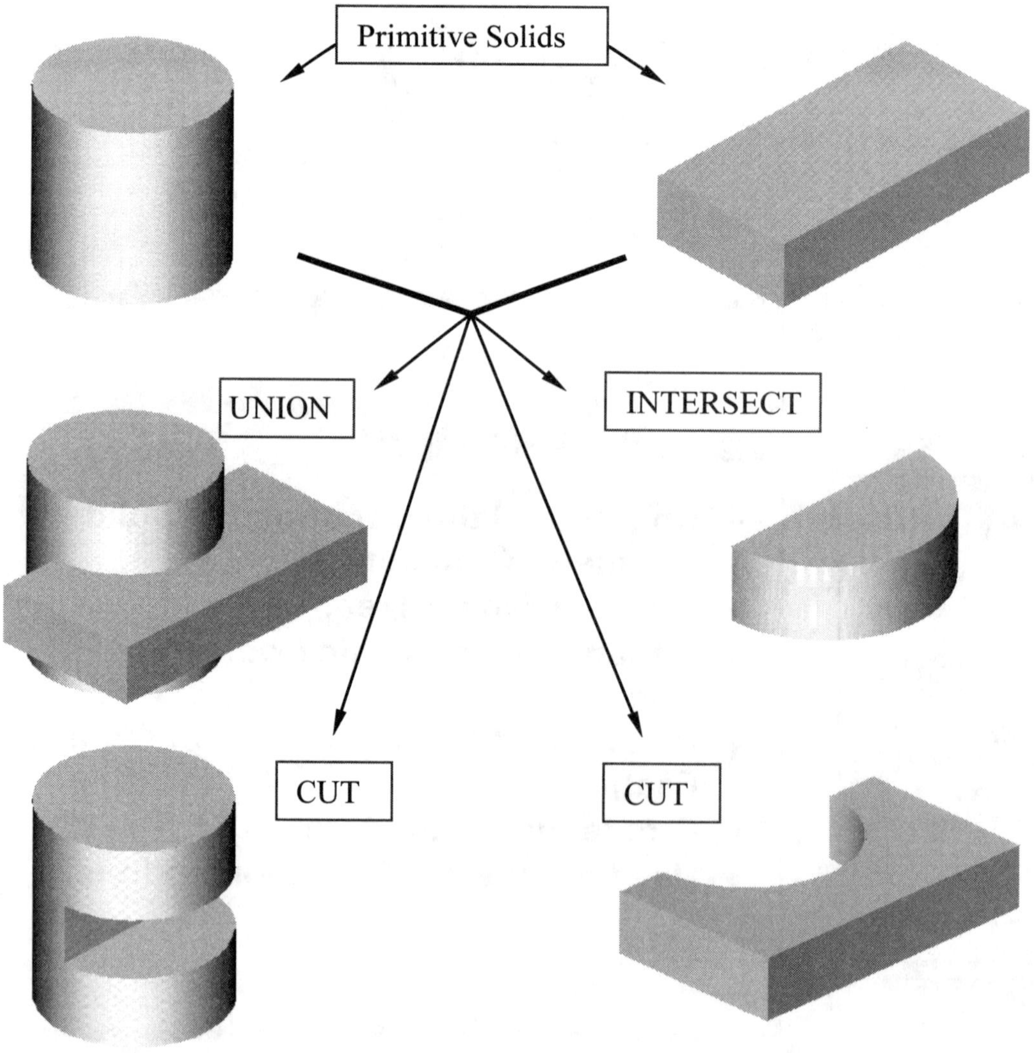

Binary Tree

Constructive Solid Geometry (CSG) is also referred to as the method used to store a solid model in the database. The resulting solid can be easily represented by what is called a **binary tree**. In a binary tree, the terminal branches (leaves) are the various primitives that are linked together to make the final solid object (the root). The binary tree is an effective way to keep track of the *history* of the resulting solid. By keeping track of the history, the solid model can be rebuilt by re-linking through the binary tree. This provides a convenient way to modify the model. We can make modifications at the appropriate links in the binary tree and re-link the rest of the history tree without building a new model.

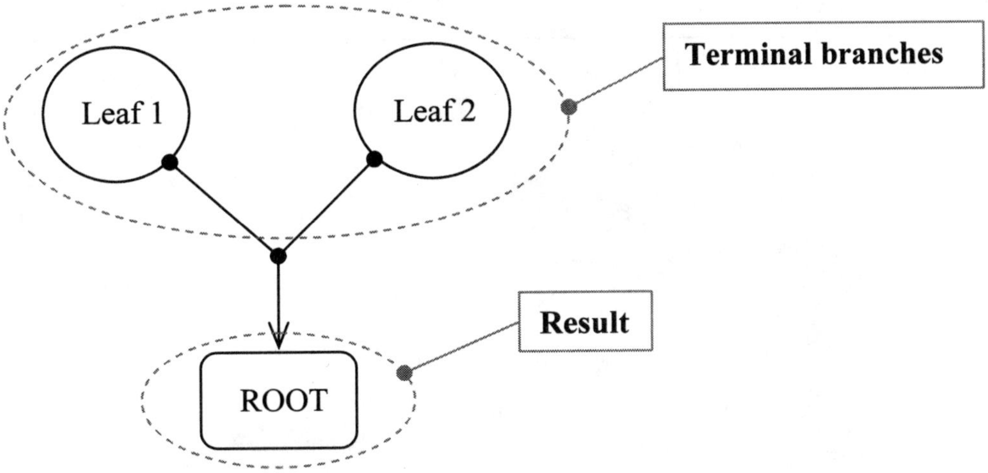

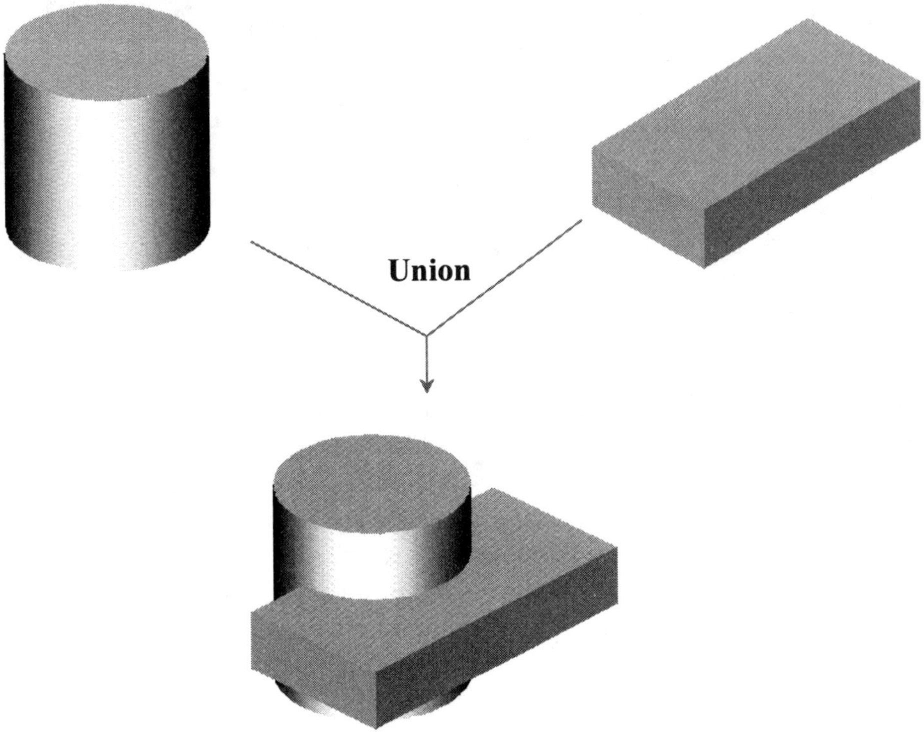

The *Locator* Design

The CSG concept is one of the important building blocks for feature-based modeling. In *Pro/ENGINEER*, the CSG concept can be used as a planning tool to determine the number of features that are needed to construct the model. It is also a good practice to create features that parallel the manufacturing process required for the design. With parametric modeling, we are no longer limited to using only the predefined basic solid shapes. In fact, any solid features we create in *Pro/ENGINEER* are used as primitive solids; parametric modeling allows us to maintain full control of the design variables that are used to describe the features. In this lesson, a more in-depth look at the parametric modeling procedure is presented. The equivalent CSG operation for each feature is also illustrated.

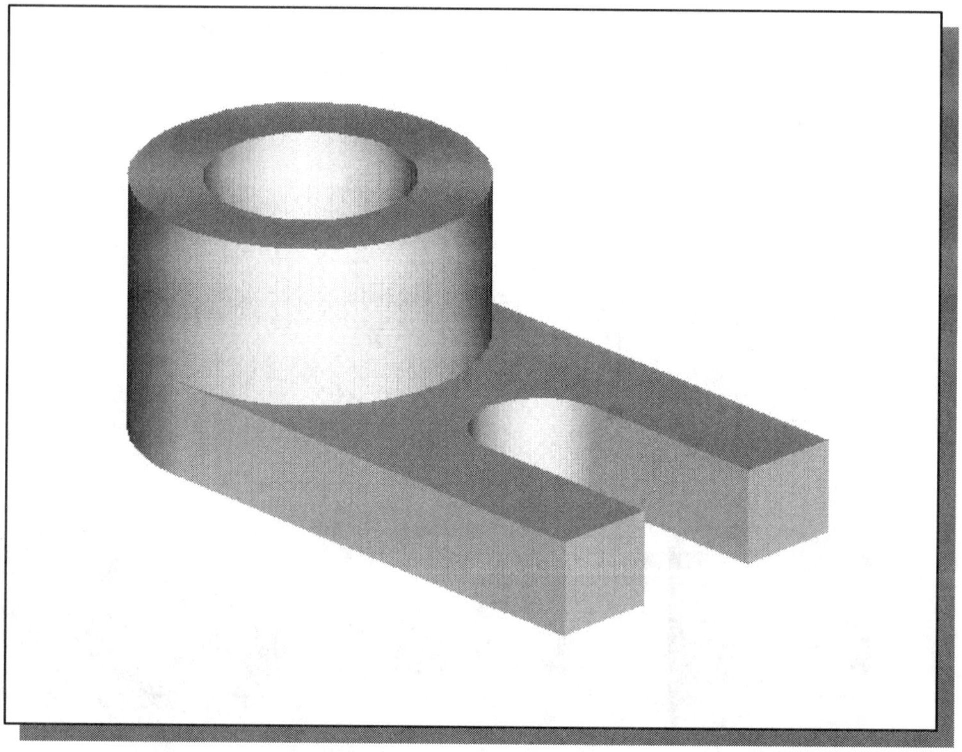

➢ Before going through the tutorial, on your own make a sketch of a CSG binary tree of the *Locator* design using only two basic types of primitive solids: cylinder and rectangular prism. In your sketch, how many *Boolean operations* will be required to create the model? What is your choice for the first primitive solid to use, and why? Take a few minutes to consider these questions and do the preliminary planning by sketching on a piece of paper. Compare the sketch you make to the CSG binary tree steps shown on the next page. Note that there are many different possibilities in combining the basic primitive solids to form the desired solid model. Even for the simplest design, it is possible to take several different approaches to creating the same solid model.

Modeling Strategy - CSG Binary Tree

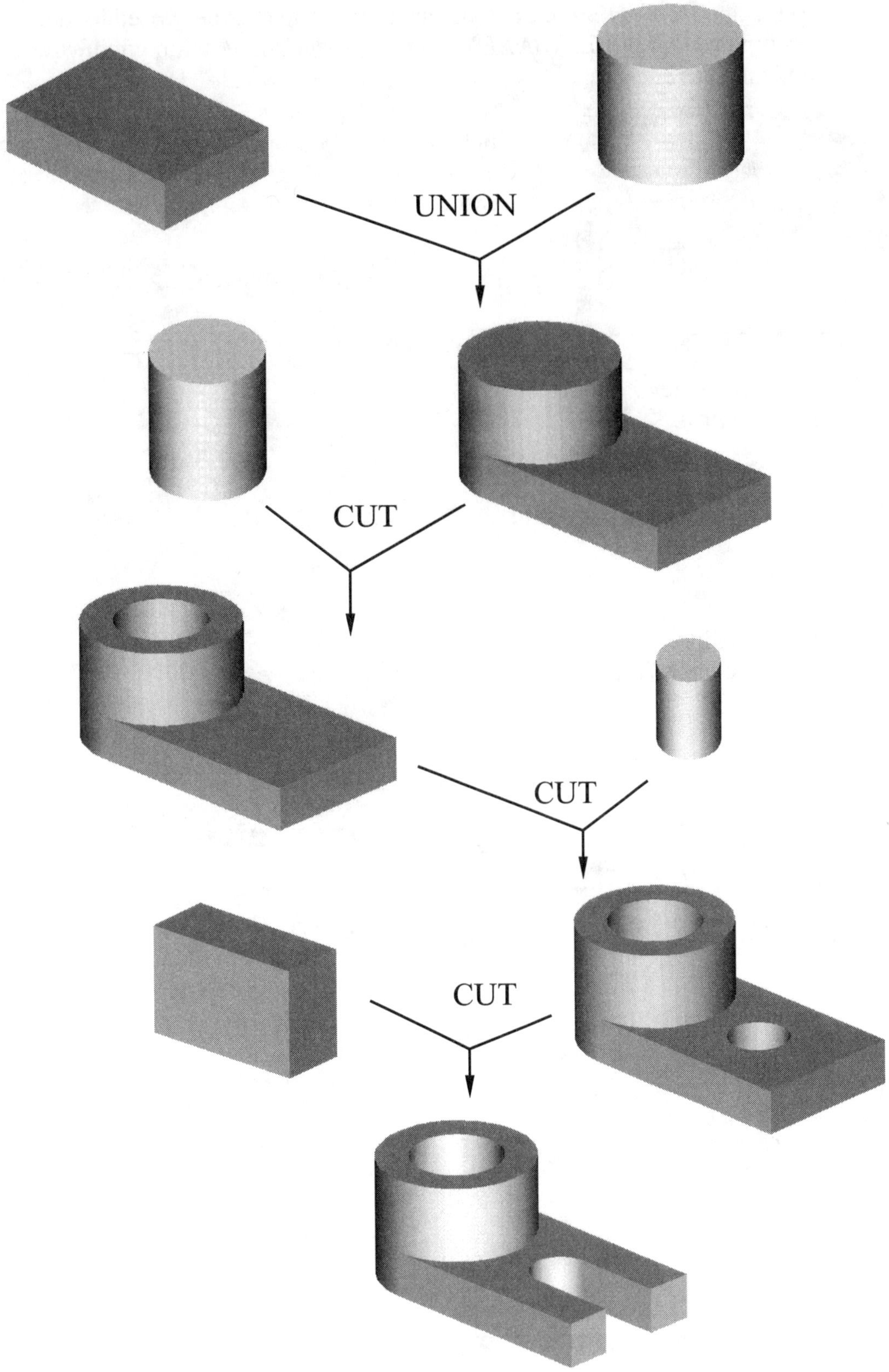

Starting *Pro/ENGINEER*

1. Select the **Pro/ENGINEER** icon on the desktop or type [***proe***] at your system prompt to start *Pro/ENGINEER*. The *Pro/ENGINEER* main window will appear on the screen.

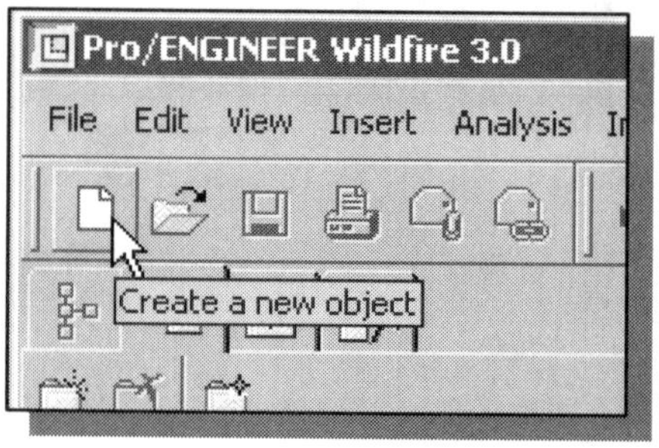

2. Click on the **New** icon as shown. Notice that we can also use the key combination **Ctrl-N** to start a new object.

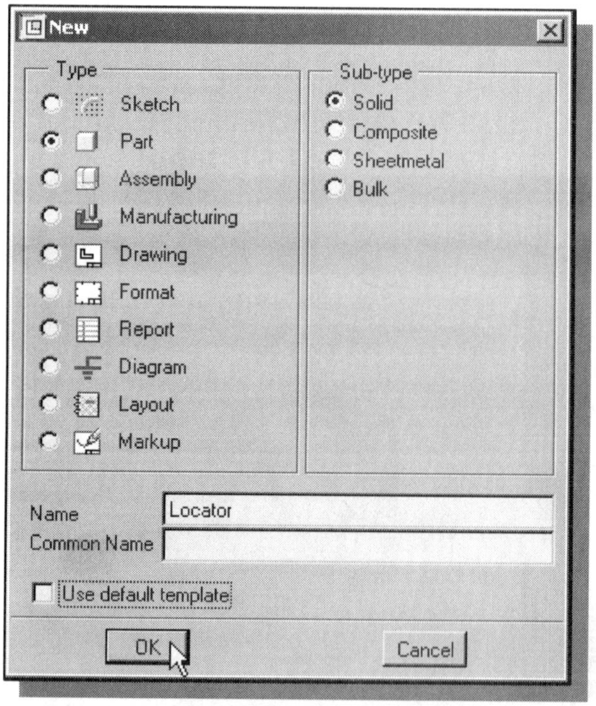

3. In the *New* form, enter ***Locator*** as the solid part file **Name**.

4. Turn ***off*** the **Use default template** option.

5. Click **OK** to continue. The *Model Tree* window and the *Feature Toolbars* appear on the screen.

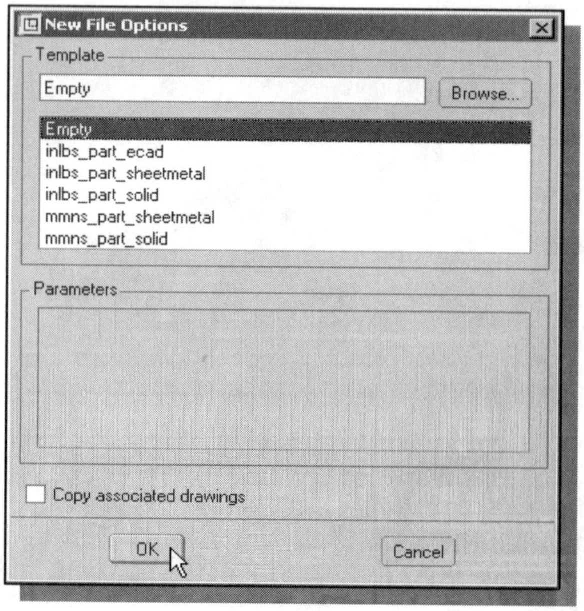

6. In the *New File Options* dialog box, select **Empty** in the option list to not use any template file.

7. Click on the **OK** button to accept the settings and enter the *Pro/ENGINEER Part Modeling* mode.

Units Setup

❖ When starting a new model, the first thing we should do is to choose the set of units we want to use.

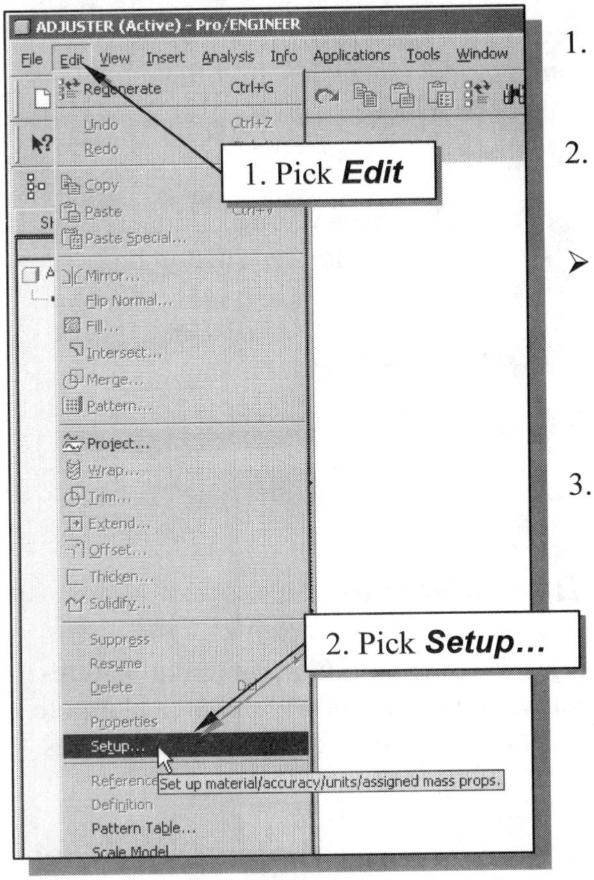

1. Use the left-mouse-button and select **Edit** in the pull-down menu area.

2. Use the left-mouse-button and select **Setup…** in the pull-down list as shown.

➢ Note that the *Pro/ENGINEER* menu system is context-sensitive, which means that the menu items and icons of the non-applicable options are grayed out (temporarily disabled.)

3. Select the **Units** option in the **Menu Manager** window that appeared to the right of the *Pro/ENGINEER* main window.

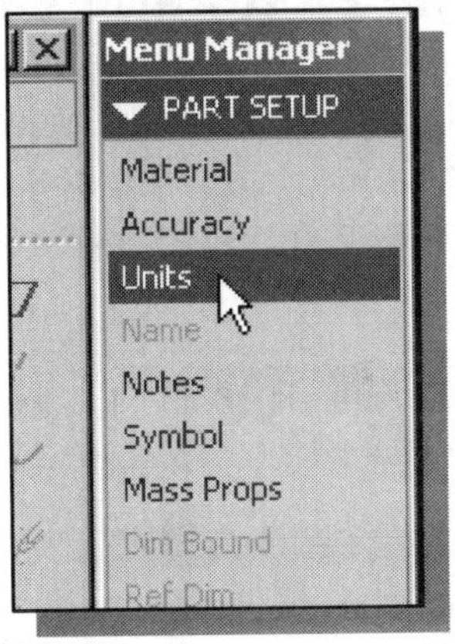

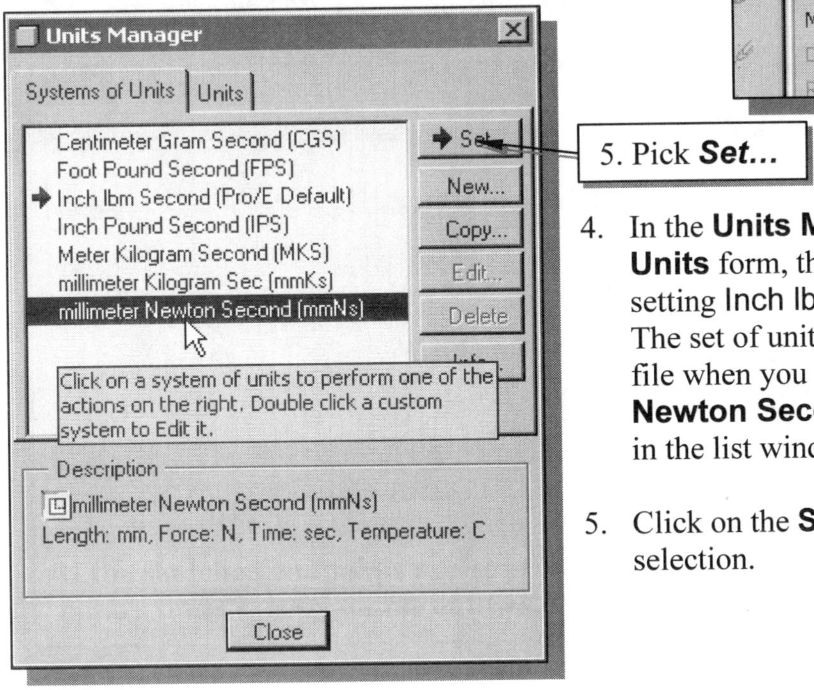

4. In the **Units Manager-System of Units** form, the *Pro/ENGINEER* default setting Inch lbm Second is displayed. The set of units is stored with the model file when you save. Pick **millimeter Newton Second (mmNs)**, by clicking in the list window as shown.

5. Click on the **Set** button to accept the selection.

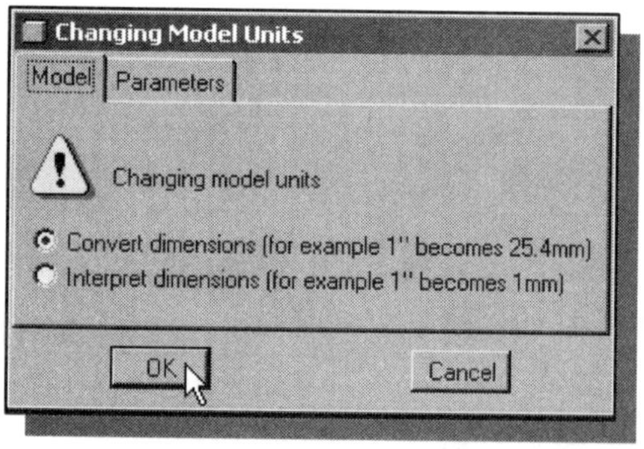

6. In the *Changing Model Units* dialog box, click on the **OK** button to accept the change of the units.

➢ Note that *Pro/ENGINEER* allows us to change model units even after the model has been constructed.

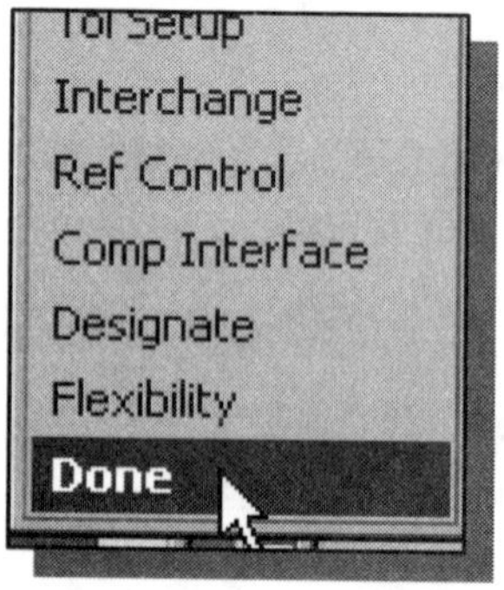

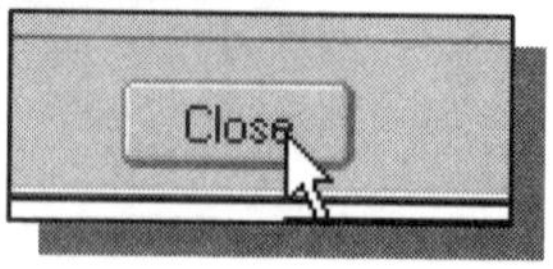

7. Click on the **Close** button to exit the *Units Manager* dialog box.

8. Pick **Done** to exit the **PART SETUP** submenu.

Adding the First Part Features — Datum Planes

❖ *Pro/ENGINEER* provides many powerful tools for model creation. In doing feature-based parametric modeling, it is a good practice to establish three reference planes to locate the part in space. The reference planes can also be used as location references in feature constructions.

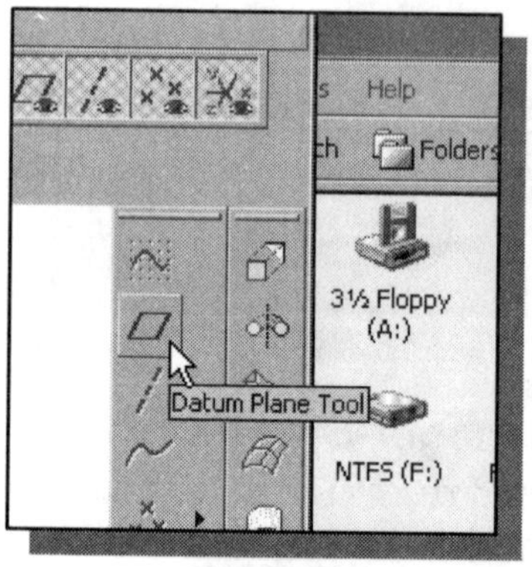

1. In the *Feature* toolbar, click on the **Datum Plane Tool** icon to create three datum planes as shown.

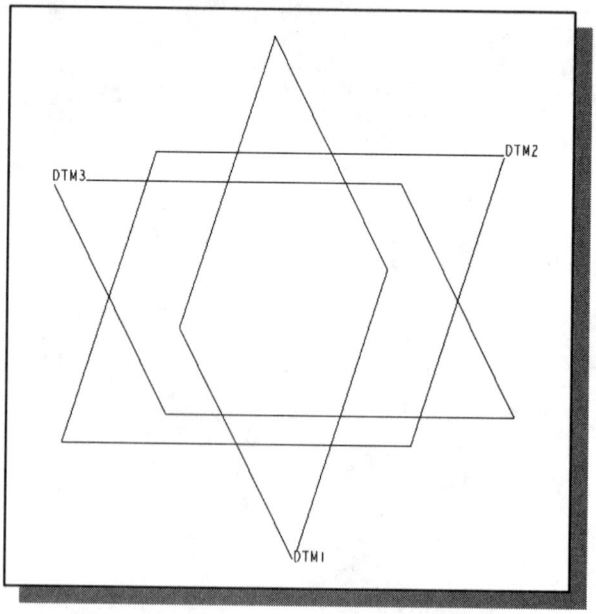

2. Click the Locator.PRT name in the navigator window to deselect DTM3.

Base Feature

In *parametric modeling*, the first solid feature is called the **base feature,** which usually is the primary shape of the model. Depending upon the design intent, additional features are added to the base feature.

Some of the considerations involved in selecting the base feature:

- **Design Intent** – Determine the functionality of the design; identify the feature that is central to the design.

- **Order of features** – Choose the feature that is the logical base in terms of the order of features in the design.

- **Ease of making modifications** – Select the feature that is more stable and is less likely to be changed.

❖ We will create a rectangular block as the base feature of the ***Locator*** design.

1. In the *Feature* toolbars (toolbars aligned to the right edge of the main window), select the **Extrude Tool** option as shown.

- The ***Feature Option Dashboard***, which contains applicable construction options, is displayed above the message area near the bottom of the *Pro/ENGINEER* main window.

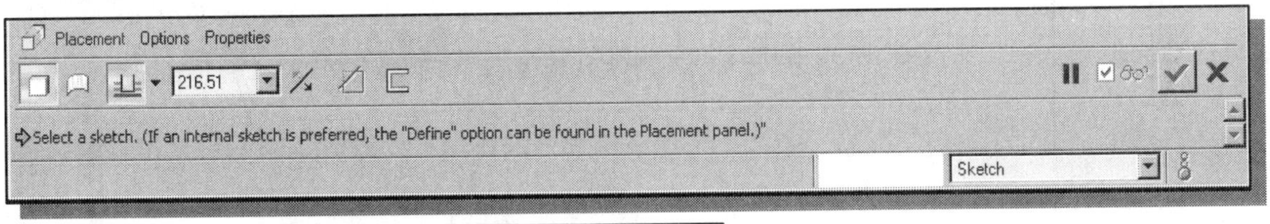

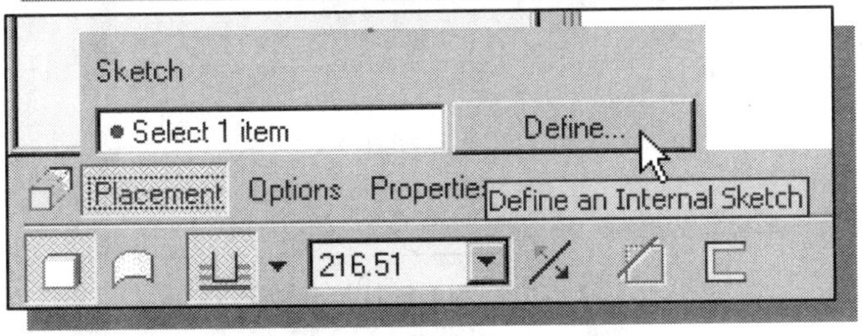

2. Click the **Placement** option and choose **Define** to begin creating a new *Internal sketch.*

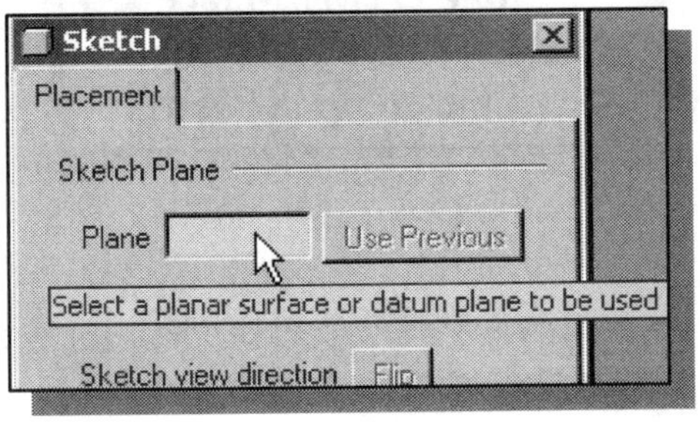

3. Move the cursor inside the **Plane** option box in the *Sketch* window as shown. The message "*Select a plane or surface to define sketch plane.*" is displayed in the message area.

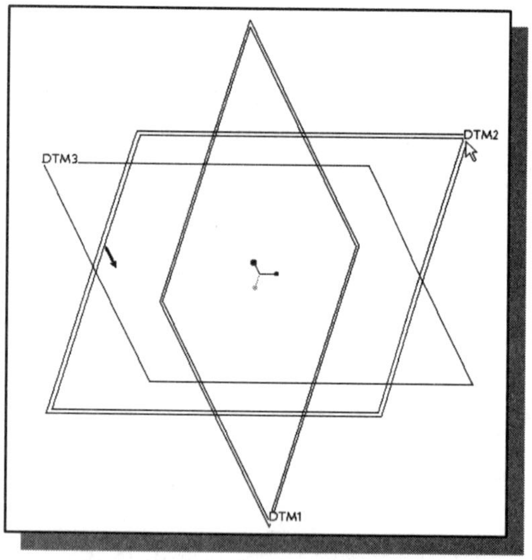

4. In the graphic area, select **DTM2** by clicking on the text DTM2 as shown.

❖ Notice an arrow appears on the edge of DTM2. The arrow direction indicates the viewing direction of the sketch plane. The viewing direction can be reversed by clicking on the **Flip** button in the **Sketch Orientation** section of the popup window.

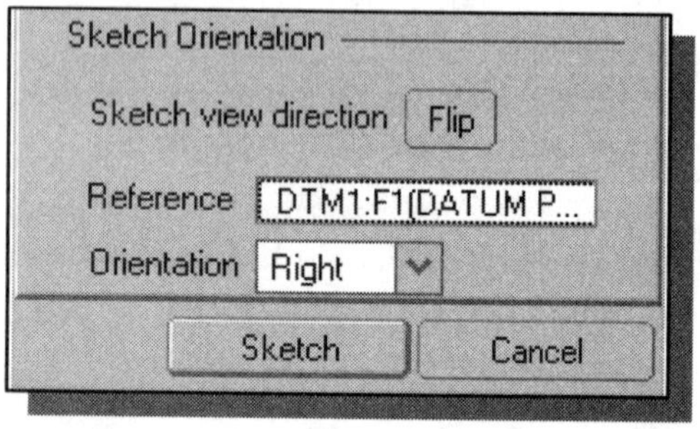

5. Confirm the **Sketch Orientation** options are set to **DTM1** and **Right**, as shown.

6. Pick **Sketch** to exit the *Section Placement* window and proceed to enter the *Pro/ENGINEER Sketcher* mode.

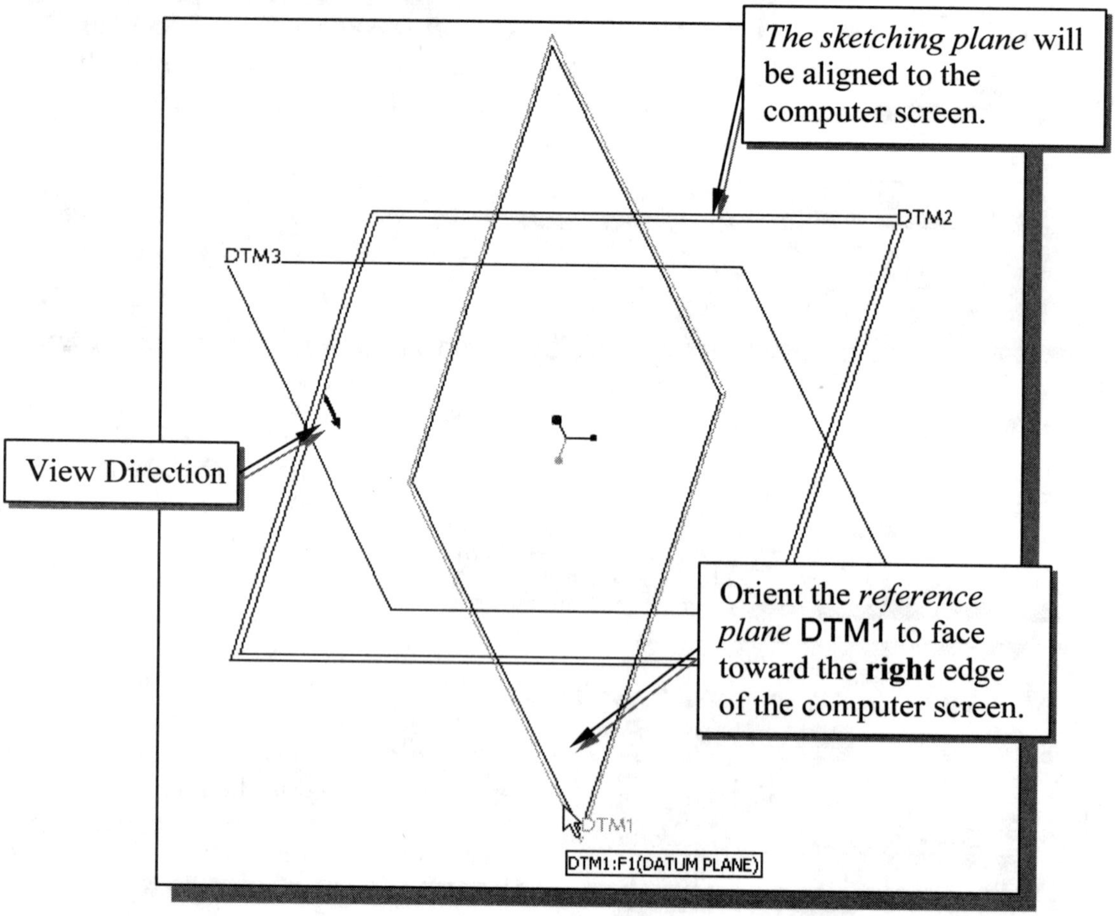

Creating a 2D parametric section

❖ The first thing that *Pro/ENGINEER Sketcher* expects us to do is specify **sketching references**. In the previous sections, we created the three datum planes to help orient the model in 3D space. Now we need to orient the 2D sketch with respect to the three datum planes. At least two references are required to orient in the horizontal direction and in the vertical direction. By default, the two planes (in our example, **DTM1** and **DTM3**) that are perpendicular to the sketching plane (**DTM2**) are automatically selected.

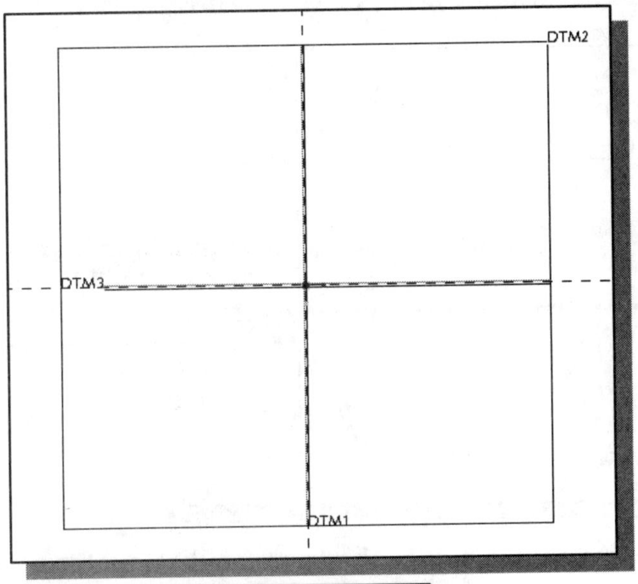

1. Note that **DTM1** and **DTM3** are pre-selected as the sketching references. In the graphics area, the two references are highlighted and displayed with two dashed lines.

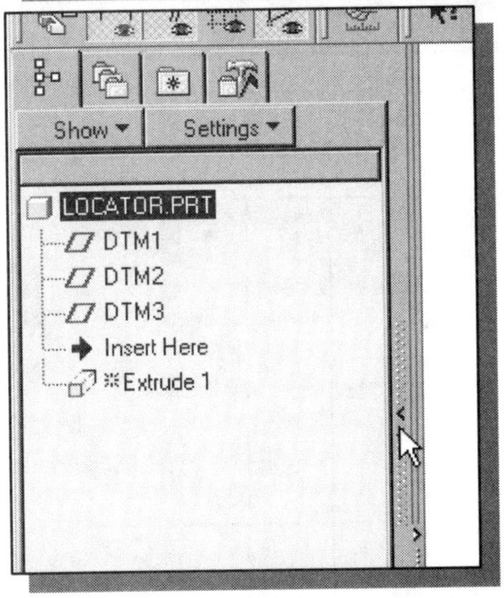

2. Click on the ***Navigator Sash*** to toggle ***off*** the display of the *Model Tree* window.

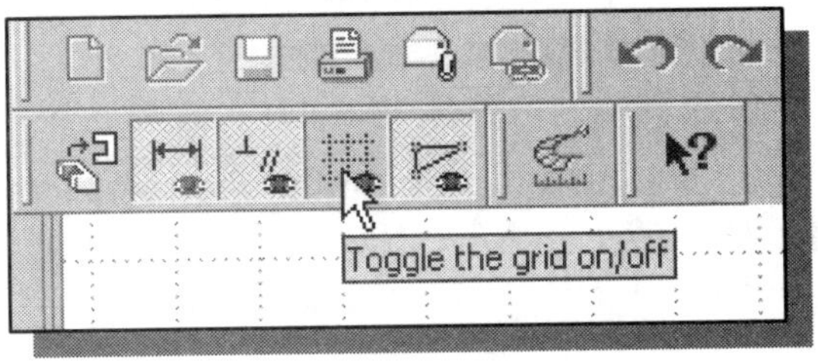

3. Click on the **Grid** icon to switch ***on*** the grid display.

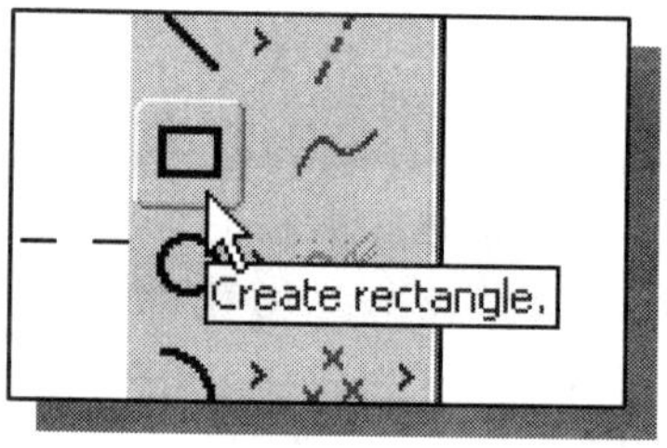

4. In the *Sketcher* toolbar, click on the **Rectangle** icon as shown to activate the Create Rectangle command.

5. Create a rectangle as shown, with one corner on the vertical axis (DTM1). (Do not be concerned if the dimensional values on your screen are different than what are shown here. Place points near the grid-points shown.)

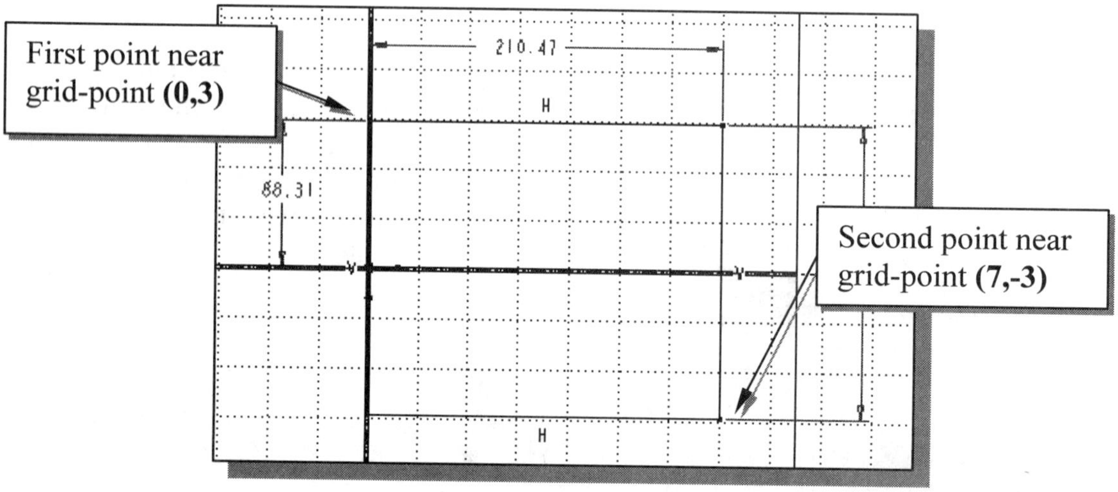

6. On your own, modify the dimensional values as shown (overall size: **75mm x 50mm**).

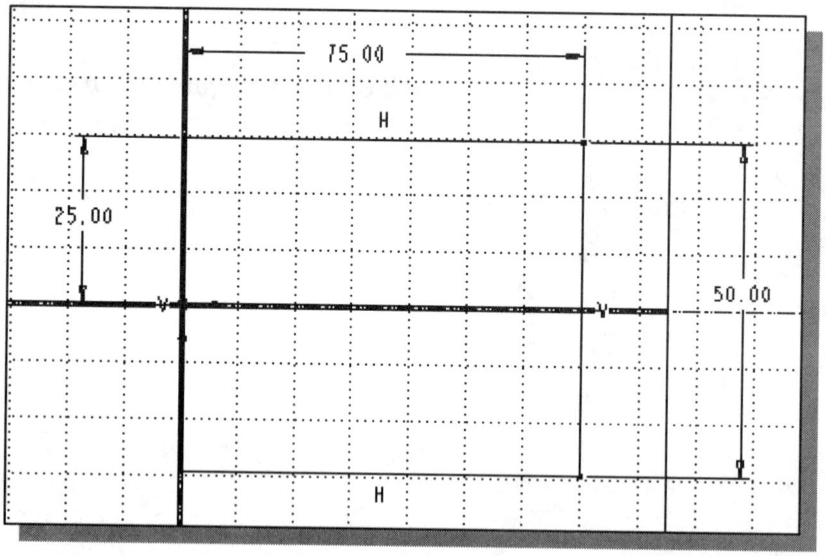

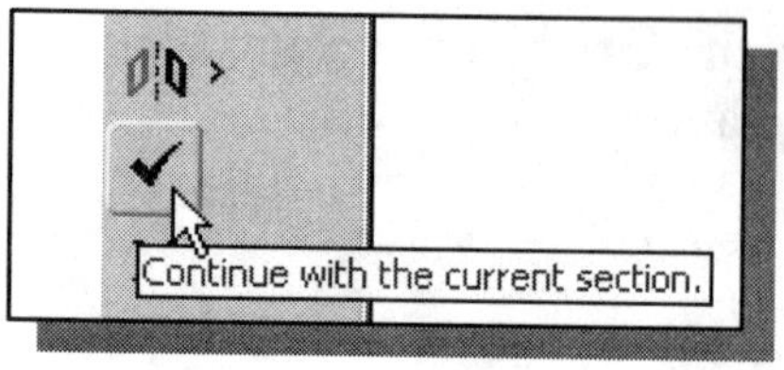

7. In the *Sketcher* toolbar, click on the **Accept** icon to exit the *Pro/ENGINEER 2D Sketcher* and proceed to the next element of the feature definition.

8. In the *Feature Options Dashboard*, confirm the **Depth Value** option is set and enter **15** as the *extrusion depth* as shown. (Hit the **Enter** key once.)

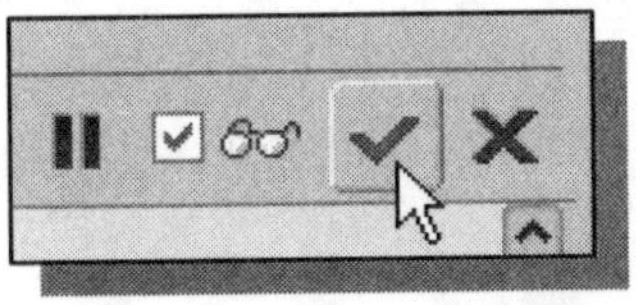

9. In the message area, click on the **Accept** button to proceed with the feature definition.

➢ Note that all dimensions disappeared from the screen. All parametric definitions are stored in the *Pro/ENGINEER* **database** and any of the parametric definitions can be displayed and edited at anytime.

10. Pick the **Standard Orientation** option [quick-key: **Ctrl-D**] in the pull-down menus.

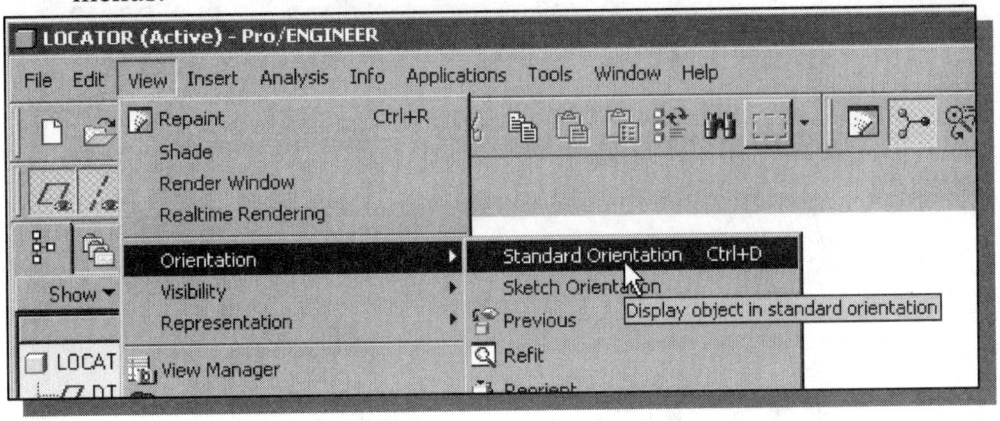

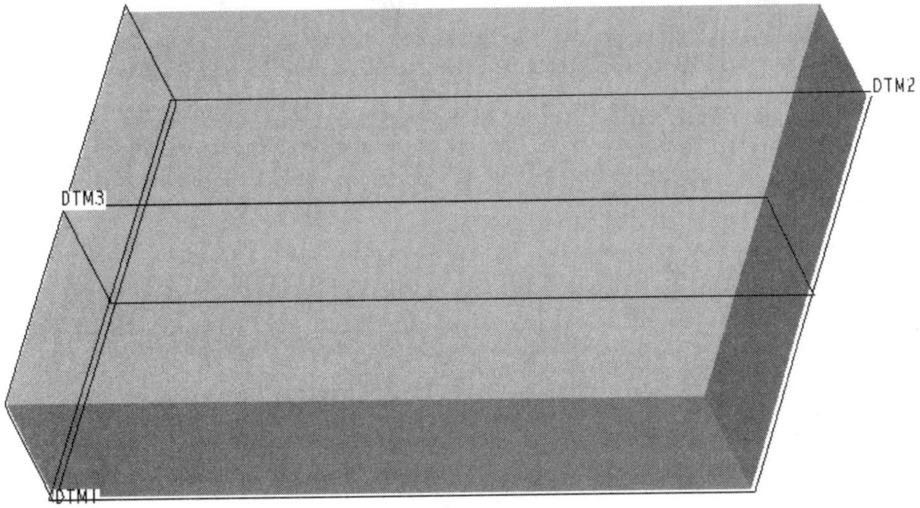

Creating the Second Solid Feature

❖ We will create a cylinder as the second solid feature of the ***Locator***.

1. In the *Feature* toolbars (toolbars aligned to the right edge of the main window), select the **Extrude Tool** option as shown.

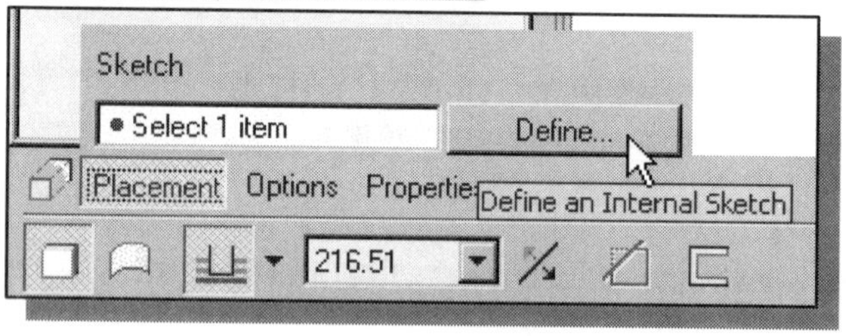

2. Click the **Placement** option and choose **Define** to begin creating a new *Internal sketch.*

3. We will use datum plane DTM2 as the *sketching plane*. Pick **DTM2**.

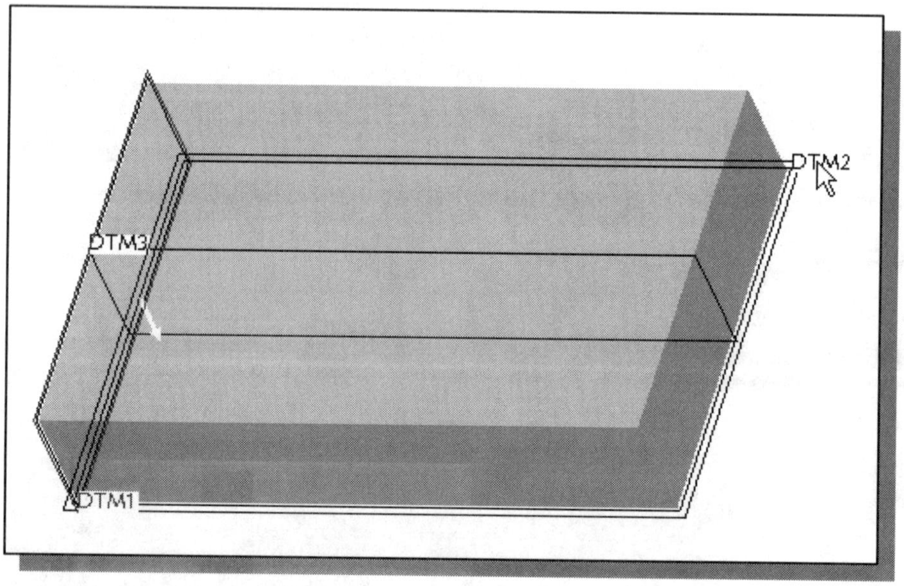

4. An arrow appears on one of the edges of DTM2 to indicate the viewing direction of the sketching plane.

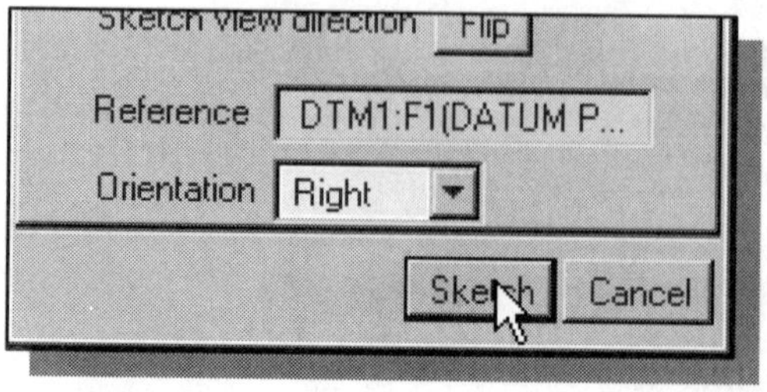

5. Confirm the **Orientation** option is set to referencing DTM1 and **Right** as shown.

6. Pick **Sketch** to exit the *Section Placement* window and proceed to enter the *Pro/ENGINEER Sketcher* mode.

Creating the 2D Sketch

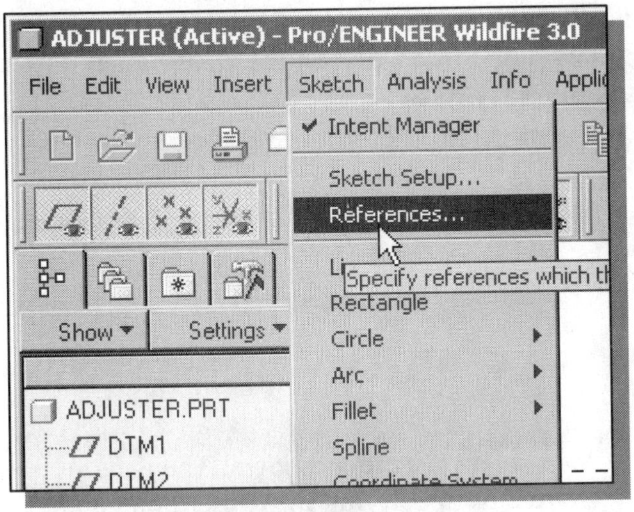

1. In the pull-down menu area, click on **Sketch** to display the option list and select the References option:

   Sketch → References...

➢ This will bring up the *References* dialog box.

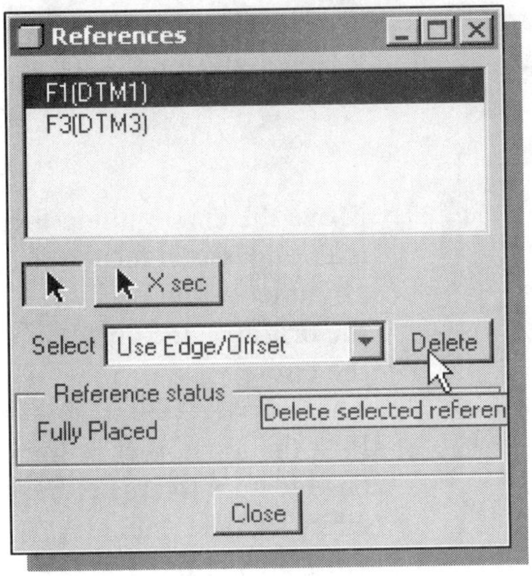

2. Note that the **DTM1** and **DTM3** are pre-selected as the sketching references. In the graphics area, the two references are highlighted and displayed with two dashed lines. Click on the **DTM1** reference in the *References* dialog box.

3. Click on the **Delete** button to remove the reference to the right surface of the base feature.

4. We will add two additional references. Click on **DTM1** and the **top edge** of the base feature as shown.

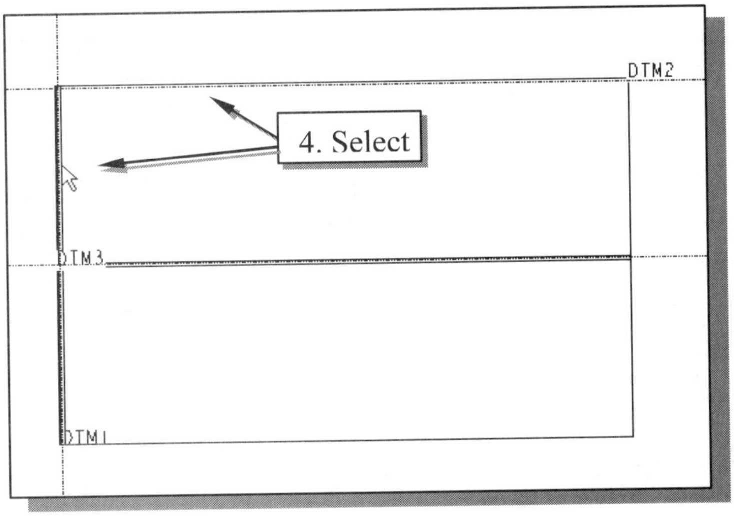

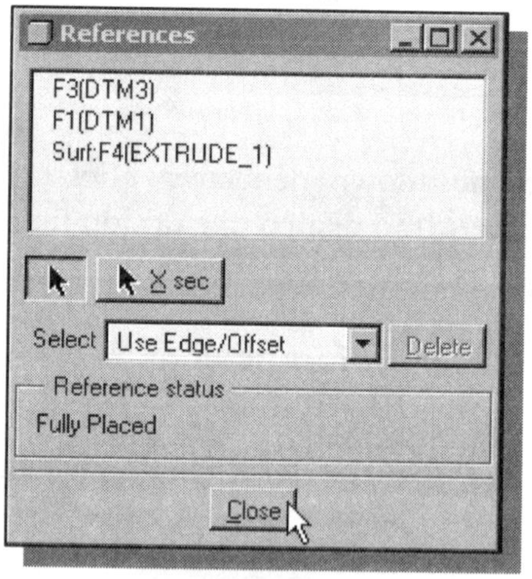

5. In the *References* dialog box, the two references are added to the list. Click on the **Close** button to accept the selections.

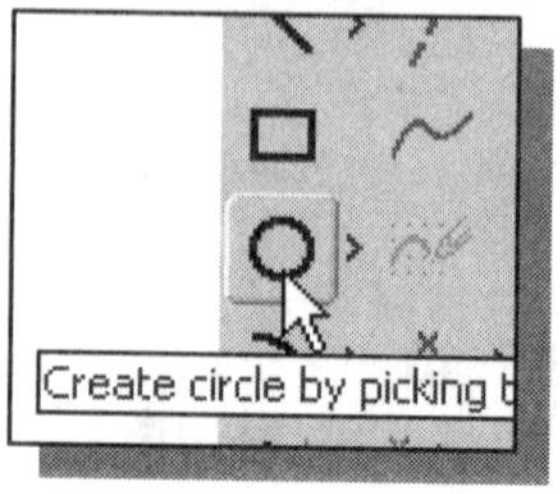

6. In the *Sketcher* toolbar, select **Circle** as shown. The default option is to create a circle by specifying the center point and a point through which the circle will pass. The message "*Select the center of a circle*" is displayed in the message area.

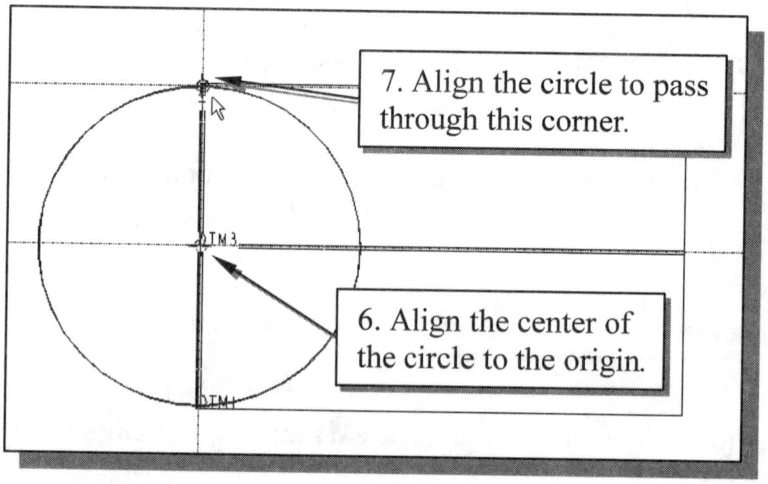

7. Move the cursor along the axes and watch for the alignment symbol. Pick the origin as the center of the circle.

8. Pick the top corner of the base feature to create the circle. (Click when the *Tangent* symbol **T** is displayed.)

- The circle is fully defined and no dimension is needed. The selected references, which are aligned to the base solid, control the size of the circle.

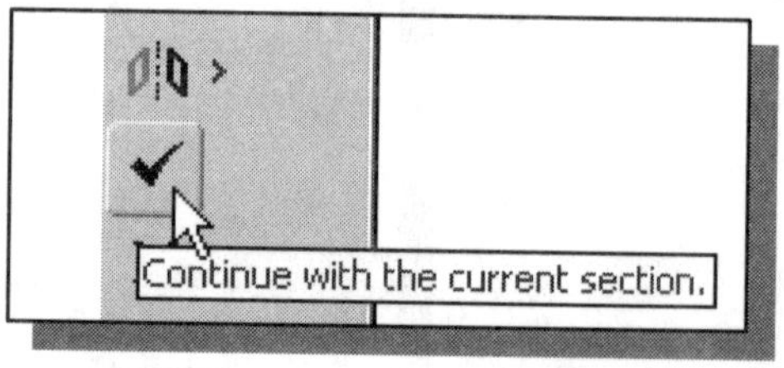

9. In the *Sketcher* toolbar, click on the **Accept** icon to exit the *Pro/ENGINEER 2D Sketcher* and proceed to the next element of the feature definition.

10. In the *Feature Option Dashboard*, confirm the **Depth Value** option is set as shown.

11. In the *depth value* box, enter **40** as the extrusion depth.

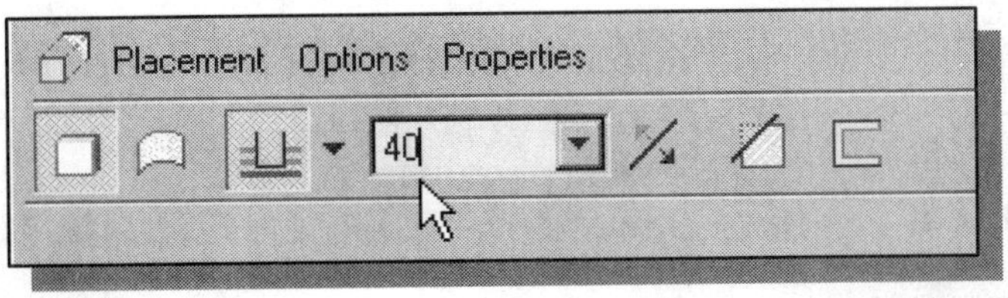

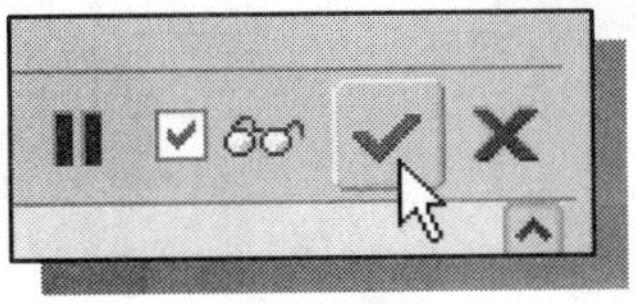

12. In the message area, click on the **Accept** button to proceed with the feature definition.

13. Use the quick-key [**Ctrl-D**] option or pick the **Standard Orientation** option or in the pull-down menus.

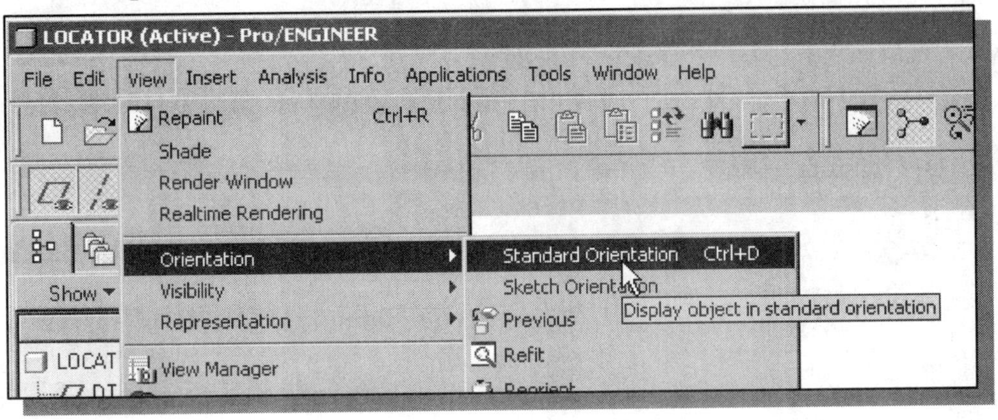

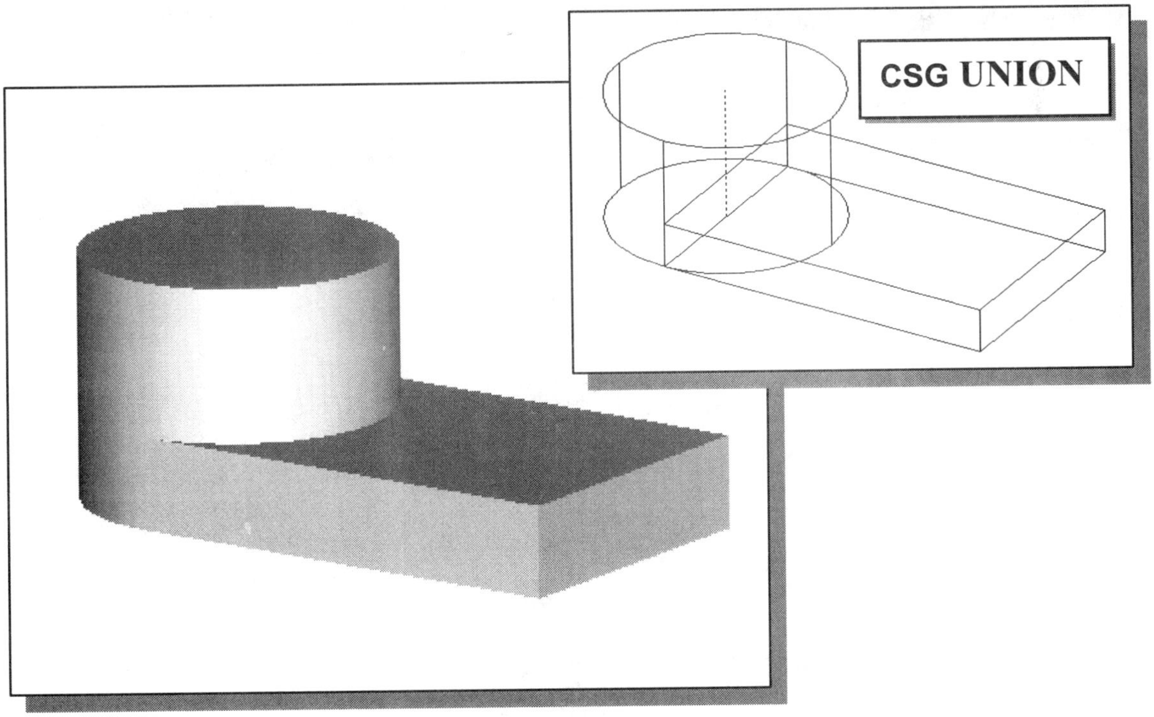

Creating a CUT Feature

- We will create a circular cut as the next solid feature of the ***Locator***.

1. In the *Feature* toolbars (toolbars aligned to the right edge of the main window), select the **Extrude Tool** option as shown.

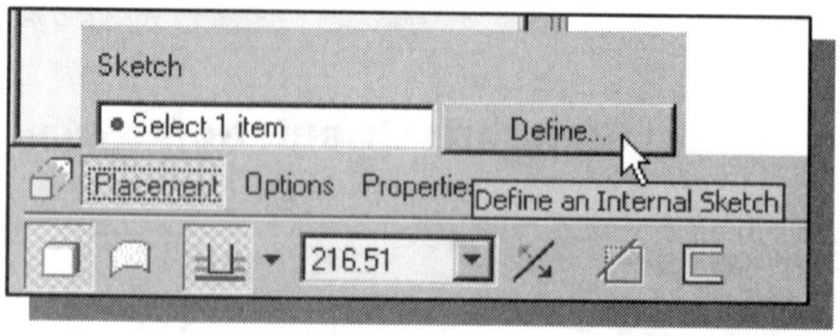

2. Click the **Placement** option and choose **Define** to begin creating a new *Internal sketch*.

3. We will use the top surface of the last feature as the sketching plane. Click once, with the left-mouse-button, inside the top surface of the circular solid feature as shown in the figure below.

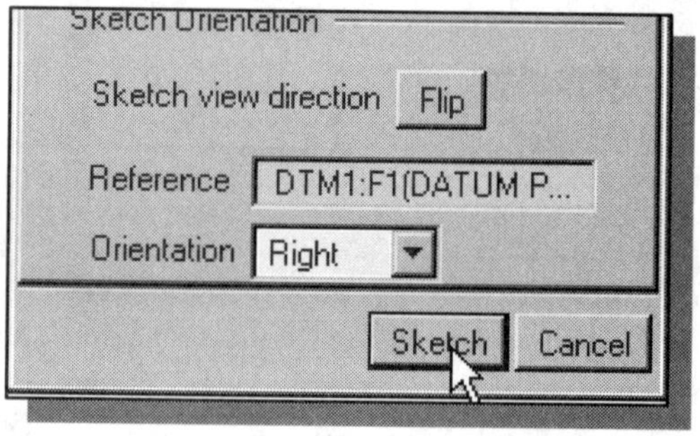

4. In the **Sketch Orientation** menu, confirm the reference plane **Orientation** is set to referencing DTM1 and orientation to **Right**.

5. Click **Sketch** to exit the *Section Placement* window and proceed to enter the *Pro/ENGINEER Sketcher* mode.

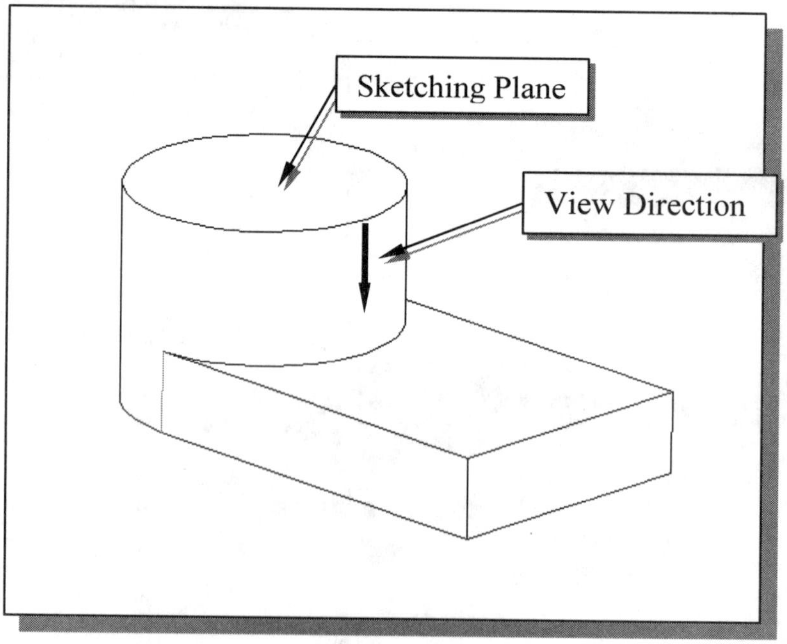

Creating a 2D Section

1. In the *Sketcher* toolbar, select **Circle** as shown. The default option is to create a circle by specifying the center point and a point through which the circle will pass. The message "*Select the center of a circle*" is displayed in the message area.

2. Move the cursor along the axes and watch for the alignment symbol. Pick the origin as the center of the circle.

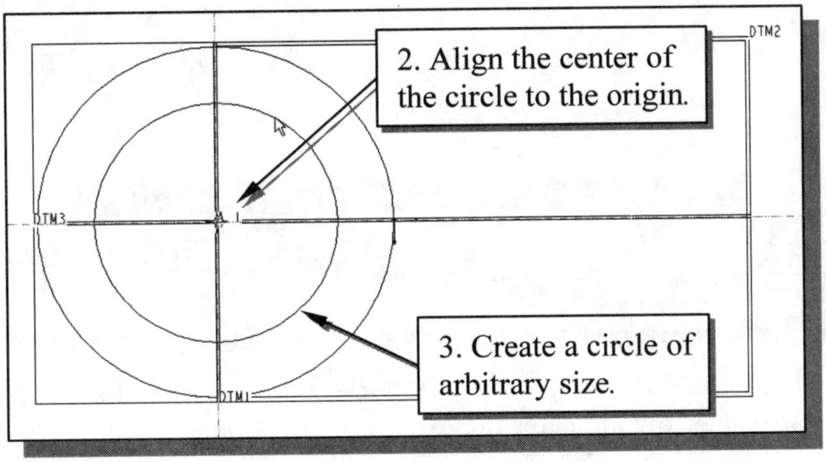

3. Pick a point inside the top surface of the cylinder and create a circle of arbitrary size.

4. On your own, modify the diameter to **40mm**.

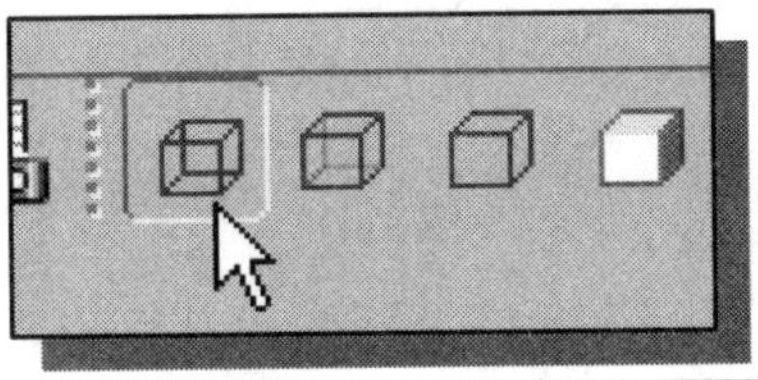

5. Click on the **Wireframe** icon to display the model in wireframe format.

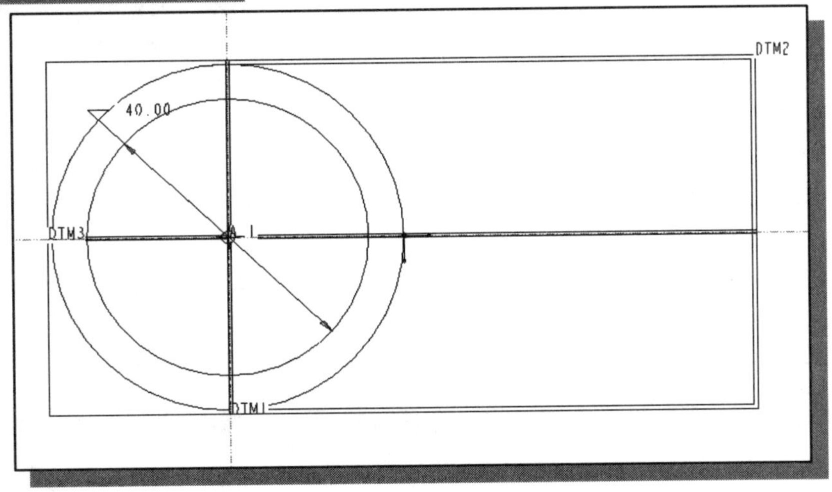

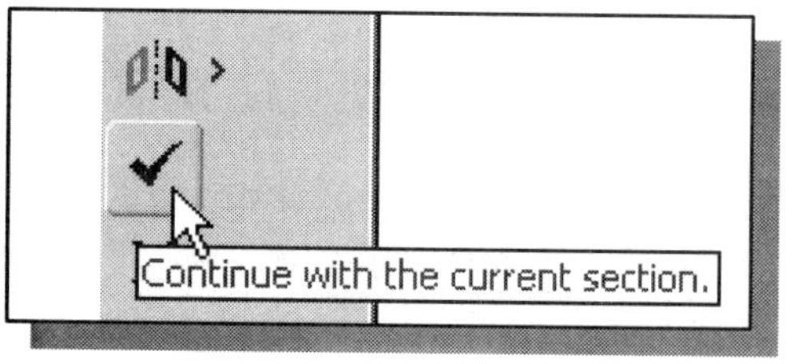

6. In the *Sketcher* toolbar, click on the **Accept** icon to exit the *Pro/ENGINEER 2D Sketcher* and proceed to the next element of the feature definition.

7. Note that the quick-key [**Ctrl-D**] or picking the **Standard Orientation** option in the pull-down menus also works in the sketcher mode.

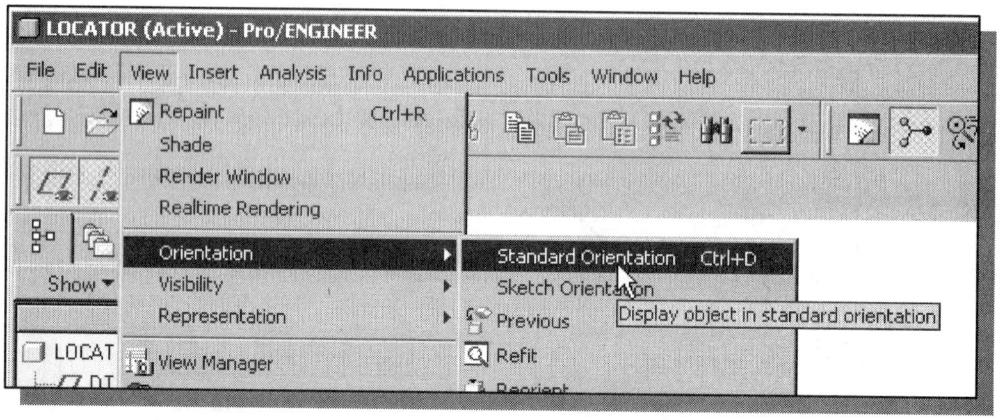

❖ Note that the default setting of the **Extrude Tool** command is set to **add material.** The arrow indicates the extrusion direction.

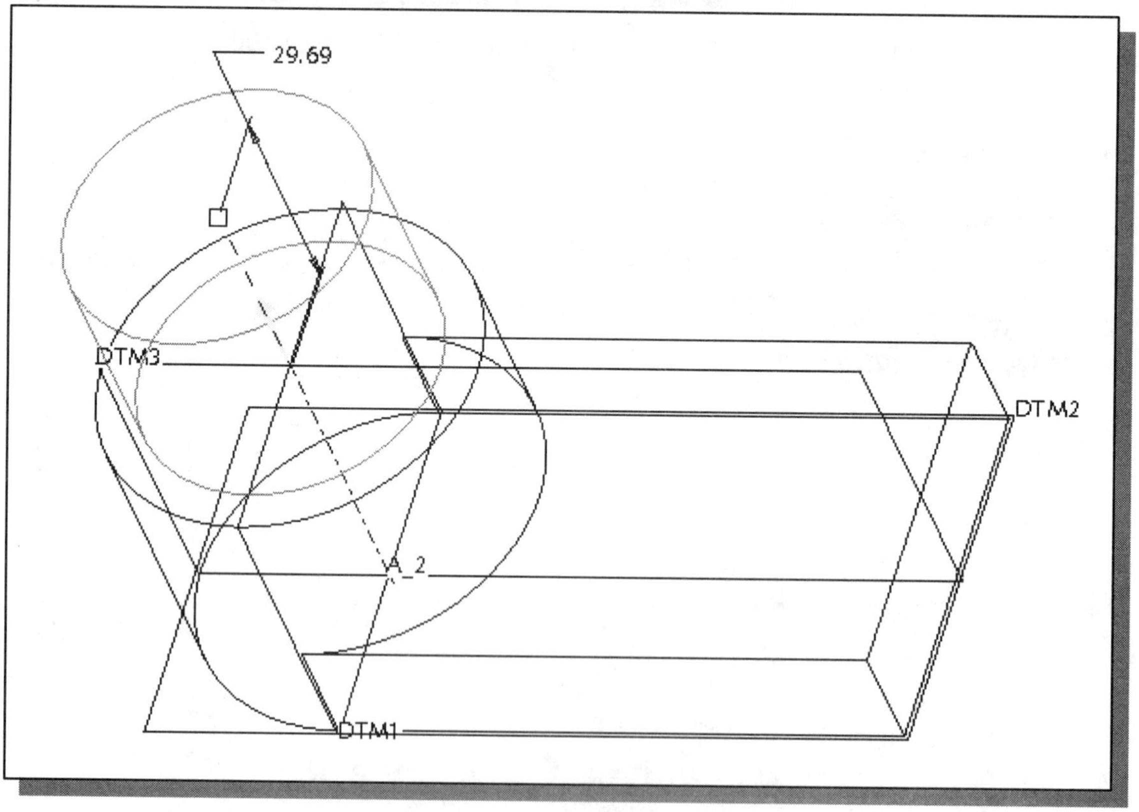

8. Click on the **Remove Material** icon as shown in the below figure.

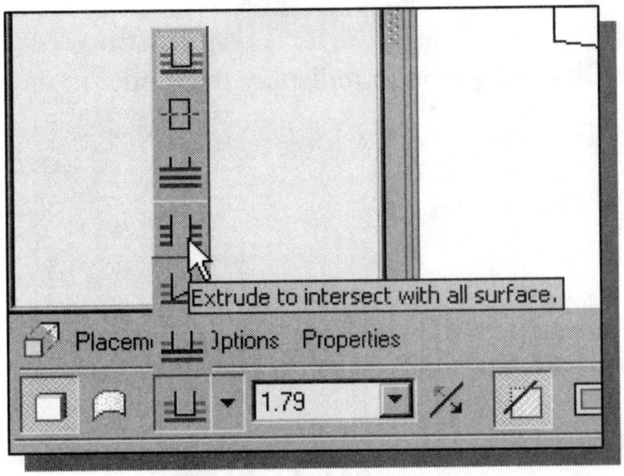

9. In the *Feature Option Dashboard*, select the **Extrude to intersect with all surface** option as shown.

- Note that the **Thru All** option does not require us to enter a value to define the depth of the extrusion; *Pro/ENGINEER* will calculate the required value to assure the extrusion is through the entire solid model.

10. Click on the **Flip direction** icon, as shown in the figure below, to set the cut direction.

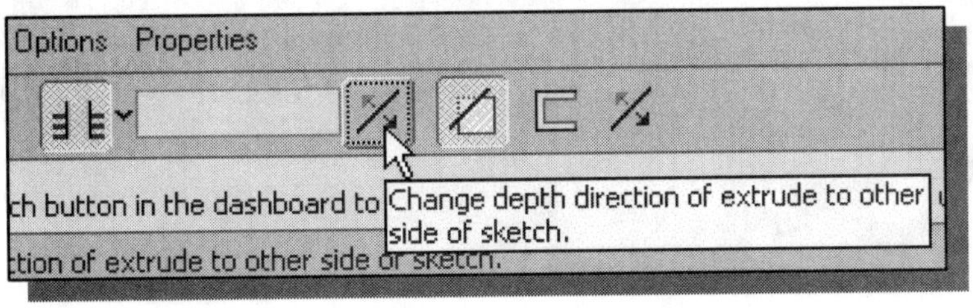

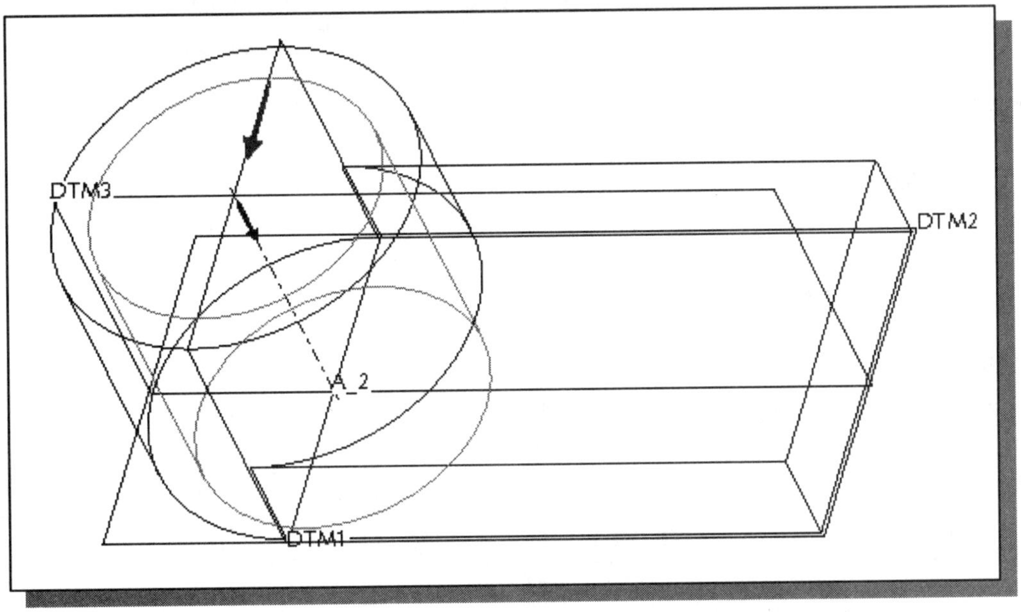

❖ We are now returned to the feature dialog box with all the required elements defined. Prior to creating the feature, we can preview and/or modify any of the elements that are currently defined.

11. Click on the **Preview** button to examine the current solid feature.

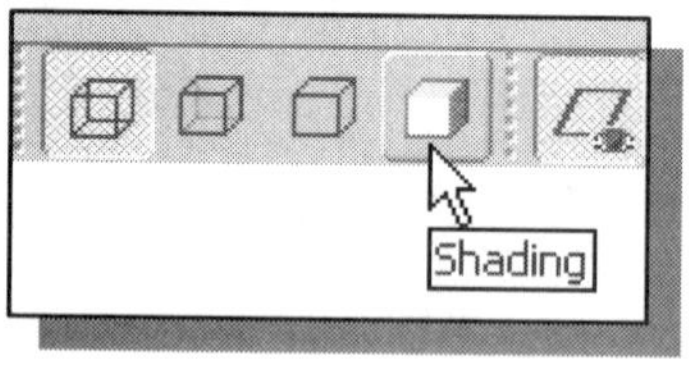

12. Click on the **Shading** icon to redisplay the model in the shaded format.

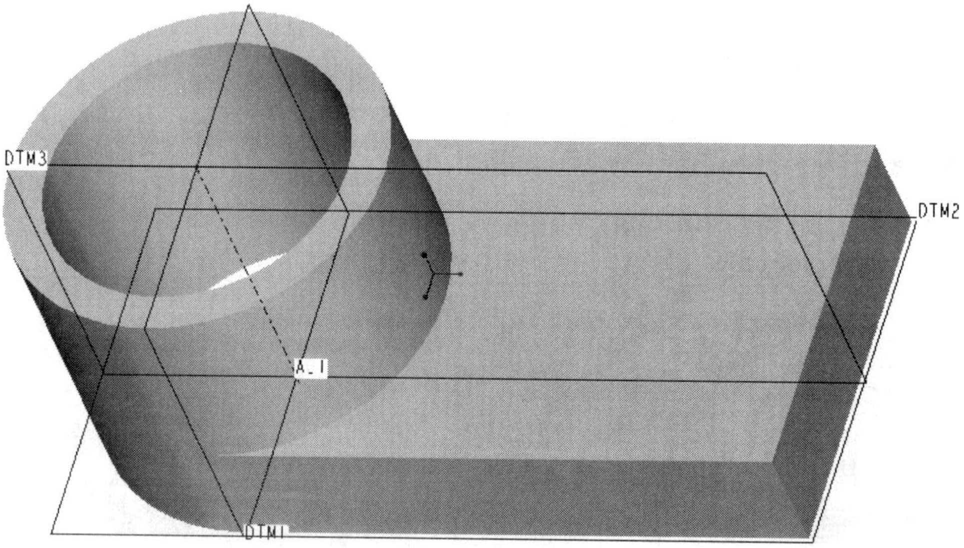

Redefine a Feature Element

❖ Quite often during the design stage, minor modifications become necessary as the model is being formed. We will illustrate the flexibility and functionality of the feature based modeling concept by redefining one of the elements of the feature.

1. Click on the **Preview** button again to exit the preview mode.

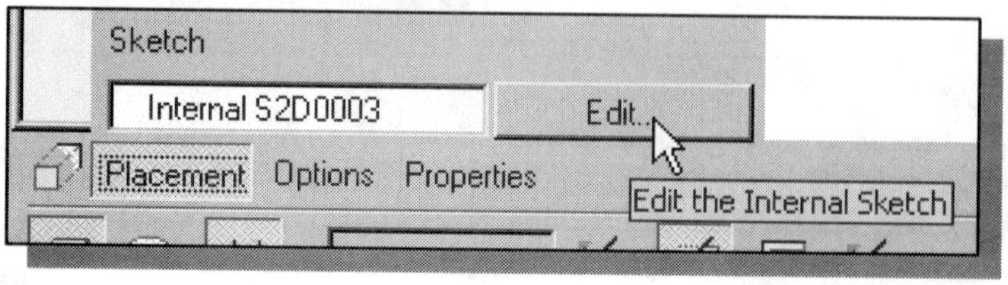

2. Click the **Placement** option and choose **Edit** to edit the current *Internal sketch.*

3. On your own, modify the diameter dimension to **30mm** as shown in the figure.

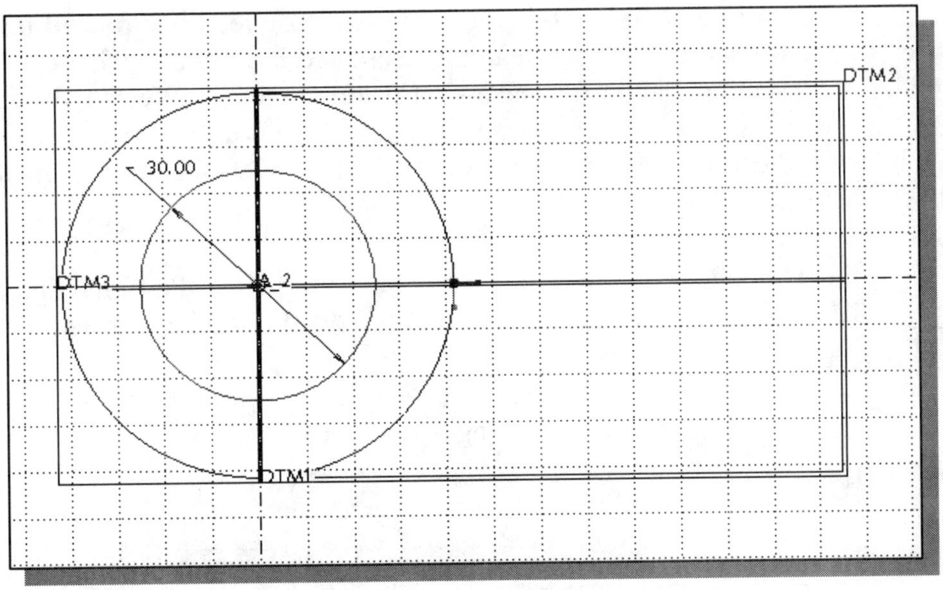

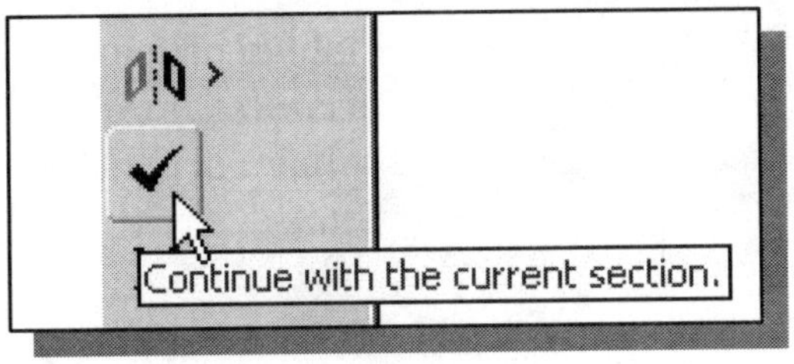

4. In the *Sketcher* toolbar, click on the **Accept** icon to end the *Pro/ENGINEER 2D Sketcher*.

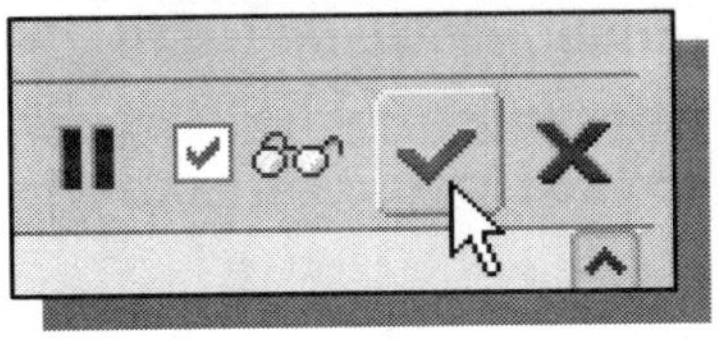

5. Click on **Accept** to proceed with the extrusion option.

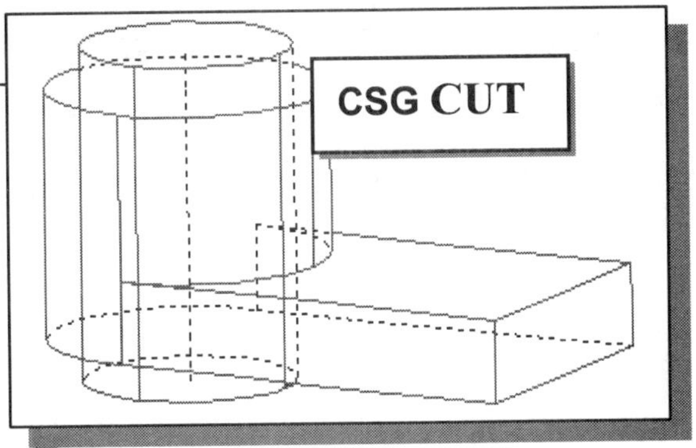

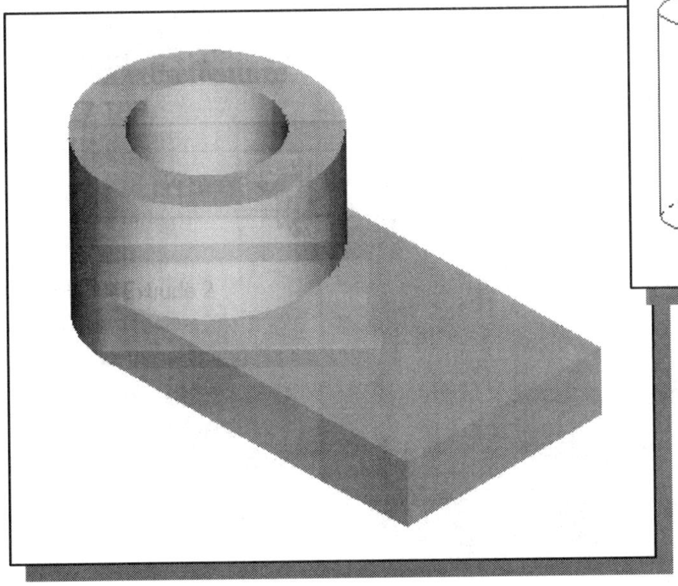

Creating a Placed HOLE Feature

- In *parametric modeling*, there are two types of geometric features: **placed features** and **sketched features**. The last cut feature we created is a *sketched feature*, where we created a 2D section and performed an extrusion operation. We can also create a hole feature, which is a *placed feature*. A *placed feature* is a feature that does not need a sketch and can be created automatically. Holes, fillets, chamfers, and shells are all placed features.

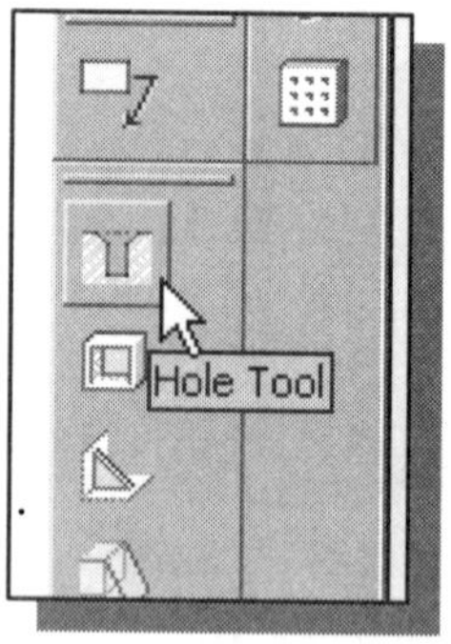

1. In the *Feature* toolbars (toolbars aligned to the right edge of the main window), select the **Hole Tool** option as shown.

❖ The message "*Select a surface, axis or point to place hole.*" is displayed in the message area.

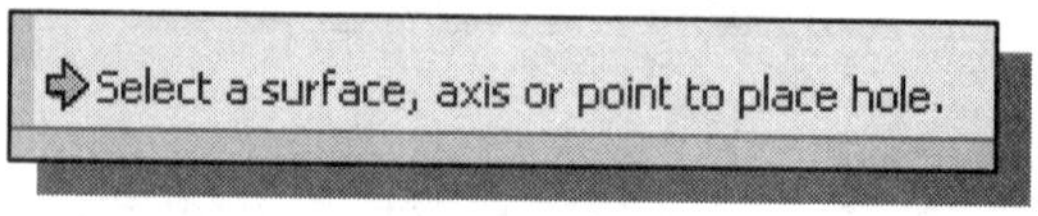

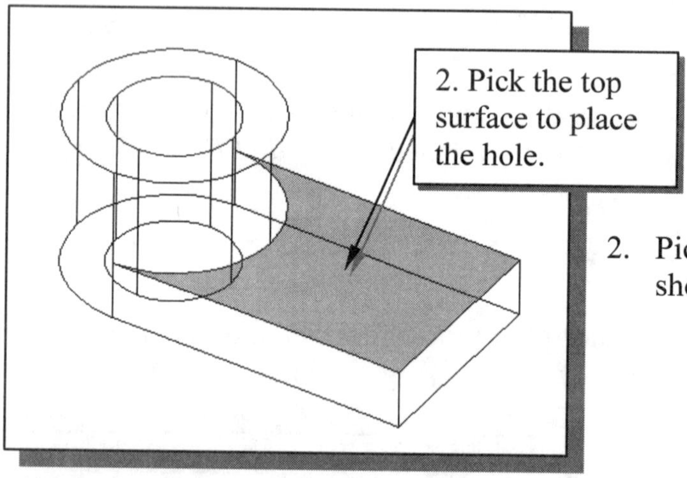

2. Pick the top face of the base feature as shown.

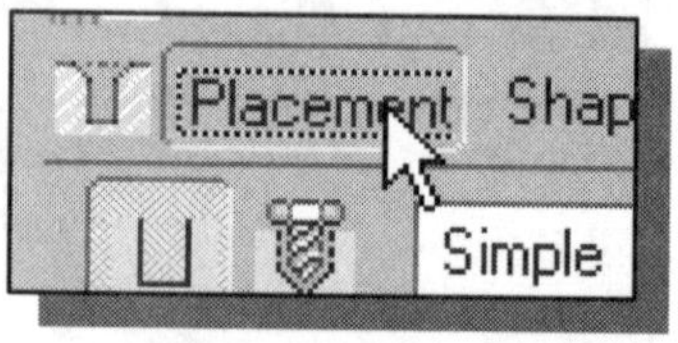

3. In the *Feature Option Dashboard*, select the **Placement** option as shown.

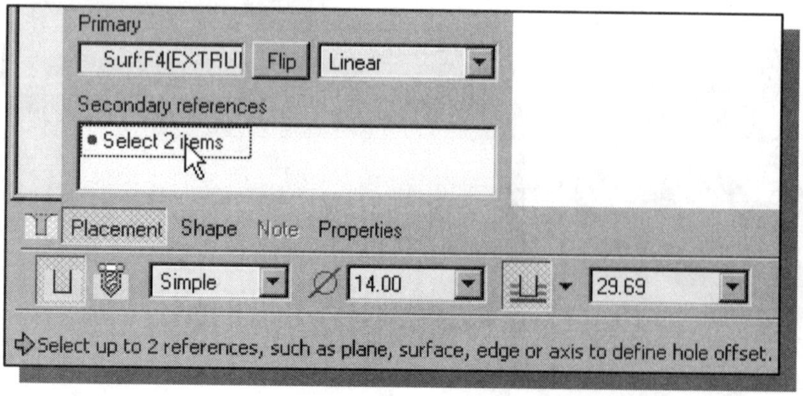

4. In the *Feature Option Dashboard*, click once with the left-mouse-button in the **Secondary references** option box as shown.

5. In the graphics area, select **DTM3** as the first secondary reference as shown in the figure below.

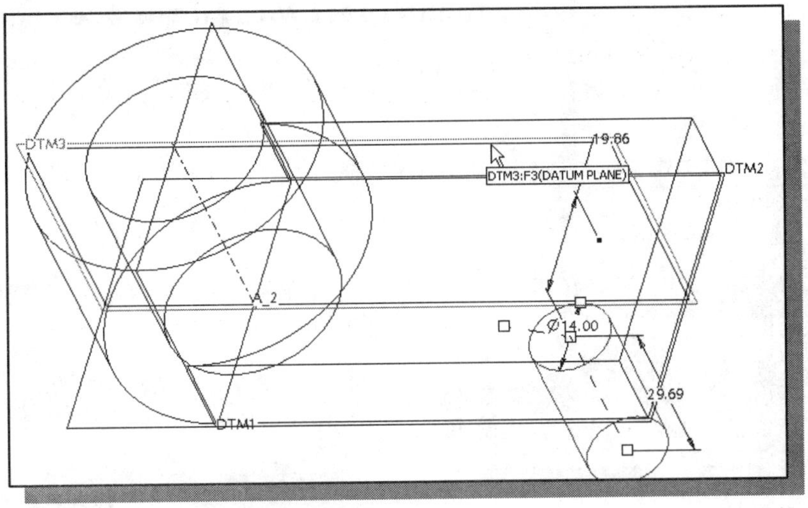

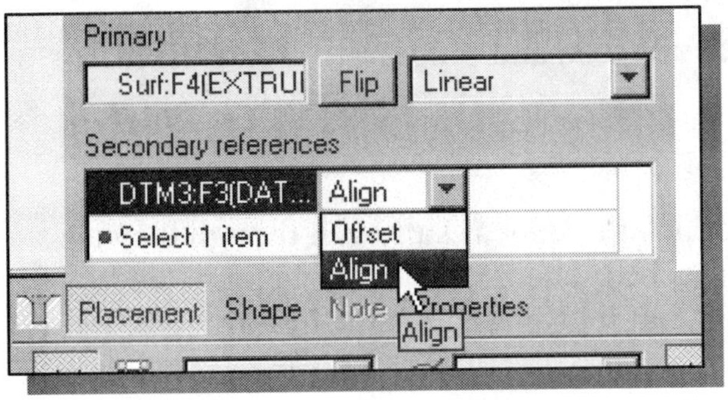

6. In the **Secondary references** section, set the alignment option to **Align** as shown.

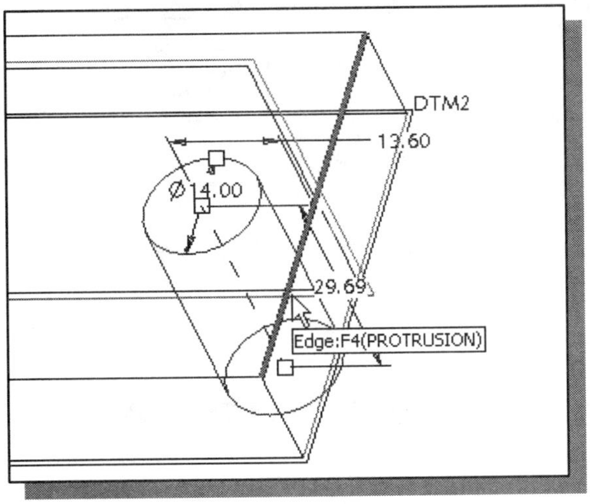

7. Hold down the [**Ctrl**] key and select the **top right edge** of the base feature as shown in the figure to the left.

8. Enter **30** for the **Offset** distance of the second reference as shown in the figure below.

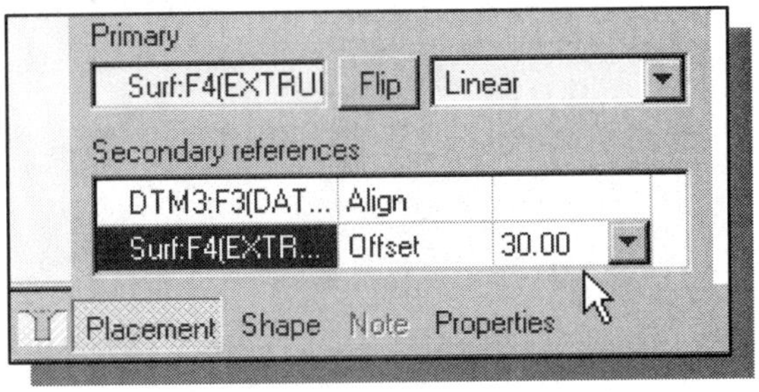

❖ Note that two references for the location of the hole are required for the **linear** placement of the **hole feature**.

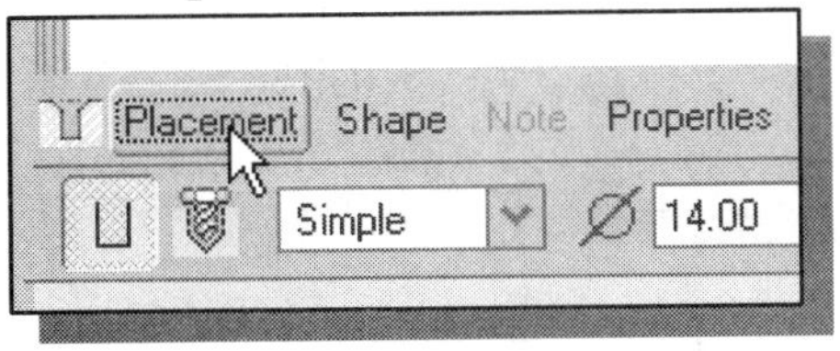

9. In the *Feature Option Dashboard*, click on the **Placement** button to close the option menu.

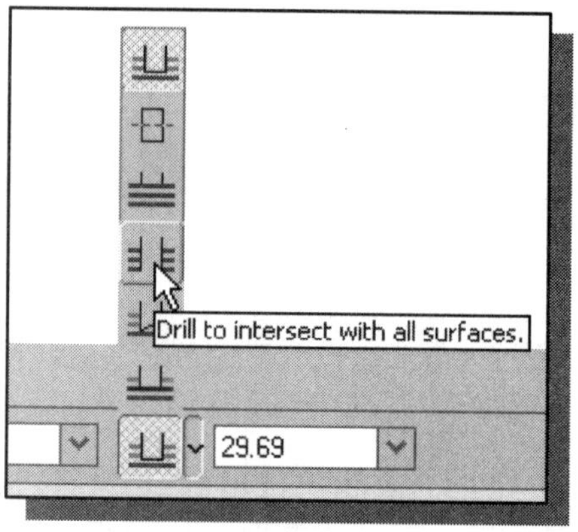

10. In the *Feature Option Dashboard*, select the **Drill to intersect with all surface** option as shown.

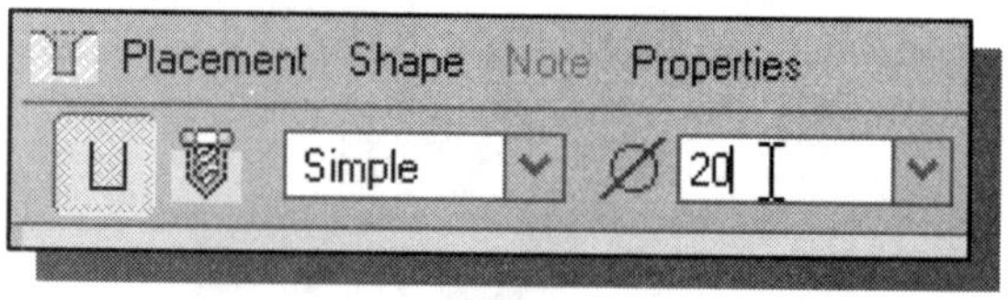

11. In the *Feature Option Dashboard*, set the *feature diameter* to **20mm** as shown.

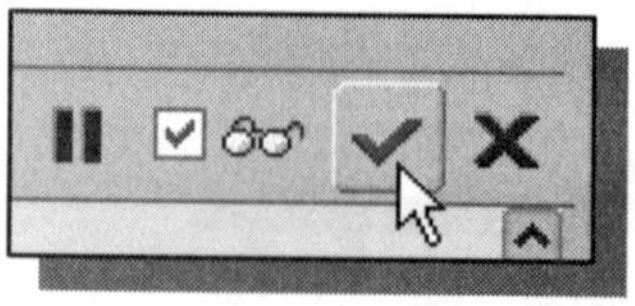

12. Click on the **Accept** icon and proceed to create the hole feature.

➢ The last two features we have created produce the same result even though they are accomplished by using two different approaches. The **Extrude Tool** command is more flexible in the shape that can be used, but it also requires the definition of more elements.

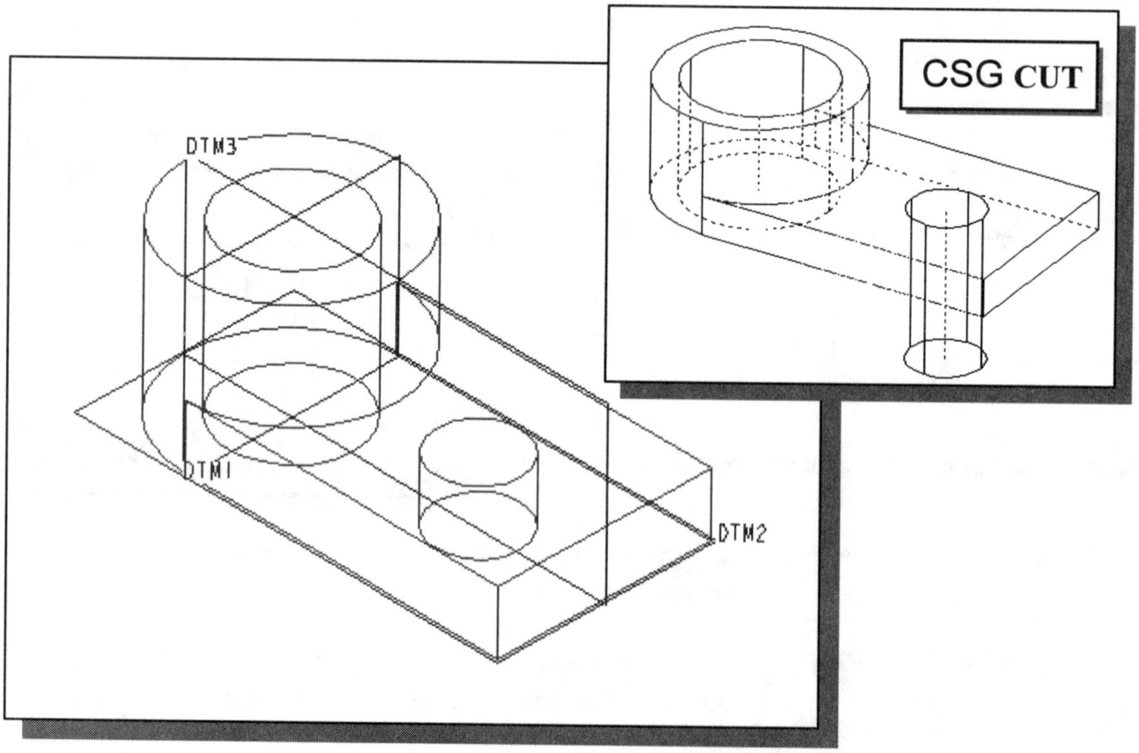

Creating the Final Feature

❖ We will create a rectangular cut as the final solid feature of the ***Locator***.

1. In the *Feature* toolbars (toolbars aligned to the right edge of the main window), select the **Extrude Tool** option as shown.

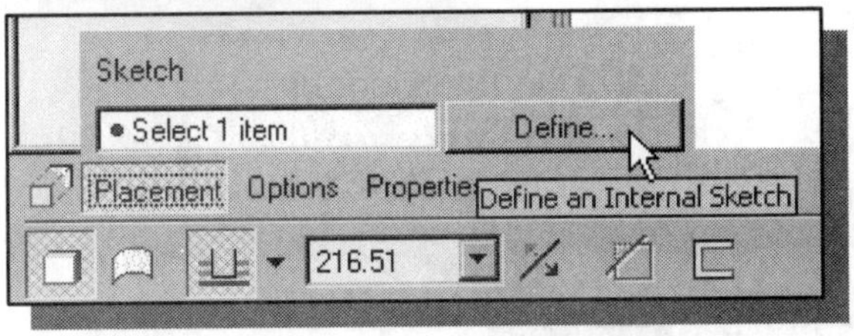

2. Click the Placement option and choose **Define** to begin creating a new *Internal sketch.*

3. We will use the right vertical surface of the base feature as the sketching plane. Click once, with the left-mouse-button, inside the right vertical surface of the base solid feature as shown in the figure below.

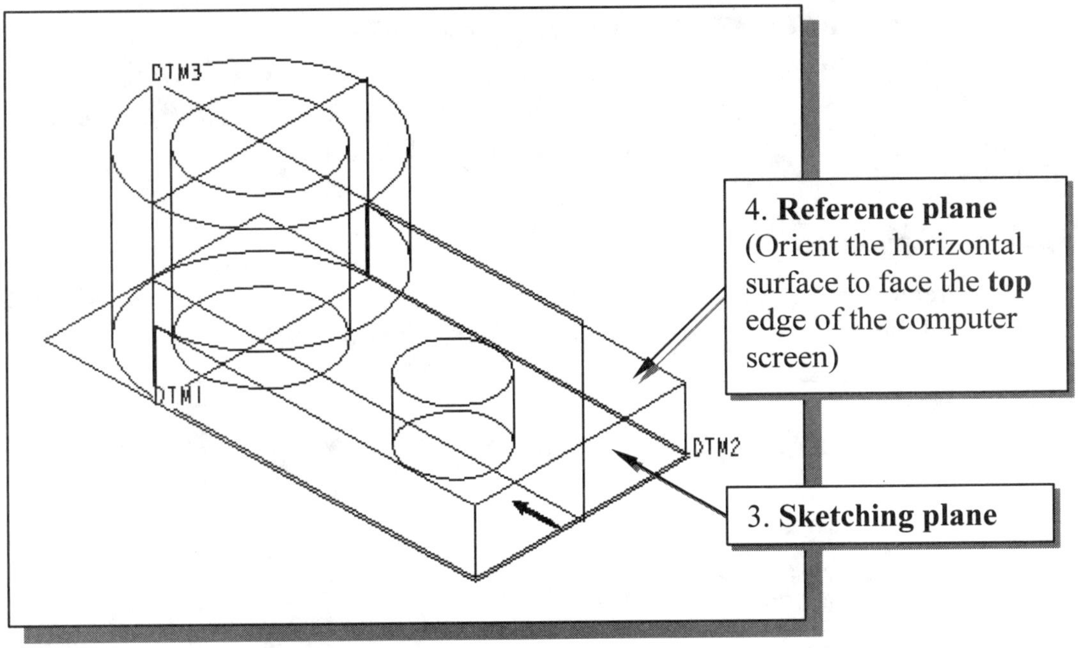

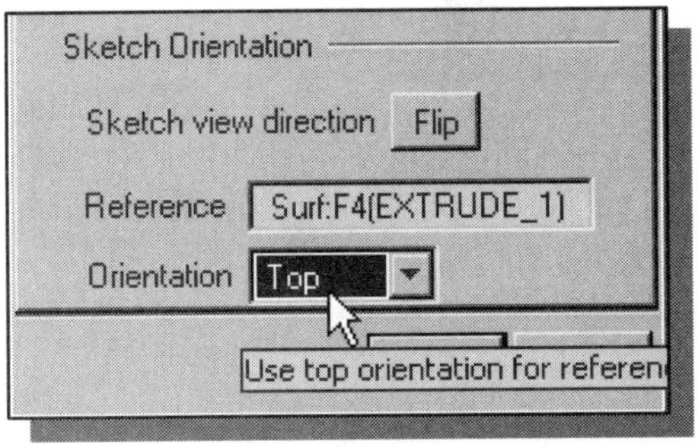

4. Select the top surface of the base feature as the orientation reference as shown in the above figure.

5. In the Sketch Orientation menu, set the reference plane Orientation to **Top**, and proceed to enter the sketcher mode.

Creating the 2D Section

1. On your own, setup to use **DTM2**, **DTM3**, and the top surface of the base feature as the references for the 2D sketch. *Pro/ENGINEER* will use these references to assure the 2D sketch is created at the correct location.

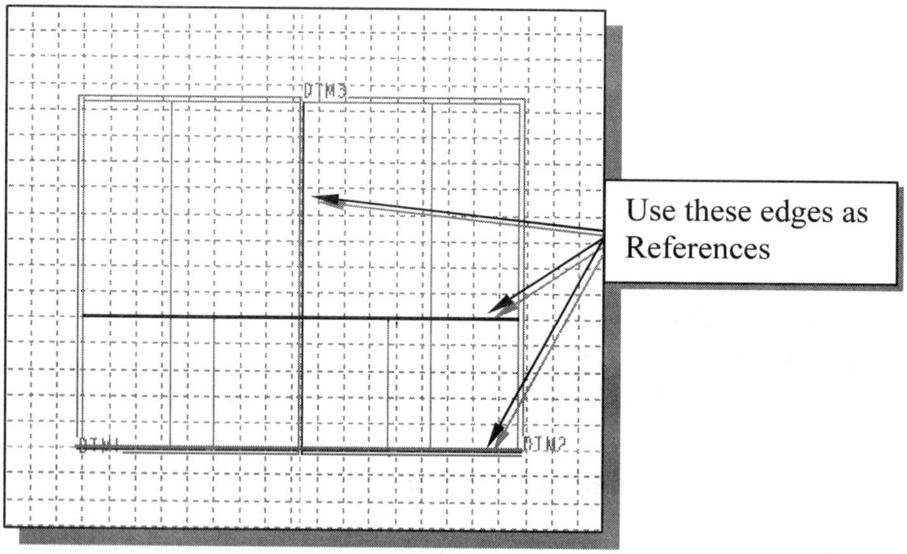

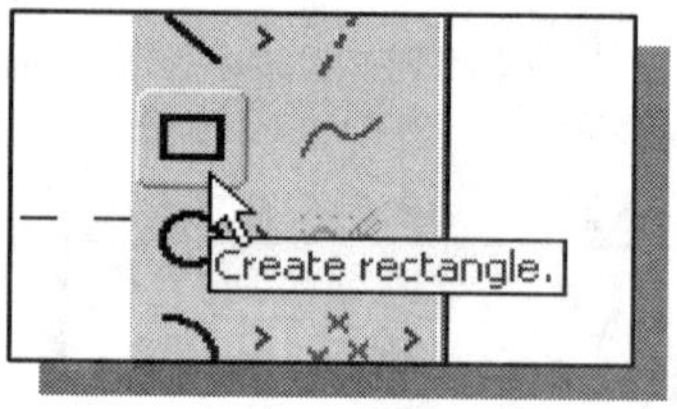

2. In the *Sketcher* toolbar, click on the **Rectangle** icon as shown to activate the **Rectangle** command.

3. Create a rectangle as shown, with one corner on the horizontal axis (**DTM2**) and the other on the top surface of the base feature.

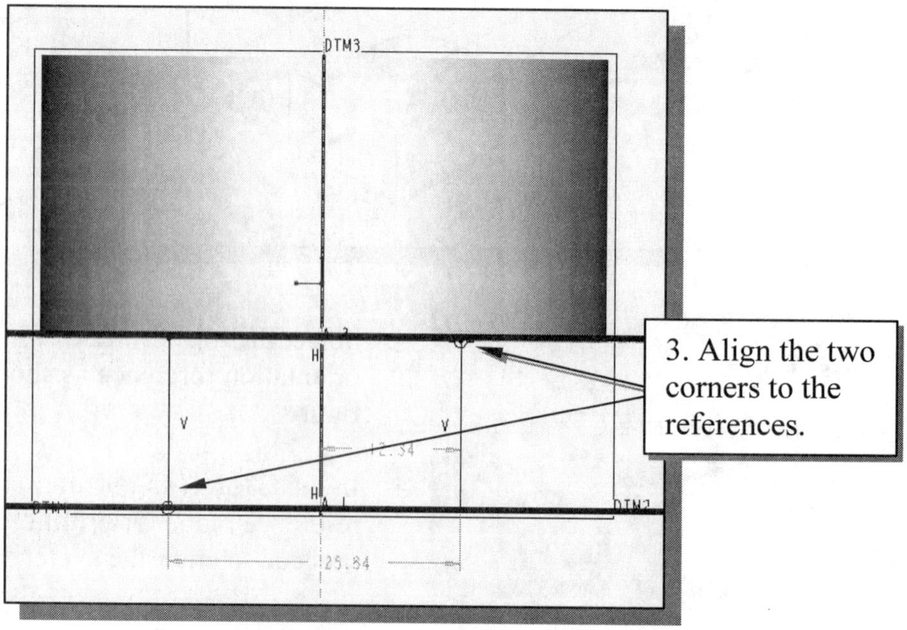

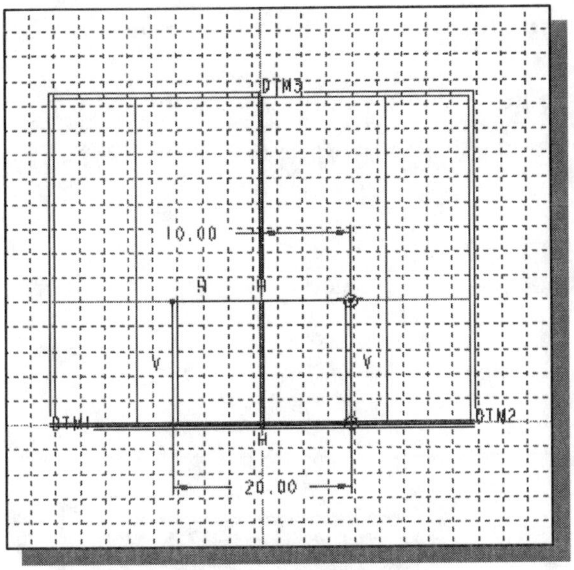

4. On your own, use the **Modify** command and adjust the rectangle to **20mm** wide and centered as shown.

5. In the *Sketcher* toolbar, click on the **Accept** button to end the *Pro/ENGINEER 2D Sketcher*.

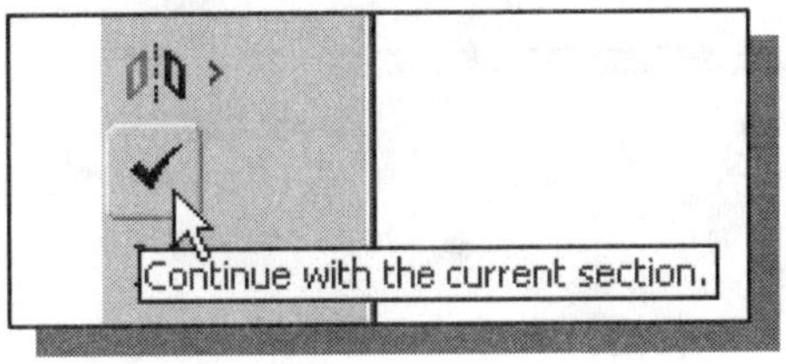

6. In the *Feature Option Dashboard*, select the **Remove Material** option as shown.

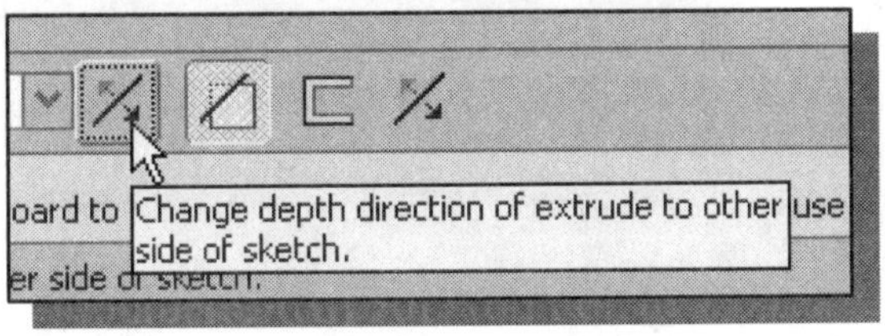

7. In the *Feature Option Dashboard*, click the **Flip Direction** icon to change the extrusion direction.

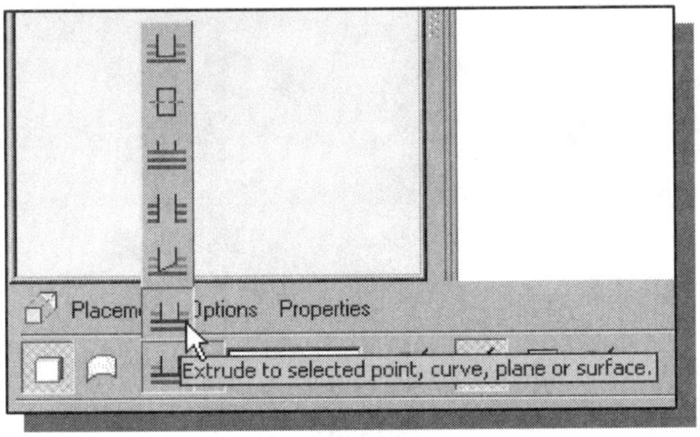

8. In the *Feature Option Dashboard*, select the **Extrude to selected point, curve, plane or surface** option as shown.

9. Choose the center axis of the hole feature as shown.

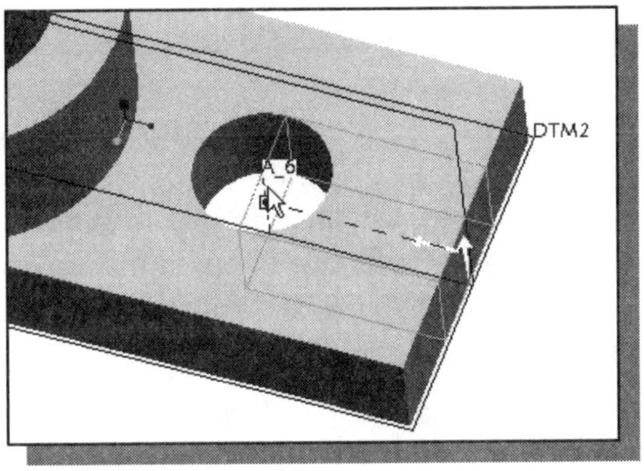

10. On your own, use the *Dynamic Viewing* functions to reorient the 3D view.

11. On your own, use the **Preview** option to examine the current feature.

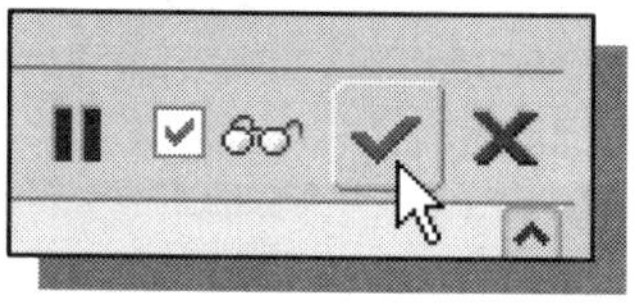

12. Click on the **Accept** icon and proceed to create the hole feature.

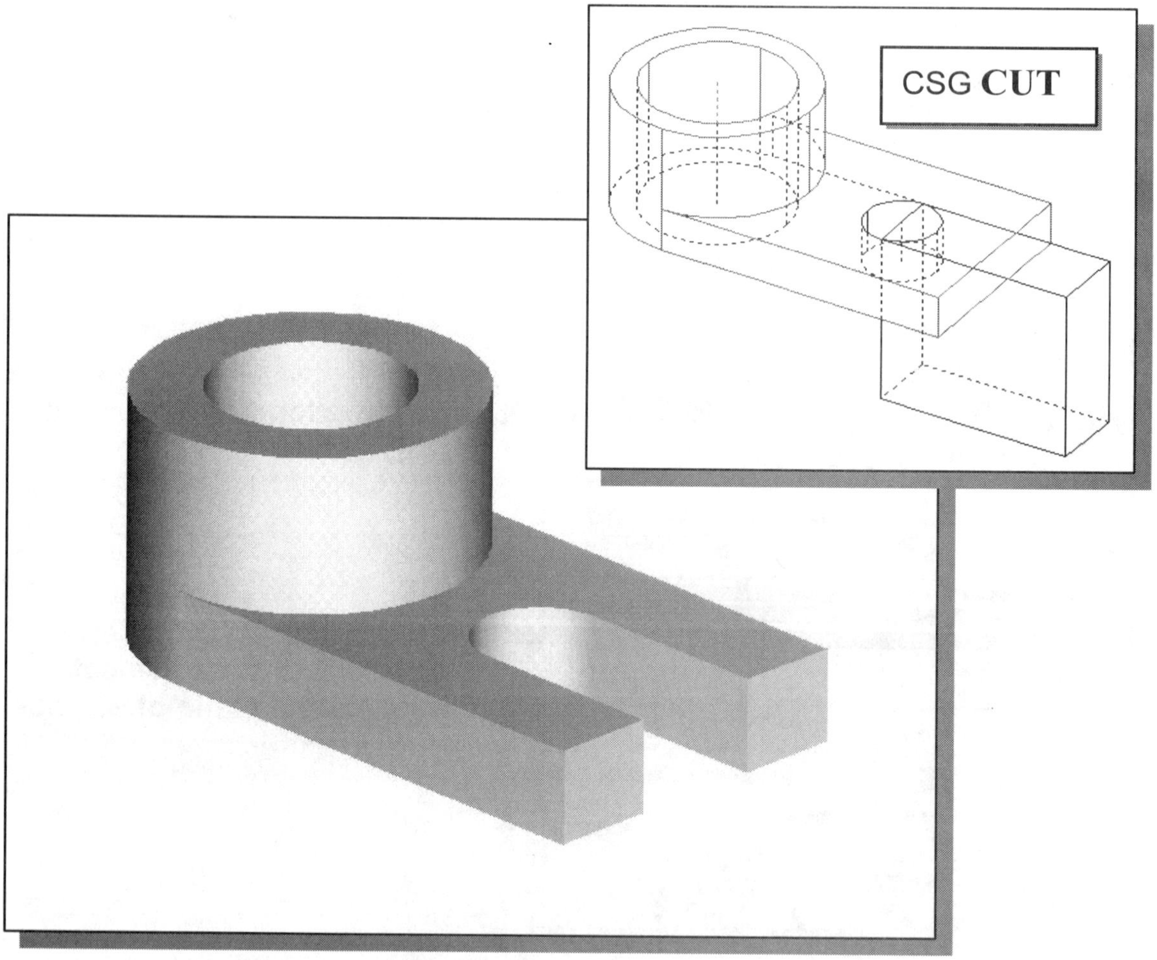

➢ Note that the above *Locator design* was constructed using rectangular blocks and cylinders to illustrate the basic CSG procedure, which are embedded in the parametric modeling system. It is also important to point out that the modern parametric modeling software is much more flexible and all parametric features created in *Pro/ENGINEER* can be treated as primitive solids.

Questions:

1. What are the three basic *Boolean operations* commonly used in computer geometric modeling software?

2. What is a *primitive solid*?

3. What does *CSG* stand for?

4. Which *Boolean operation* keeps only the volume common to the two solid objects?

5. What are the differences in creating a CUT feature and creating a HOLE feature in *Pro/ENGINEER*?

6. Using the CSG concept, create Binary Tree sketches showing the steps you plan to use to create the two models shown on the next page:

Ex.1)

Ex.2)

Exercises: (All dimensions are in inches.)

1.

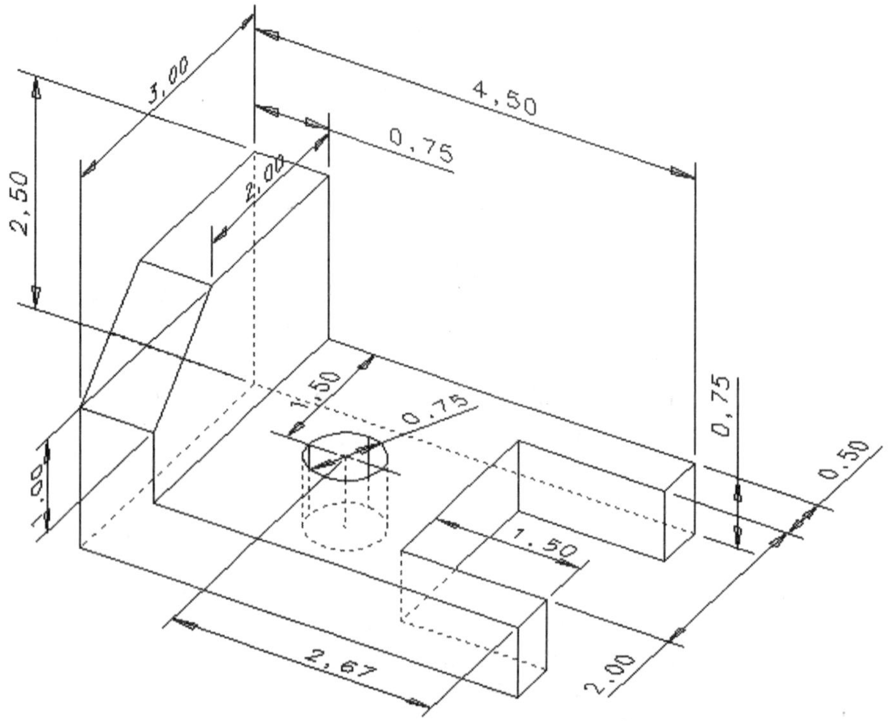

2.

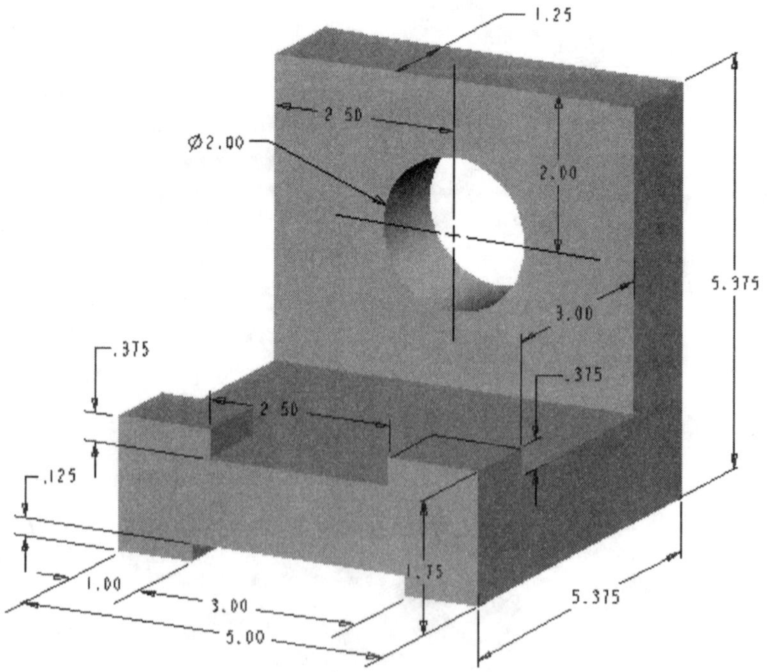

3.

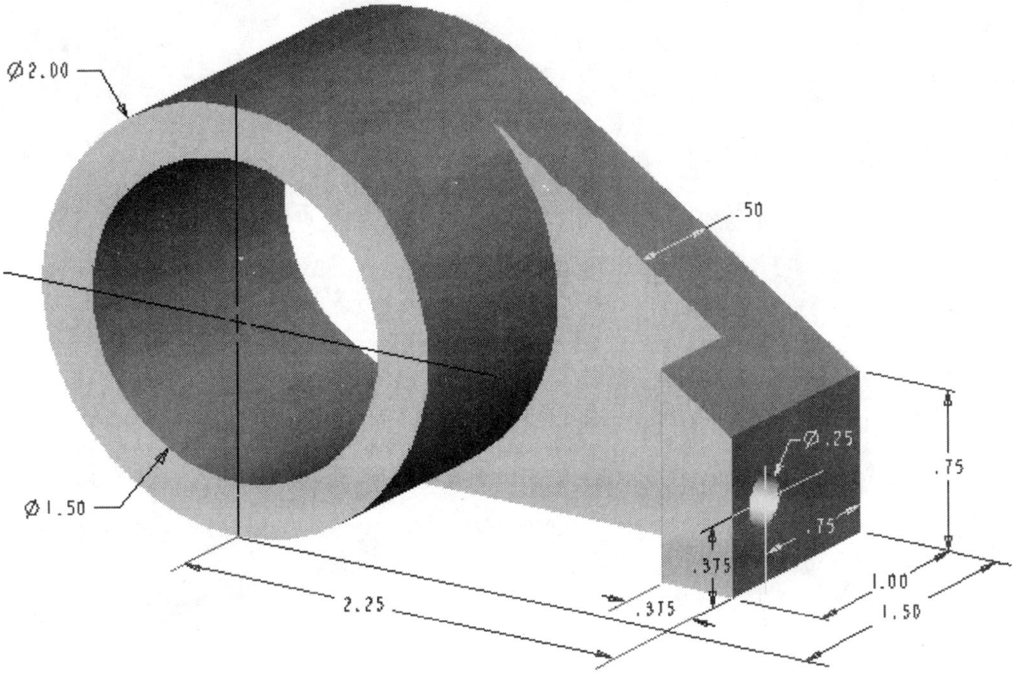

Notes:

Chapter 3
Model History Tree

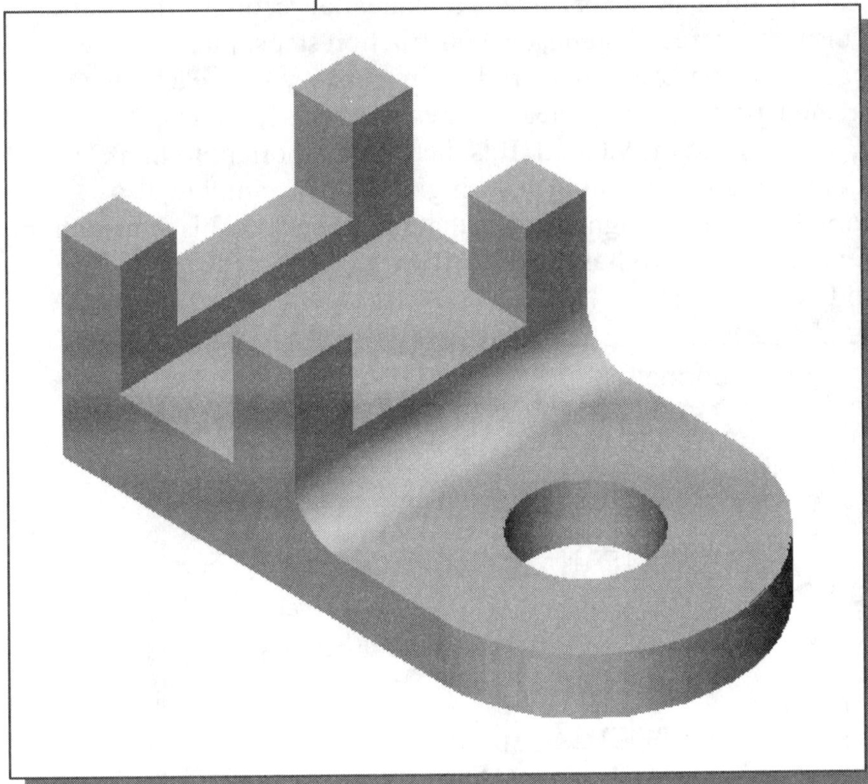

Learning Objectives

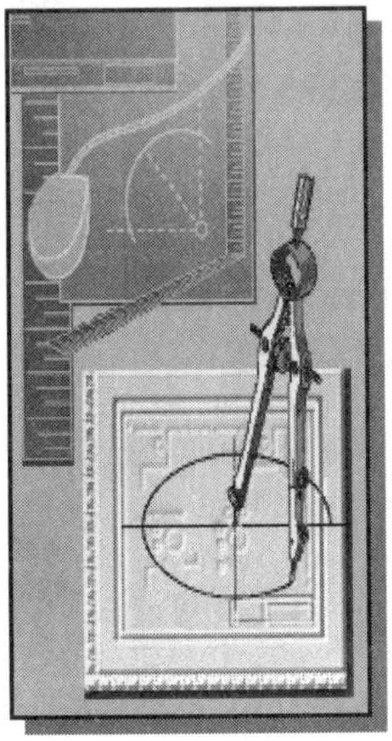

- ♦ **Understand Feature Interactions.**
- ♦ **Use the Part Browser.**
- ♦ **Modify and Update Feature Dimensions.**
- ♦ **Perform History-Based Part Modifications.**
- ♦ **Change the Names of created Features.**
- ♦ **Perform Basic Design Changes.**

Introduction

In *Pro/ENGINEER*, the *design intents* are stored as features in the **history tree**. The structure of the model history tree resembles that of a **CSG binary tree**. A CSG binary tree contains only Boolean relations, while the *Pro/ENGINEER* **history tree** contains all features, including Boolean relations. A history tree is a sequential record of the features used to create the part. This history tree contains the construction steps, plus the rules defining the design intent at each construction operation. In a history tree, each time a new modeling event is created, previously defined features can be used to define information such as size, location and orientation. It is therefore important to think about your modeling strategy before you start creating anything. It is important, but also difficult, to plan ahead for all possible design changes that might occur. This approach in modeling is a major difference of **feature based cad software**, such as *Pro/ENGINEER*, from previous generation CAD systems.

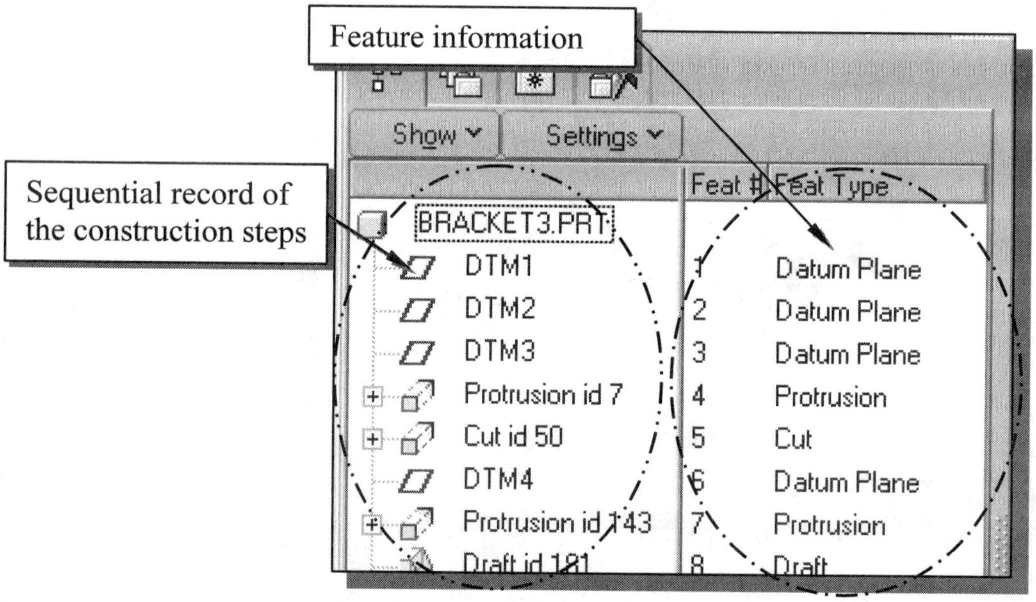

Feature based parametric modeling is a cumulative process. Every time a new feature is added, a new result is created and the feature is also added to the history tree. The database also includes parameters of features that were used to define them. All of this happens automatically as features are created and manipulated. At this point, it is important to understand that all of this information is retained, and modifications are done based on the same input information.

In *Pro/ENGINEER*, the model tree gives information about modeling order and other information about the feature. Part modifications can be done through accessing the features in the history tree. It is important to understand the concept of the history tree when you modify parts. *Pro/ENGINEER* remembers the history of a part, including all the rules that were used to create it, so that changes can be made to any operation that was performed to create the part. To modify a feature in *Pro/ENGINEER*, you may select the feature by picking it in the display area or selecting the name of the feature in the ***Model Tree*** window.

The *L-Support* Design

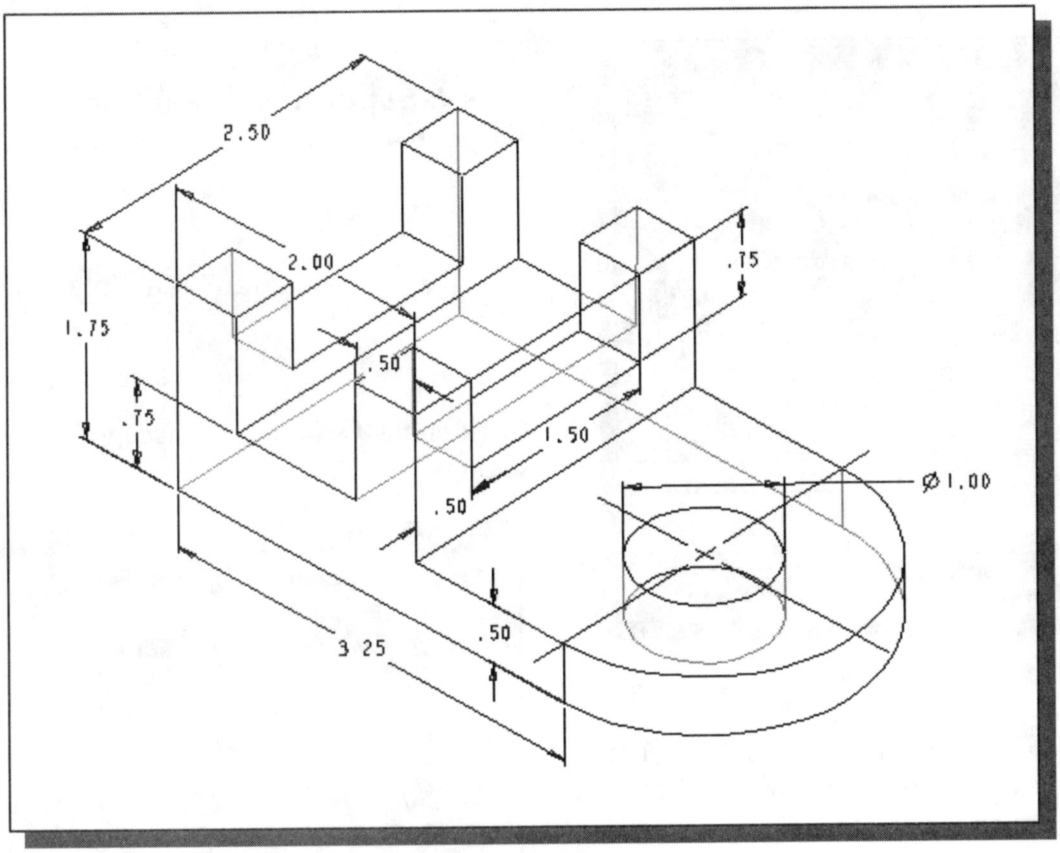

❖ Based on your knowledge of *Pro/ENGINEER* so far, how many features would you use to create the design? Which feature would you choose as the **base feature**, the first feature, of the model? What is your choice in arranging the order of the features? Would you organize the features differently if additional fillets were to be added in the design? Take a few minutes to consider these questions and do preliminary planning by sketching on a piece of paper. You are also encouraged to create the model on your own prior to following through the tutorial.

Starting Pro/ENGINEER

1. Select the **Pro/ENGINEER** option on the *Start* menu or select the **Pro/ENGINEER** icon on the desktop to start *Pro/ENGINEER*. The *Pro/ENGINEER* main window will appear on the screen.

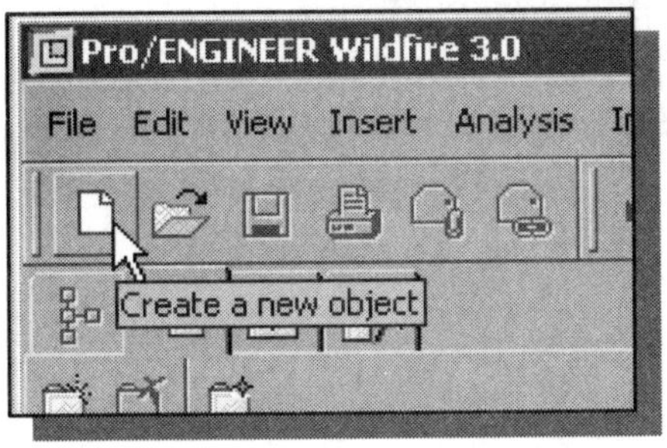

2. Click on the **New** icon, located in the *Standard* toolbar as shown.

3. In the *New* dialog box, confirm the model Type is set to **Part** (**Solid** Sub-type).

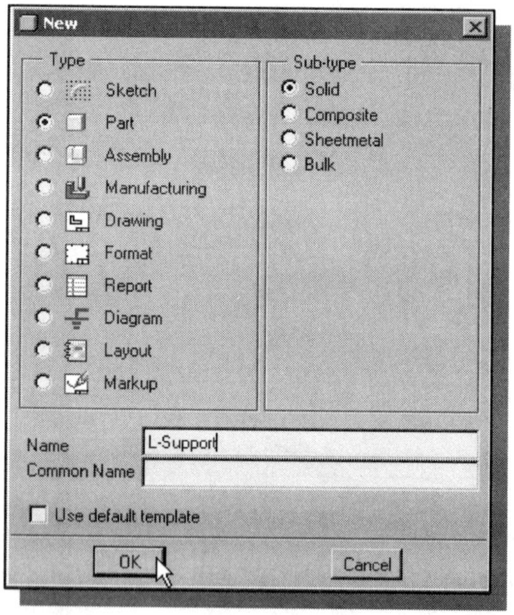

4. Enter ***L-Support*** as the part Name as shown in the figure.

5. Turn *off* the Use default template option.

6. Click on the **OK** button to accept the settings.

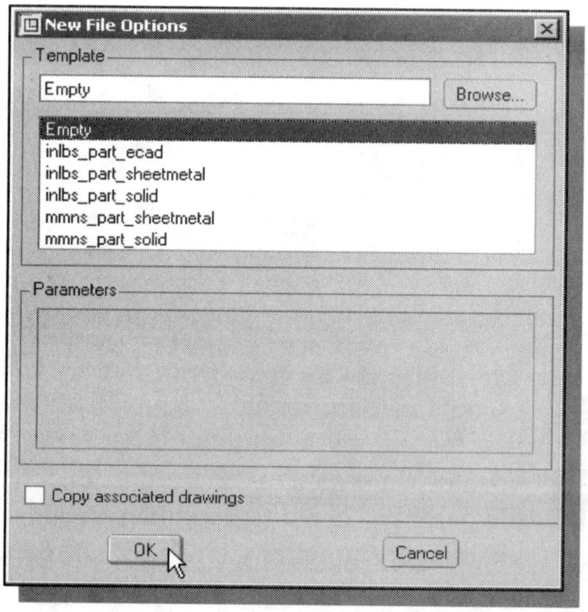

7. In the *New File Options* dialog box, select **Empty** in the option list to not use any template file.

8. Click on the **OK** button to accept the settings and enter the *Pro/ENGINEER Part Modeling* mode.

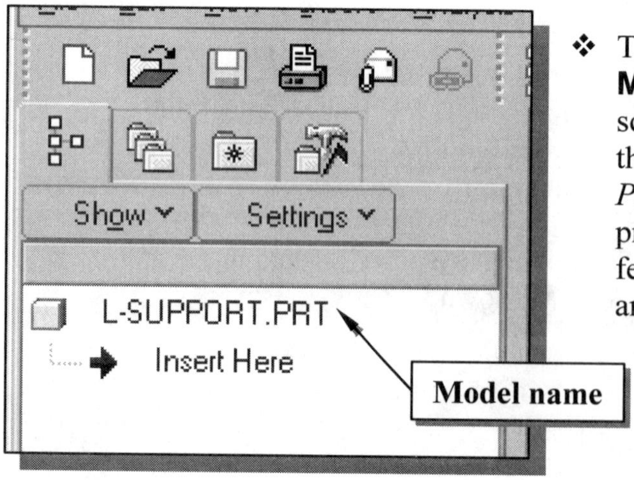

❖ The *Model Tree* window and the **Menu Manager** window appeared on the screen. Notice in the *Model Tree* window, the part name is displayed. The *Pro/ENGINEER Model Tree* window presents the model structure, feature by feature, in the order in which the features are created.

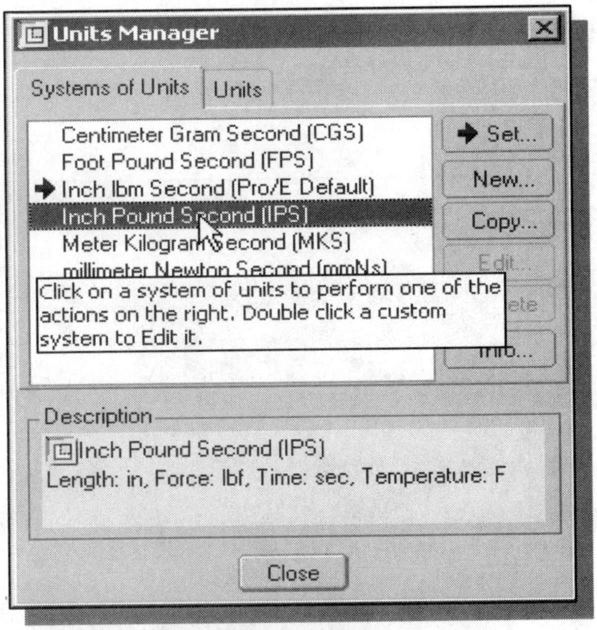

9. On your own, set up the **Units** to **Inch-Pound-Second (IPS)**.

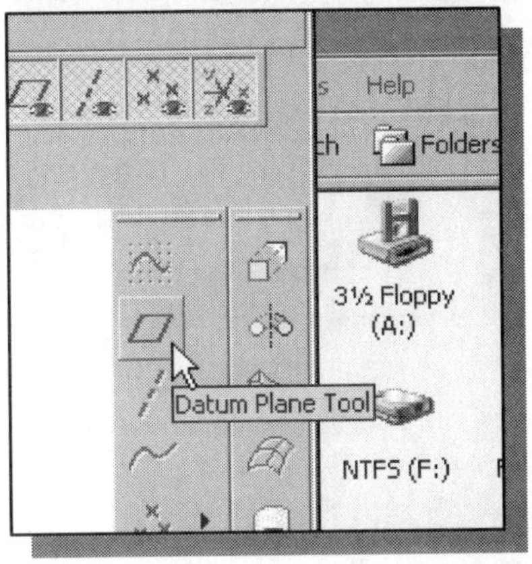

10. On your own, click on the **Datum Plane** icon and create the basic set of datum planes.

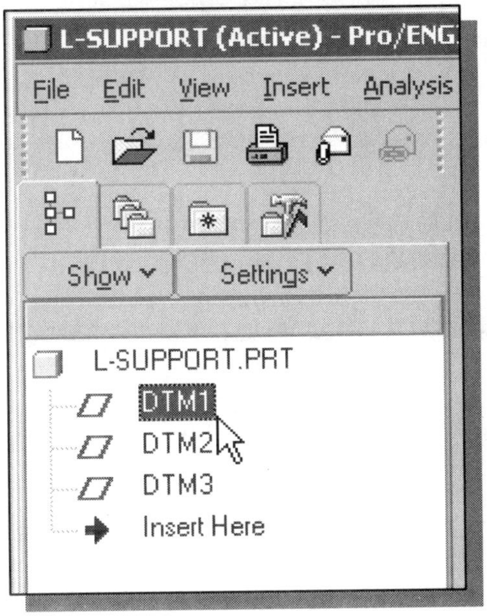

- In the *Model Tree* window, the datum planes are new features added into the model structure.

11. Click on **DTM1** in the *Model Tree* window and notice the selected feature is highlighted in the display area.

12. Click on **DTM2**; the selected feature is highlighted in the display area.

13. Select **DTM3** as the pre-selected datum plane before moving on to the next section.

- Note that the main advantage of the Model Tree is: **Any selection of features can also be done through the *Model Tree* window**.

Modeling Strategy

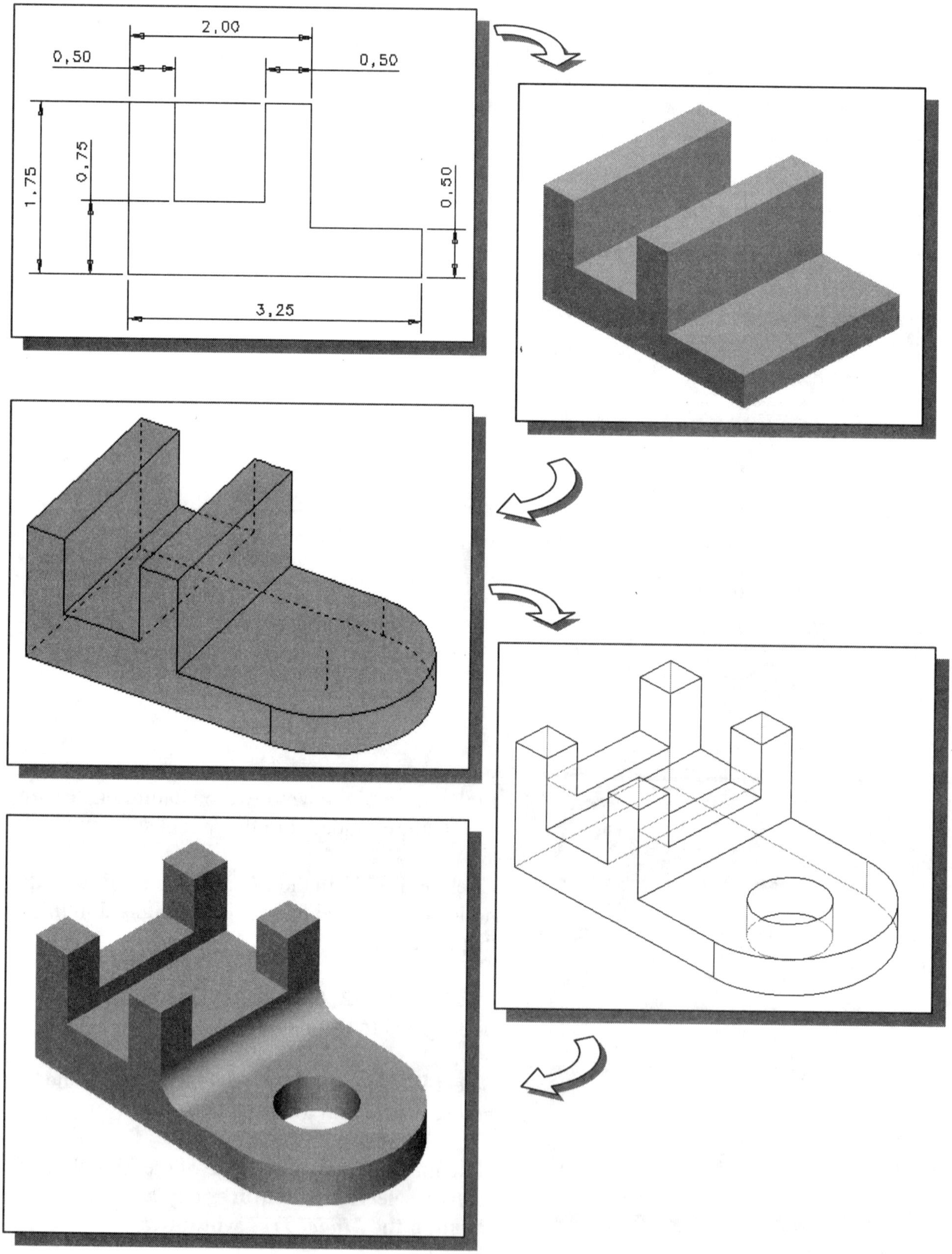

Creating the Base Feature

1. In the *Feature* toolbars (toolbars aligned to the right edge of the main window), select the **Extrude Tool** option as shown.

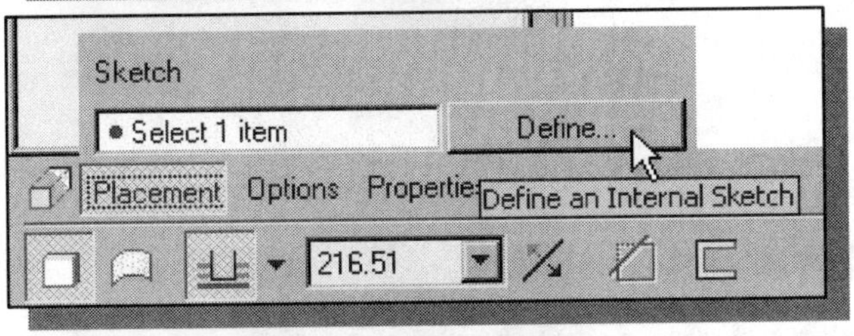

2. Click the **Placement** option and choose **Define** to begin creating a new *Internal sketch.*

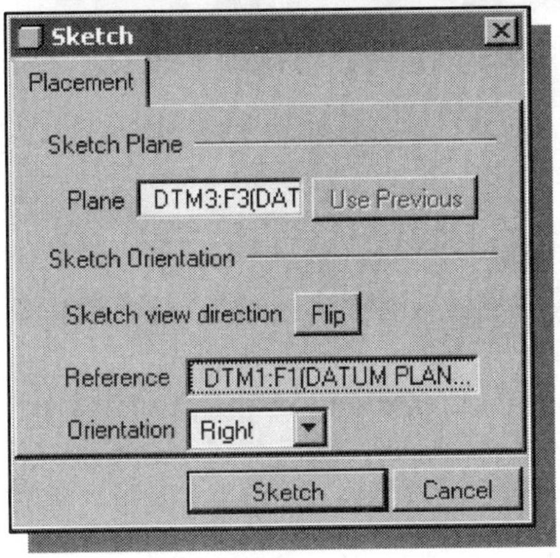

3. In the *Sketch* dialog box, confirm that **DTM3** is set as the Sketch Plane and the Sketch Orientation is set to **DTM1-Right** orientation as shown.

4. Click **Sketch** to exit the *Sketch* window and proceed to enter the *Pro/ENGINEER Sketcher* mode.

2D Sketching of the Base Feature

1. Create the 2D section as shown. (The lower left corner is aligned to the origin.)

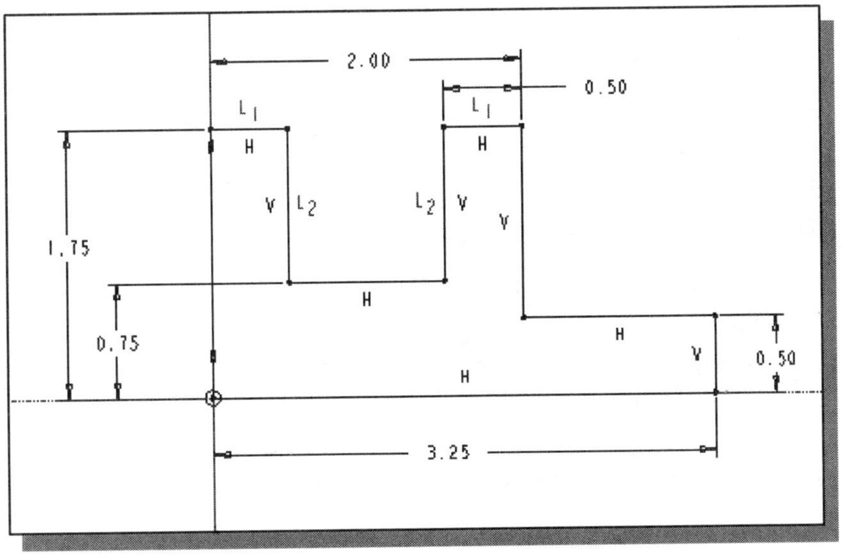

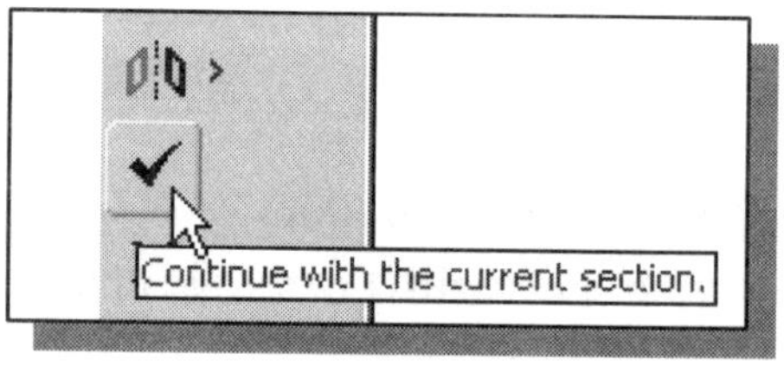

2. Now that the 2D sketch is completed, we will proceed to the next *element.* In the *Sketcher* toolbar, click on the **Accept** button to end the *Pro/ENGINEER* Sketcher command.

3. In the *Feature Option Dashboard*, choose the **Both Side** option as shown. This option sets the extrusion of the section to **Extrude on both sides of sketch plane**.

4. In the *depth value* box, enter **2.5** as the extrusion depth.

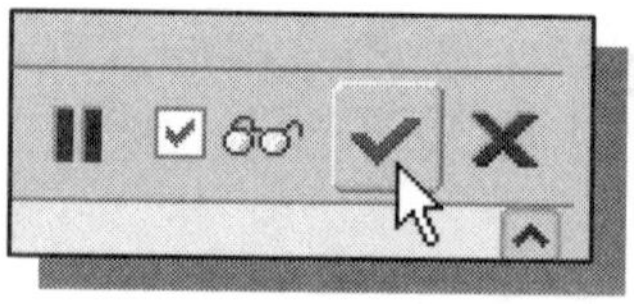

5. In the *Extrude Option Dashboard*, click on the **Accept** button to proceed with the feature creation.

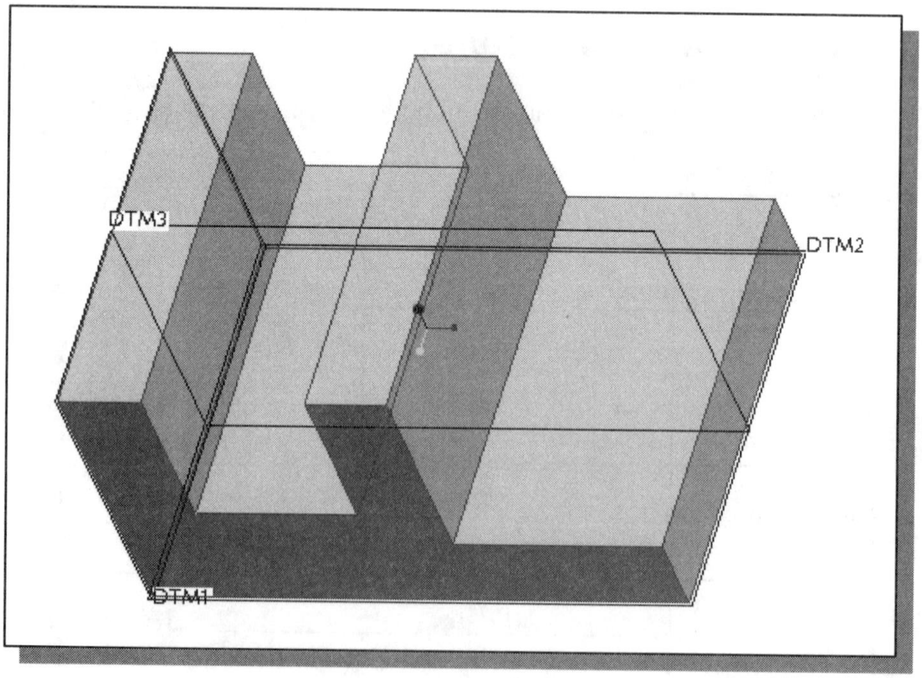

Set up the *Pro/ENGINEER* Default View

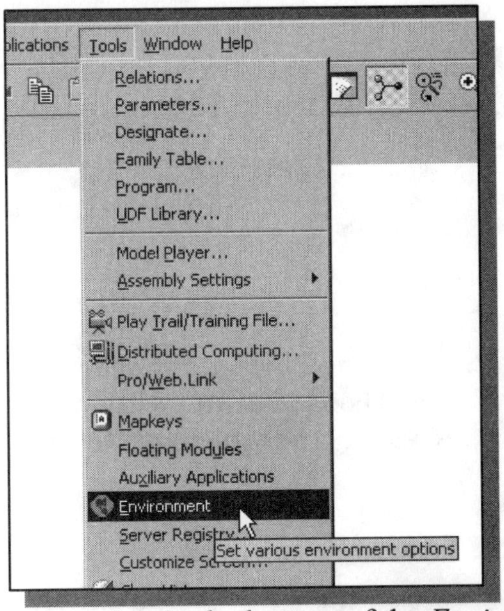

1. Select **Tools** in the pull-down menus. Pick the **Environment** option. The environment option allows us to adjust various environment settings.

2. At the bottom of the *Environment* form, three groups of display settings are available. Pick **Isometric** as the *default model orientation.*

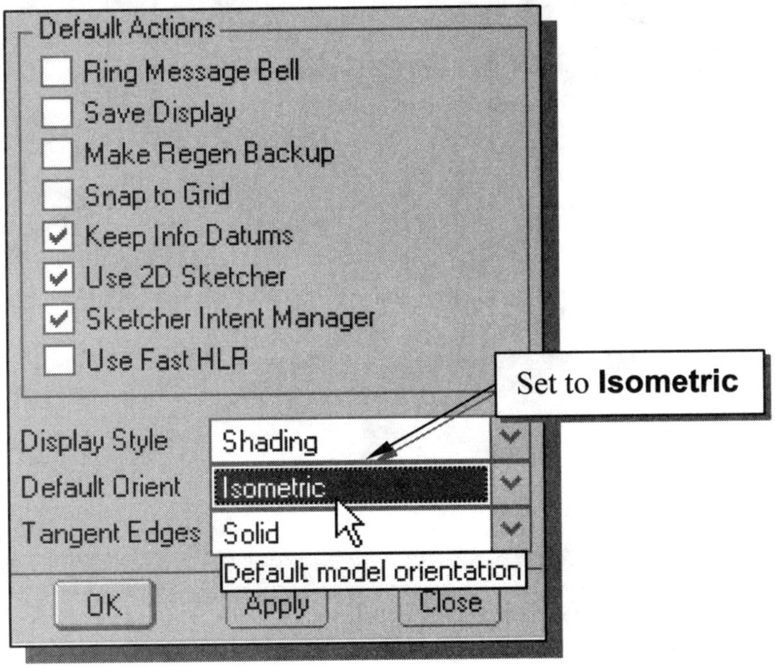

3. Pick **OK** to exit the form.

4. Rotate the Model and then use the quick key combination **[Ctrl-D]** to change the view orientation to the Isometric orientation in the display area.

➢ Note that DTM3 passes through the center of the base feature and the extrusion feature is added as the last item in the *Model Tree* window.

Add the Second Solid Feature

1. In the *Feature* toolbars (toolbars aligned to the right edge of the main window), select the **Extrude Tool** option as shown.

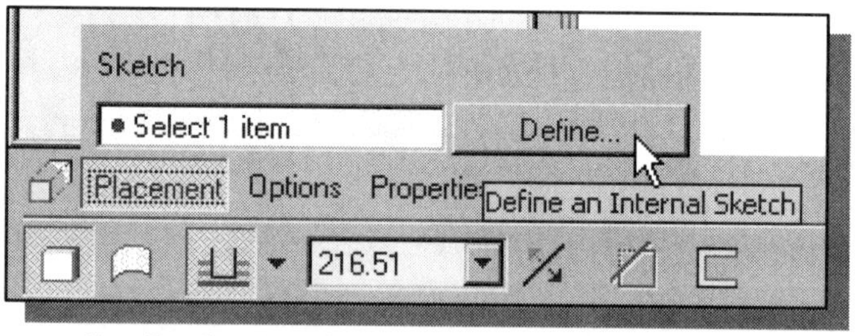

2. Click the **Placement** option and choose **Define** to begin creating a new *Internal sketch*.

3. In the *Model Tree* window, pick **DTM2** as the sketching plane as shown.

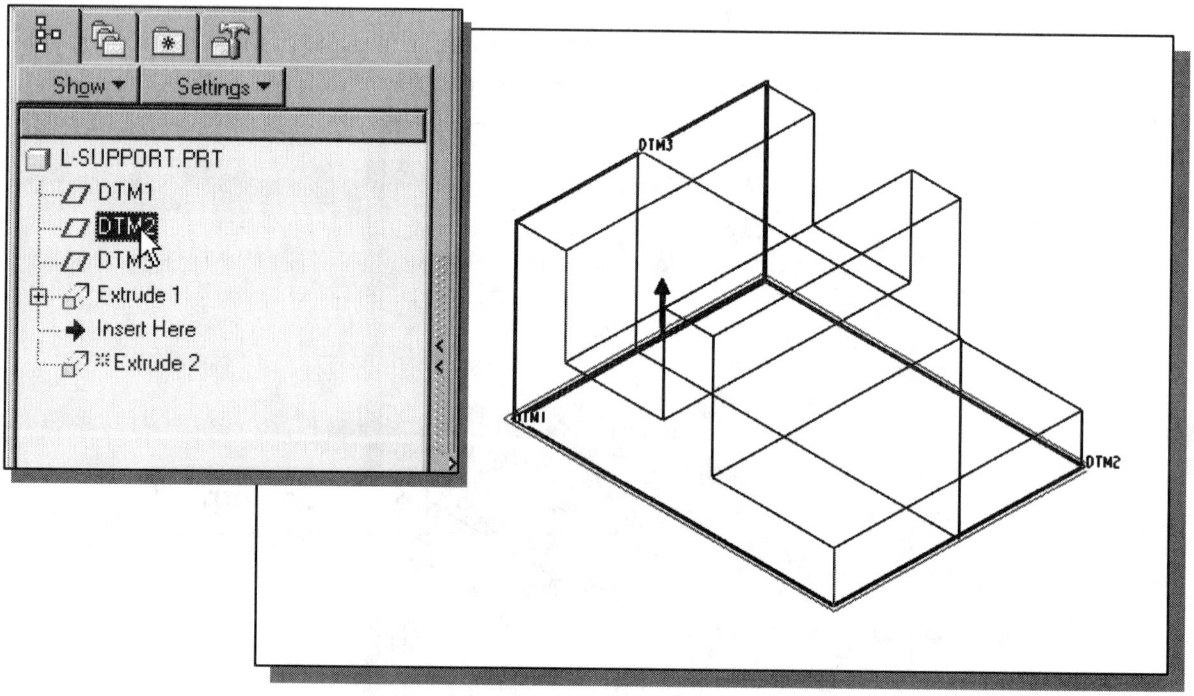

❖ As more features are created, selecting through the *Model Tree* is an extremely convenient option. As can be seen, *Pro/ENGINEER* provides us a variety of tools to aid the selection and construction of the desired features.

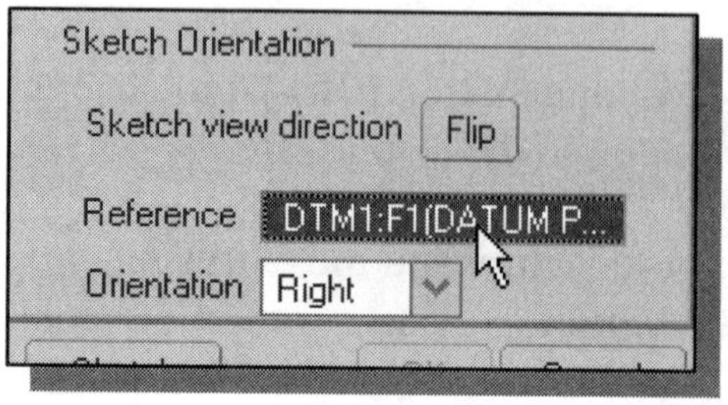

4. In the Sketch Orientation options, select the **Reference** option box (currently DTM1 is listed) by clicking once with the left-mouse button. Note that the Orientation option is set to **Right**.

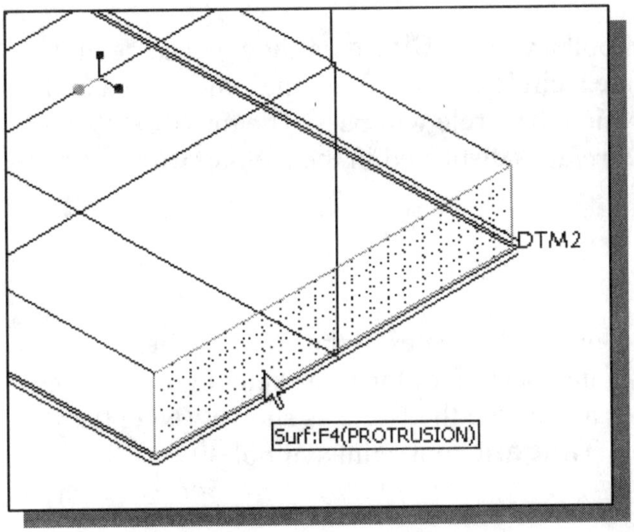

5. Pick the right vertical face of the base feature as shown.

6. Pick **Sketch** to enter the *Pro/ENGINEER Sketcher* mode.

7. Click inside the display area to set the display area as the active window.

8. Use the key combination [Ctrl-D] to change the view orientation in the display area.

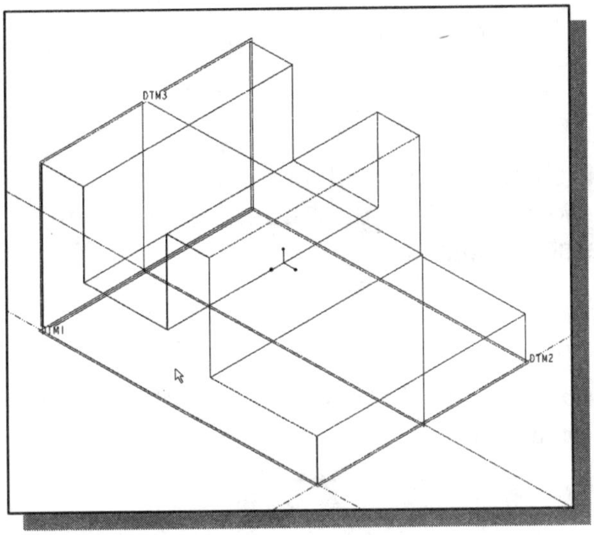

9. On your own, activate the Reference dialog box.

10. Select the front vertical surface of the base feature as an additional sketching reference as shown in the figure.

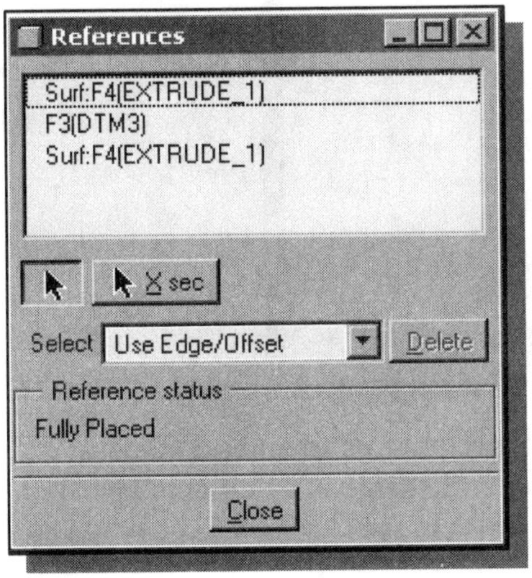

11. In the *References* dialog box, three references are listed as shown. Click on the **Close** button to accept the selections.

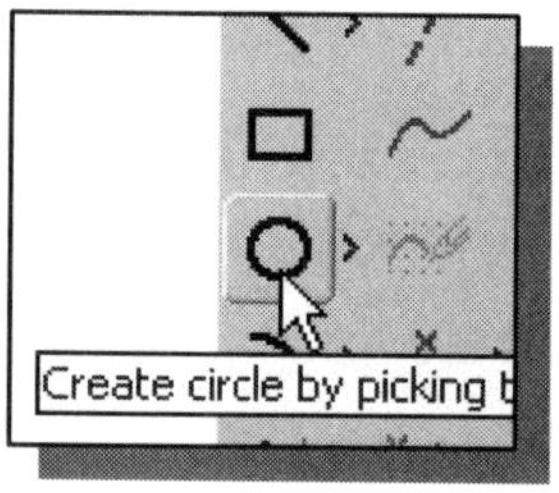

12. In the *Sketcher* toolbar, select **Circle** as shown. The default option is to create a circle by specifying the center point and a point through which the circle will pass. The message "*Select the center of a circle*" is displayed in the message area.

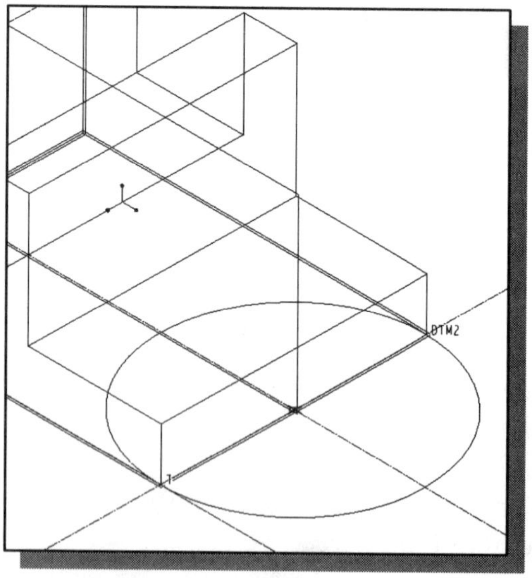

13. Align the center of the circle to the intersection of the two sketching references and create the circle as shown. (Note the **Tangent** constraint symbol.)

- The circle is fully defined and no dimension is needed. The base solid controls the size of the circle.

14. Now that the 2D sketch is completed, we will proceed to the next *element.* In the *Sketcher* toolbar, click on the **Accept** button to end the *Pro/ENGINEER* Sketch command.

15. In the *Feature Option Dashboard*, select the **Extrude to selected point, curve, plane or surface** option as shown.

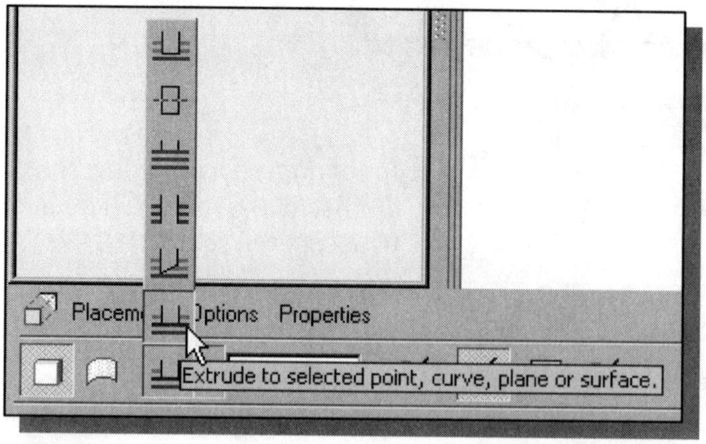

➢ Note that the **Until Selected** option does not require us to enter a value to define the depth of the extrusion; the extrusion distance is calculated based on the termination entity.

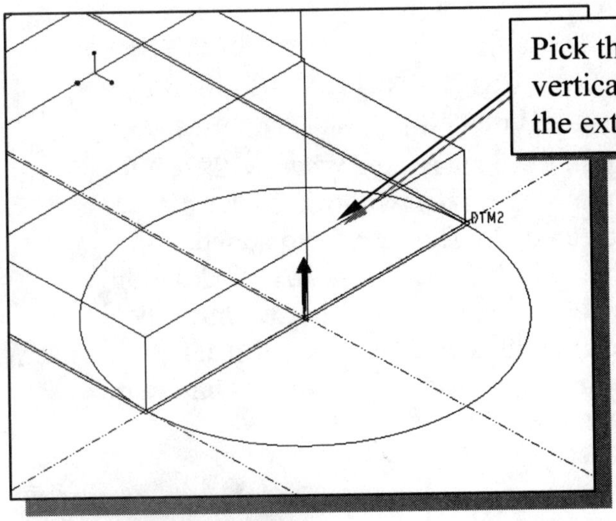

16. Pick the top edge of the vertical surface of the base to define the extrusion distance as shown.

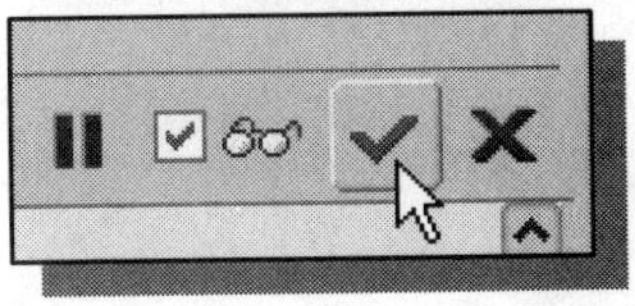

17. Click on **Accept** to proceed with the extrusion option.

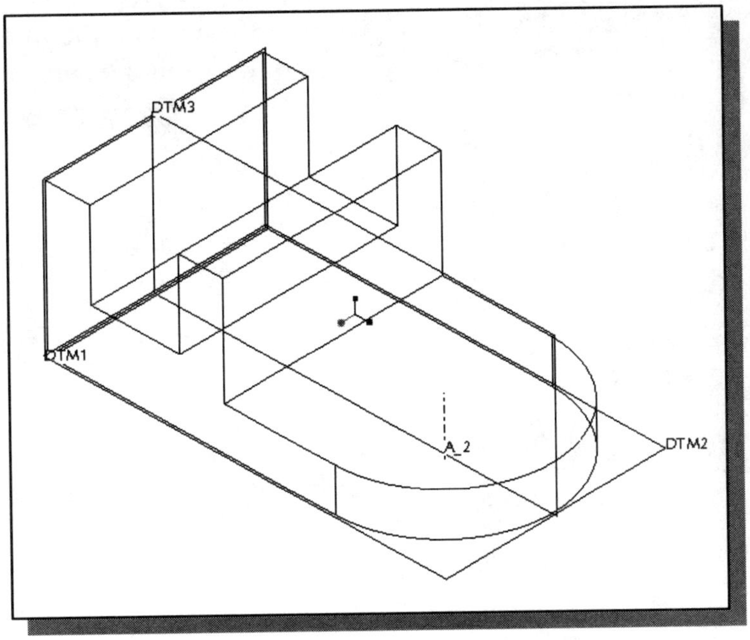

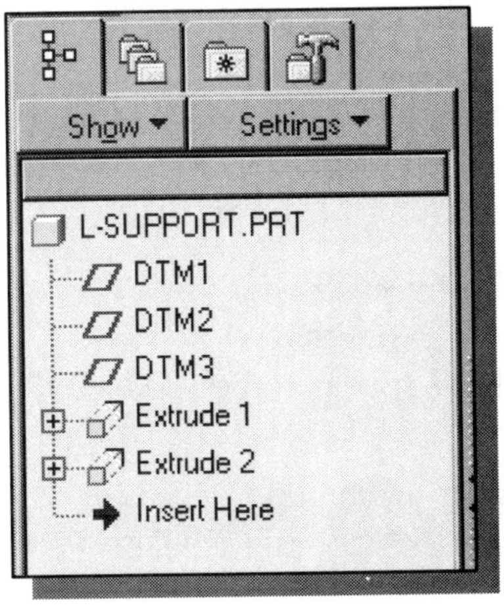

➢ Notice the created feature is added to the model history and the feature is also added in the *Model Tree* window.

Renaming the Part Features

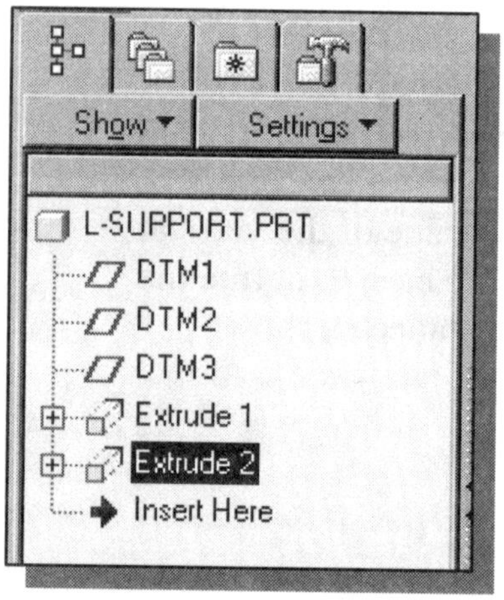

♦ Currently, our model contains five features: three datum planes and two protrusions. The feature is highlighted in the display area when we select the feature in the *Model Tree* window. Each time a new feature is created, the feature is also added in the *Model Tree* window. By default, *Pro/ENGINEER* will use generic names for part features, but when we begin to deal with parts with a large number of features, it will be much easier to identify the features using more meaningful names.

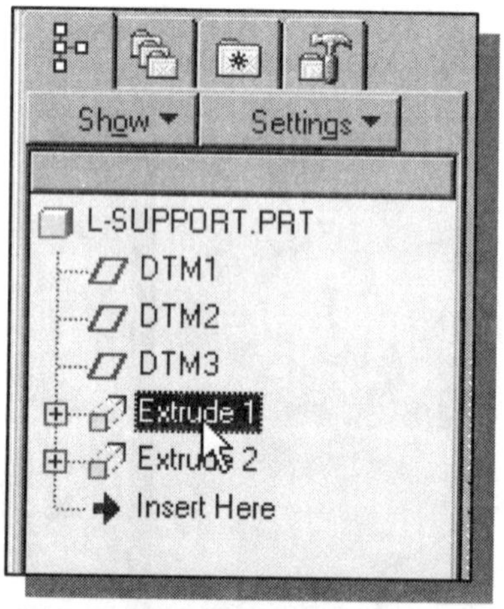

1. Inside the *Model Tree* window, click twice with the **left-mouse-button** on the first protrusion feature, **Extrude 1**, to edit the name of the feature.

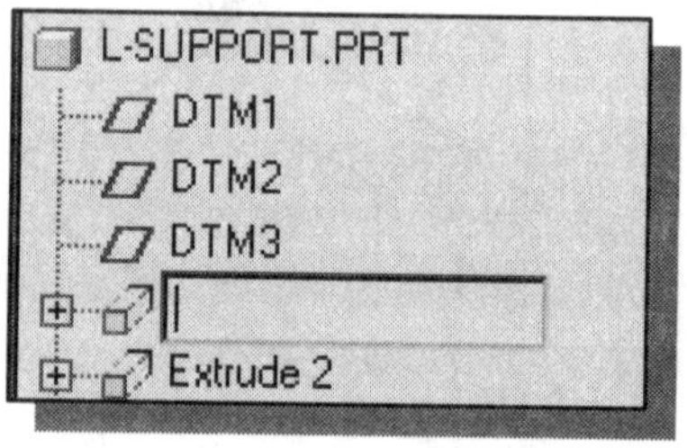

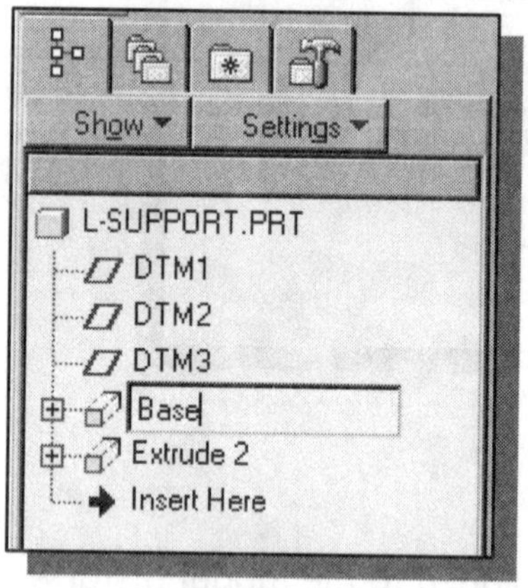

2. Enter **BASE** as the new feature name as shown.

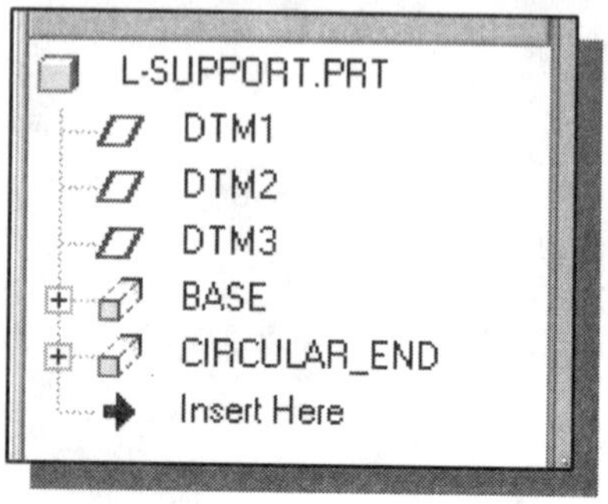

3. On your own, rename the second protrusion to **CIRCULAR_END** as shown. (Note that *Pro/ENGINEER* does not allow spaces to be used as part of the feature names.)

• In the *Model Tree* window, the new feature names are displayed as shown.

Expanding the Model Tree Listing

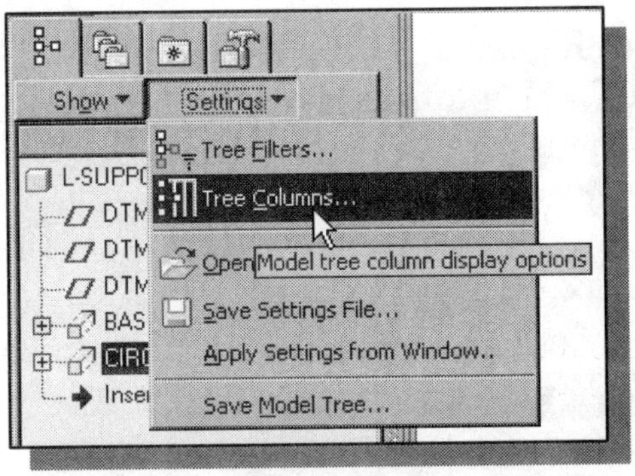

1. In the *Model Tree* window, click on the **Tree Columns** icon to set the **column display options**.

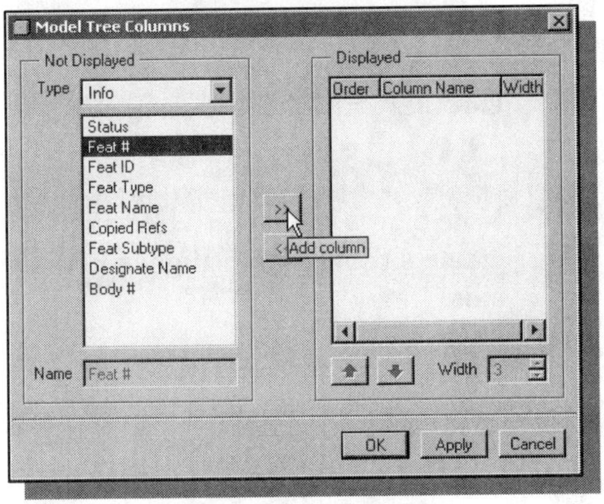

2. In the *Model Tree Columns* form, select

   [Feat #] → [>>]

3. In the *Model Tree Columns* form, select

   [Feat Type] → [>>]

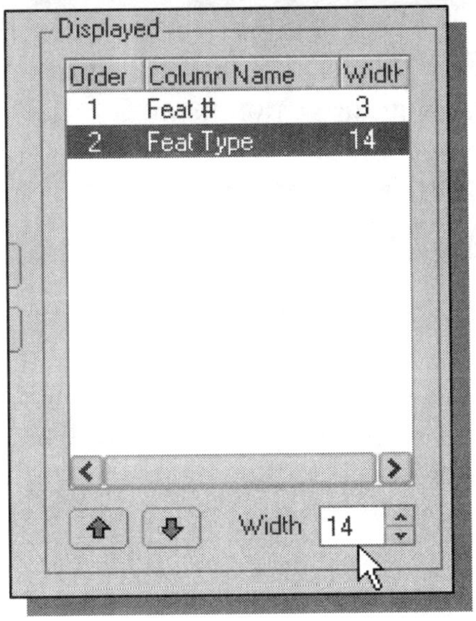

4. In the **Displayed** section, set the column **Width** of **Feat #** to **3** and **Feat Type** to **14** as shown in the figure.

5. Click on the **OK** button to apply the settings and exit the form.

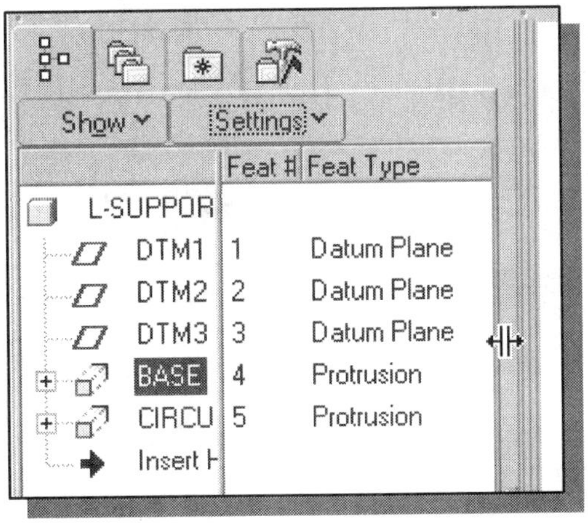

6. Resize the *Model Tree* window by dragging the edge of the window with the left-mouse-button to look at the different feature information.

Creating a Placed HOLE Feature

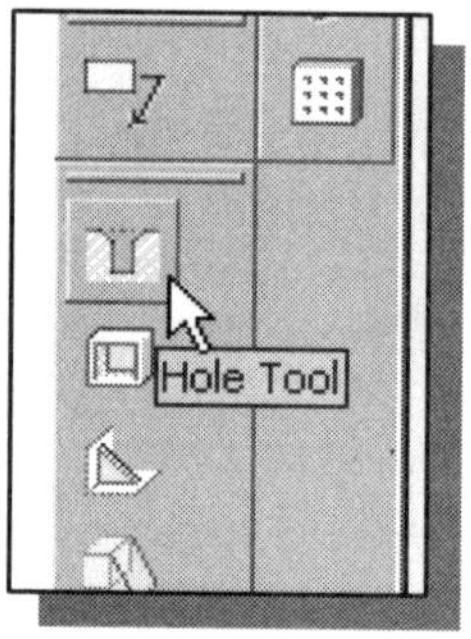

1. In the *Feature* toolbars (toolbars aligned to the right edge of the main window), select the **Hole Tool** option as shown.

❖ The message, "*Select a surface, axis or point to place hole*" is displayed in the message area.

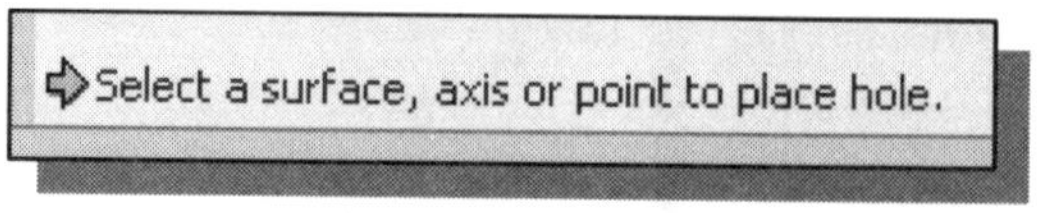

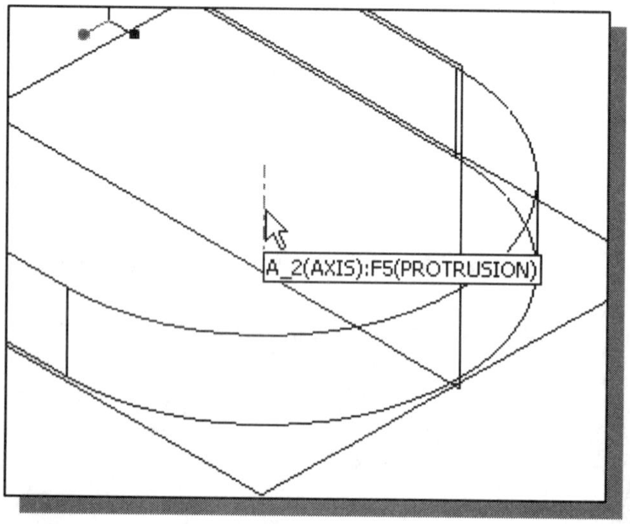

2. Pick the **center axis** of the Circular_End feature as shown.

❖ Note that selecting an axis will create a coaxial hole aligned to the axis.

3. Inside the graphics area, click once with the **right-mouse-button** to bring up the option menu.

4. Select the **Secondary Reference Collector** option as shown.

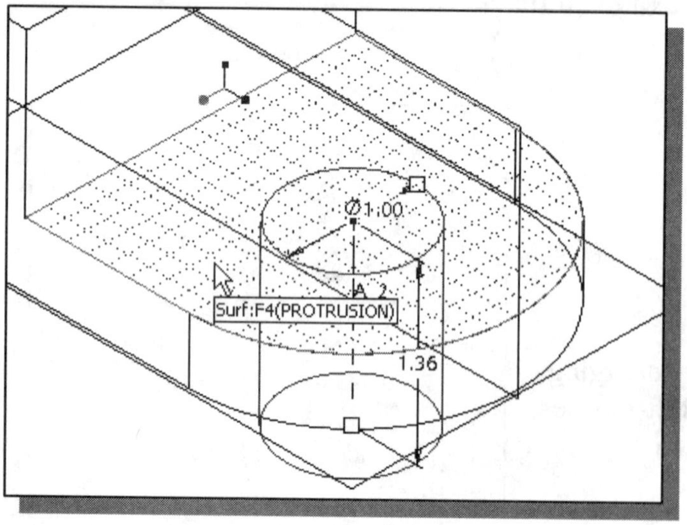

5. Pick the **top horizontal** surface of the Circular_End feature as the *secondary placement reference* for the hole feature.

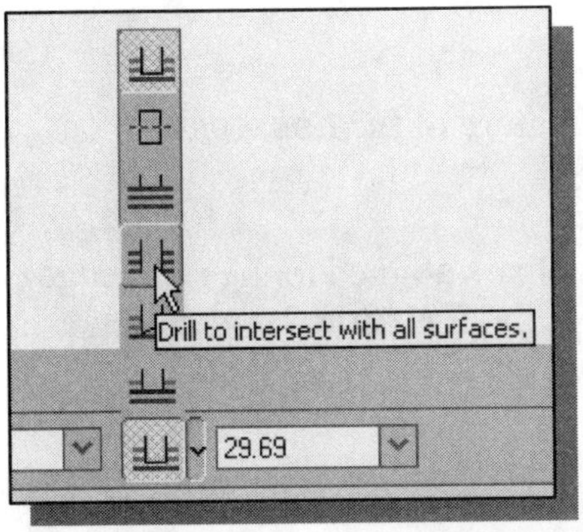

6. In the *Feature Option Dashboard*, select the **Drill to intersect with all surface** option as shown.

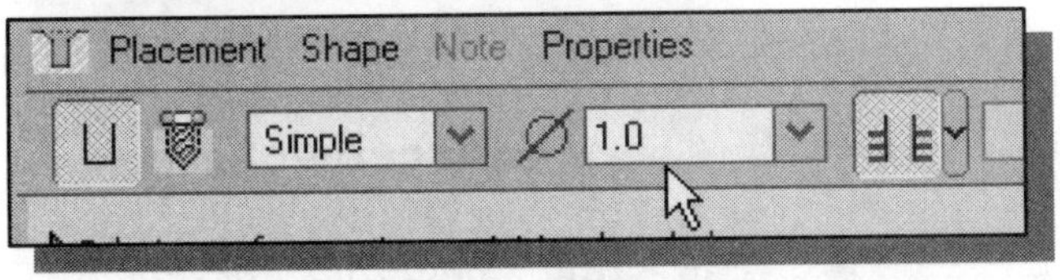

7. In the *Feature Option Dashboard*, set the *feature diameter* to **1.0** as shown.

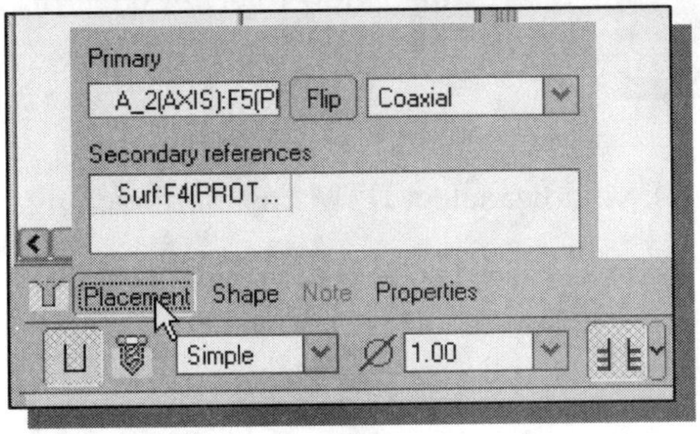

8. In the *Feature Option Dashboard*, click the **Placement** option and examine the placement settings. Note the **Coaxial** option is being used.

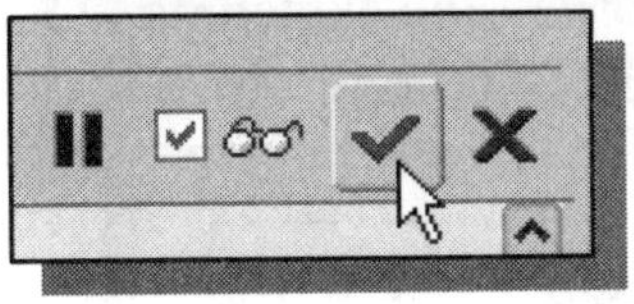

9. Click on the **Accept** icon and proceed to create the hole feature.

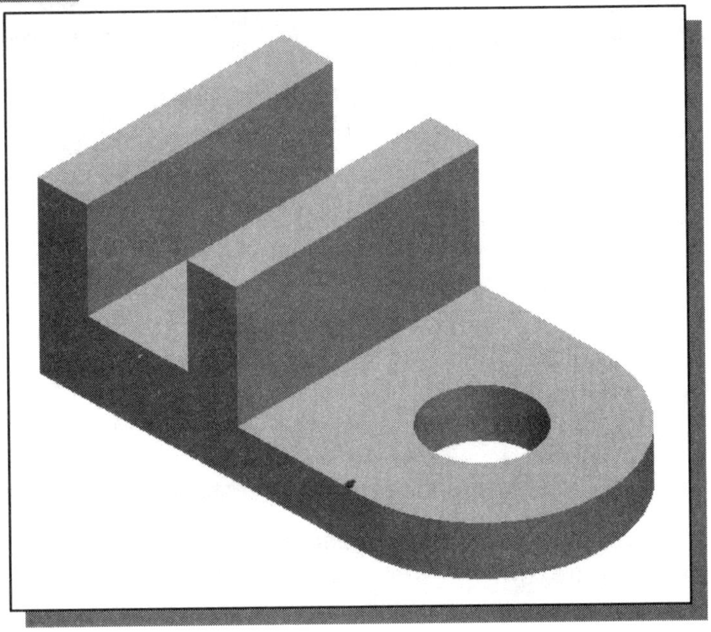

Creating a Rectangular Cutout

- We will create a rectangular cutout as the final solid feature of the *Locator.*

1. In the *Feature* toolbars (toolbars aligned to the right edge of the main window), select the **Extrude Tool** option as shown.

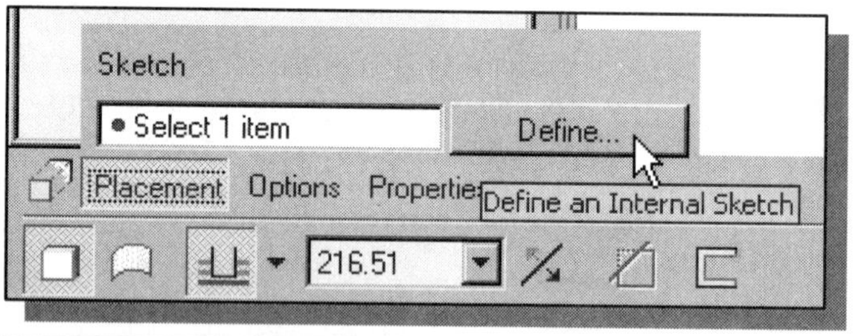

2. Click the **Placement** option and choose **Define** to begin creating a new *Internal sketch.*

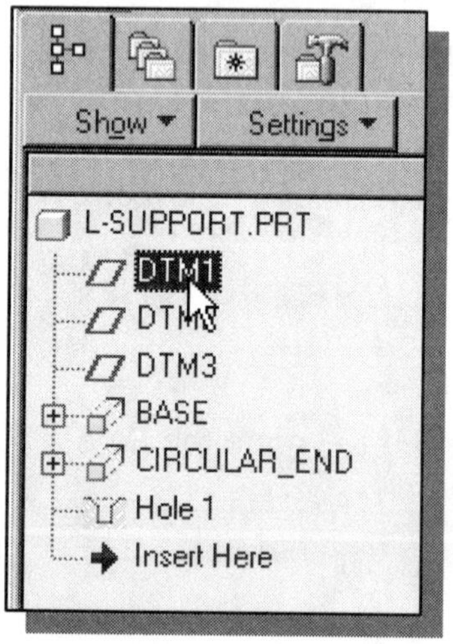

3. In the *Model Tree* window, select **DTM1** as the sketching plane.

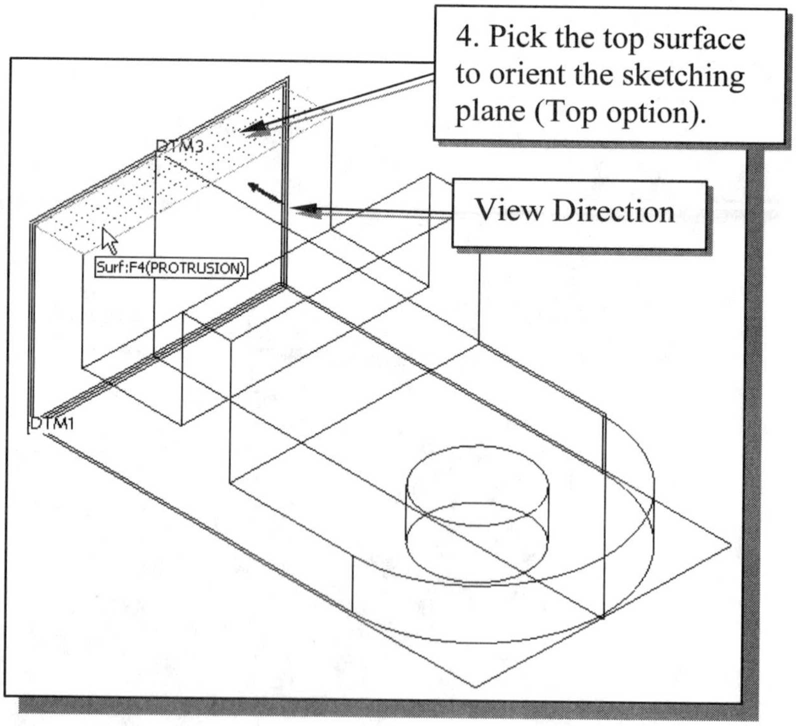

4. Select the top surface of the base feature as the orientation reference as shown in the above figure.

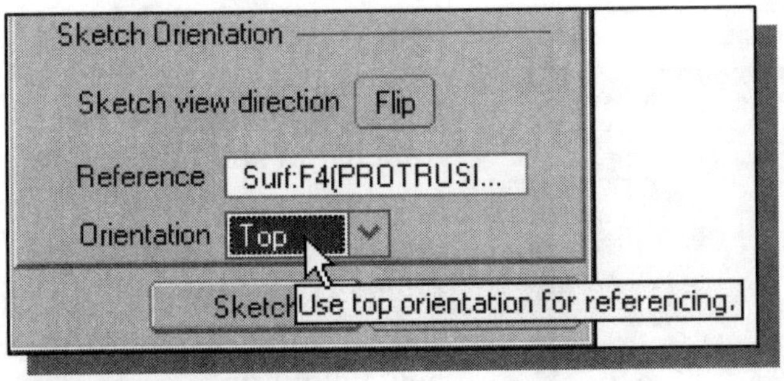

5. In the **Sketch Orientation** menu, confirm the reference plane **Orientation** is set to **Top**.

6. Pick **Sketch** to exit the *Section Placement* window and proceed to enter the *Pro/ENGINEER Sketcher* mode.

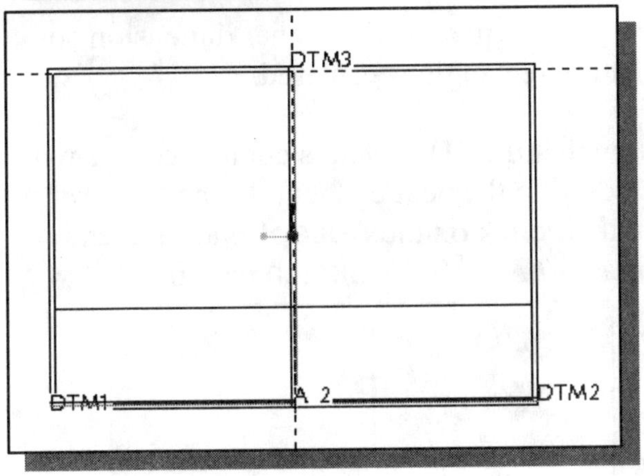

7. Note that **DTM3** and the **top horizontal plane** are set as the *sketching references* as shown.

8. Create a rectangle of arbitrary size, with the top edge aligned to the top surface of the solid model, as shown in the figure below.

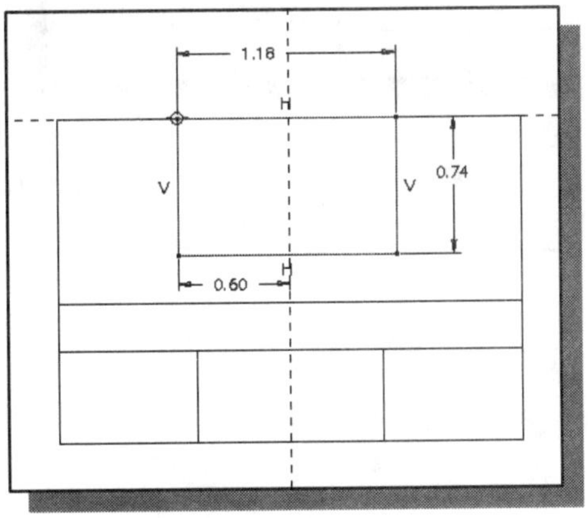

9. In the *Sketcher* toolbar, click on the **Modify** icon as shown.

10. Click on the width dimension and enter **1.5** as the new value, as shown.

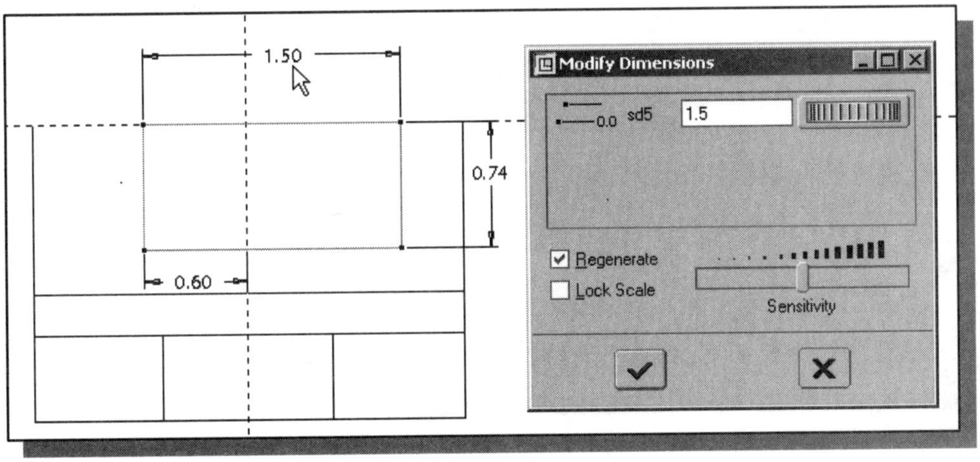

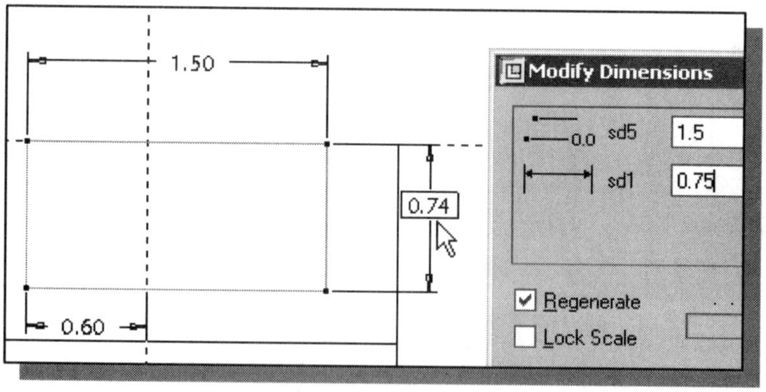

11. Click on the height dimension and notice that the dimension value appears in the *Modify Dimensions* box as shown. Enter **0.75** as the new height dimension.

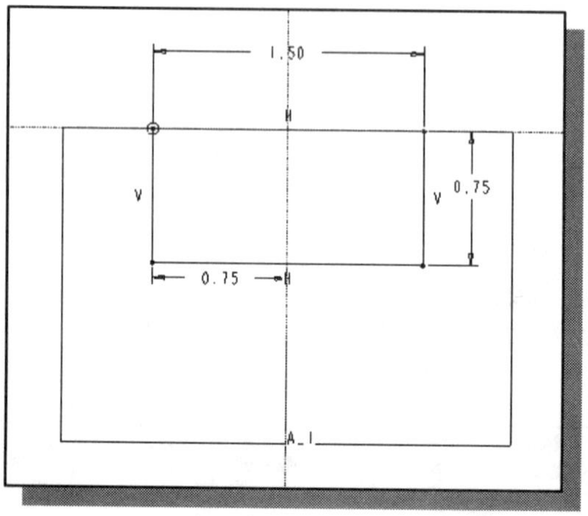

12. On your own, adjust the other dimension so that the rectangle is centered.

13. Now that the 2D sketch is completed, we will proceed to the next element. In the *Sketcher* toolbar, click on the **Accept** button to end the *Pro/ENGINEER* **Sketch** command.

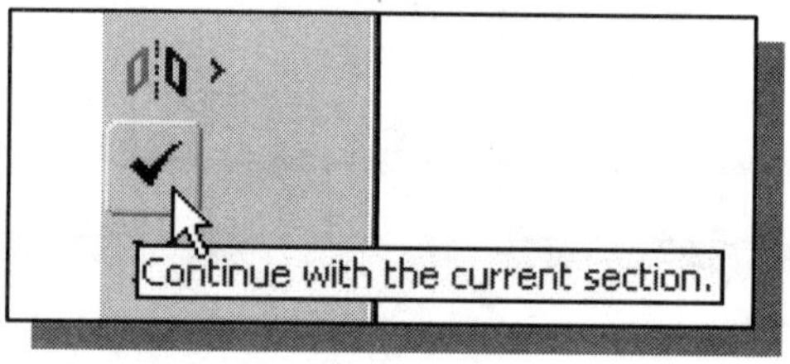

14. Reset the display to the default orientation [quick-key: **Ctrl-D**].

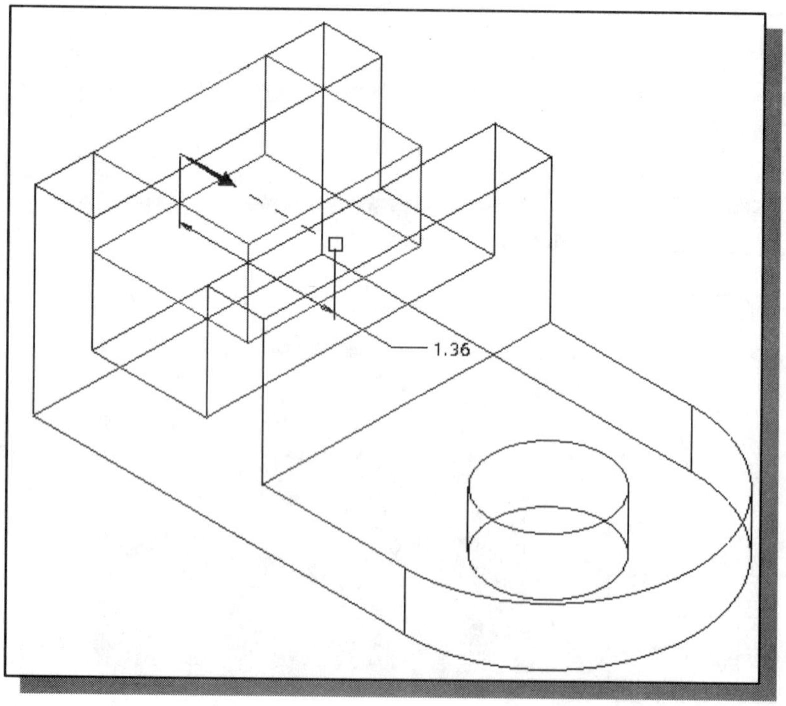

❖ Note that the default setting of the **Extrude Tool** command is set to **add material**. The displayed arrow indicates the extrusion direction.

15. Click on the **Remove Material** icon as shown in the figure below.

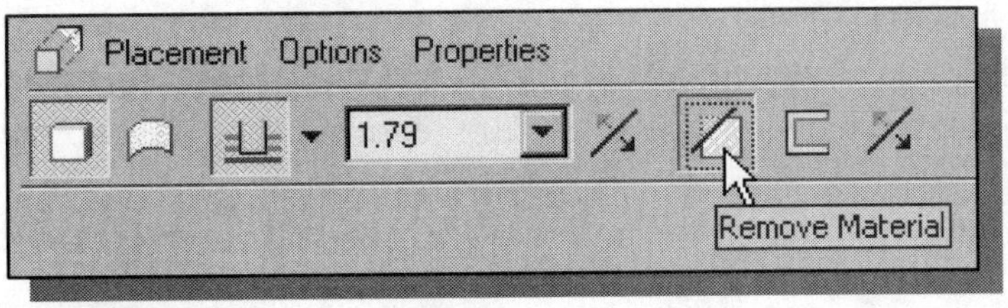

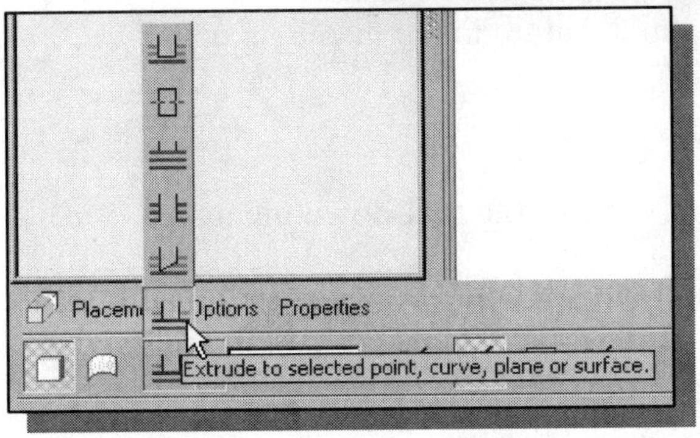

16. In the *Feature Option Dashboard*, select the **Extrude to selected point, curve, plane or surface** option as shown.

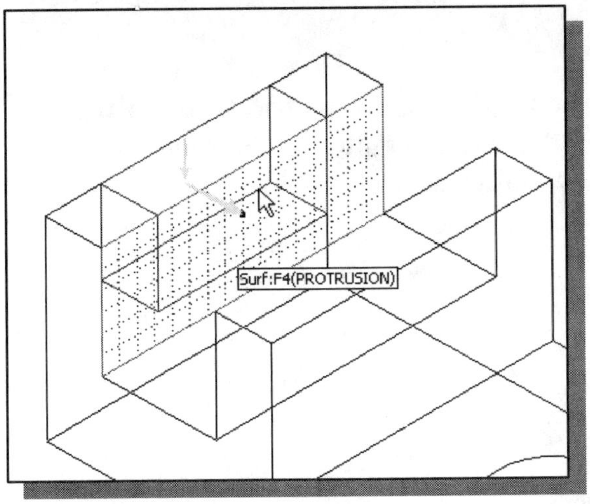

17. Select the vertical surface as shown.

18. Click **Accept** and proceed to create the feature.

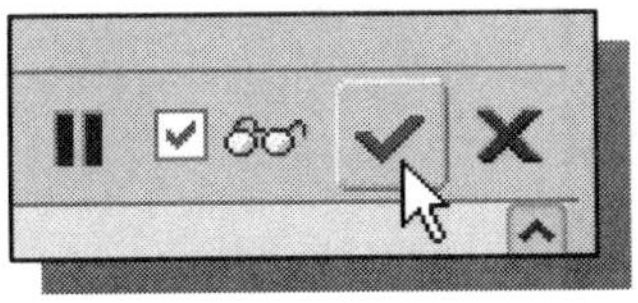

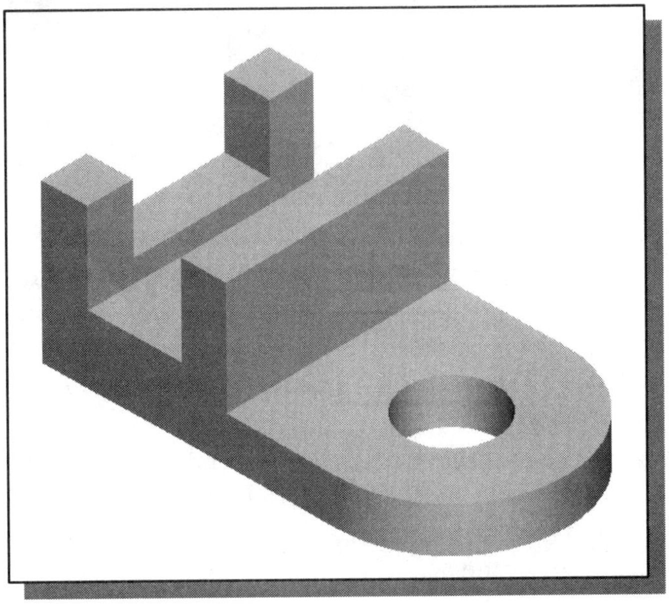

Examining the Model History

1. In the *Model Tree* window, pick **DTM1**. Notice the feature is highlighted in the display area when we select the feature in the *Model Tree* window.

2. On your own, change the name of the last feature to **Rect_CUT.**

3. Pick **Hole 1** and observe the highlighted feature in the display area.

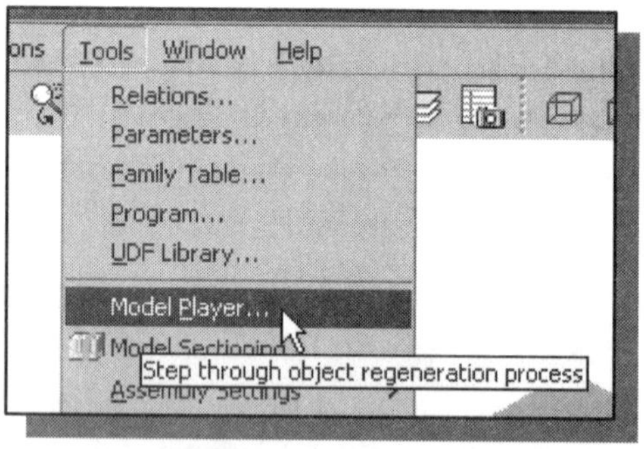

4. Select **Tools** in the pull-down menus.

5. Pick the **Model Player** option.

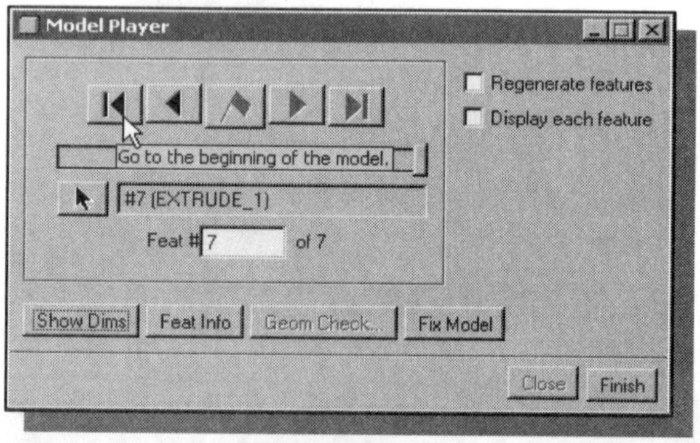

6. Click on the **Go to the beginning of the model** button in the *Model Player* window as shown.

7. Click on the **Step Forward** button to look at the first feature created.

◆ Notice that we have literally gone back in time. We are at the first feature of the model.

8. On your own, click the **Step Forward** button and review the features used to create the model.

9. Click **Finish** to exit the *Model Player*.

❖ Note that the *Model Player* simply redisplays the features recorded in the *Model Tree*. The *Pro/ENGINEER Model Tree* is a sequential record of the features used to create the part. We can review and make modifications to assure the accuracy of our model at any time.

History-based Part Modifications

❖ *Pro/ENGINEER* uses the *history-based part modification* approach, which enables us to make modifications to the appropriate features in the *Model History Tree* and re-link the rest of the *History Tree.* We can think of it as going back in time and modifying some aspects of the modeling steps used to create the part. We can modify any feature that we have created. As an example, we will adjust the depth of the rectangular cutout.

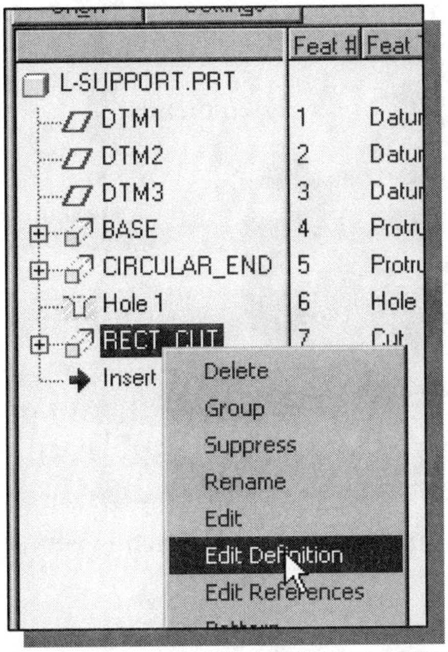

1. In the *Model Tree* window, move the cursor on top of **Rect_CUT**, then click once with the right-mouse-button. A pop-up menu is displayed.

2. Select **Edit Definition**. Notice the feature dialog box appears near the bottom of the main window.

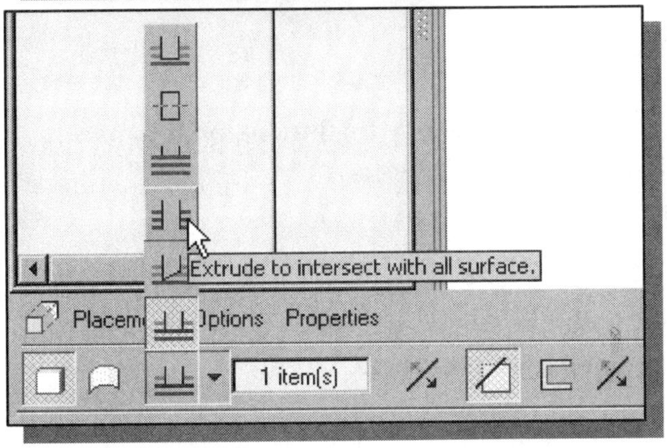

3. Change the *depth* option to **Extrude to intersect with all surface** as shown.

4. Click on the **Preview** button to examine the effect of the modification.

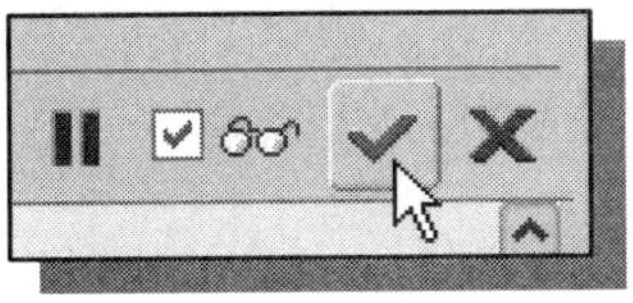

5. Click on the **Accept** icon and proceed to create the hole feature.

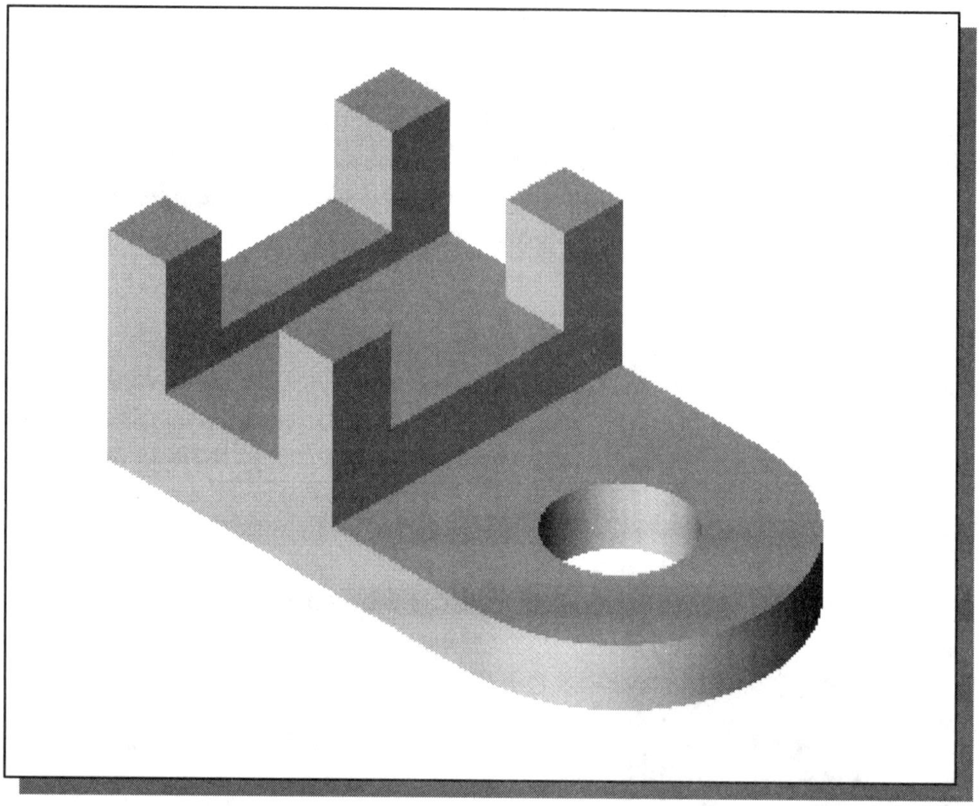

❖ As can been seen, the *Pro/ENGINEER* history-based modification approach is very straightforward and it only took a few seconds to do this modification.

Modifying Dimensional Values

❖ Once a feature has been created, it is very easy to modify its dimensional values in *Pro/ENGINEER*.

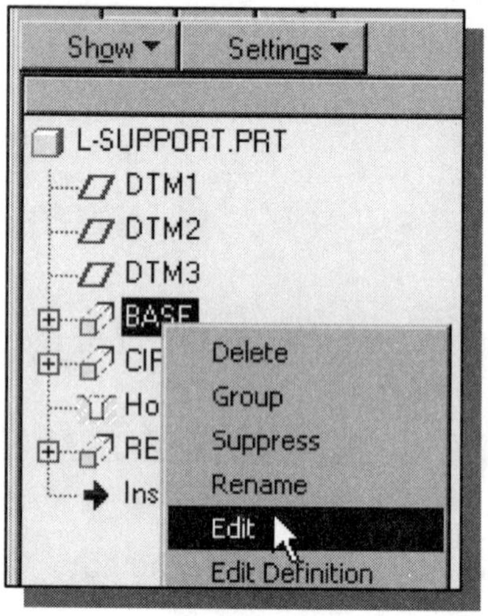

1. In the *Model Tree* window, move the cursor on top of **BASE**, then click once with the right-mouse-button to bring up the option menu.

2. Select **Edit** in the option menu. Notice the 2D section, along with all the dimensions, appears in the graphics area.

3. Pick the width dimension of the upper section of the part (**2.0**) by double-clicking with the left-mouse-button.

4. Enter **2.5** as the new dimension value.

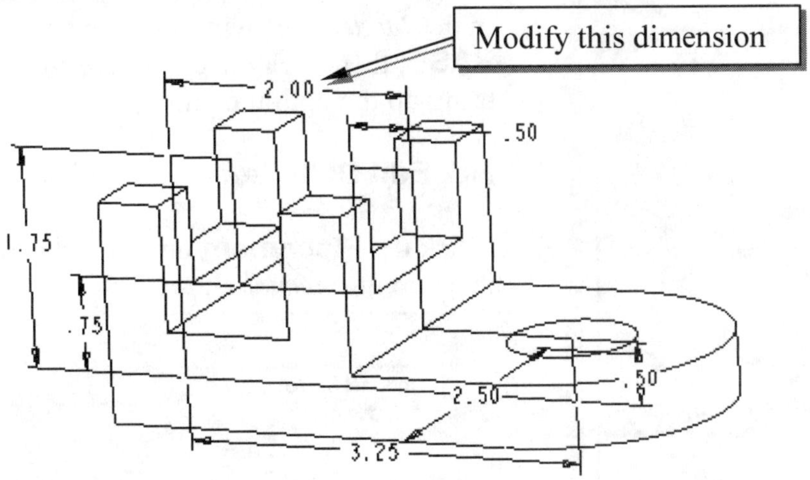

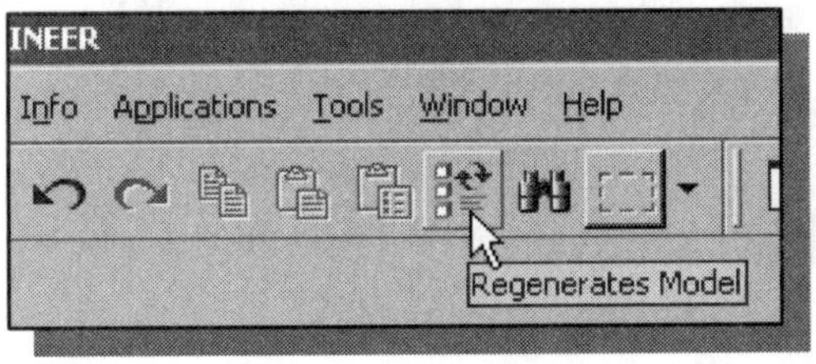

5. Pick **Regenerate** in the *Standard* toolbar to apply the modification.

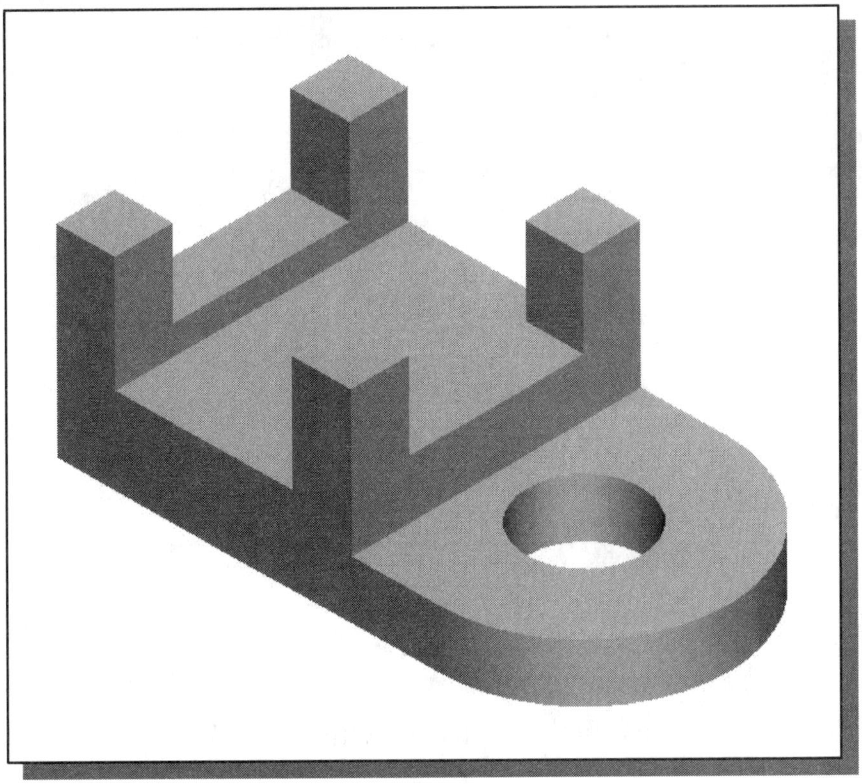

Modifying the 2D Section of the Base Feature

❖ We can also make changes to the original 2D sections we have created.

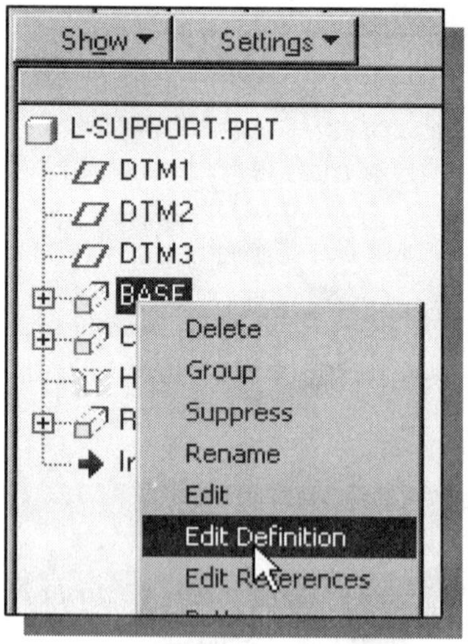

1. In the *Model Tree* window, move the cursor on top of **BASE,** then click once with the right-mouse-button to bring up the option menu.

2. Pick **Edit Definition** in the option menu as shown.

3. Click the **Placement** option and choose **Edit** to edit the current *Internal sketch.*

❖ The *Pro/ENGINEER Sketcher* toolbar appears on the screen and the original 2D section is displayed in the display area. We have literally gone back in time to the point where we first created the 2D section.

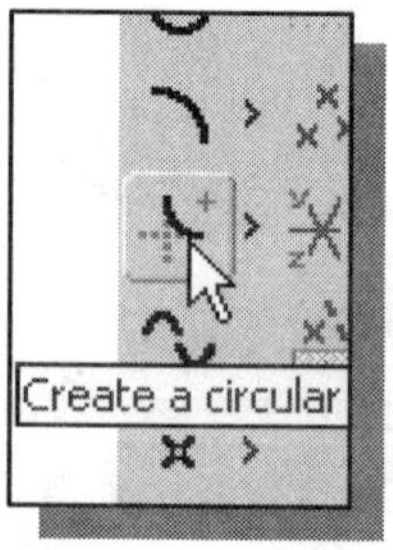

4. In the *Sketcher* toolbar, select the **Fillet** command.

5. Pick the two lines on the right, as shown in the below figure, to create a fillet.

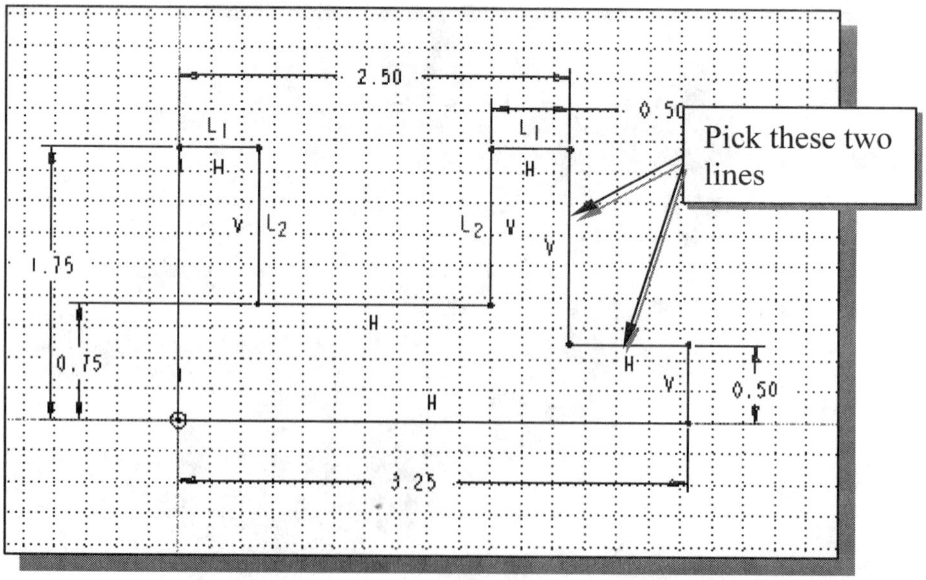

6. On your own, modify the radius of the fillet to **0.25** inches.

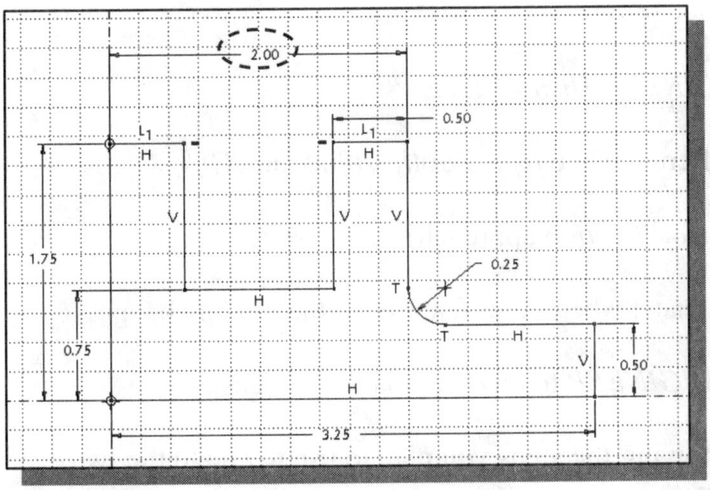

7. On your own, adjust the width dimension of the upper section of the part from **2.5** to **2.0** as shown in the figure.

8. In the *Sketcher* toolbar, click on the **Accept** button to accept the modification and exit the *Pro/ENGINEER* **Sketch** command.

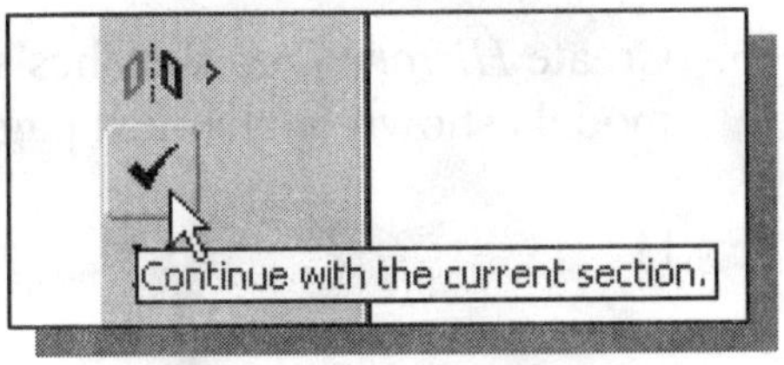

9. Pick **OK** in the section dialog box.

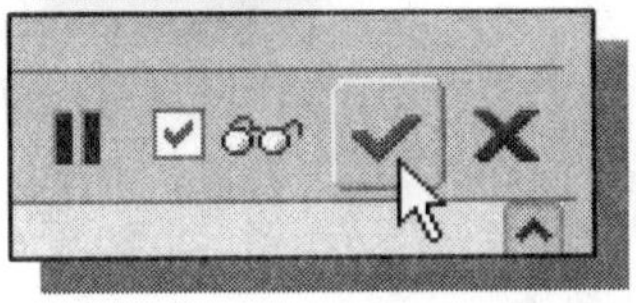

10. Pick **Accept** to complete the modification.

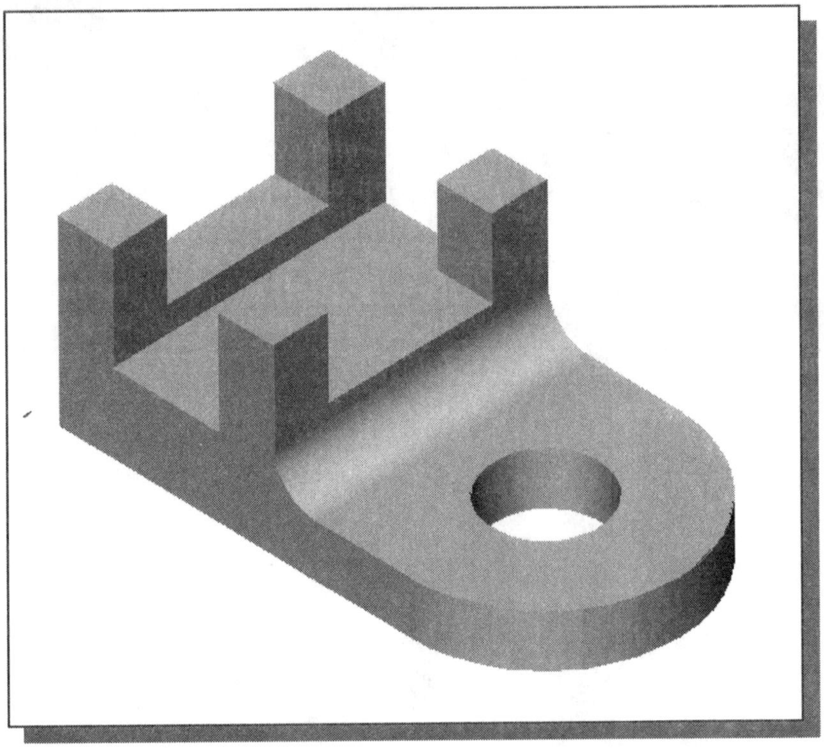

❖ In a typical design process, the initial design will undergo many analyses, testing and reviews. The *history-based part modification* approach is an extremely powerful tool that enables us to quickly update the design. At the same time, it is quite clear that PLANNING AHEAD becomes more important in doing feature-based modeling.

Questions:

1. What are stored in the *Pro/ENGINEER History Tree*?

2. When extruding, what is the difference between ***Depth Value*** and ***Thru All***?

3. What is the *history-based part modification* approach?

4. What determines how a model reacts when other features in the model change?

5. Describe two methods in *Pro/ENGINEER* to change existing dimensions.

6. Create *History Tree* sketches showing the steps you plan to use to create the two models shown on the next page:

Ex.1)

Ex.2)

Exercises: (All dimensions are in inches.)

1.

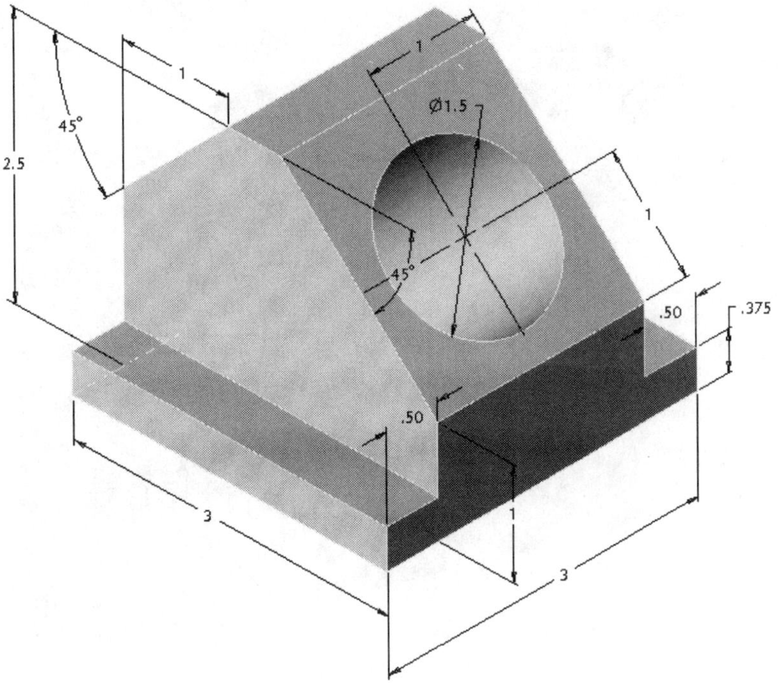

2.

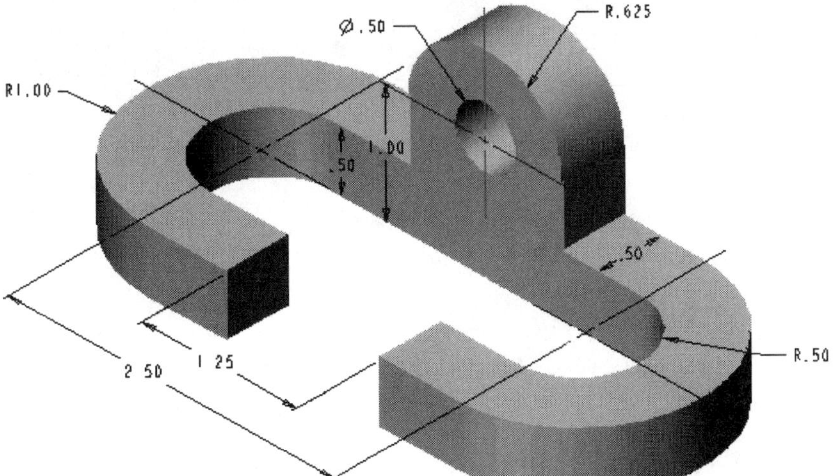

3.

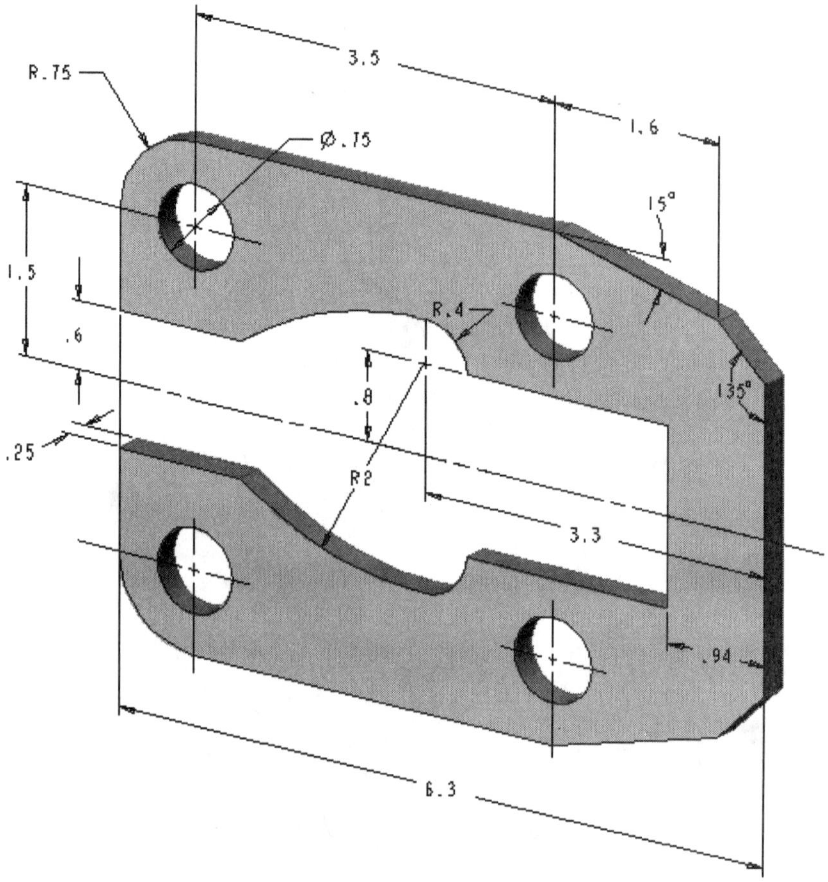

Chapter 4
Constraints and Parametric Relations

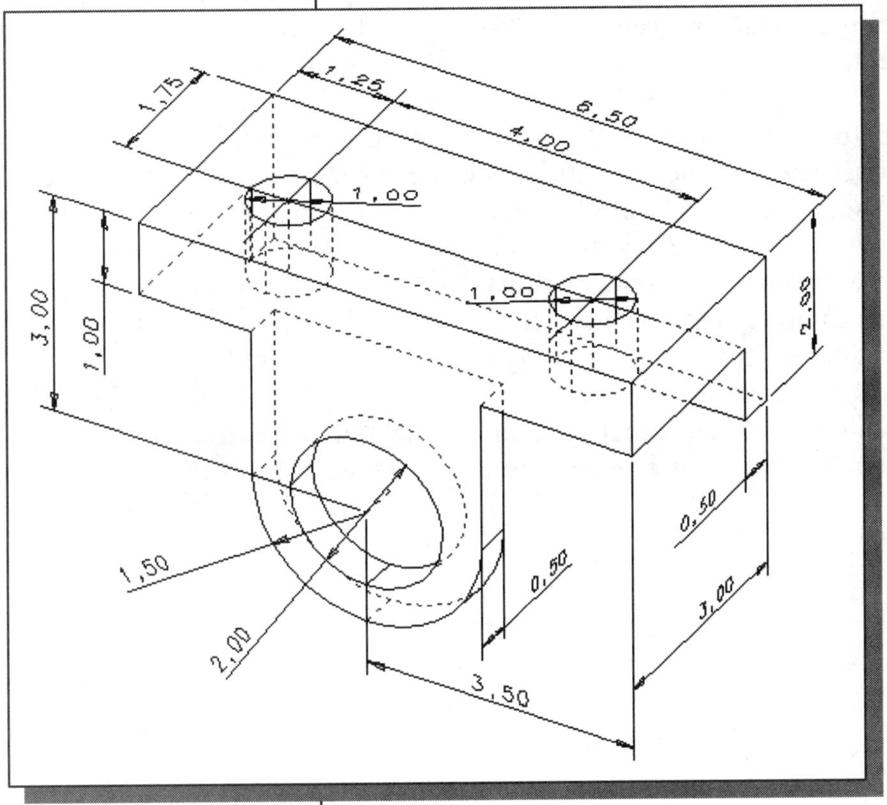

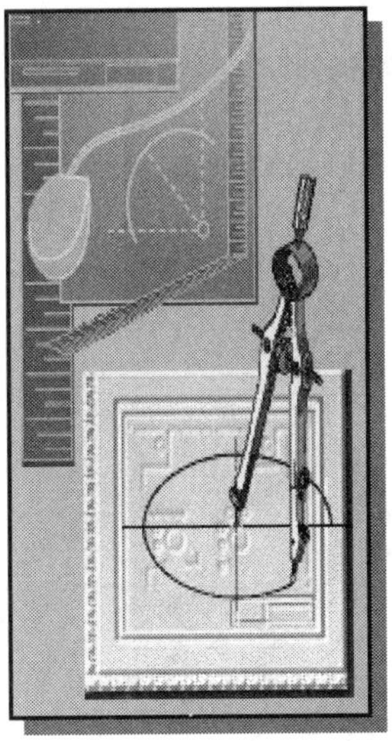

Learning Objectives

- **Understand the importance of Fully Constrained Sketches.**
- **Apply Constraints manually.**
- **Create Parametric Relations.**
- **Creating Sketch Files.**
- **Editing by Drag-and-Drop.**
- **Display and Modify Parametric Relations.**
- **Using the Basic Editing commands.**
- **Create Offset Features.**

Geometric Construction and Constraints

The main characteristics of solid modeling are the accuracy and completeness of the geometric database of the three-dimensional objects. However, working in three-dimensional space using input and output devices that are largely two-dimensional in nature is potentially tedious and confusing. The ***Pro/ENGINEER 2D Sketcher*** provides an assortment of two-dimensional construction tools to make the creation of 2D geometry easier and more efficient.

In doing geometric constructions, dimensional values are necessary to describe the **SIZE** and **LOCATION** of constructed geometric entities. Besides using dimensions to define the geometry, we can also apply geometric rules to control geometric entities. This set of geometric rules is known as **geometric constraints** in parametric modeling. The geometric constraints are **geometric restrictions** that can be applied to geometric entities, such as tangent, parallel, perpendicular, etc. In *parametric modeling*, **parametric relations** and **geometric constraints** are commonly used in 2D sketches to assist the capture of individual design intent. In this next example, the use of *Pro/ENGINEER Sketcher's* geometric construction tools, including the application of **geometric constraints** is illustrated. The parametric relation is also illustrated in a later section of this chapter.

The *Gasket* Design

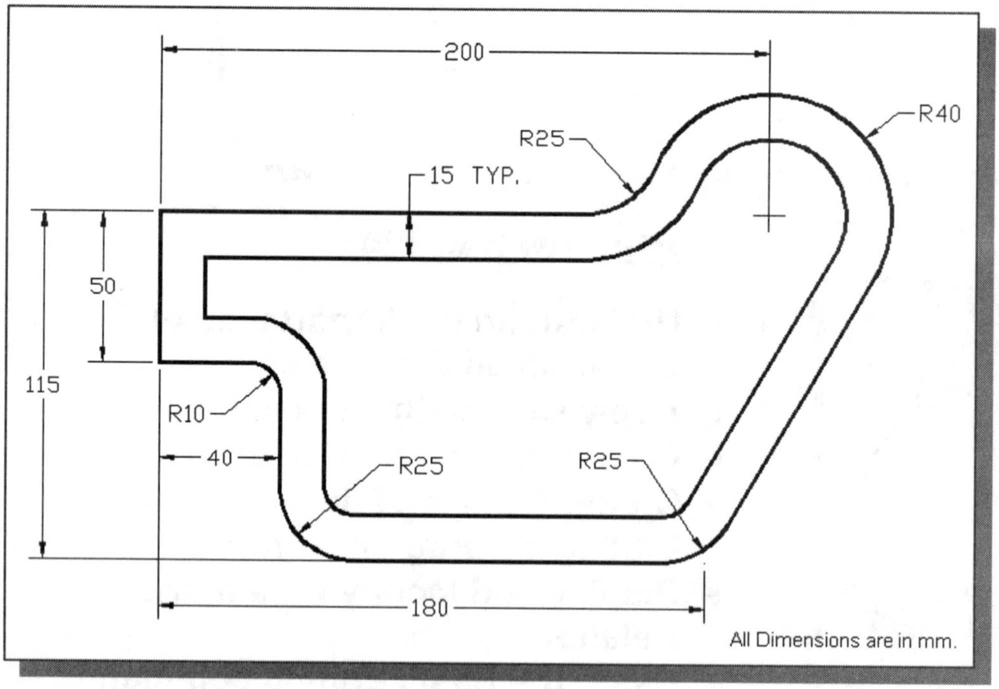

❖ Based on your knowledge of *Pro/ENGINEER* so far, how would you create this design? What are the more difficult geometry involved in the design? Take a few minutes to consider a modeling strategy and do preliminary planning by sketching on a piece of paper. You are also encouraged to create the design on your own prior to following through the tutorial.

Modeling Strategy

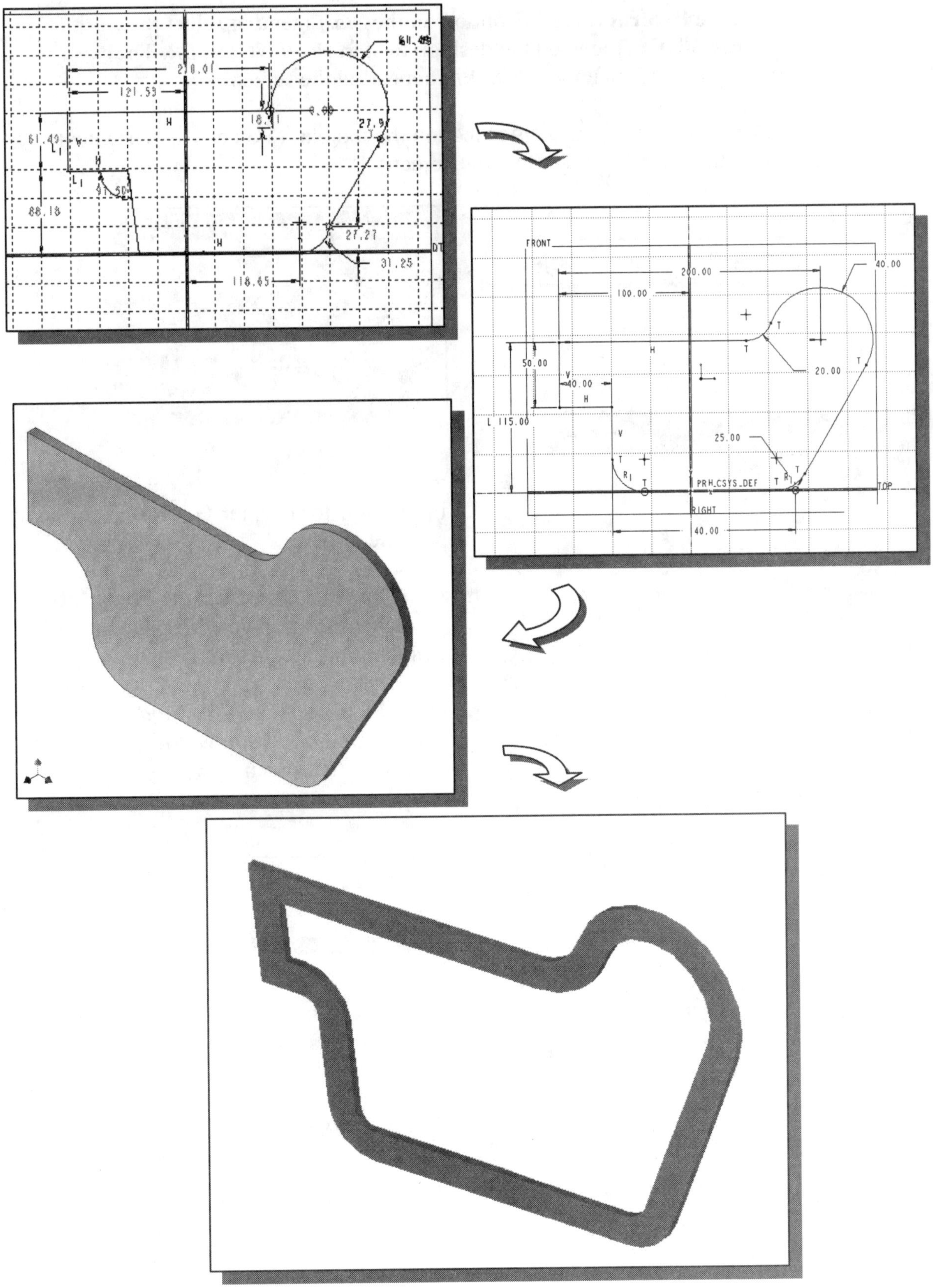

Starting *Pro/ENGINEER* and using the Metric Template

1. Select the **Pro/ENGINEER** option on the *Start* menu or select the **Pro/ENGINEER** icon on the desktop to start *Pro/ENGINEER*. The *Pro/ENGINEER* main window will appear on the screen.

2. Pick the **Create New Object** icon in the toolbar. (You can also use the key combination **Ctrl-N** to start a new object.)

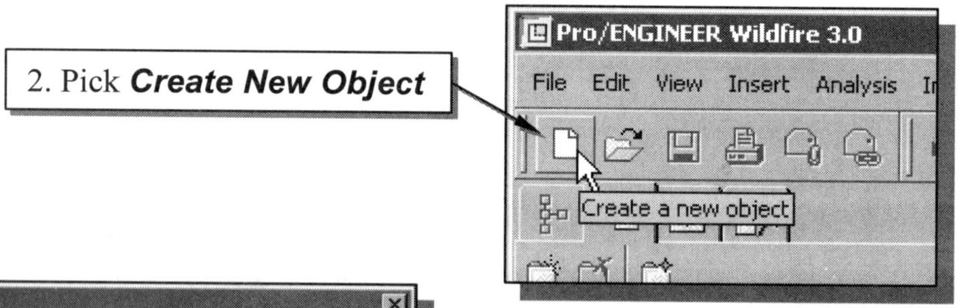

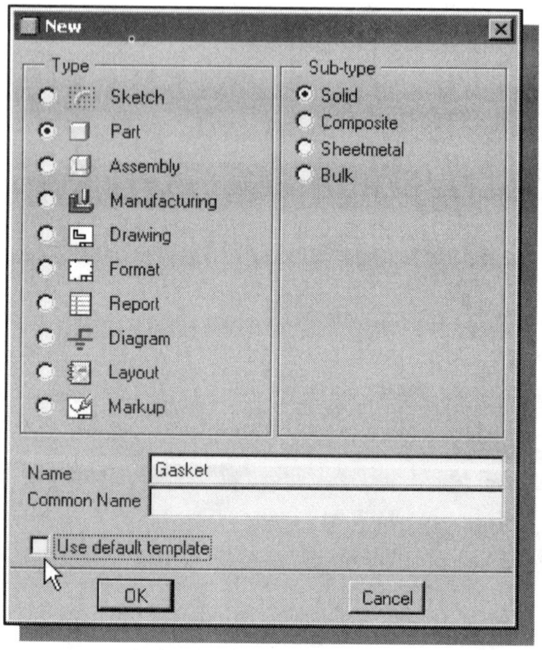

3. In the *New* form, enter ***Gasket*** as the solid part file Name.

4. Switch ***off*** the **Use Default Template** option as shown. (We will select the metric template for the *Gasket* design.)

5. Pick **OK** to continue. The *Model Tree* window and the Menu Manager window appear on the screen.

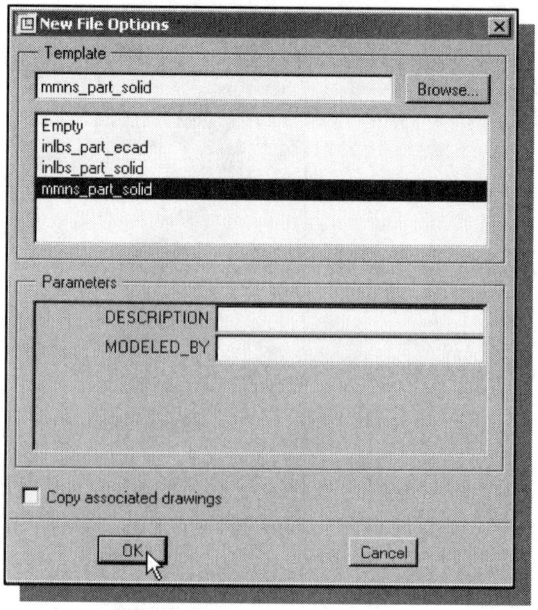

6. In the *New File Options* window, select **mmns_part_solid** as the template to use. This is the template for using with the metric system (**millimeter**).

❖ In Pro/E, a template file can contain a set predefined settings, such as the datum planes, datum axes and units setup. The template files allow us to reduce startup repetitions.

Using the 2D Sketcher in 3D Orientation

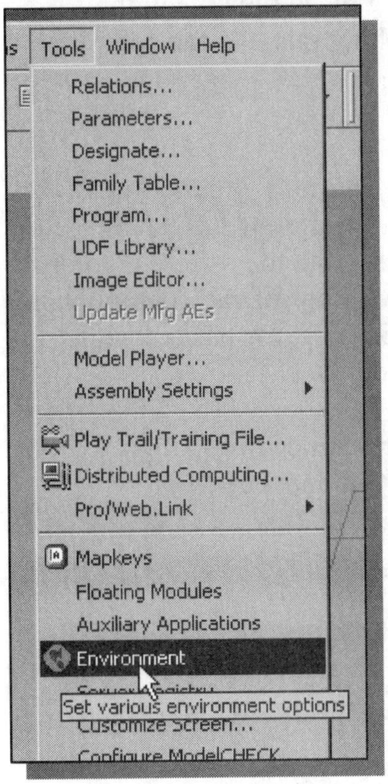

1. Select **Tools** in the pull-down menu.

2. Pick **Environment** in the pull-down list.

3. Toggle **off** the ***Use 2D Sketcher*** option. The Use 2D Sketcher option determines if the sketching plane will be oriented to the computer screen or will remain in the default view angle.

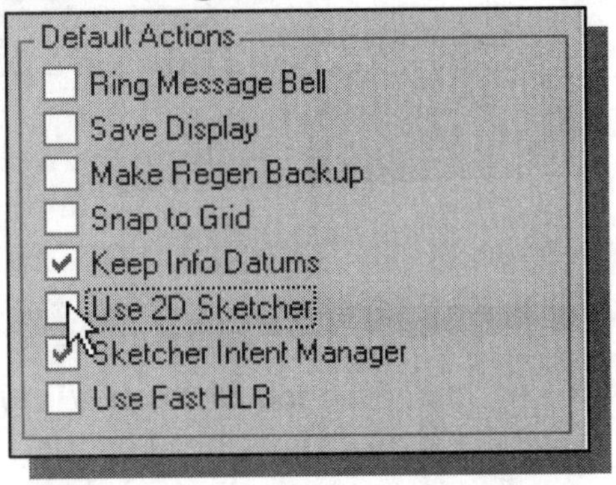

4. Pick **OK** to exit the *Environment* form.

Creating the Base Feature

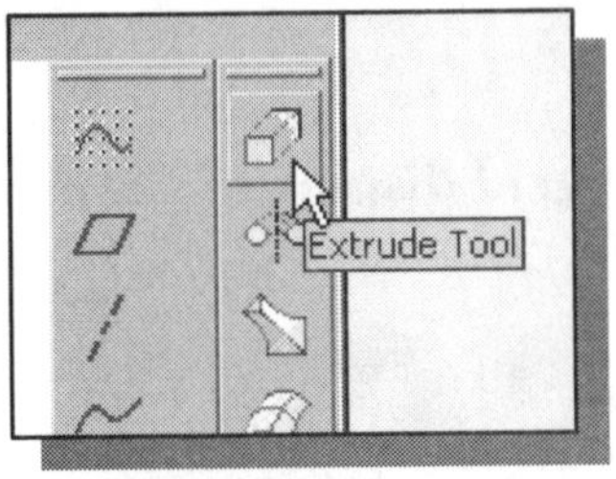

1. In the *Feature* toolbars (toolbars aligned to the right edge of the main window), select the **Extrude Tool** option as shown.

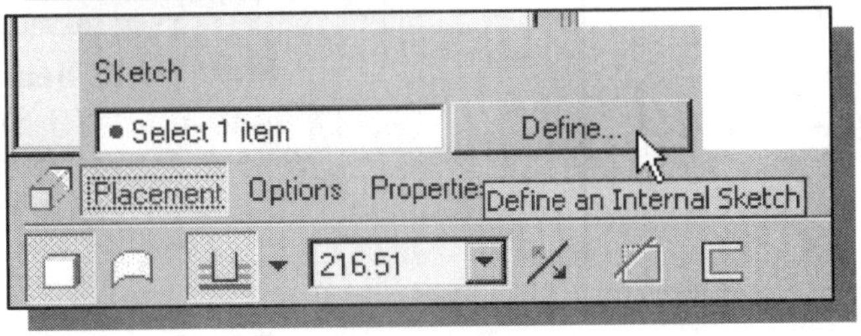

2. Click the **Placement** option and choose **Define** to begin creating a new *Internal sketch.*

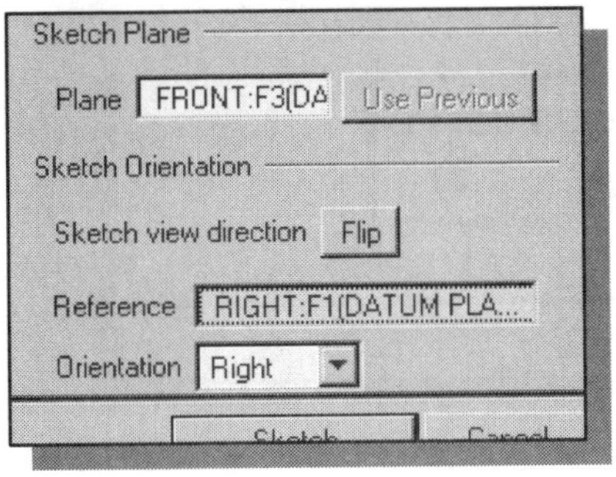

3. On your own, set up the **Front** datum plane as the sketch plane with the **Right** datum plane facing the **right** edge of the computer screen.

❖ Note that the display of the datum planes remains in the 3D orientation. The **Use 2D Sketcher** option was turned *off* and the sketching plane remains in the default view angle. In most cases, orienting the sketching plane to the computer screen simplifies the construction of 2D sketches.

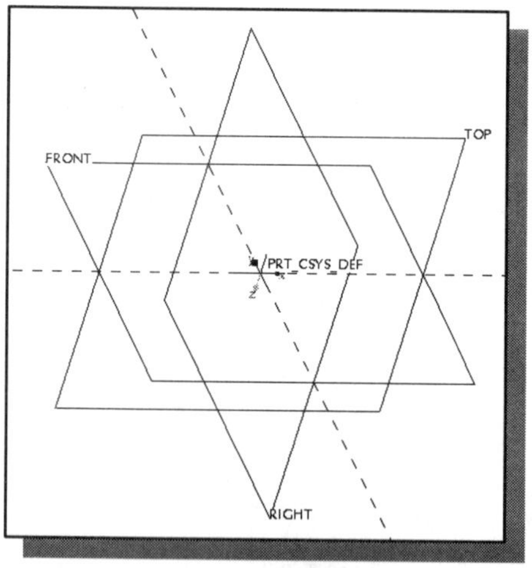

4. On your own, use the *Dynamic Rotation* quick key [middle-mouse-button] and view the established datum planes and axes. Note the different colors associated with the two sides of each datum plane. Each surface in *Pro/ENGINEER* has a positive side and a negative side.

➢ Note the two dashed reference lines, aligned to the two datum planes, **Right** and **Top**.

5. Pick **Sketch View** in the **Sketcher** menu. This will align the sketching plane to the computer screen. Also note the four display-control icons to the right side of the **Sketch View** icon.

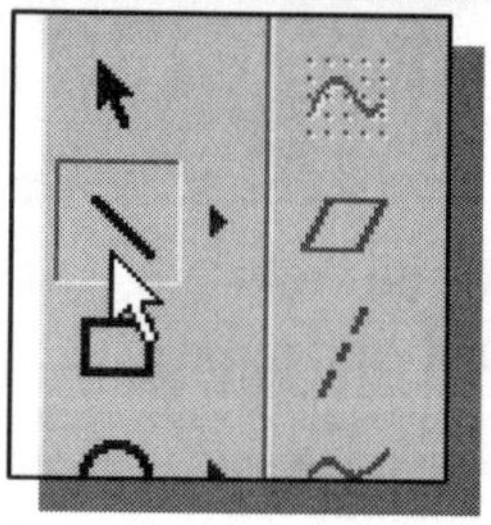

6. Choose the **Line** command in the *Sketcher* toolbar.

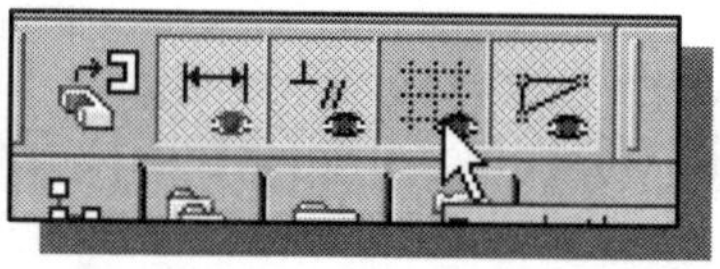

7. Switch ***on*** the **grid display** as shown.

8. Create the geometry, with the geometric constraints shown by selecting points near the grid-points indicated.

❖ Do not end the **Line** command. We will continue to sketch after the following discussion of the implicit geometric relationships.

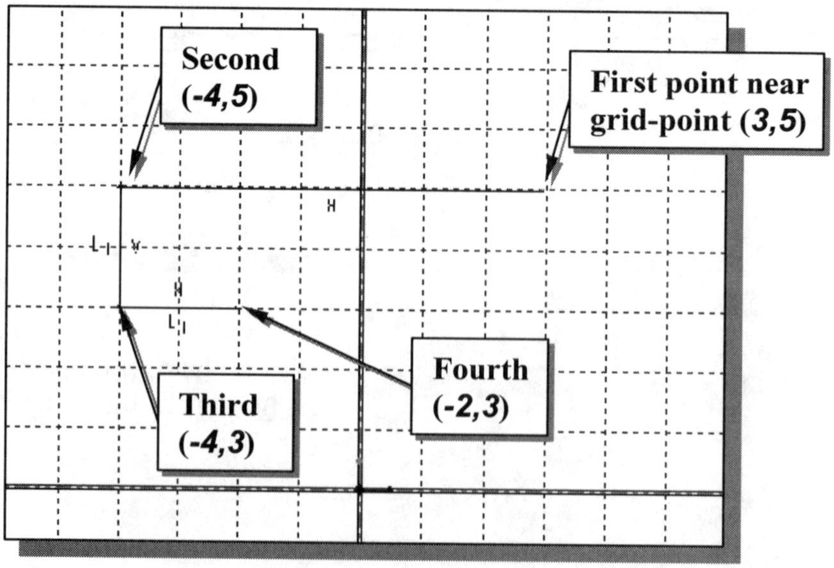

❖ Note that the *Intent Manager* displays different constraint symbols to show perpendicularities, verticals, equal length, etc. The three geometric constraint symbols shown on the screen represent three types of geometric constraints: *vertical*, *horizontal* and *equal length.*

V	Vertical	indicates a segment is vertical
H	Horizontal	indicates a segment is horizontal
L	Equal Length	indicates two segments are of equal length

Implicit Geometric Relationships

In parametric modeling software, such as *Pro/ENGINEER*, **geometric constraints** can be applied to geometric entities to ensure the shape of a part behaves predictably as changes are made. By default, *Pro/ENGINEER* automatically adds constraints to geometry as they are constructed. The following is a list of assumptions about geometric relationships in *Pro/ENGINEER* sketches.

- **If the sketched lines are approximately horizontal or vertical, *Pro/ENGINEER* makes them exactly horizontal or vertical.**

- **If the sketched lines are approximately parallel or perpendicular, *Pro/ENGINEER* makes them exactly parallel or perpendicular.**

- **If the sketched lines are approximately equal to the length of an existing line, *Pro/ENGINEER* makes them equal to the length of the existing line.**

- **If the sketched arcs or circles are approximately equal to the radius of an existing arc or circle, *Pro/ENGINEER* makes them equal to the radius of the existing arc or circle.**

- **If the sketched entities are approximately symmetrical about a centerline, *Pro/ENGINEER* makes them symmetrical about the centerline.**

- **If the sketched entities are approximately tangent to each other, *Pro/ENGINEER* makes them tangent to each other.**

- **If the sketched entities are approximately collinear to each other, *Pro/ENGINEER* makes them collinear to each other.**

- **If the sketched endpoints are approximately on other entities, *Pro/ENGINEER* makes them exactly on the entities.**

Note that we can override these implicit rules by adding appropriate dimensions and/or geometric constraints. We can also disable the implicit constraints during geometric constructions. A single click of the right-mouse-button (in the display area) will toggle on/off the displayed constraint. We can always modify the constraints to ensure the design intent.

A good rule of thumb to follow in using an implicit-geometric-constraint-based modeler, such as *Pro/ENGINEER*, is to exaggerate the feature during the initial sketch. For example, if we want to create an angle between two lines, we should exaggerate when sketching (that is, sketch an eighty-five degree angle at sixty degrees) so that *Pro/ENGINEER* will not make the lines perpendicular.

Disable/Enable Constraints

Geometric constraints can be modified while sketching; the rules to **disable** or **enable** constraints are as follows:

- The current constraint is highlighted in red.
- Disable the current constraint by pressing the **right-mouse-button**. To enable it again, press the **right-mouse-button** again.
- Lock in the current constraint by holding the [**SHIFT**] key and pressing the **right-mouse-button**. To unlock the constraint, press the [**SHIFT**] key and the **right-mouse-button** again.
- When more than one constraint is active, you can walk through them all by using the [**TAB**] key. Each one will be made current in turn, and can then be disabled or locked.

➢ We will next demonstrate the usage of the rules in the current sketch.

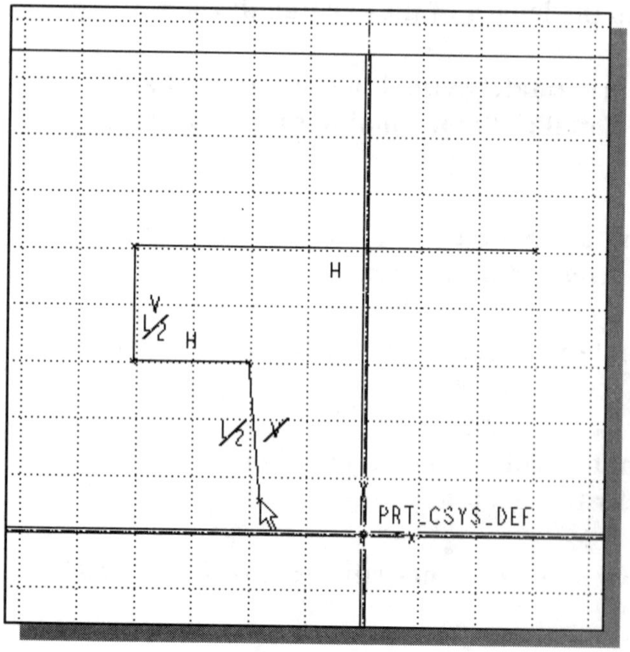

1. Move the cursor below the last endpoint we created, near grid point **(-2,3)**; click the **right-mouse-button** once when both the **vertical** and **equal length** constraint symbols are displayed. The right-mouse-click disables the current constraint, the *equal length constraint* in this case.

2. Hit the [**TAB**] key once and click the **right-mouse-button** again. Both of the constraint symbols are displayed with an inclined line to signify the constraints have been disabled.

3. Click the **right-mouse-button** again and the current constraint, the *vertical constraint*, is enabled.

4. Hit the **[TAB]** key once and click the right-mouse-button again. Both of the constraints are reactivated.

5. Move the cursor near the horizontal axis (datum plane **TOP**). Notice the cursor will *SNAP* on to the horizontal axis.

6. Pick near grid point (**-1.5,0**) and create an inclined line as shown.

7. Pick near grid point (**4,0**) to create a horizontal line on the horizontal axis.

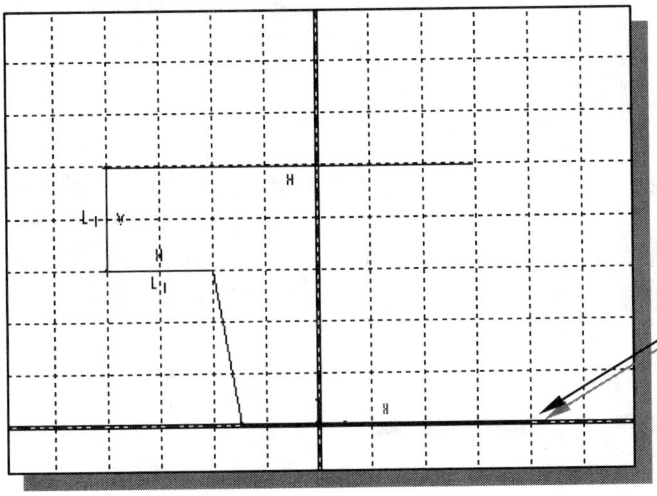

8. In the display area, click the **middle-mouse-button** twice to end the Line command.

❖ Notice that dimensions are added automatically to describe the locations of the lines.

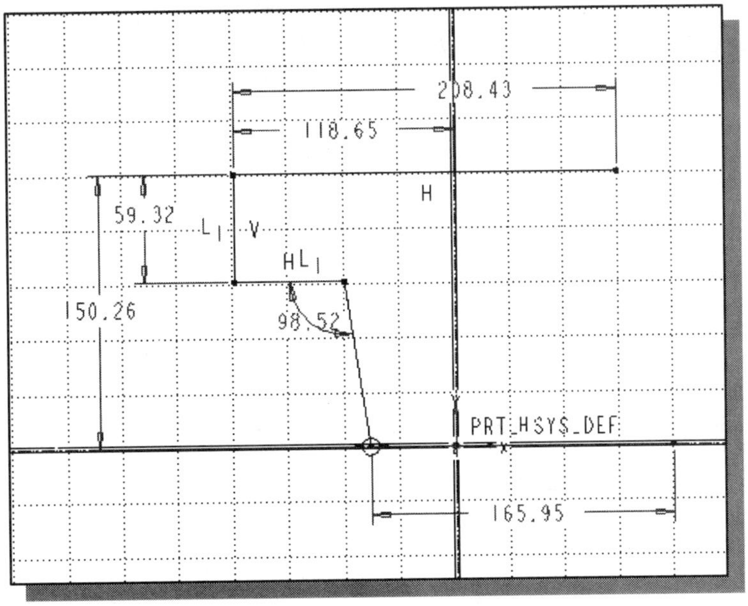

Create and Divide a Circle

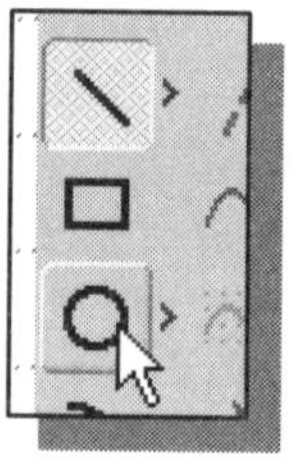

1. In the *Sketcher* toolbar, select **Circle** as shown. The default option is to create a circle by specifying the center point and a point through which the circle will pass.

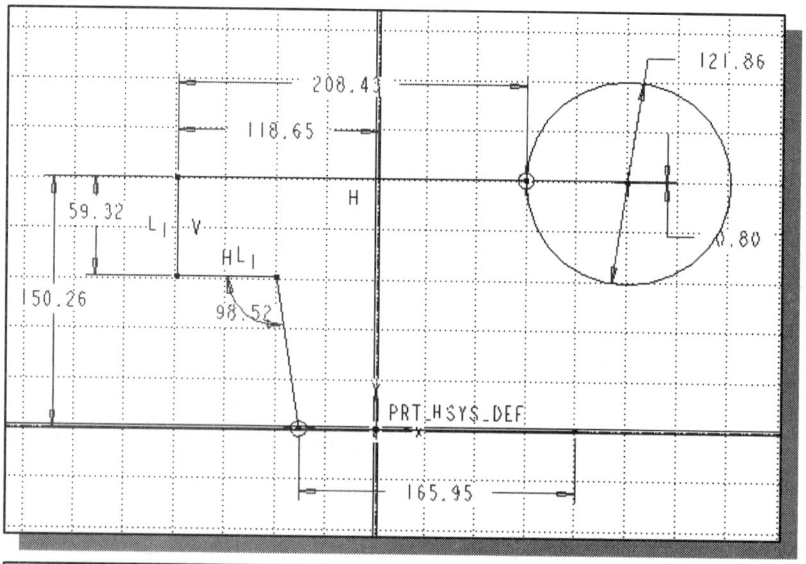

2. Place the center of the circle toward the right side of the sketch, approximately at grid point (**5,5**).

3. Pick near the right endpoint (**3,5**) of the top horizontal line; an implicit alignment constraint will applied automatically.

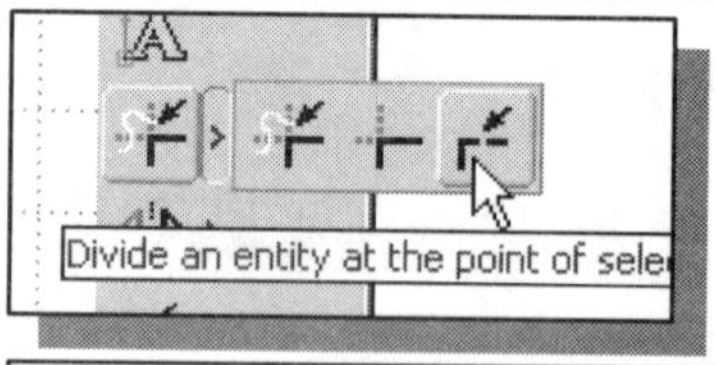

4. In the *Sketcher* toolbar, activate the Divide command by selecting the icon in the icon stack.

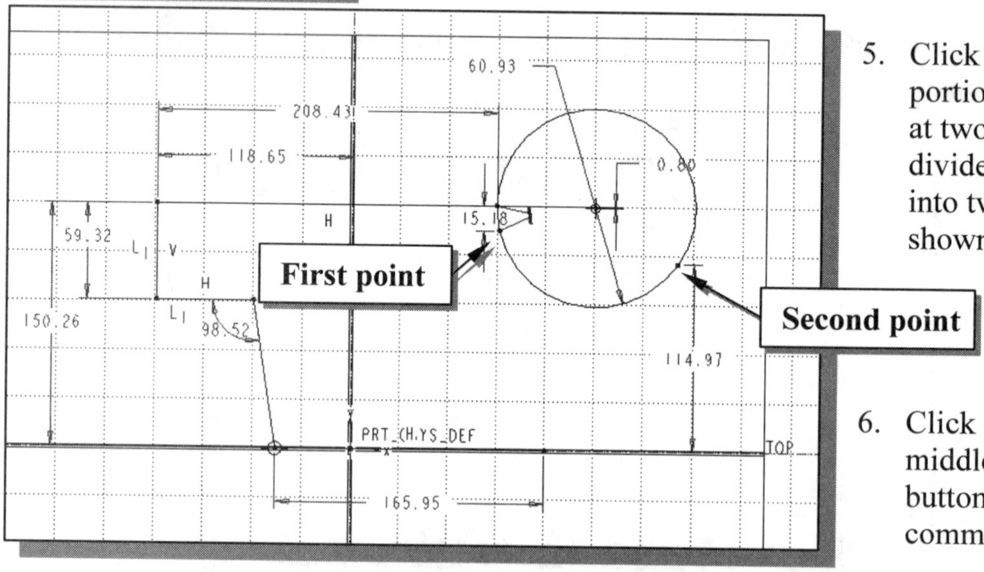

5. Click on the lower portion of the circle at two locations to divide the circle into two arcs, as shown in the figure.

6. Click once with the middle-mouse-button to end the command.

❖ Note the *Intent Manager* automatically added additional dimensions as the circle is divided into two arcs.

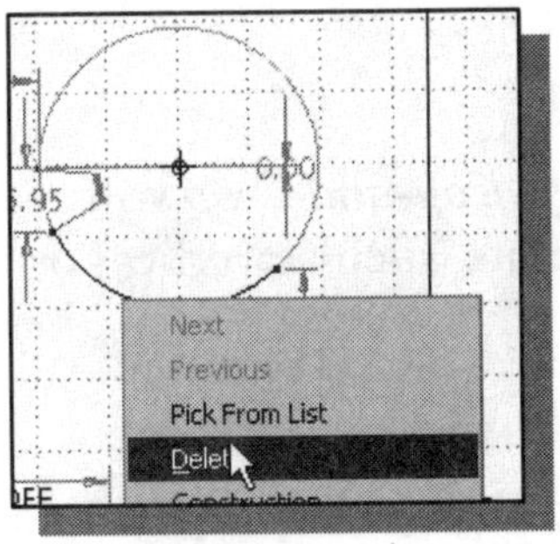

7. Select the lower arc by clicking with the **left-mouse-button**. Notice the selected entity is highlighted.

8. Inside the graphics area, press down the right-mouse-button to display the option menu and select **Delete** as shown.

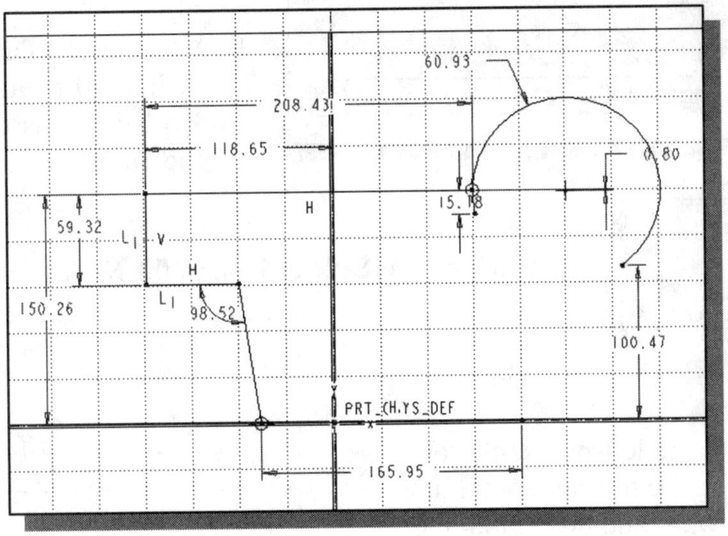

Create a closed region Sketch

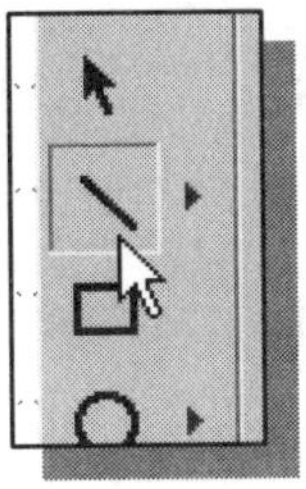

1. Pick **Line** in the *Sketcher* toolbar.

2. Pick the right endpoint of the lower horizontal line to attach the first endpoint of the new line.

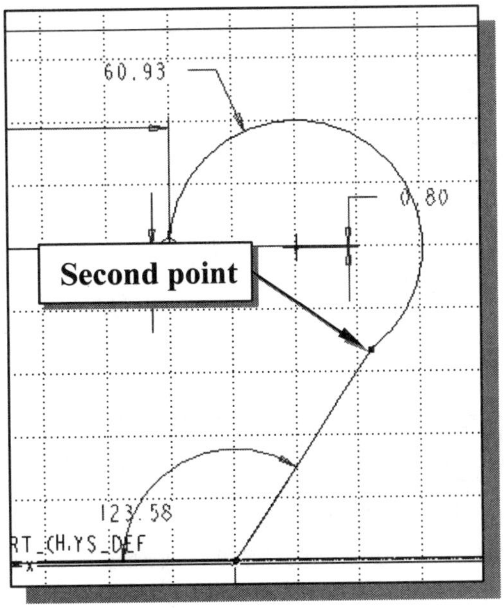

3. Move the cursor along the arc and note that different types of constraints are feasible at different locations. Attach the end of the line to the end of the arc. (Note that a *tangent* constraint may be applied automatically depending upon the size of the arc you have created.)

4. Inside the display area, click the **middle-mouse-button** once to end the Line command.

❖ Note the *Intent Manager* automatically adds and removes additional dimensions and/or constraints to maintain a fully described sketch.

Modify Geometry by Drag-and-Drop

- In *Pro/ENGINEER*, geometric entities can also be modified by *drag-and-drop* with the left-mouse-button. Note that the applied geometric constraints, such as vertical or horizontal, are maintained by the system during the editing.

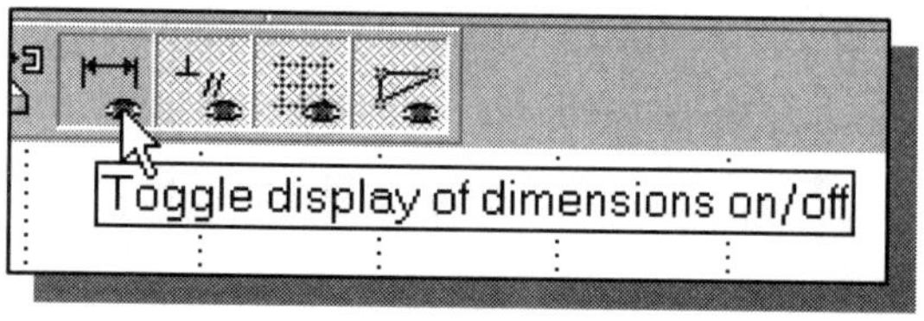

1. Click on the **Dimensions display** icon to switch ***off*** the display of the dimensions. Note the other display icons available in the *Standard* toolbar area.

2. Click on the **Select** icon in the *Sketcher* toolbar.

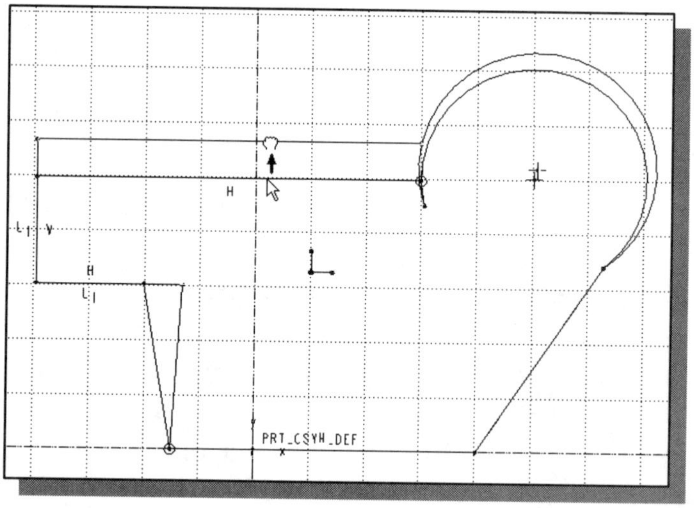

3. Use the left-mouse-button to drag the top-horizontal line upward and notice that the geometry is adjusted while maintaining the applied geometric constraints.

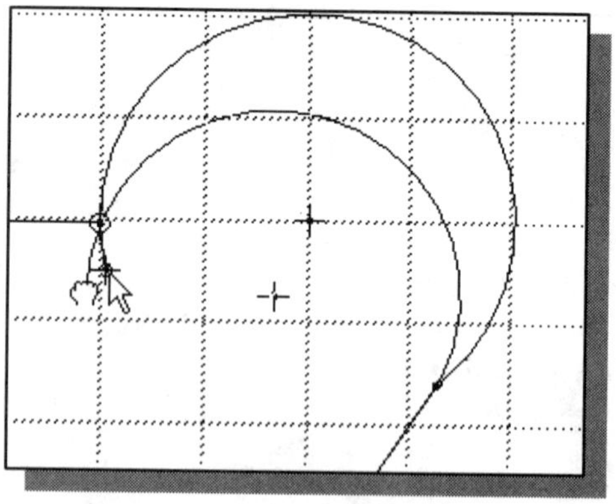

4. On your own, experiment with adjusting the locations of the endpoint of the arc, the center point of the arc, etc.

- We can dynamically modify geometric entities using drag-and-drop. The drag-and-drop capabilities provide a more flexible modeling environment, especially when exact dimension values are not known.

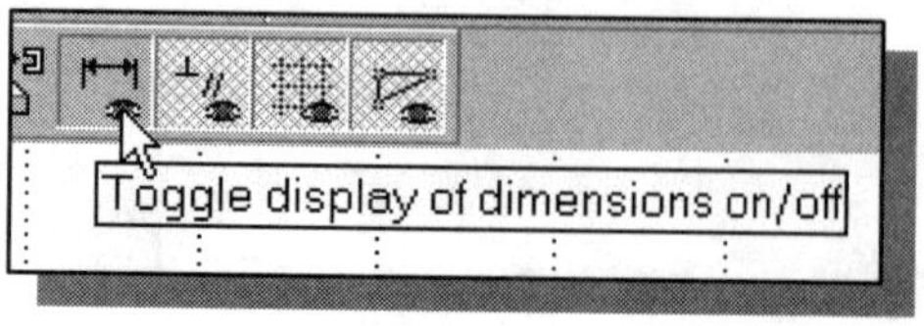

5. Click on the **Dimensions display** icon to switch ***on*** the display of the dimensions.

Controlling Geometric Constraints

❖ In *Pro/ENGINEER*, a set of rules can be applied to geometric entities to ensure the capture of the design intent. This set of rules is called the **geometric constraints**. As we create geometric entities such as lines and circles in the *Sketcher* mode, the *Intent Manager* automatically add dimensions and/or geometric constraints. The application of different constraints will affect the geometry differently, and the design intent can be embedded into the model file through the proper use of constraints.

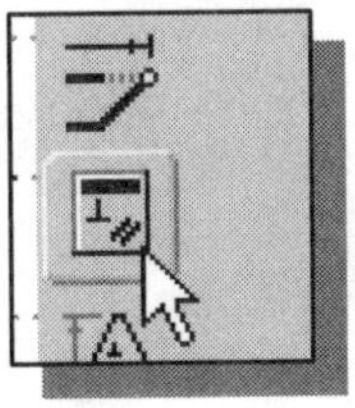

1. Pick **Constraints** in the *Sketcher* toolbar.

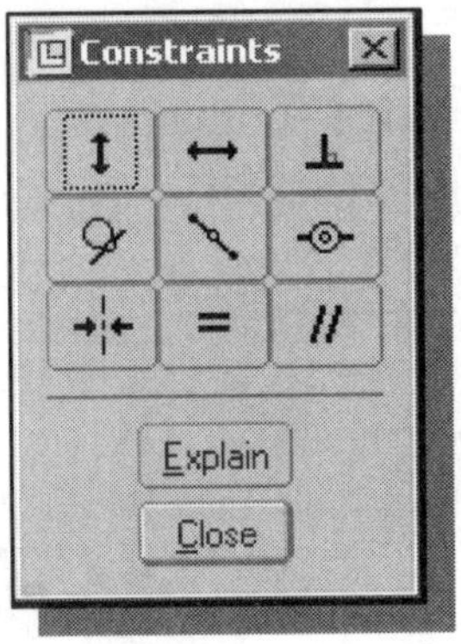

❖ The **Constraints** submenu appears on the screen. Note that the nine constraint icons actually provide more than a dozen of the applicable constraint options.

Vertical — Make an entity vertical
Line Up Vertical — Line up two points vertically

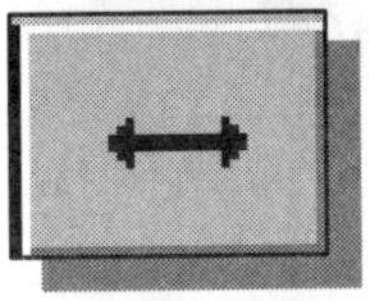

Horizontal — Make an entity horizontal
Line Up Horizontal — Line up two points horizontally

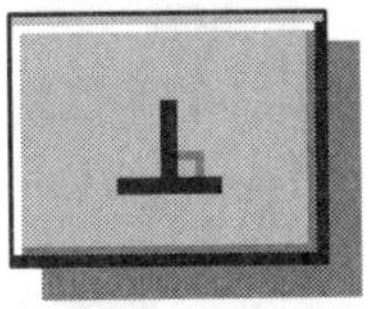

Perpendicular — Make entities perpendicular

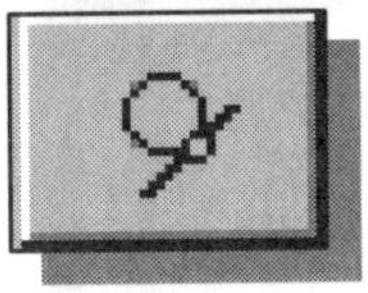

Tangent — Make entities tangent to each other

Midpoint — Place point on the middle of the line

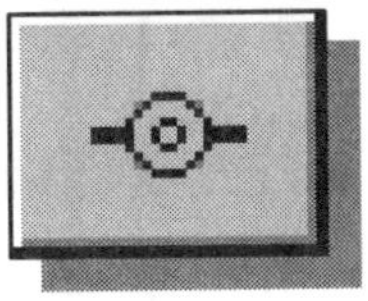

Collinear — Make two lines collinear
Point On Entity — Move a point on entity
Same Points — Make two points coincident

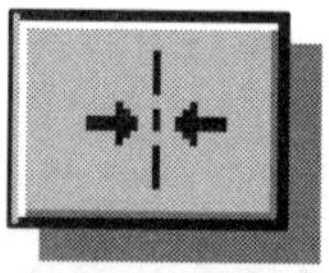

Symmetric — Make two points symmetric about a centerline

Equal Radii — Make arcs or circles equal radii

Equal Lengths — Make lines equal length

Parallel — Make entities parallel

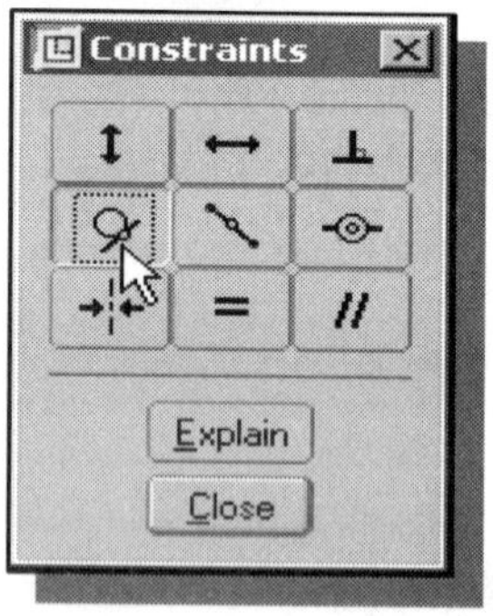

2. Pick **Tangent** in the **Constraints** submenu.

3. Pick the arc and the line as shown.

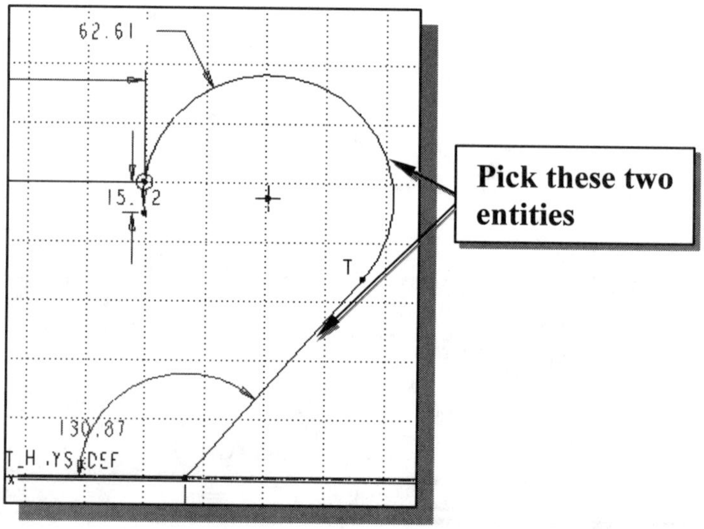

❖ Note that *Pro/ENGINEER* adjusts the two entities to satisfy the applied tangent constraint.

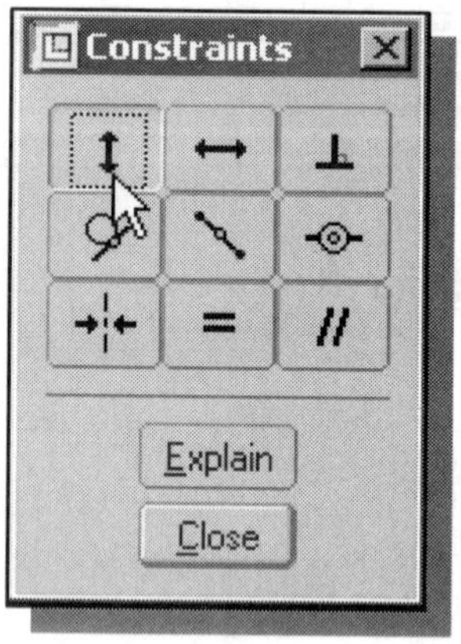

4. Pick **Vertical** in the **Constraints** submenu.

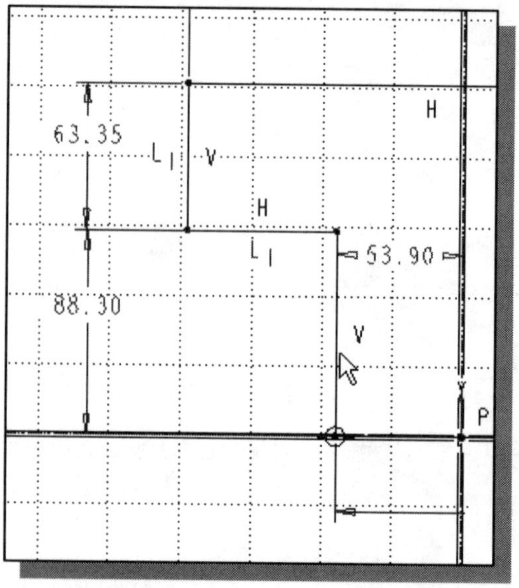

5. Pick the inclined line on the left. Notice the **vertical symbol** is displayed as the constraint is applied.

REDO and UNDO

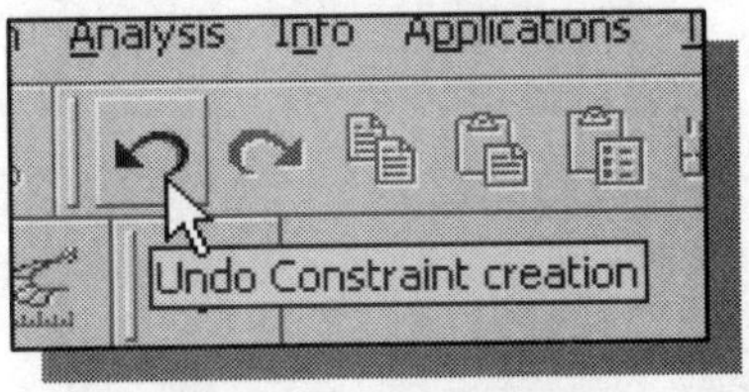

1. Pick **Undo** in the *Sketcher* toolbar.

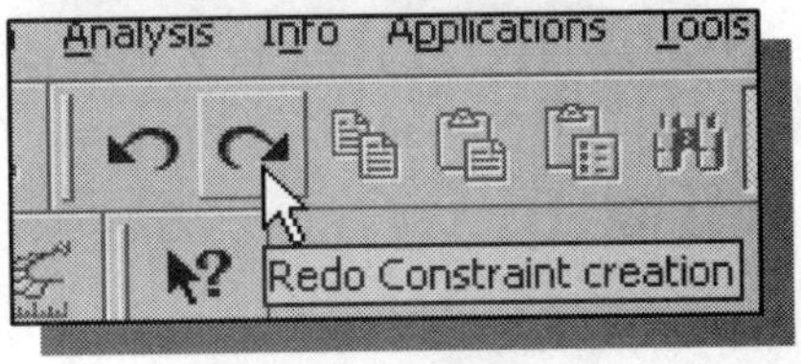

2. Pick **Redo** to reapply the vertical constraint.

❖ *Pro/ENGINEER* records all the steps used while we are in *Sketcher* mode. All *Sketcher* operations can be undone with the **Undo** command. We can use **Undo** and **Redo** to experiment with different designs we have in mind.

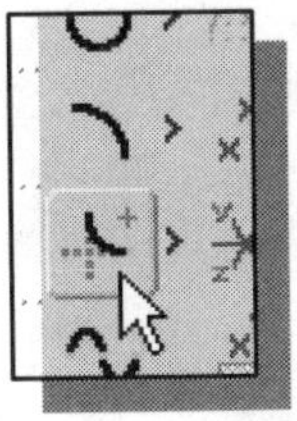

3. In the *Sketcher* toolbar, select **Fillet**.

4. The **Fillet** command allows us to create a fillet by picking two intersecting entities. Create the three fillets as shown.

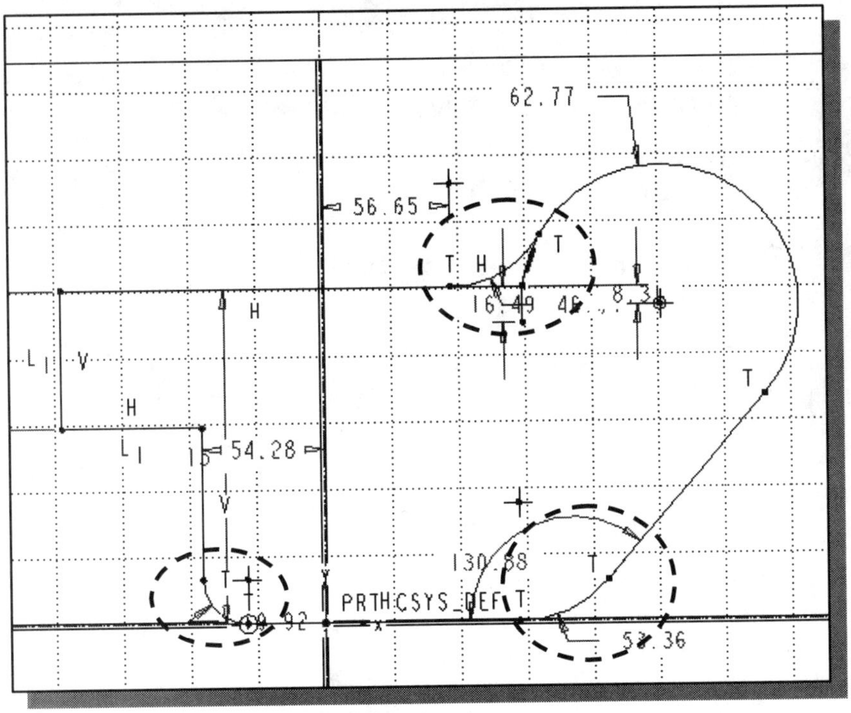

❖ Note that in *Pro/ENGINEER*, the **Fillet** command automatically trims two lines and creates break points when arcs or circles are involved.

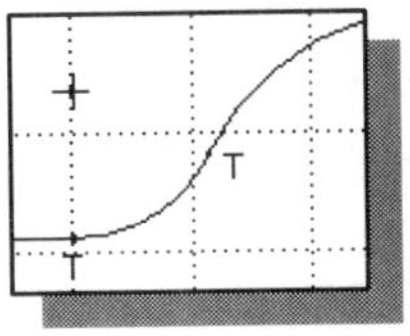

5. On your own, **Delete** the two extra segments near the top fillet.

Using Sketcher Points

We will next create a *Sketcher* **Point** to define the theoretical intersection of two lines; the point can then be used for dimensioning purposes.

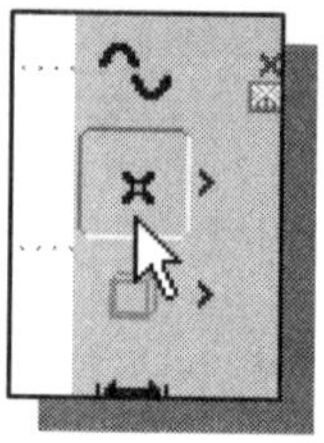

1. Select **Point** in the *Sketch* toolbar as shown.

2. Move the cursor near the lower right corner of the sketch. Place the point so that it is aligned to the inclined line as shown.

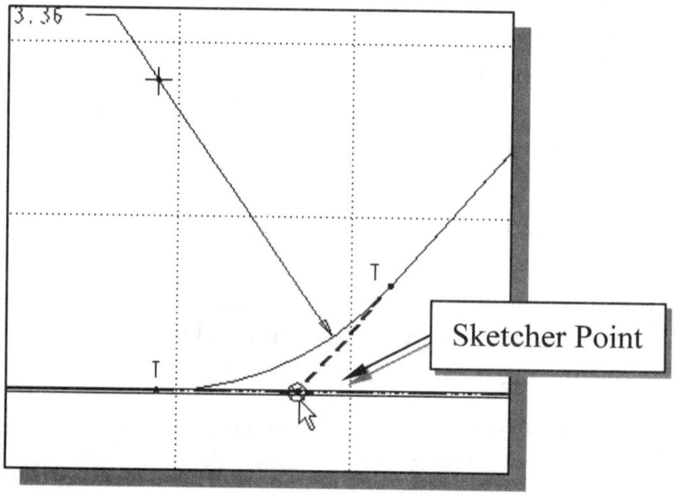

3. On your own, use the **Line Up Horizontal** option to align the center of the top arc and the top left corner of the sketch.

4. On your own, use the **Equal Radii** option to make the two lower arcs equal radii.

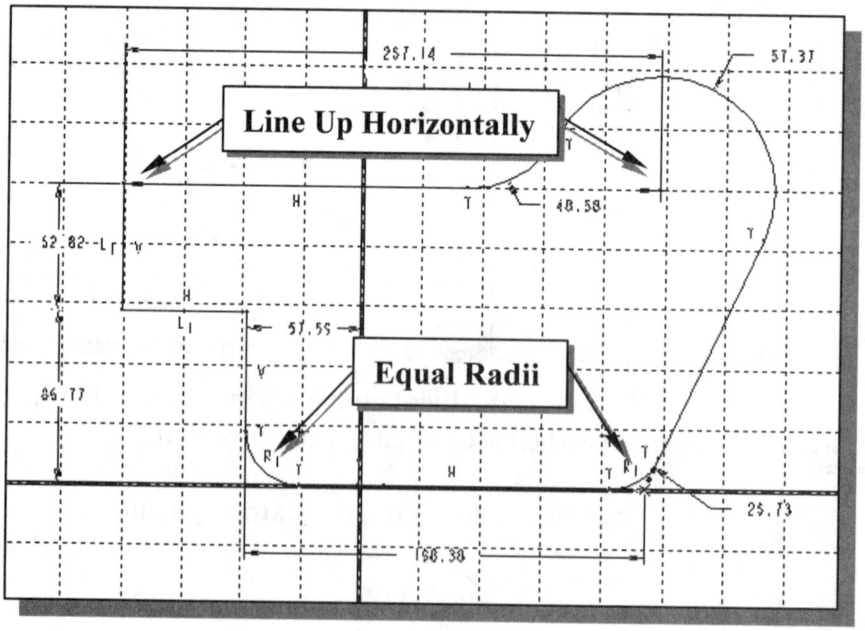

5. Use the **Dimension** command and add the two dimensions as shown.

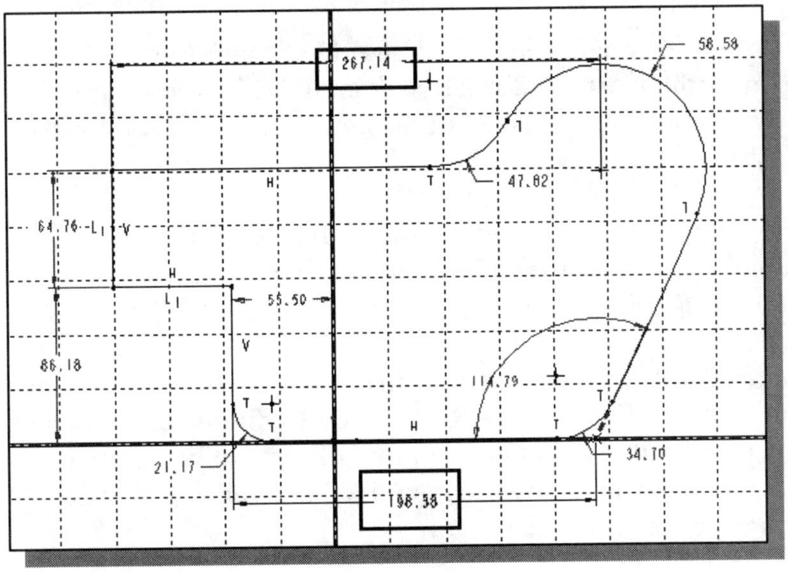

Deleting Geometric Constraints

1. Choose **Select** in the *Sketcher* toolbar.

2. Pick one of the equal length symbols (**L1**).

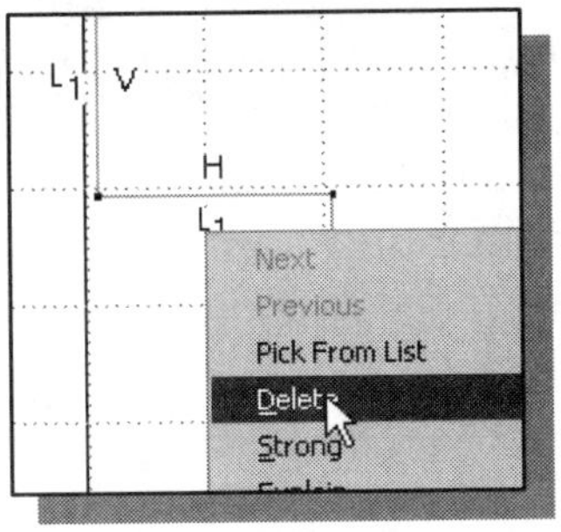

3. Press and hold down the right-mouse-button to display the option menu and select **Delete** to remove the *equal length constraint*.

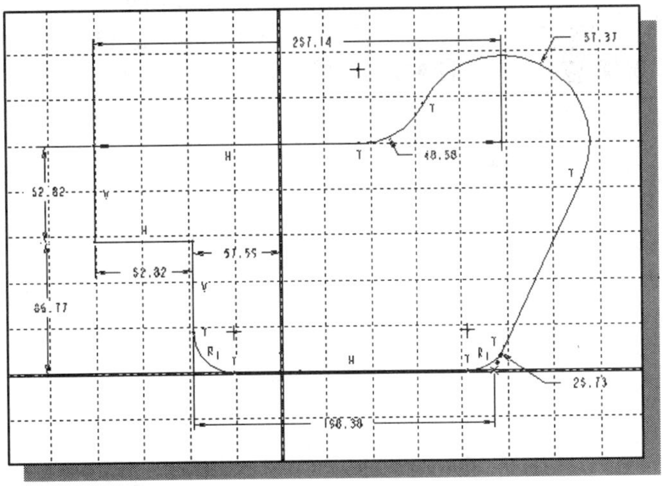

❖ Notice *Pro/ENGINEER* removed the equal length constraint and a horizontal dimension is added to the sketched section.

❖ In *Pro/ENGINEER*, completed 2D sketches (**sections**) are always fully constrained. *Pro/ENGINEER's Intent Manager* incorporates the ***Dynamic Dimensioning*** feature, which means that *Sketcher* adds or removes dimensions and constraints *on the fly* to keep sections fully constrained. Even in the middle of sketching. 2D sections are never under-dimensioned or over-dimensioned.

Weak Dimensions and Constraints

❖ Dimensions and constraints are called "***weak***" if *Pro/ENGINEER Sketcher* can remove them without any confirmation from us. All dimensions added dynamically by *Sketcher* are *weak*, as are many of the constraints. Weak dimensions and constraints appear gray.

There is no need to remove undesired weak dimensions and constraints explicitly. Instead, simply add the desired dimensions and constraints. The weak ones will be removed automatically as appropriate.

If you attempt to add a constraint or dimension that conflicts with other *strong* dimensions or constraints, the conflicting items will immediately be highlighted and you will be prompted to select one or more of them to be removed. This keeps the section from ever becoming over-constrained or over-dimensioned.

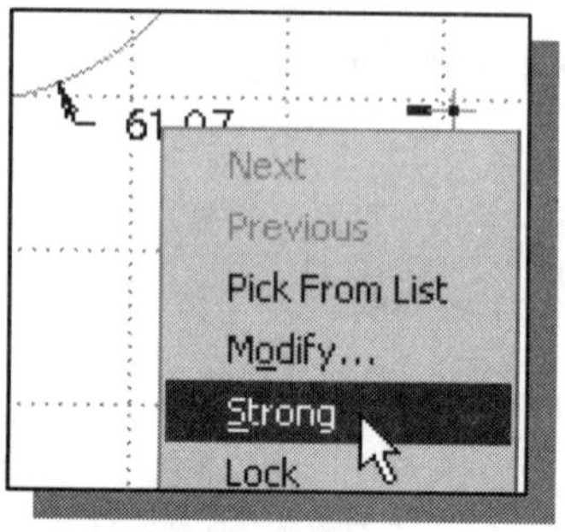

- Weak constraints and dimensions can be "***strengthened***" by choosing **Strong** in the popup option menu. This will ensure that they are never removed without confirmation. Modifying weak dimensions also strengthens them.

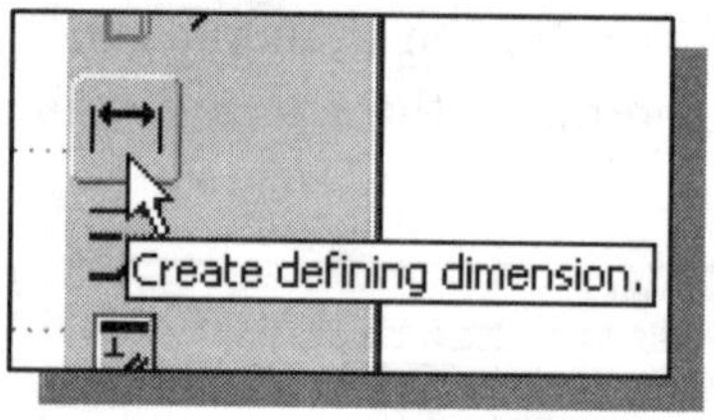

➢ On your own, create/strengthen additional dimensions so that the sketch appears as shown in the figure below. (Do not be overly concerned with the dimension values; concentrate on applying proper constraints and dimensions.)

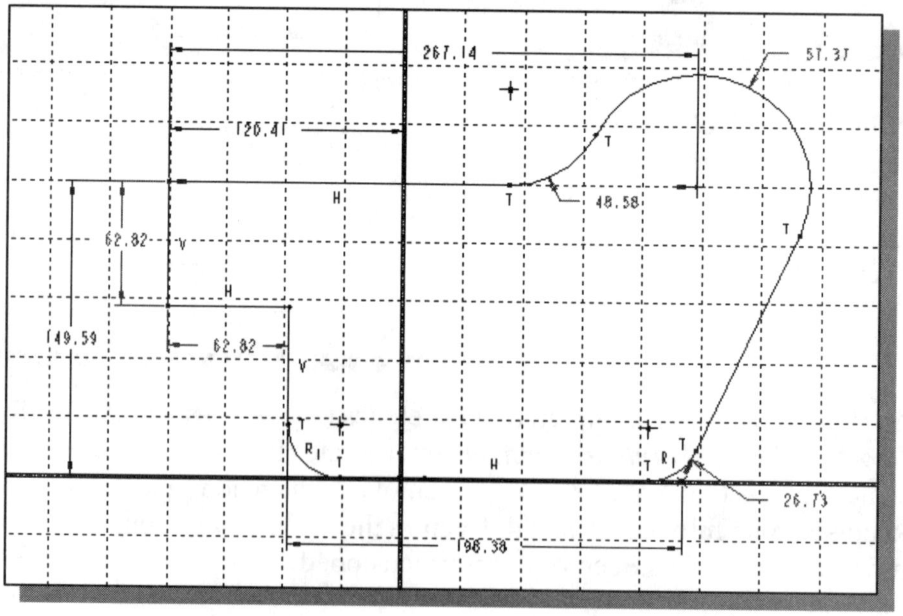

Completing the Base Feature

1. On your own, adjust the dimensions as shown in the figure below.

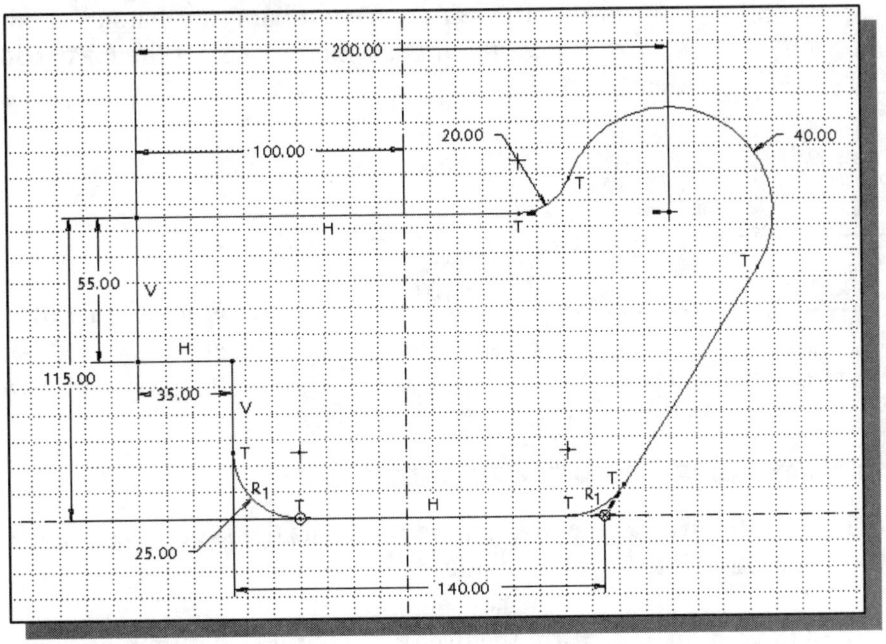

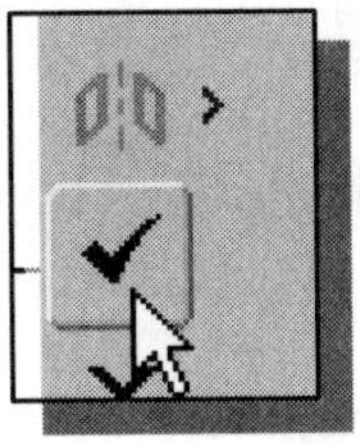

2. In the *Sketcher* toolbar, click on the **Accept** icon to end the *Pro/ENGINEER 2D Sketcher* and proceed to the next element of the feature definition.

3. In the *Extrude Options Dashboard*, confirm the ***depth option*** is set and enter **5** as the *extrusion depth* as shown in the figure.

4. Pick **OK** in the feature dialog box to create the solid.

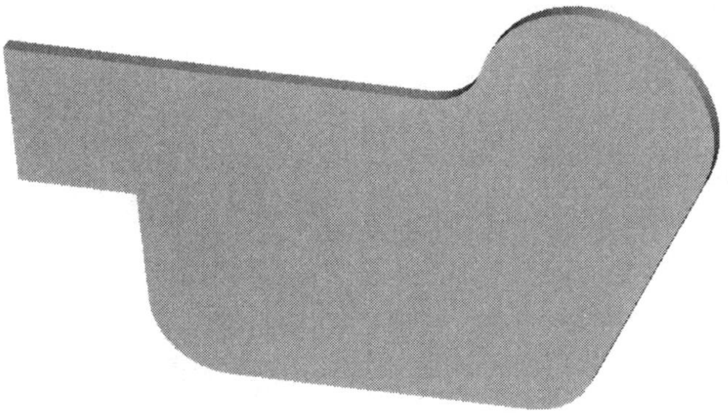

➢ On your own, use the *Dynamic Viewing* functions to view the completed 3D solid model.

Using the Offset Loop Option to create a CUT feature

1. In the *Feature* toolbars (toolbars aligned to the right edge of the main window), select the **Extrude Tool** option as shown.

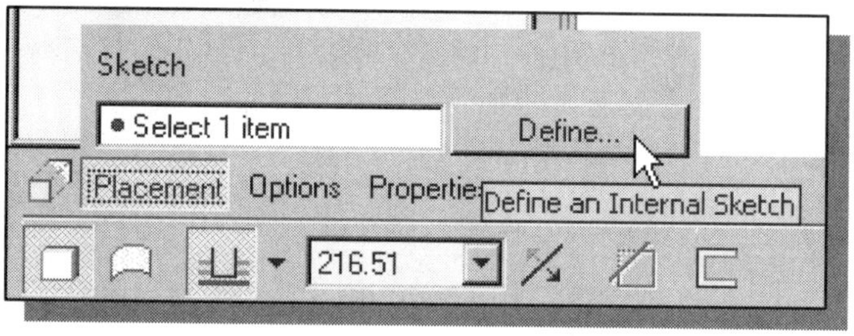

2. Click the **Placement** option and choose **Define** to begin creating a new *Internal sketch.*

3. Pick the **front face** of the solid model as the sketching plane and set the view direction into the solid model.

4. On your own, confirm the use of the **RIGHT** datum plane as the *reference plane* oriented toward the **right** edge of the computer screen.

➢ Note that we are using the **RIGHT** and **TOP** datum planes as sketching references.

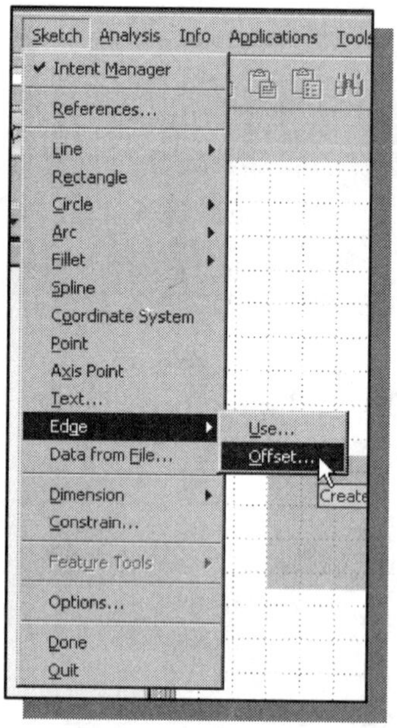

5. In the **Sketch** pull-down menu, select

   Sketch → Edge → Offset

6. Choose **Loop** in the **Type** menu as shown.

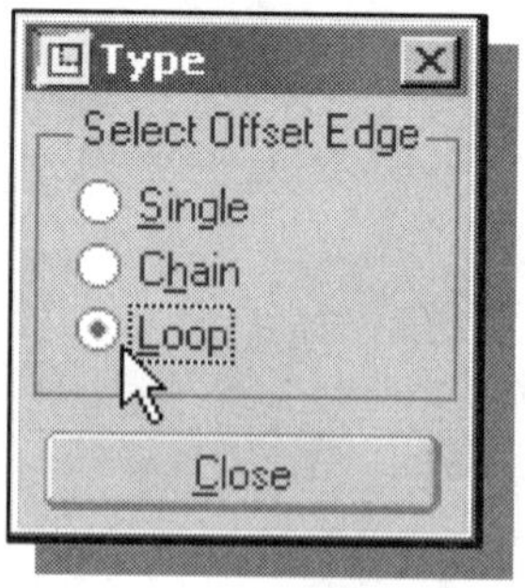

7. Click on the inside of the front surface of the base feature. The contour of the surface is automatically selected. An arrowhead is displayed pointing away from the center of the base feature.

8. The message "*Enter offset in the direction of the arrow*" is displayed in the message area. Enter **-15**. The negative sign will place the offset curve on the inside of the base feature.

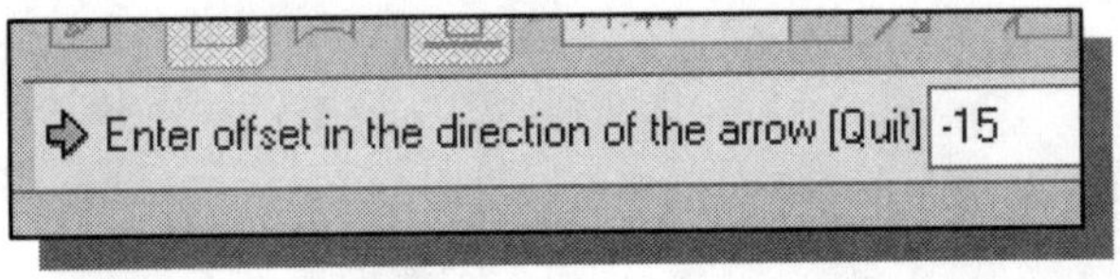

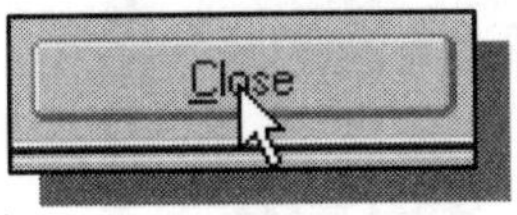

9. Click the **Close** button to accept the created offset geometry.

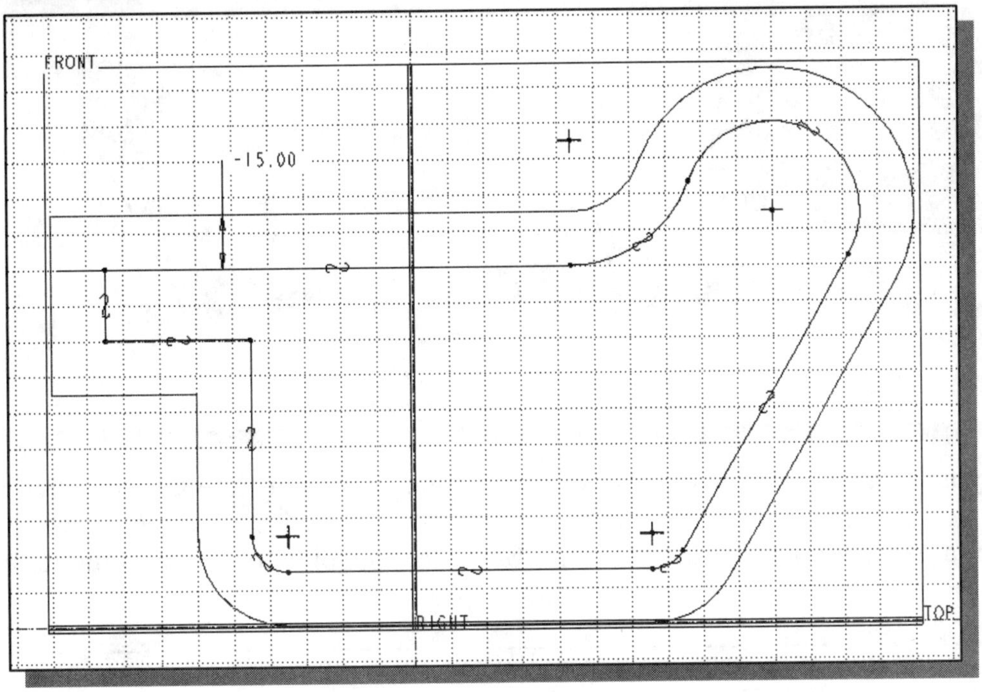

➢ Notice there is only the offset dimension needed to define the geometry. The **Offset Edge** command creates parent/child relationships with the referenced feature. The offset entities are updated automatically as the base geometry is modified.

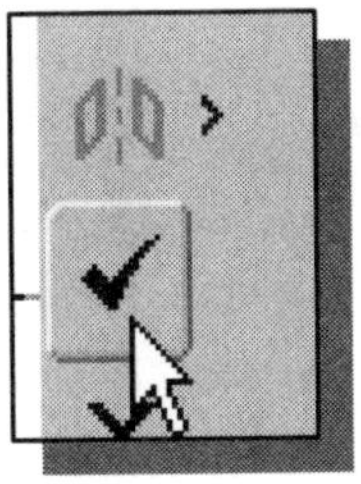

10. In the *Sketcher* toolbar, click on the **Accept** icon to end the *Pro/ENGINEER 2D Sketcher* and proceed to the next element of the feature definition.

11. On your own, set extrude options to **Remove Material** and **Thru All**. Confirm the cut direction and complete the offset cut feature.

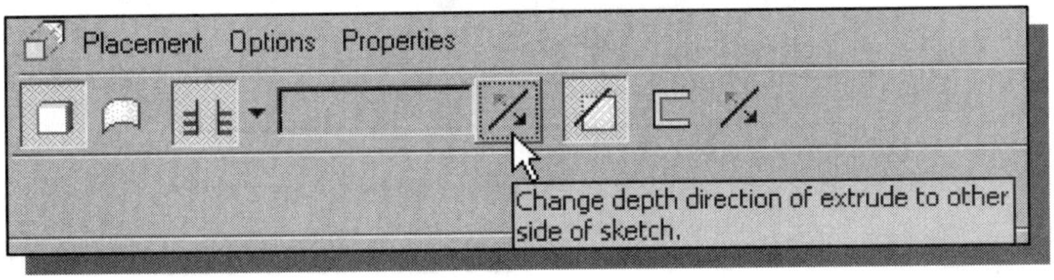

➤ Use the *Dynamic Viewing* functions to view the completed 3D model.

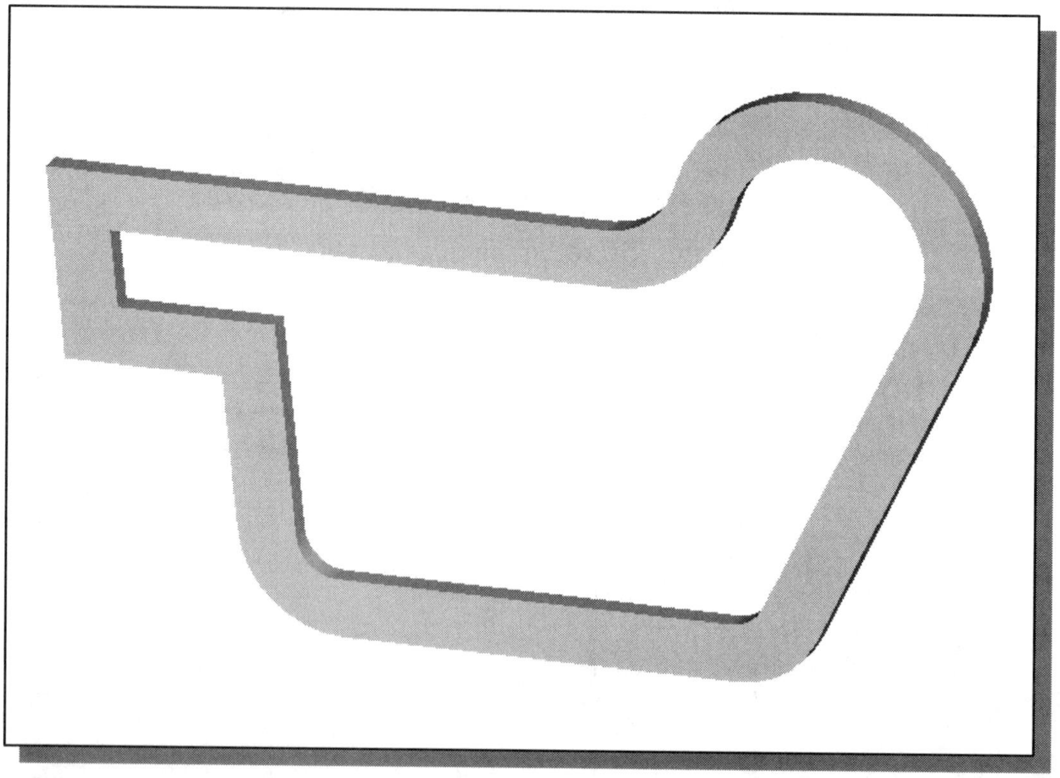

➤ On your own, modify the design by creating one additional (Radius: **10**) fillet on the left side as shown, adjust some of the dimensions and notice the offset curve is maintained and updated automatically. **Save** the model before proceeding to the next section.

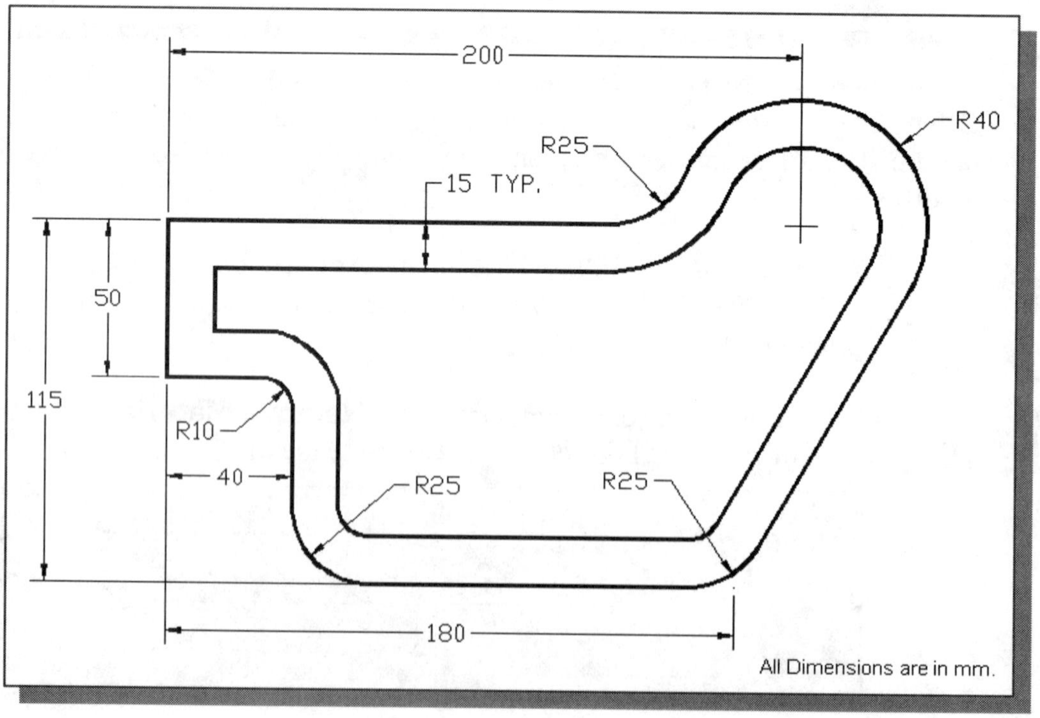

Parametric Relations and Geometric Constraints

A primary and essential difference between parametric modeling and previous generation computer modeling is that parametric modeling captures the *design intent*. In the previous lessons, we have seen that the design philosophy of "*shape before size*" is implemented through the use of *Pro/ENGINEER's* ***Intent Manager*** and ***Sketcher***. In performing geometric constructions, dimensional values are necessary to describe the **size** and **location** of constructed geometric entities. Besides using dimensions to define the geometry, we can also apply geometric rules to control geometric entities. More importantly, *Pro/ENGINEER* can capture design intent through the use of **geometric constraints**, **dimensional constraints**, and **parametric relations.** In *Pro/ENGINEER*, there are two types of constraints: **geometric constraints** and **dimensional constraints**. For part modeling in *Pro/ENGINEER*, constraints are applied to *2D sketches*, as they were illustrated in the previous section. **Geometric constraints** are **geometric restrictions** that can be applied to geometric entities; for example, *horizontal*, *parallel*, *perpendicular*, and *tangent* are commonly used *geometric constraints* in parametric modeling. **Dimensional constraints** are used to describe the SIZE and LOCATION of individual geometric shapes. In *Pro/ENGINEER*, **parametric relations** are user-defined mathematical equations composed of *dimensional variables* and/or *design variables*. In parametric modeling, features are made of geometric entities with both relations and constraints describing individual design intent. In this section, we will discuss the use of parametric relations.

Create a Simple *PLATE* Design

- In parametric modeling, dimensions are design parameters that are used to control the sizes and locations of geometric features. Dimensions are more than just values; they can also be used as feature control variables. This concept is illustrated by the following example.

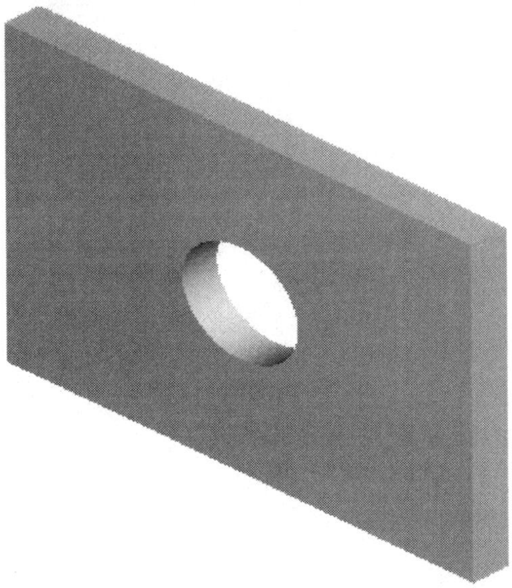

Using a Sketch file

1. Pick the **Create New Object** icon in the toolbar. (We can also use the key combination Ctrl-N to start a new object.)

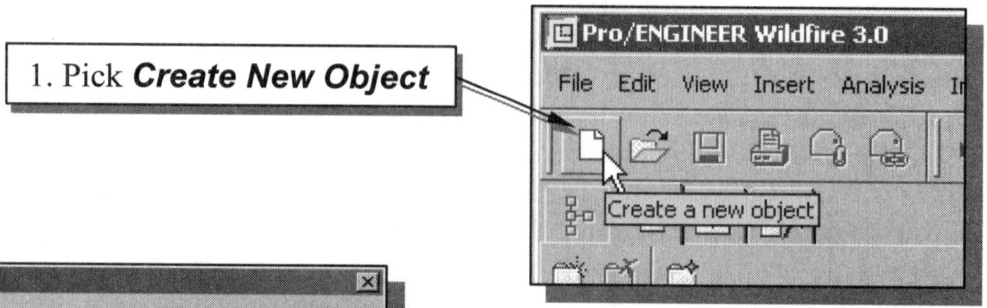

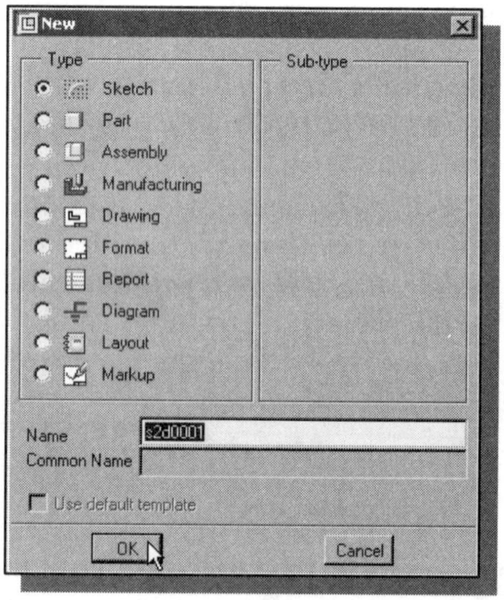

2. In the ***New*** form, pick **Sketch** in the **Type** list.

3. Pick **OK** to accept the default filename **(s2d0001.sec)**.

❖ The *Pro/ENGINEER* ***Sketcher*** appears. We have entered *Pro/ENGINEER's 2D Sketcher* mode directly. The concept of this option is to allow users to utilize *Pro/ENGINEER* in all aspects of the design process. In this case, we are using *Pro/ENGINEER* as a 2D sketching pad.

4. On your own, use the **Rectangle** and **Circle** commands and create a sketch as shown. (Do not be concerned with the displayed dimensional values; we will adjust them in the following sections.)

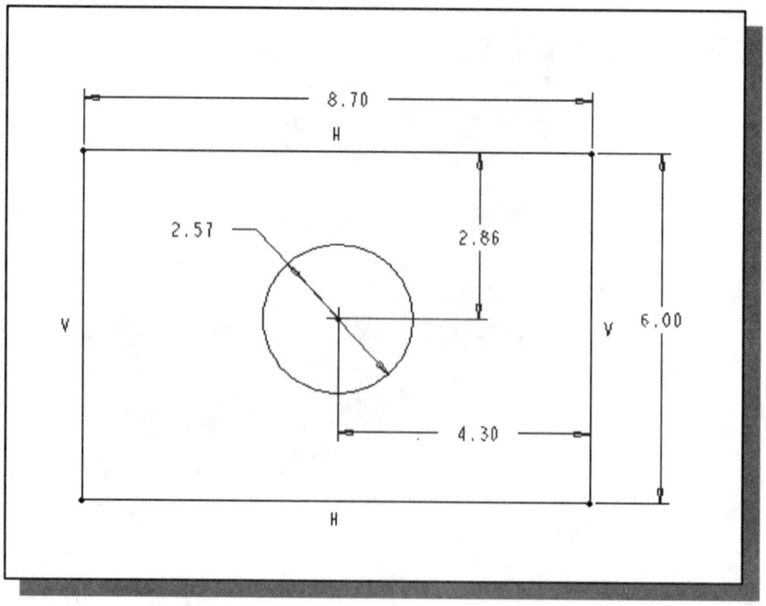

Dimensional Values and Dimensional Variables

❖ Initially in *Pro/ENGINEER*, dimensional values are used to reflect the size or distance between geometric entities created in the *Sketcher*. Each dimension is also assigned a name that allows the dimension to be used as a control variable. The default format is "**sdxx**," which stands for *Sketcher dimension* and the "xx" is a number that *Pro/ENGINEER* increments automatically.

Let us look at our current design, which represents a plate with a hole at the center. The dimensional values describe the size and/or location of the plate and the hole. If a modification is required to change the width of the plate, the location of the hole will remain the same as described by the two dimensional values. This is okay if that is the *design intent*. On the other hand, the *design intent* may require (1) keeping the hole at the center of the plate and (2) maintaining the size of the hole to be one-third of the height of the plate. We will establish a set of parametric relations using the dimensional variables to capture this design intent.

1. Pick **Relations** in the **Tools** pull-down menu.

❖ Notice in the display area the *dimensional variables* are displayed instead of the *dimensional values*. (The dimensional variables are in the **sdxx** format, which represents the dimensions are *2D Sketcher* dimensions.)

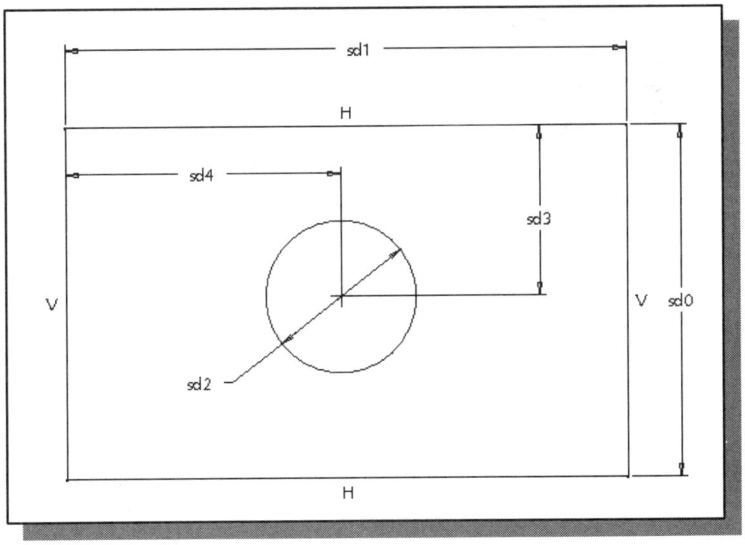

2. In the *Relations* window, enter **sd3=sd0/2** as the first equation. This equation sets the vertical location dimension (**sd3**) of the hole to be one-half of the height of the rectangle (**sd0**). The variable names on your sketch might be different than what is displayed here, use the corresponding variable names to establish the parametric relation.

3. Inside the *Relations* window, enter **sd4=sd1/2** as the second equation. This equation makes sure the horizontal location dimension (**sd2**) of the hole is always one-half of the height of the rectangle (**sd0**). Use the corresponding variable names in your sketch to establish the parametric relation.

4. Enter **sd2=sd0/3** as the third equation. This equation sets the size of the hole (**sd2**) to be always one-third of the height of the rectangle (**sd0**). The variable names on your sketch might be different than what is displayed here. Use the corresponding variable names to establish the parametric relation.

5. Click the **Execute/Verify** button in the *Relations* toolbar as shown.

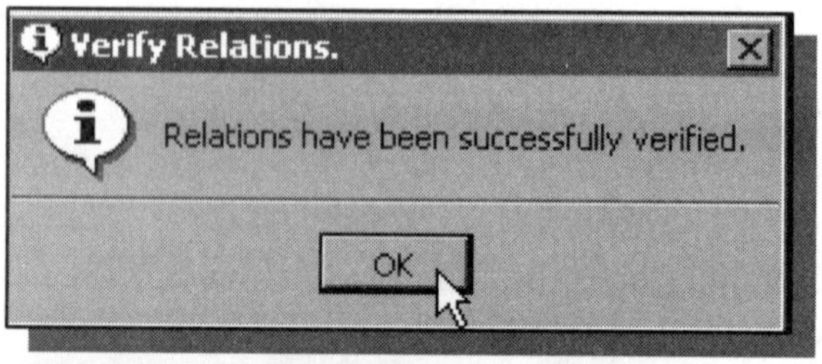

6. Click OK to close the verify relations window.

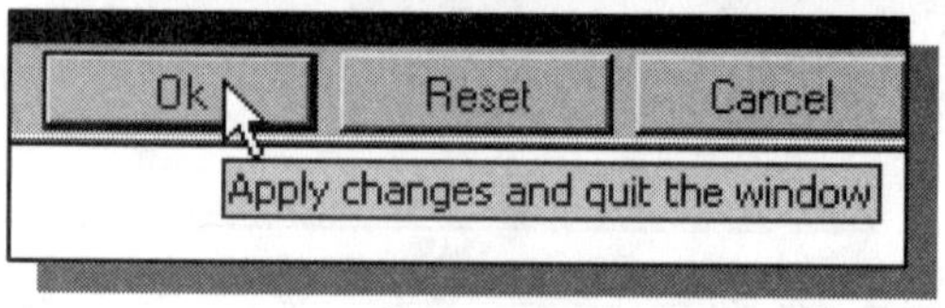

7. Click **OK** to close the *Relations* window.

8. On your own, modify the width of the rectangle to **8**, and the height of the rectangle to **6**.

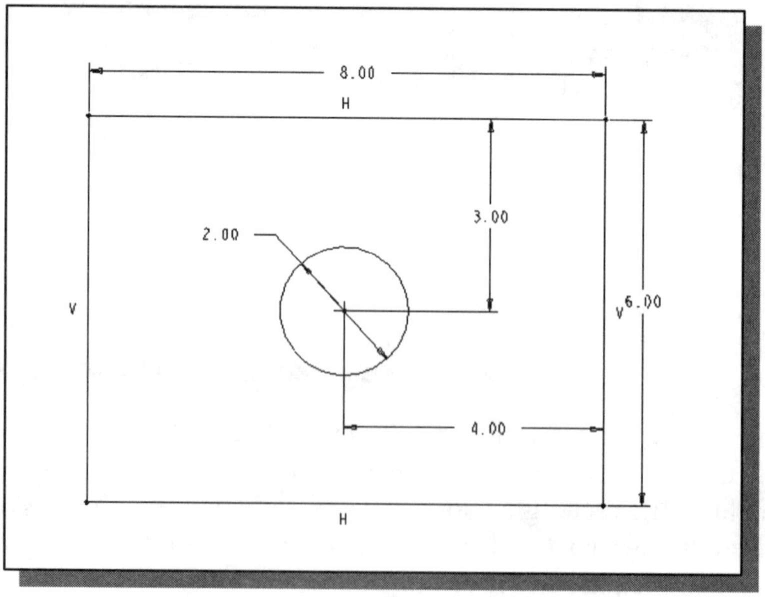

9. On your own, modify the width of the rectangle to **10**, and the height of the rectangle to **7.5**.

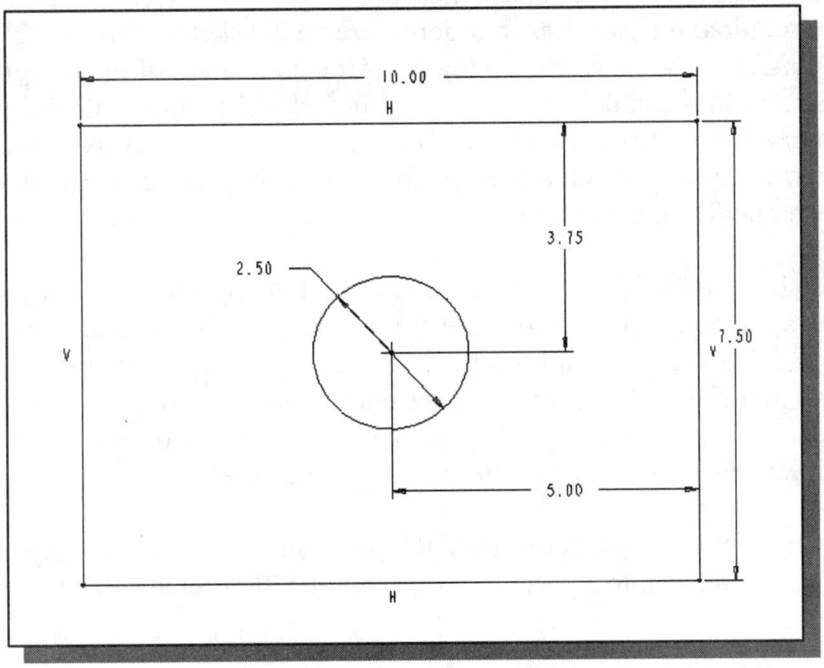

❖ *Pro/ENGINEER* automatically adjusts the dimensions of the design, and the parametric relations we entered are also applied and maintained. The dimensional constraints are used to control the size and location of the hole. The design intent, previously expressed by statements (1) and (2) at the beginning of this section, is now embedded into the model.

Save the Sketch File and Exit the Sketcher Mode

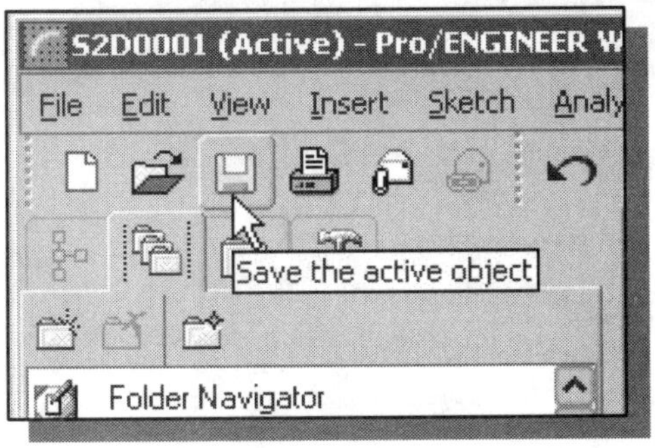

1. From the icon panel, select the **Save** icon. You can also use **Ctrl-S** to save the model.

2. Click on the **OK** button or hit the **ENTER** key to accept the filename and proceed to save the file.

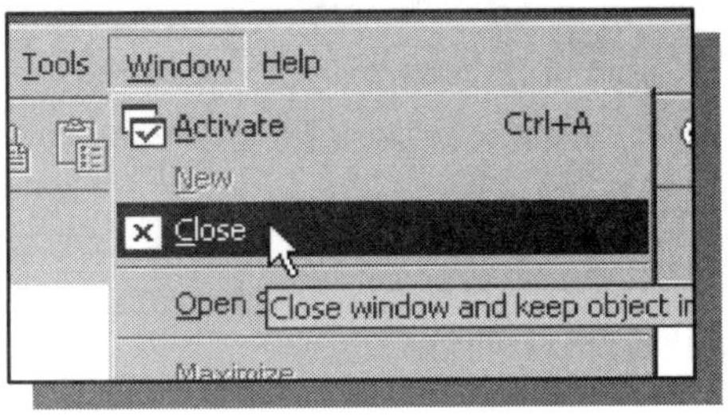

3. Click the **Close** button in the **Window** pull-down menu to end the *Pro/ENGINEER Sketcher*.

Sketch Files vs. Part Files

The main advantage of creating sketch files is that we do not need to go through all the steps that are required for part files in order to create 2D sketches in *Pro/ENGINEER*. The Pro/*ENGINEER Sketcher* can be viewed as the equivalent of the designer's sketchpad. Sketch files can be used to create rough sketches during the initial design stage. Modifications and changes can be done quickly and effectively without worrying about the interactions between features or parts. The use of sketch files allows us to concentrate on one feature at a time.

Once the sketches are finalized, we can then incorporate the finalized sketches into any part file. This approach allows us to reuse any of the 2D sketches that have been created. Note that *Pro/ENGINEER* also allows us to export any of the 2D sketches that we have created in any part file. In fact, each time we entered the *2D Sketcher*, *Pro/ENGINEER* created a sketch file automatically. The automatically generated sketch file is discarded when we exit *Pro/ENGINEER* or when we issue the **Erase** command.

In the following section, we will illustrate the procedure for incorporating an existing sketch from a sketch file into a part model, so that the 2D sketch can be used to create a solid feature.

Starting a new *Pro/ENGINEER* Part

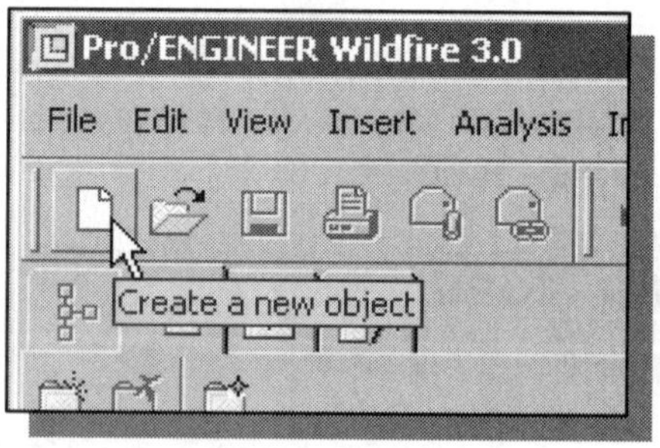

1. Select **File** in the pull-down menus. Pick the **New** option or use the quick-key combination **Ctrl-N** to start a new object.

2. In the *New* form, enter **Plate** as the solid part file **Name**.

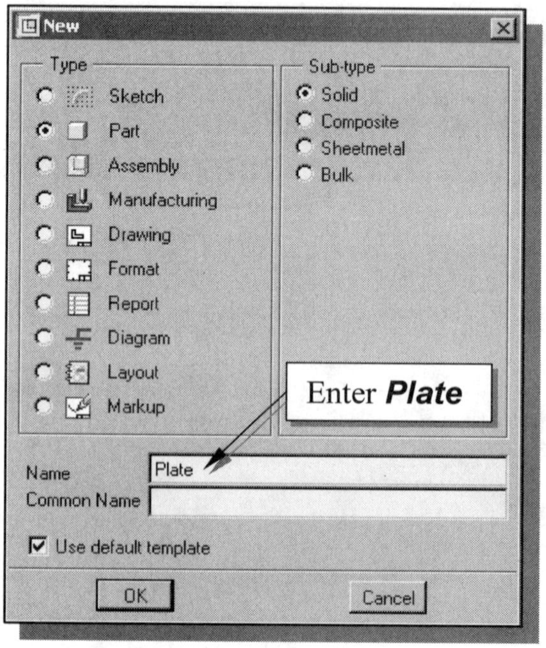

❖ Note the **Use default template** option is switched ***on***. This will use whatever default units that was selected during installation. If you are unsure about the default settings, use the procedure described on page 5-4 instead.

3. Pick **OK** to continue. A set of predefined features, three datum planes and a *coordinate system*, appeared on the screen. The template can help reducing a consider amount of repetitions that are performed each time we start a new drawing.

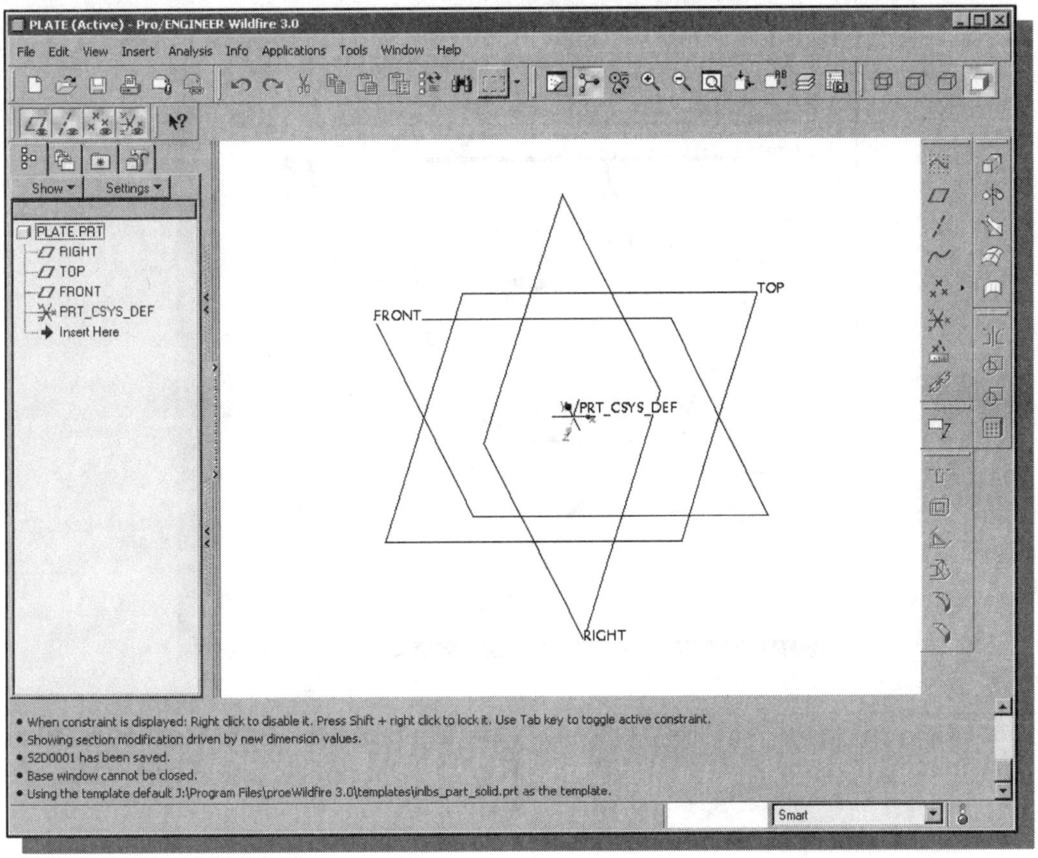

Creating the Base Feature

1. In the *Feature* toolbars (toolbars aligned to the right edge of the main window), select the **Extrude Tool** option as shown.

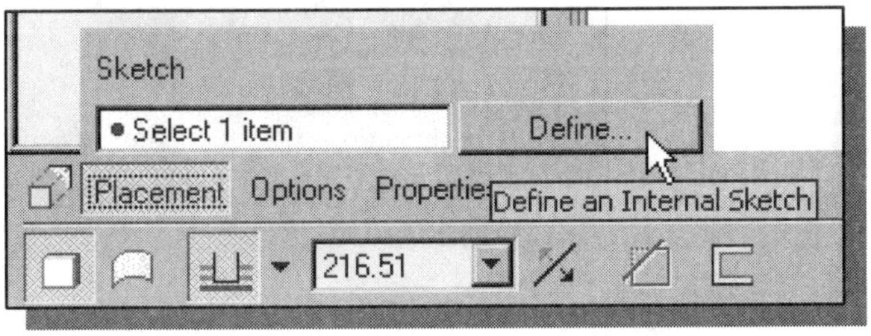

2. Click the **Placement** option and choose **Define** to begin creating a new *Internal sketch.*

3. The message "*Select a plane or surface to define sketch plane.*" is displayed in the message area. Pick the **Front** datum plane in the sketching plane as shown.

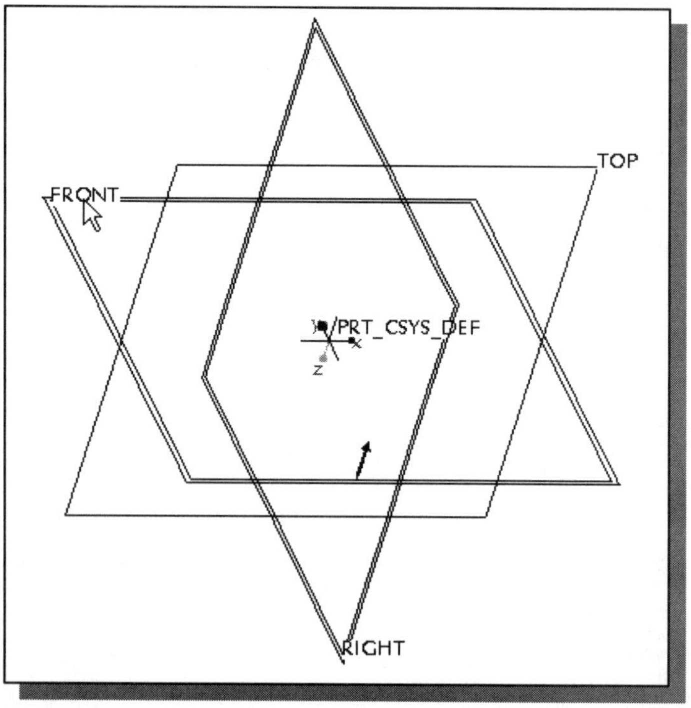

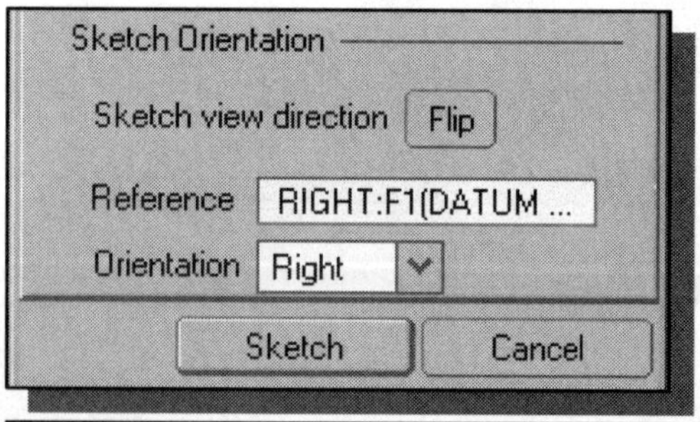

4. Confirm that the sketch **Reference** is set to **Right** datum plane and **Right Orientation** as shown.

5. Pick **Sketch** to exit the *Section Placement* window and proceed to enter the *Pro/ENGINEER Sketcher* mode.

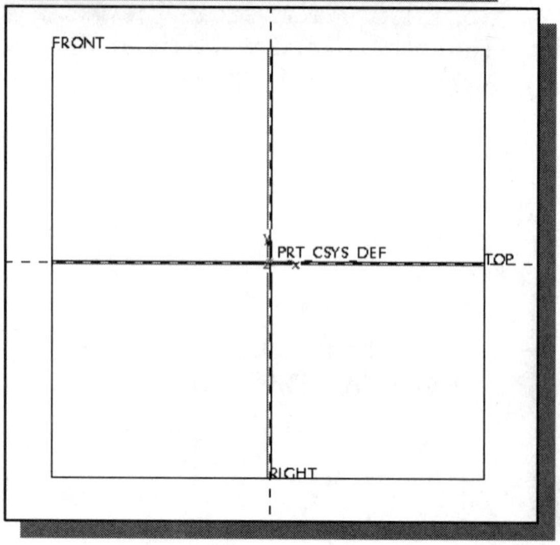

6. Pick **Sketch View** in the **Sketcher** menu. This will align the sketching plane to the computer screen. Also note the four display-control icons to the right side of the **Sketch View** icon.

Placing the 2D Sketch

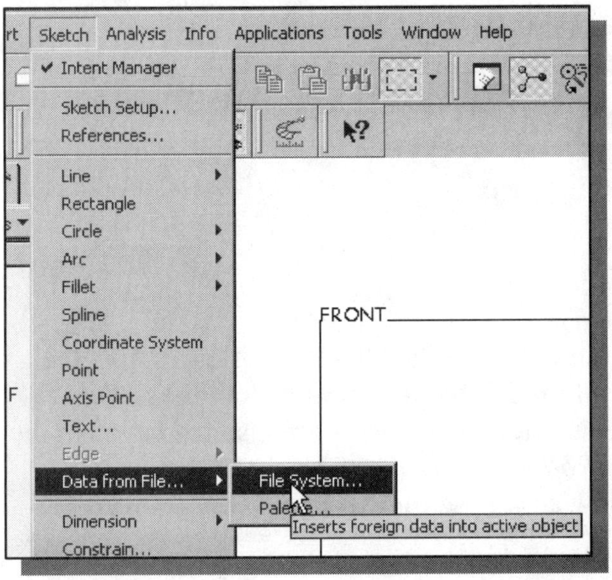

1. Pick **Data from file > File System** in the **Sketch** pull-down menu as shown.

2. In the *Open* form, select **s2d0001.sec** in the file list.

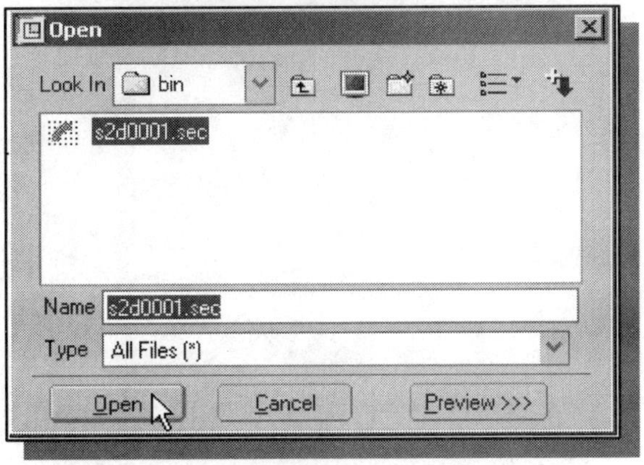

3. Pick **Open** to proceed with importing the 2D sketch into the current model file.

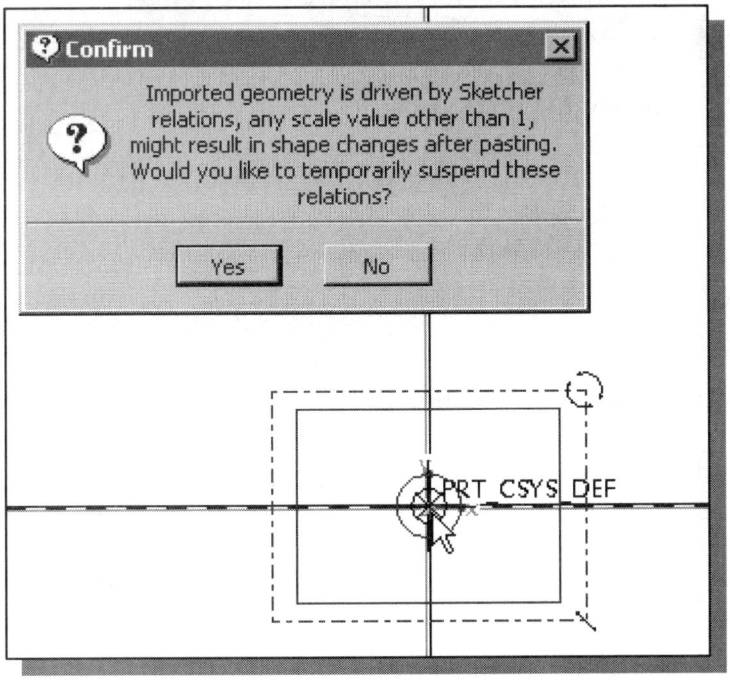

4. Click near the intersection of the two references to place a copy of the 2D sketch into the sketcher.

5. Click **No,** in the confirm window, to keep the relations that were established in the 2D sketch.

❖ The copy of the selected **sketch file** is placed into the sketcher with three control handles: the **Locate**, **Scale** and **Rotate** handles.

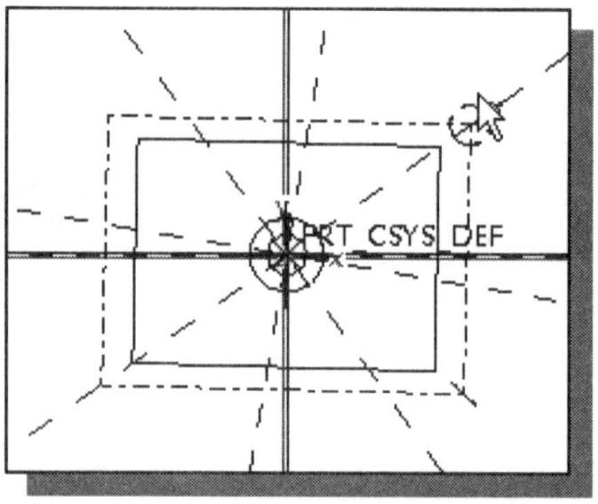

6. Drag with the left-mouse-button on the **Rotate** handle to rotate the 2D section.

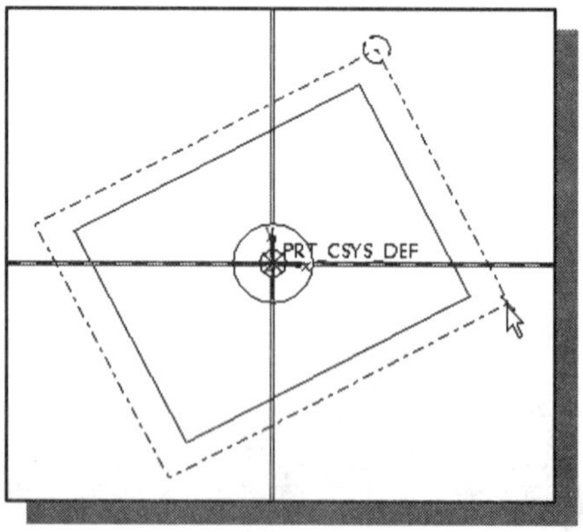

7. On your own, use the **Scale** handle to adjust the scale of the 2D section. Notice both the rotated angle and new scale factor are also displayed in the *Scale Rotate* window to the right of the main window as we use the *adjusting handles* with the mouse.

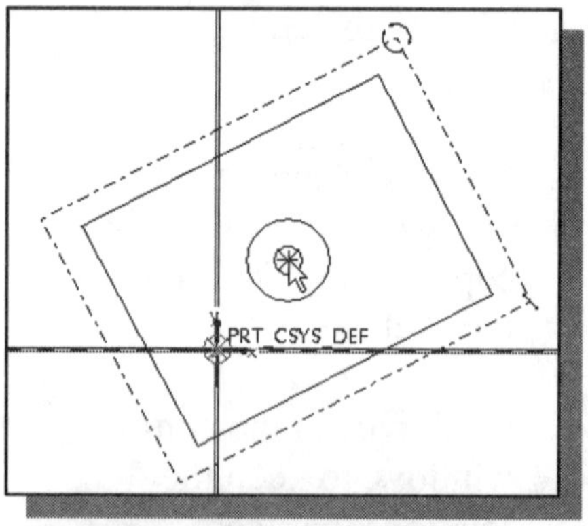

8. The **Locate** handle, the circle mark at the center of the 2D section, can be used to move the 2D section. Note that the cursor will automatically snap to the references.

9. On your own, experiment with the three handles, but place the 2D sketch back at the origin before continuing to the next step.

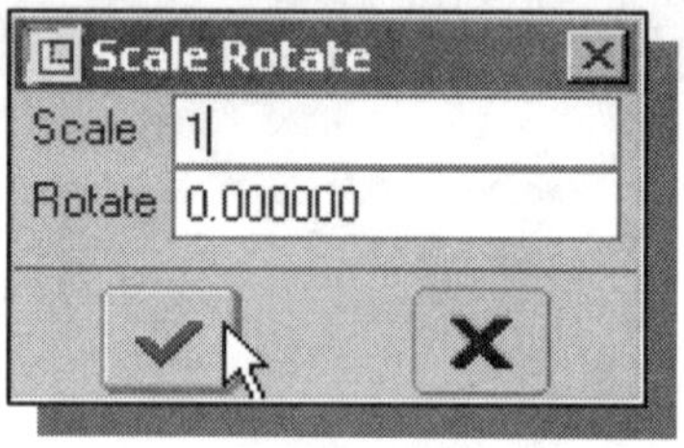

10. In the *Scale Rotate* window, enter **1** as the new scale factor and set the rotating angle to **0.00** as shown.

11. Click on the **Accept** button to proceed with the insertion of the 2D section.

12. Use the **Refit** command to have *Pro/ENGINEER* adjust the zoom scale so that all objects are displayed on screen.

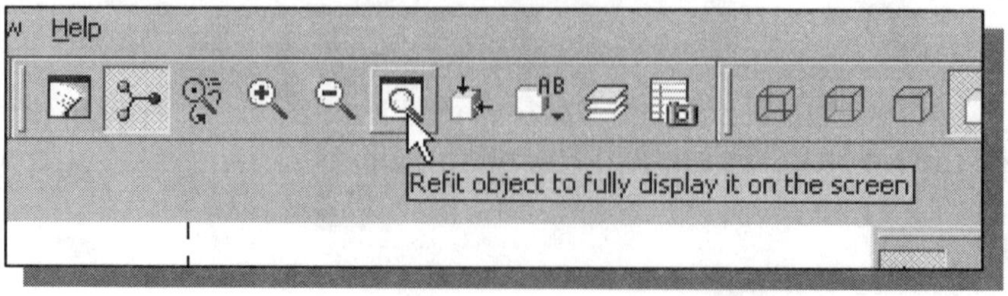

➢ We have successfully transferred in the 2D sketch done previously. We remain in the *Pro/ENGINEER Sketcher* as if we had just created the 2D sketch.

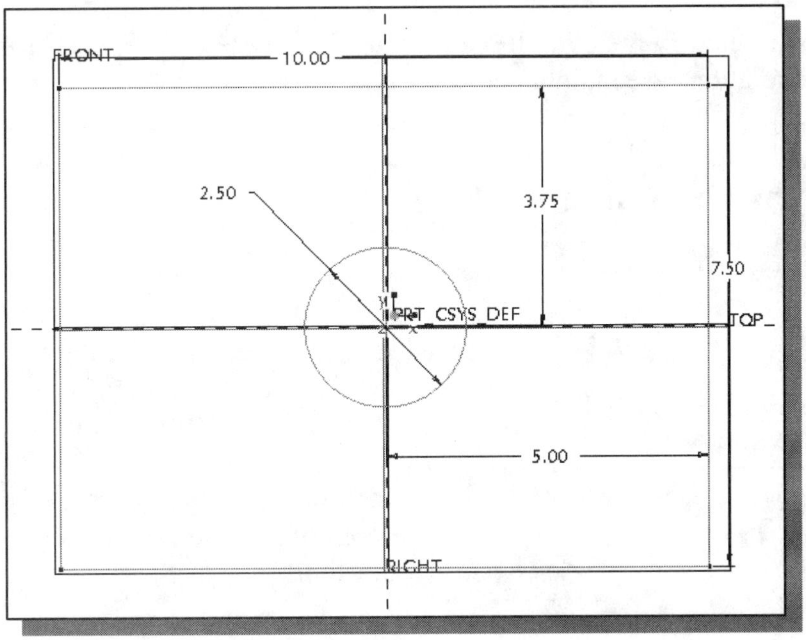

➢ On your own, try to adjust the hole diameter dimension (**2.5**). And notice the message area displays "*This dimension is governed by a relation and can not be modified.*" The set of parametric relations we have established in the sketch file is also transferred into the current part file.

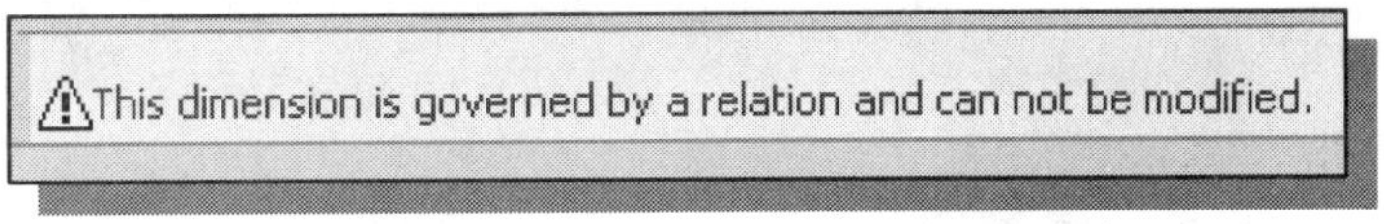

Display the Parametric Relations

1. Pick **Relations** in the **Tools** pull-down menu.

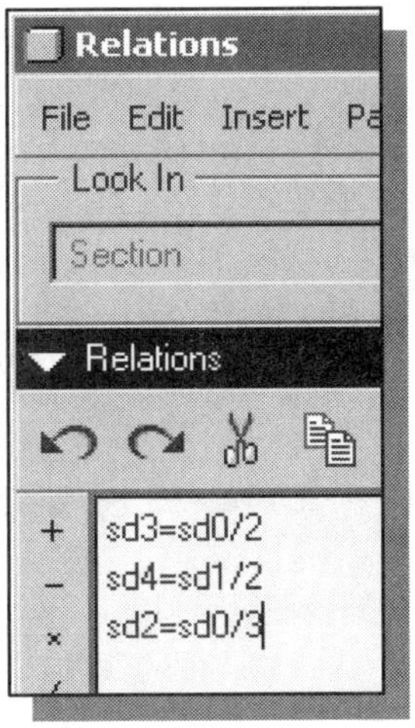

➢ The *Relations* window appears and the parametric equations defined are displayed. Note that the parametric relations are imported from the saved sketch file.

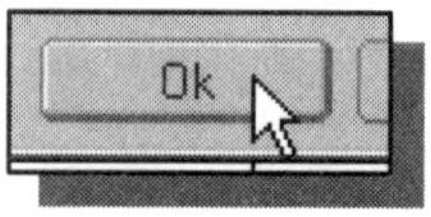

2. Pick **OK** to exit the *Relations* window and return to the ***Sketcher***.

❖ On your own, modify the overall width of the rectangle to **8** and confirm the parametric relations are maintained. Reset the value back to **10** before continuing to the next section.

Complete the Extrusion Feature

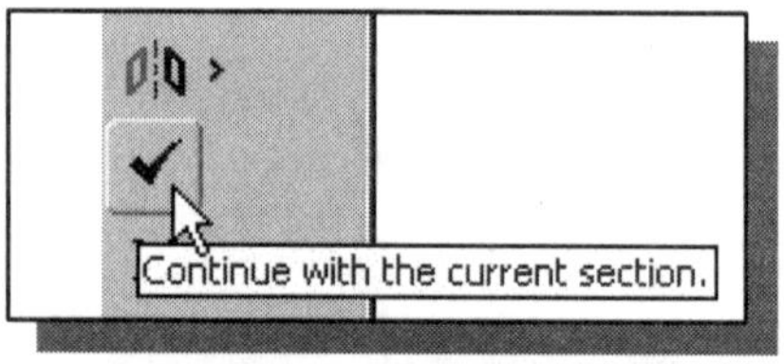

1. Now that the 2D sketch is completed, we will proceed to the next element. In the *Sketcher* toolbar, click on the **Accept** button to end the *Pro/ENGINEER* Sketcher command.

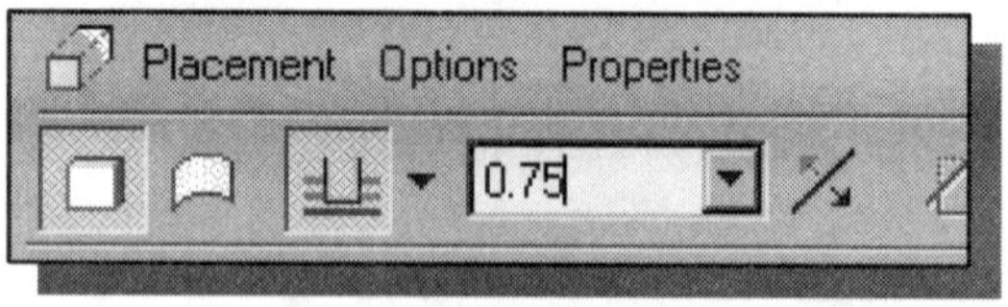

2. In the feature option area, enter **0.75** as the depth of extrusion and click **OK** to create the solid.

3. On your own, use the *Dynamic Rotate* function [middle-mouse-button] to view the 3D model.

Parametric Relations at different levels

In *Pro/ENGINEER*, **parametric relations** can also be defined within a part or among parts of an assembly. So far, we have demonstrated the use of parametric **section relations**, which are applied to the 2D section within a solid feature. Similar procedures can also be applied to the feature, part, and assembly levels to establish specific parametric relations.

In *Pro/ENGINEER*, five types of **parametric relations** are available:

- **Assembly Relations** relate different part parameters to each other in an assembly model.
- **Part Relations** relate different feature parameters to each other in a single model. Part relations can include feature relations which are within one feature in the model.
- **Feature Relations** relate specific parameters within one feature in the model. A feature relation can be for the 2D section only or for the entire feature.
- **Section Relations** relate specific parameters within one 2D sketch in the model.
- **Pattern Relations** relate specific parameters within a pattern in the model.

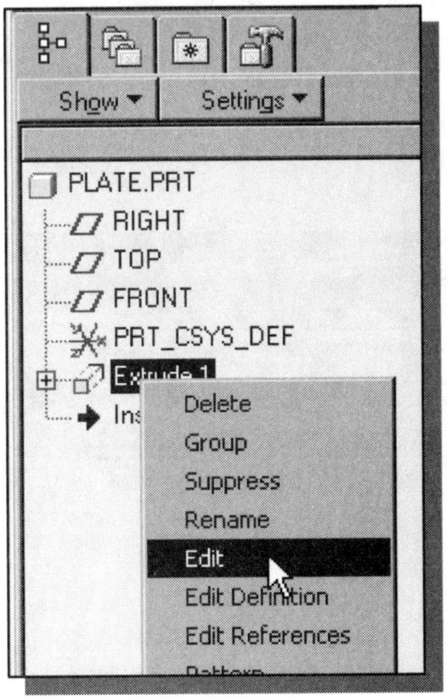

1. In the *Model Tree* window, move the cursor on top of **Extrude 1** and click once with the right-mouse-button to bring up the option menu.

2. In the option menu, select **Edit.**

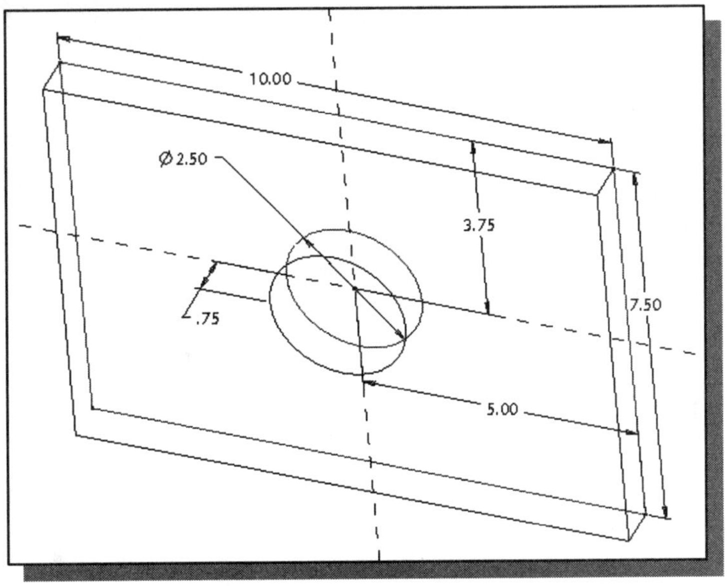

➢ Notice that all dimensions related to the feature are displayed, including the extrusion depth (0.75).

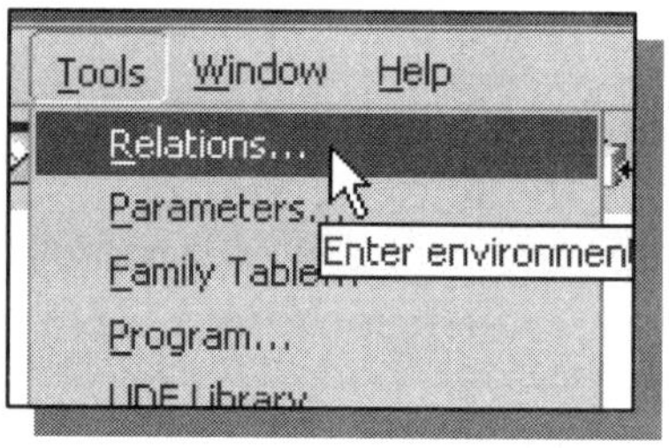

3. Pick **Relations** in the **Tools** pull-down menu.

❖ The **Relations** window appears and in the graphics area the *dimensional names* are displayed instead of the *dimensional values*. (The dimensional variables are in the **dxx** format, which represents the dimensions are in *Part level* or *Assembly level.*)

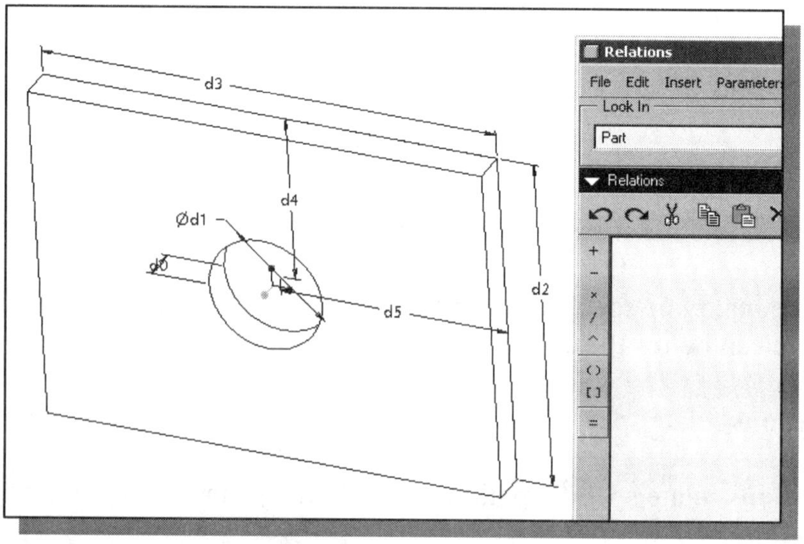

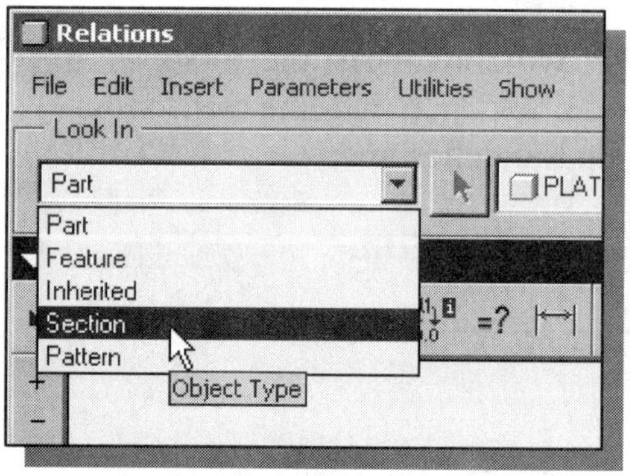

4. Choose **Section** in the **Look In** option of the *Relations* window as shown.

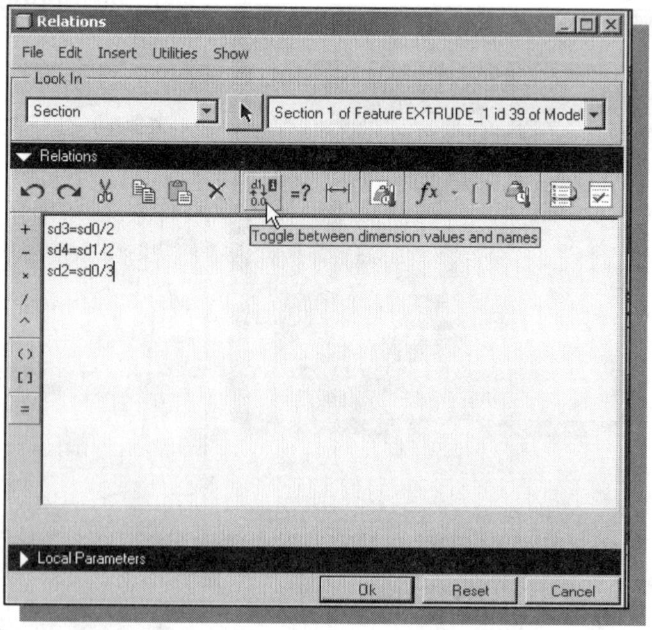

5. Choose any edge of the *Plate* model. (There is only one *section* in the model.) Note the parametric relations we set up in the sketch file are displayed.

6. On your own, use the **Switch Dim** option to examine the existing ***section relations***.

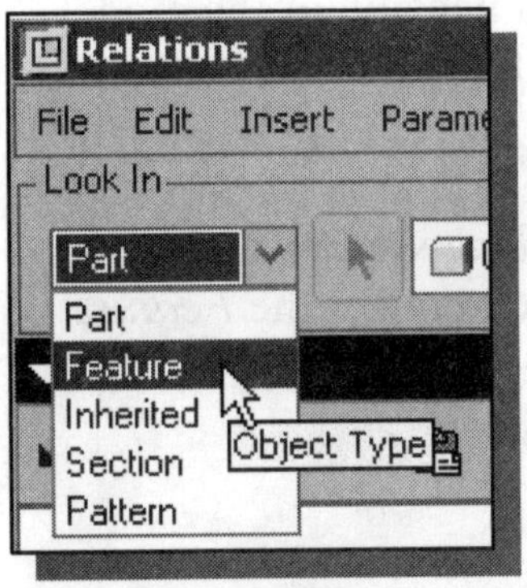

7. Choose **Feature** in the **Look In** option of the *Relations* window as shown.

8. Choose any edge of the ***Plate*** model. Note that there is only one solid feature in the model, but the datum planes are also considered as features. Also note the selected feature information is displayed in the *Relations* window.

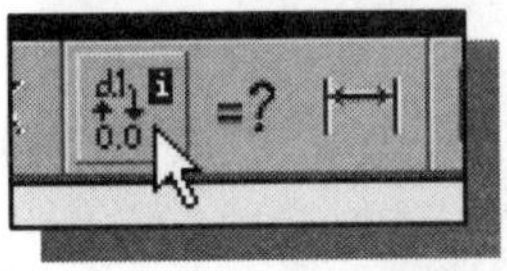

9. Use the **Switch Dim** option to toggle the display of dimensional values and names.

10. In the message area, *Pro/ENGINEER* expects us to input mathematical equations. Enter **d0=d2/10**, where **d0** is the depth of extrusion and **d2** is the height dimension of the ***Plate***. The variable names on your screen might be different than what is displayed here. Use the corresponding variable names to establish the parametric relation.

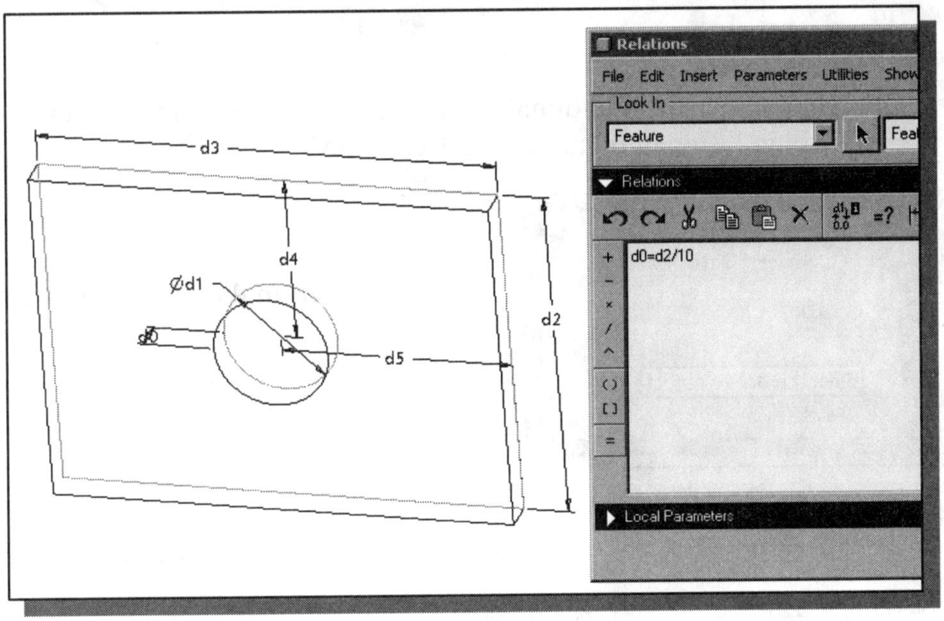

11. Click the **OK** icon to accept the entered equation and close the *Relations* window.

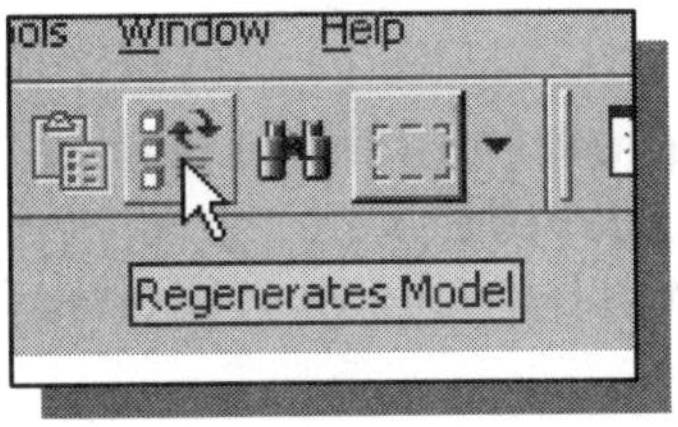

12. On your own, observe the changes of the model by adjusting the height of the ***Plate*** and **Regenerate** the model.

❖ The thickness of the part is updated based on the height dimension of the model. The parametric relations are maintained by the system so that an intelligent part is created.

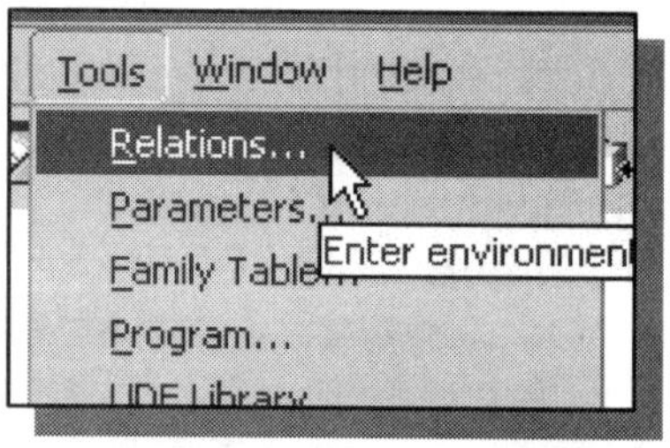

13. On your own, open the *Relations* window and delete the equation (**d0=d2/10**) we entered at the *Feature level*.

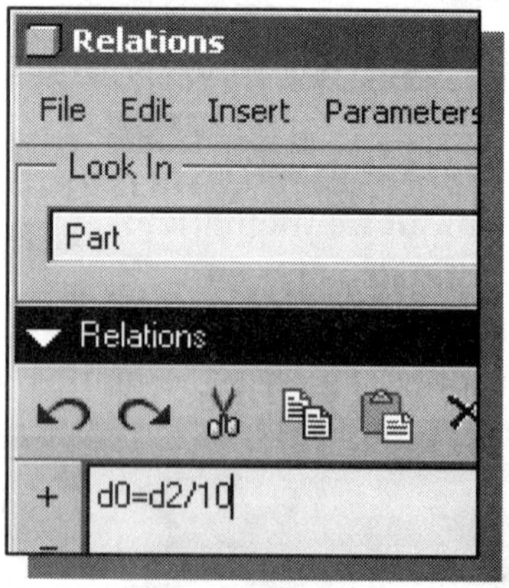

14. Place the same equation (**d0=d2/10**) at the ***Part level*** as shown.

15. On your own, **Edit** the height dimension of the model and observe the relations established within the part.

16. On your own, create additional solid features and experiment with establishing different relations at different levels.

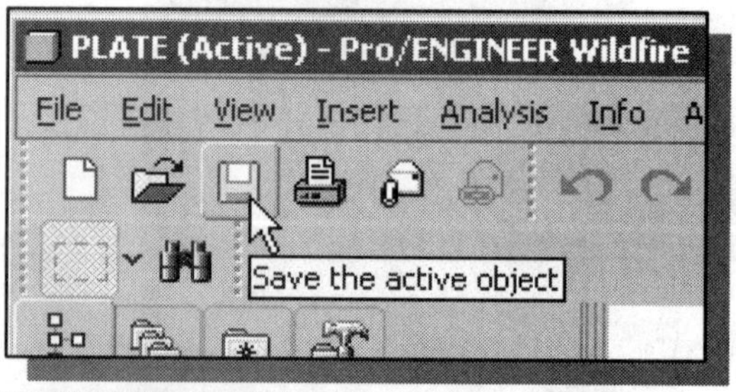

17. Select **Save** in the *Standard* toolbar.

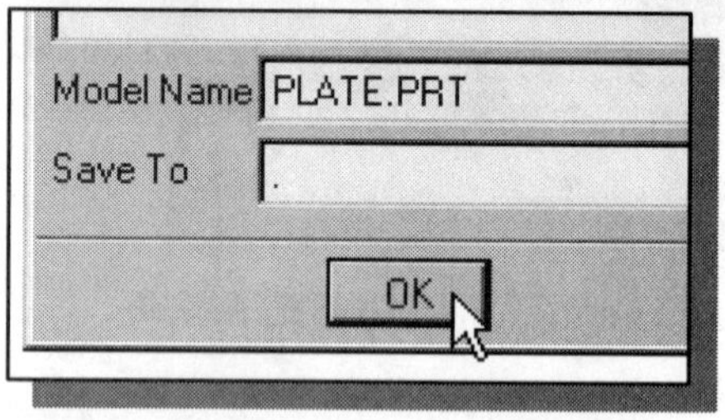

18. Click **OK** to proceed with saving the model.

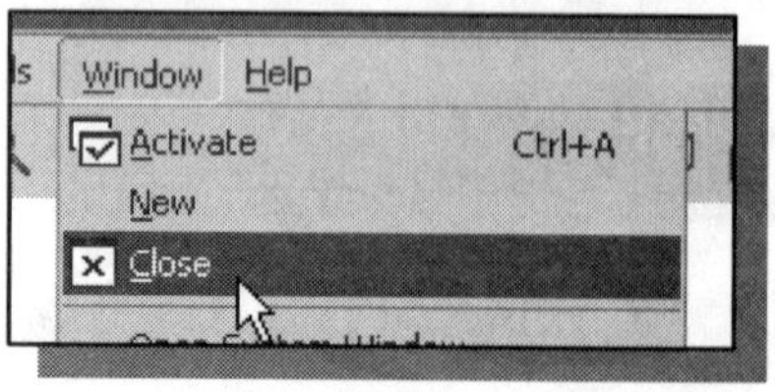

19. Choose **Close** in the **Window** pull-down menu to end and close the *Plate* model file. Note the different options to switch in between the different open windows.

Questions:

1. What is the difference between *dimensional values* and *dimensional variables*?

2. How do we **disable** *geometric constraints* while sketching?

3. How are *geometric constraints* applied in *Pro/ENGINEER Sketcher*?

4. Describe the different types of **parametric relations** available in *Pro/ENGINEER*.

5. How do we create parametric relations in *Pro/ENGINEER*?

6. How do we assure two points are ***lined up horizontally***?

7. Create binary tree sketches showing the steps you plan to use to create the two models shown on the next page:

Ex.1)

Ex.2)

Exercises:

(Establish three parametric relations for each of the designs and experiment with the different modification techniques illustrated in this lesson.)

1. Dimensions are in inches.

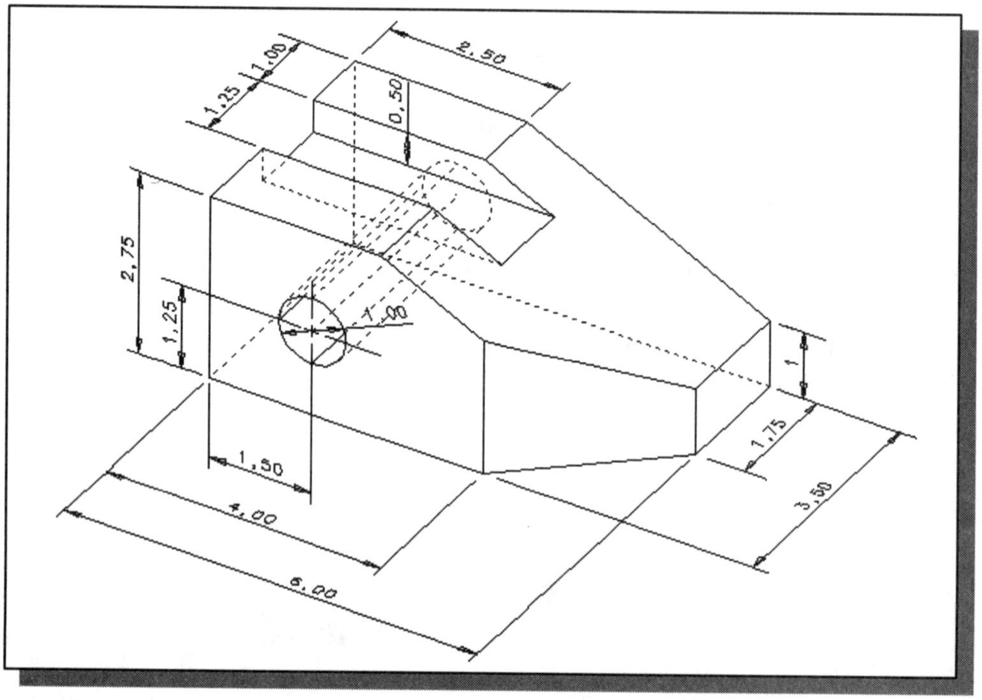

2. Dimensions are in inches.

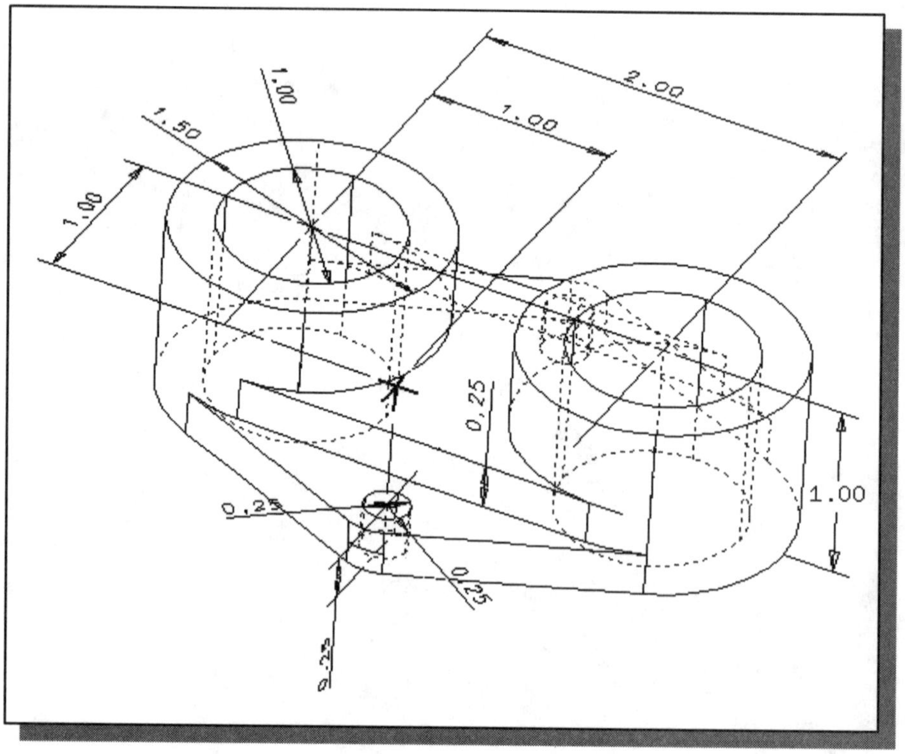

3. Dimensions are in inches.

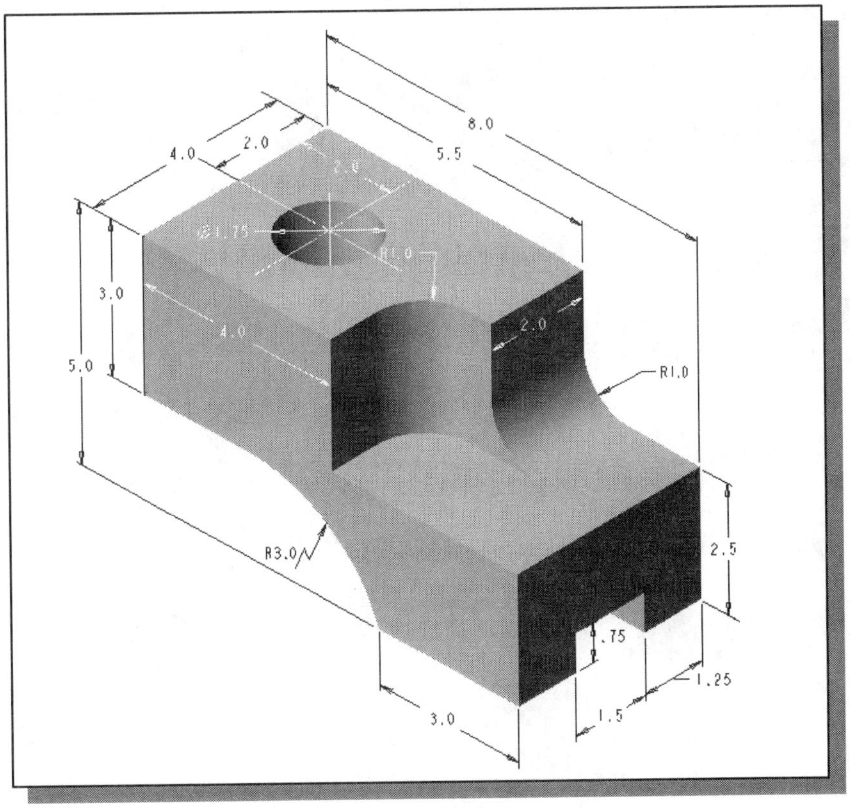

4. Dimensions are in millimeters. (Base thickness: 10 mm., Boss height: 20 mm.)

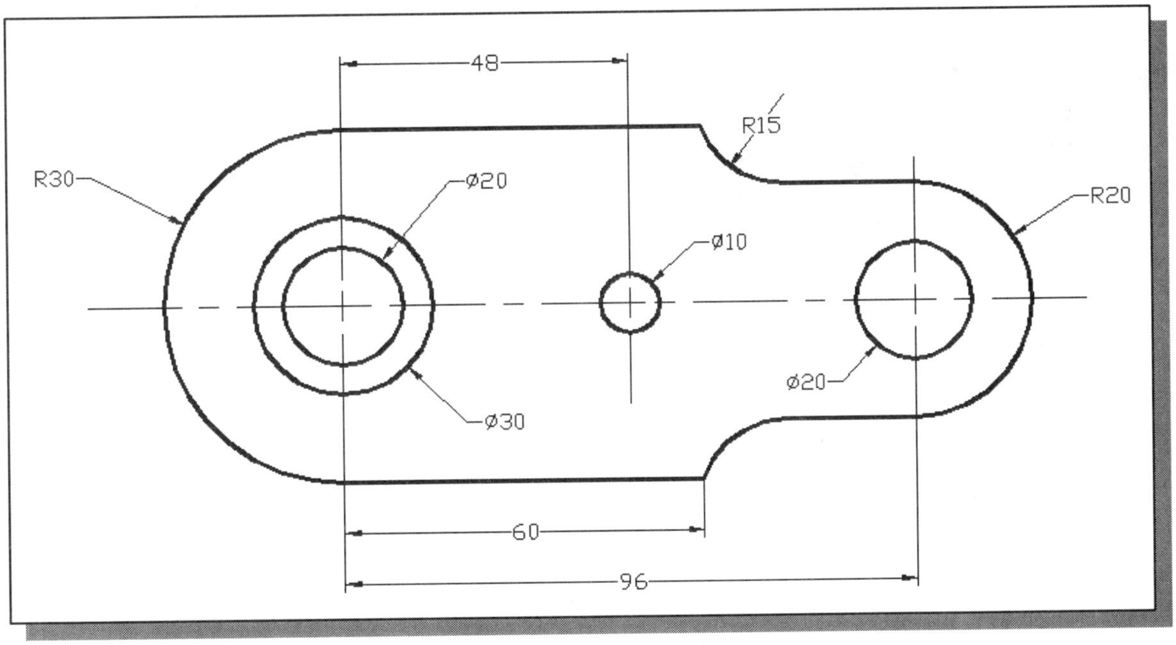

Notes:

Chapter 5
Parent/Child Relationships

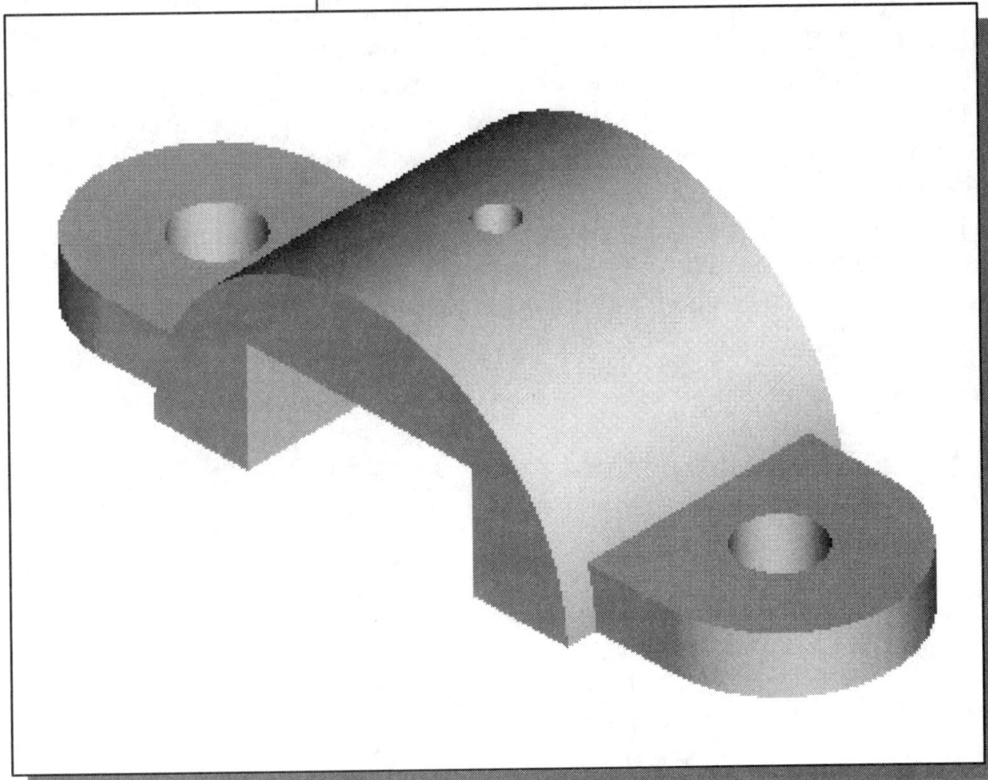

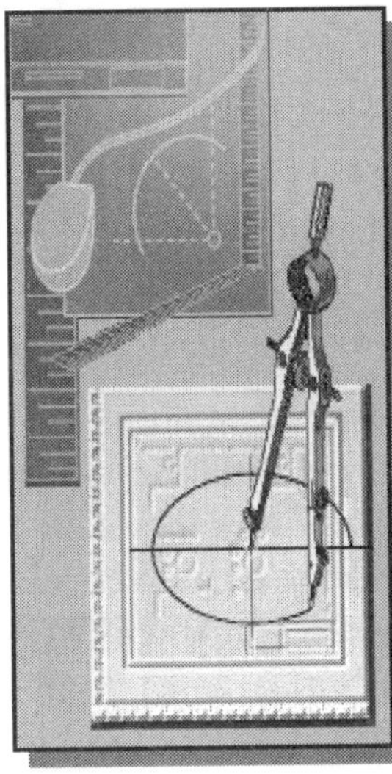

Learning Objectives

- **Understand the Parent/Child relations between Features.**
- **Create Swept Features.**
- **Set up Multiple Work planes.**
- **Create Lofted Features.**
- **Use the Shell Command.**
- **Create 3D Rounds and Fillets.**

Introduction

The parent/child relationship is one of the most powerful aspects of ***parametric modeling***. In *Pro/ENGINEER*, each time a new modeling event is created, previously defined features can be used to define information such as size, location, and orientation. The referenced features become **PARENT** features to the new feature, and the new feature is called the **CHILD** feature. The parent/child relationships determine how a model reacts when other features in the model change, thus capturing design intent. It is crucial to keep track of these parent/child relations. Any modification to a parent feature can change one or more of its children. As one might expect, parent/child relationships can become quite complicated when the features begin to accumulate. It is therefore important to think about modeling strategy before we start to create anything. The main consideration is to try to plan ahead for all possible design changes that might occur which would be affected by the existing parent/child relationships.

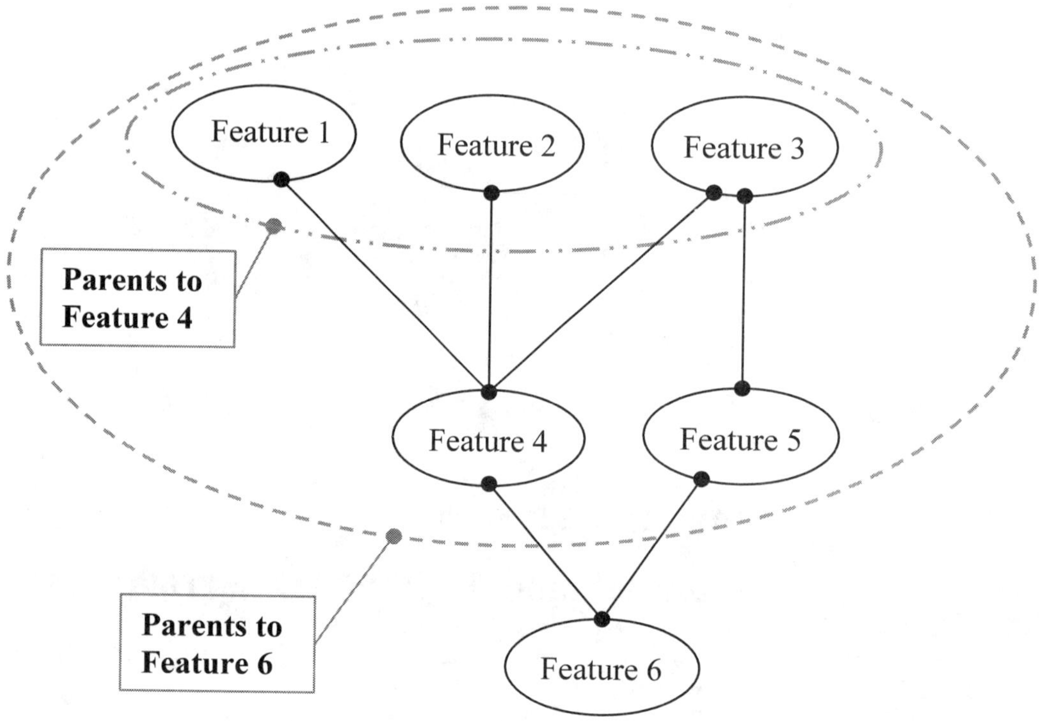

Parent/child relationships can be created *implicitly* or *explicitly*; implicit relationships are implied by the feature creation method and explicit relationships are entered manually by the user. We have been using implicit parent/child relationships in the previous lessons. For example, before creating a 2D section in the *Pro/ENGINEER Sketcher* mode, we select a sketching plane and a reference plane. Both of these become parents of the new feature. If the sketching plane is moved, the child feature will move with it. The reference plane determines the orientation of the section on the sketching plane. If the orientation is changed, the orientation of the child feature is changed as well. The use of *explicit parent/child relationships* will be demonstrated in the following lessons.

The *U-Bracket* Design

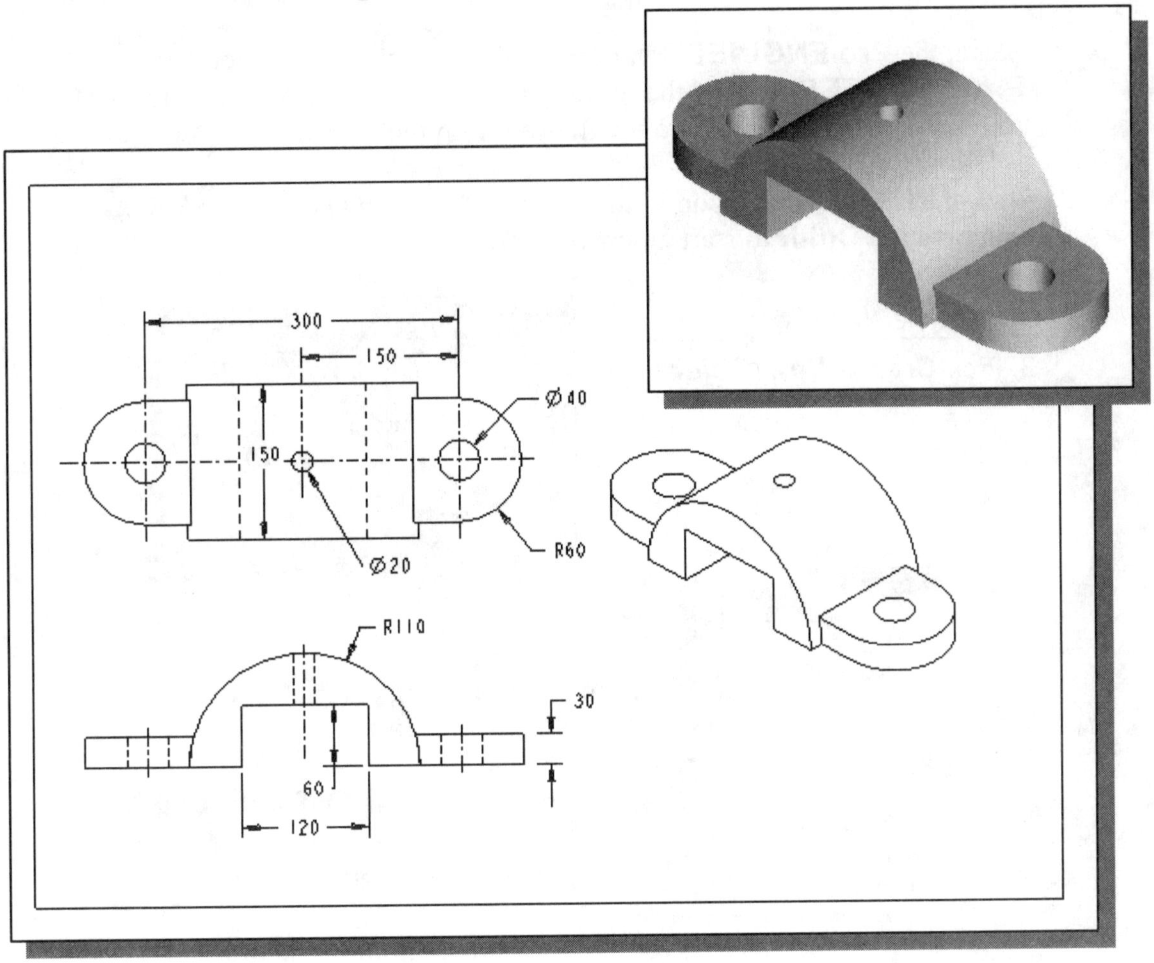

Preliminary Modeling Considerations

Prior to creating the model, a few minutes of planning and sketching on paper will save many hours of redefining the design on the computer. Two of the main modeling considerations are:

- **Design Intent** – Determine the functionality of the design; identify features that are central to the design.

- **Order of Features** – Consider the parent/child relationships necessary for all features.

➢ Based on your knowledge of *Pro/ENGINEER* so far, how many features would you use to create the model? Which feature would you choose as the **base feature**? What is your choice for arranging the order of the features? Would you organize the features differently if the rectangular cut at the center is changed to a circular shape (Radius: 80mm)? Take a few minutes to consider these questions and do preliminary planning by sketching on a piece of paper.

Starting *Pro/ENGINEER* and using the Metric Template

1. Select the **Pro/ENGINEER** option on the *Start* menu or select the **Pro/ENGINEER** icon on the desktop to start *Pro/ENGINEER*. The *Pro/ENGINEER* main window will appear on the screen.

2. Pick the **New Object** icon in the toolbar. (You can also use the key combination **Ctrl-N** to start a new object.)

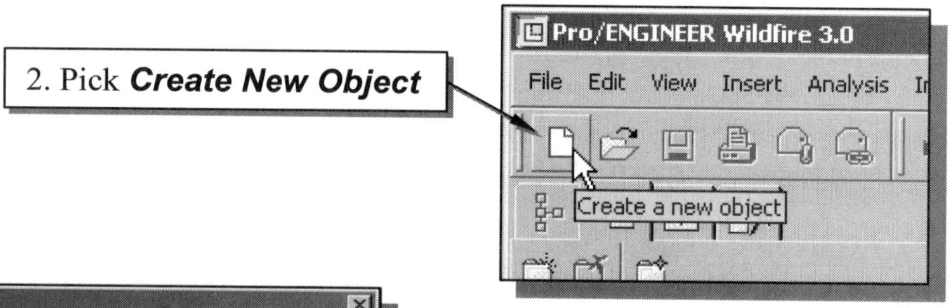

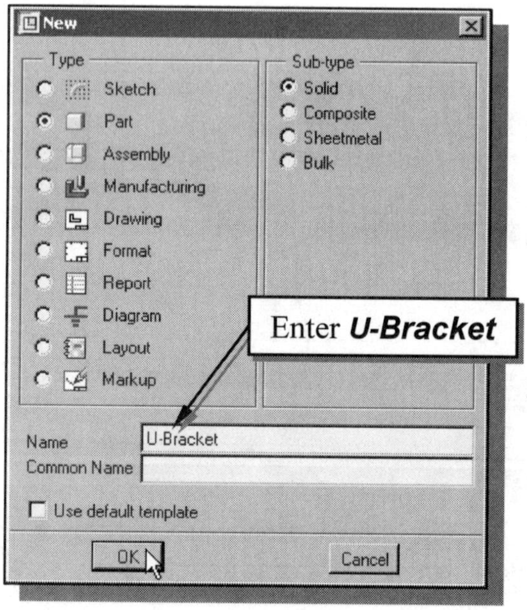

3. In the *New* form, enter ***U-Bracket*** as the solid part file Name.

4. Switch ***off*** the **Use Default Template** option as shown. (We will select the metric template for the *Gasket* design.)

5. Pick **OK** to continue. The *Model Tree* window and the Menu Manager window appear on the screen.

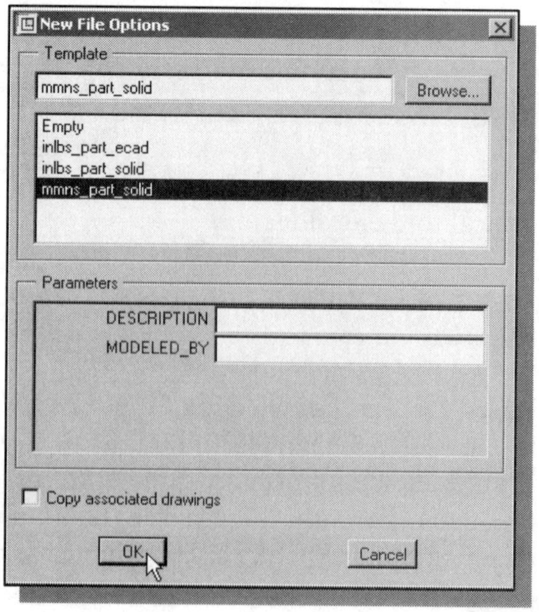

6. In the *New File Options* window, select **mmns_part_solid** as the template to use. This is the template for using with the metric system (**millimeter**).

❖ In Pro/E, a template file can contain a set predefined settings, such as the datum planes, datum axes and units setup. The template files allow us to reduce startup repetitions.

Creating the Base Feature

❖ We will create the model using four features and the base feature will be the bottom section of the model. We will use Pro/E as a sketching tool for quickly changing the sketched geometric entities. Note that this is a common practice during the initial conceptual design stage.

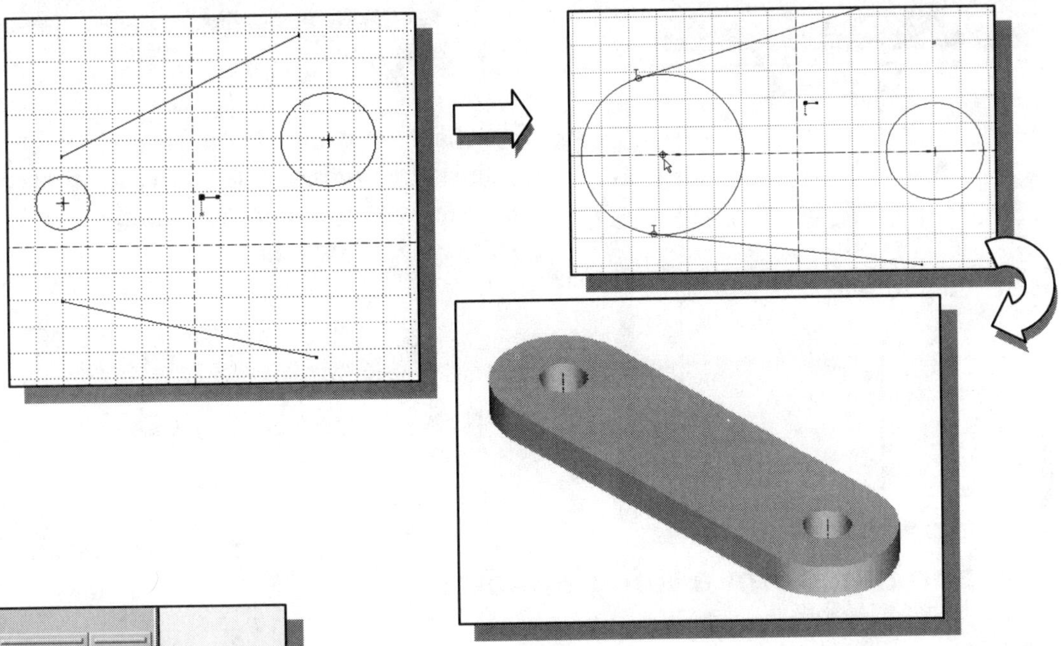

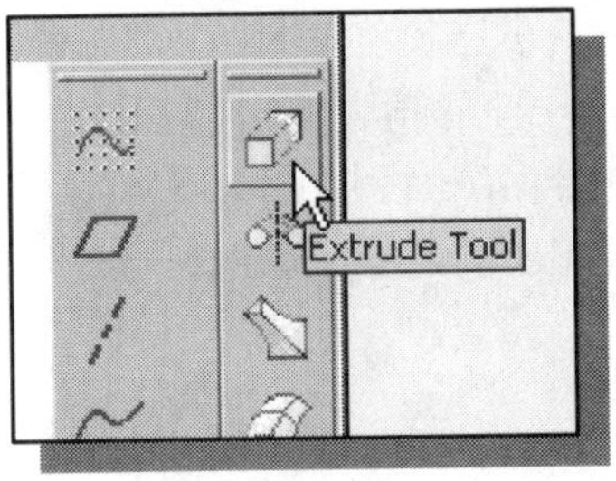

1. In the *Feature* toolbars (toolbars aligned to the right edge of the main window), select the **Extrude Tool** option as shown.

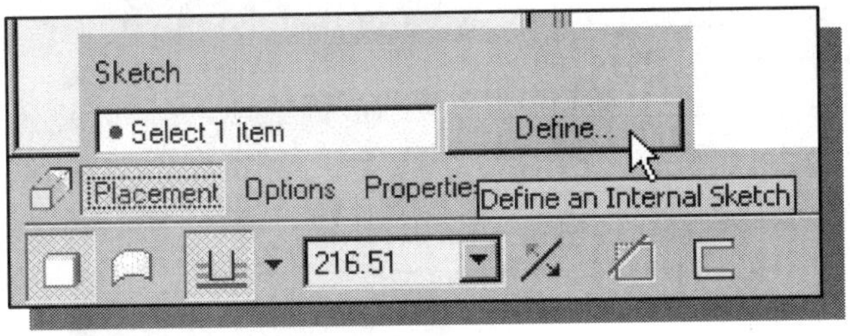

2. Click the **Placement** option and choose **Define** to begin creating a new *Internal sketch.*

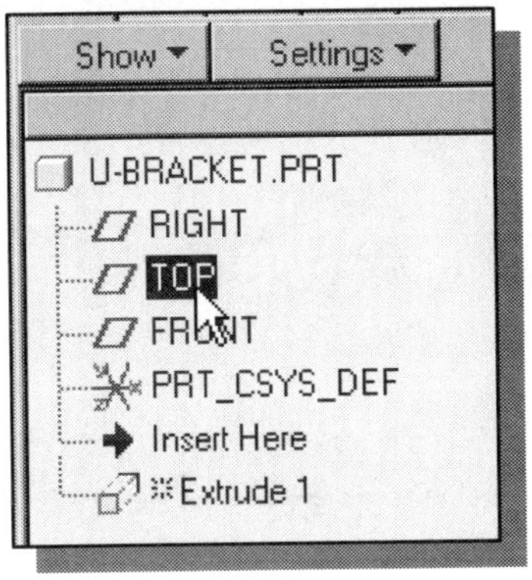

3. Select **TOP** by clicking on the **Datum Plane** name in the model tree window as shown.

❖ The feature names, listed in the Model Tree window, can be selected just like selecting the features in the graphics window.

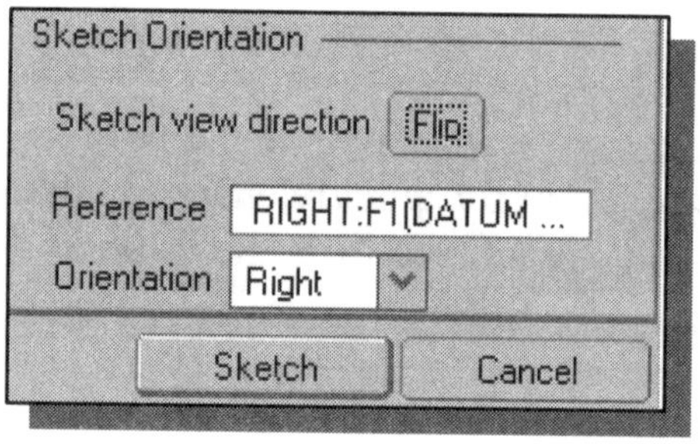

4. Confirm the **Right** datum plane is facing the **right** edge of the computer screen and pick **Sketch** to enter the *Sketcher* mode.

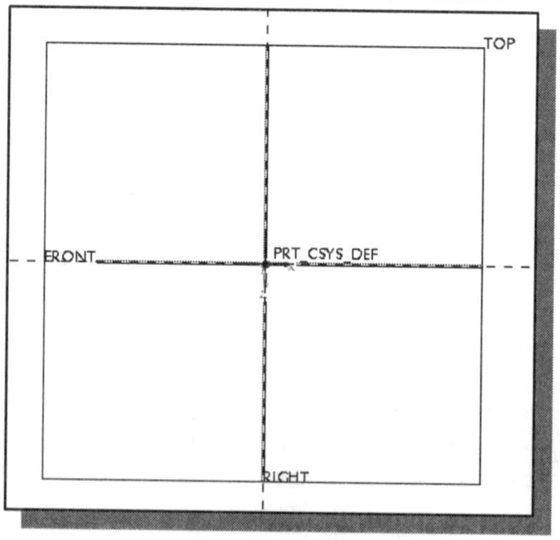

➢ Note that Front and Right are pre-selected as the sketching references. In the graphics area, the two references are highlighted and displayed with two dashed lines.

Creating and transform a Rough Sketch

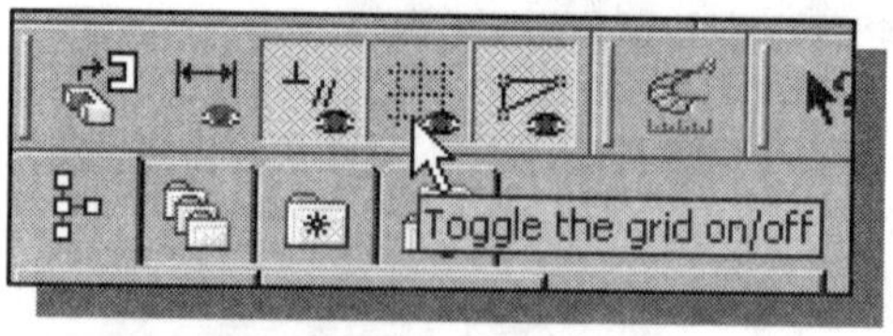

1. Click on the **Grid** icon to switch ***on*** the grid display and turn **off** the **Dimension display** as shown.

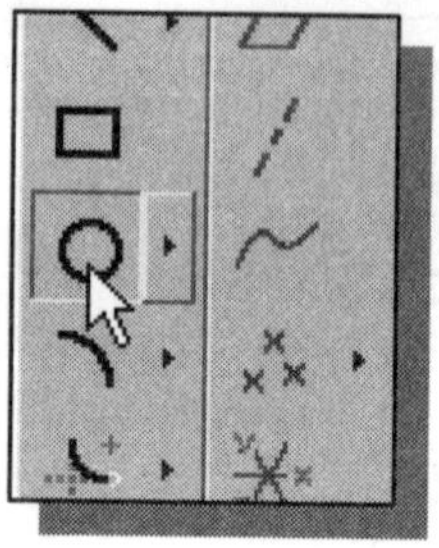

2. In the *Sketcher* toolbar, select **Circle** as shown. The default option is to create a circle by specifying the center point and a point through which the circle will pass.

3. On your own, create two circles of different sizes, with the centers intentionally not aligned to the reference datum planes as shown.

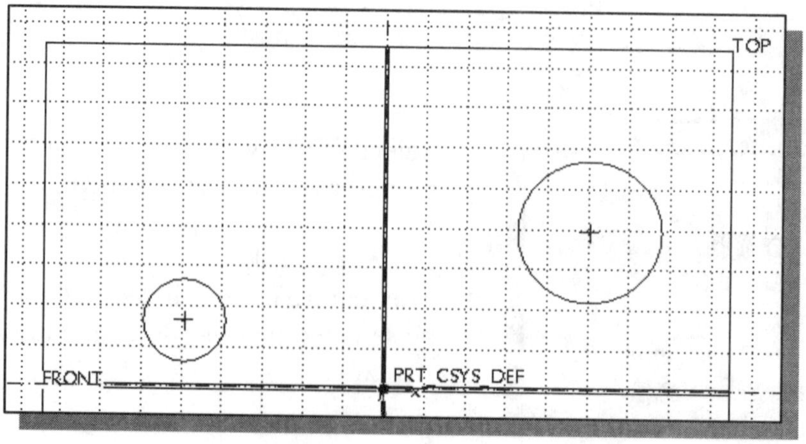

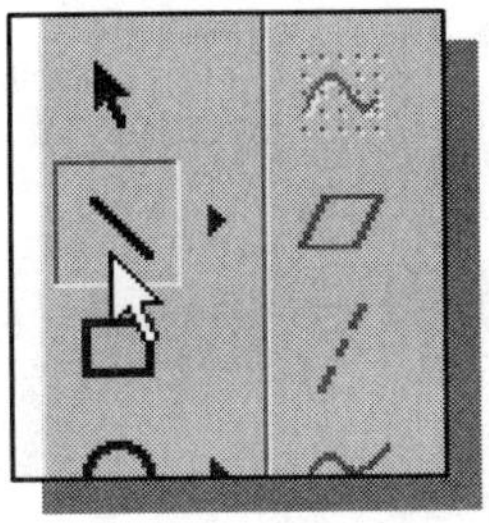

4. Choose the **Line** command in the *Sketcher* toolbar.

5. Create two lines of arbitrary length as shown in the below figure. Again, we will intentionally not align the lines to the reference lines or the circles.

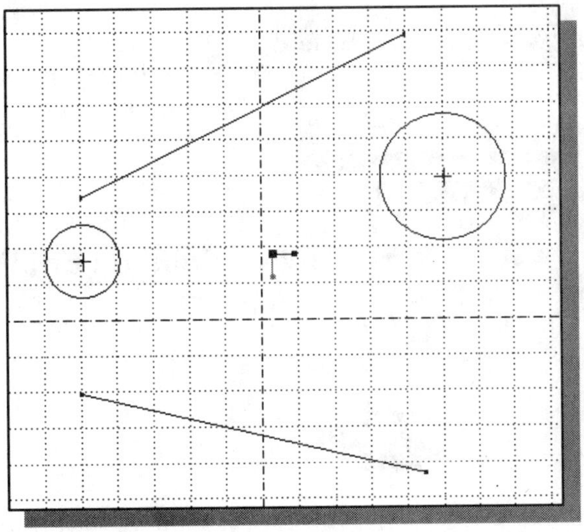

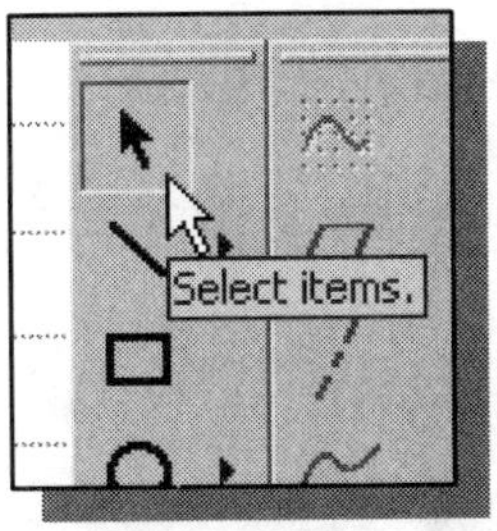

6. In the *Sketcher* toolbar, click on the **Select Items** icon.

7. On your own, move the created objects around by using the **drag and drop method**, with the **left-mouse-button**.

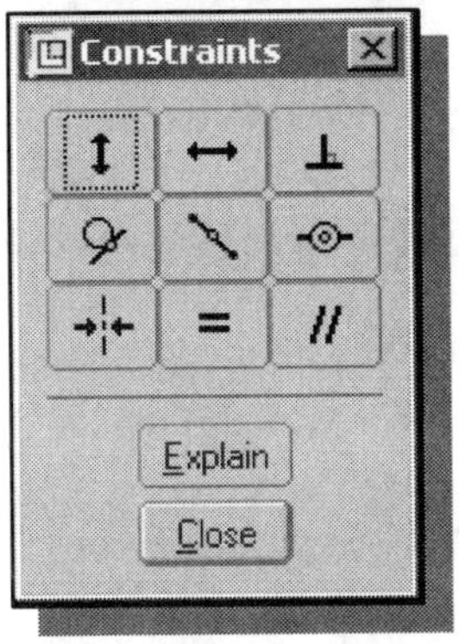

8. Pick **Constraints** in the *Sketcher* toolbar.

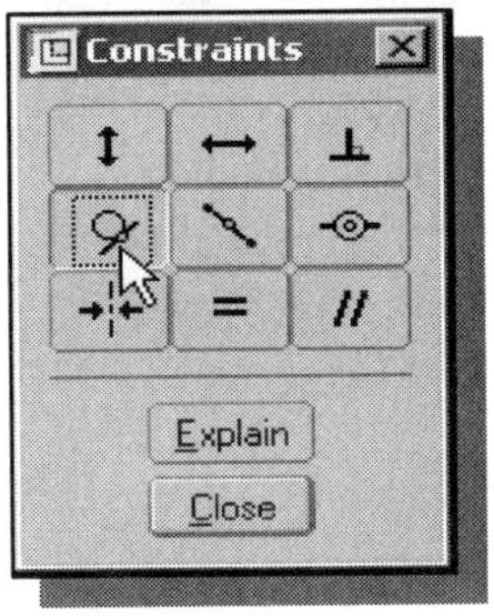

9. Pick **Tangent** in the **Constraints** submenu.

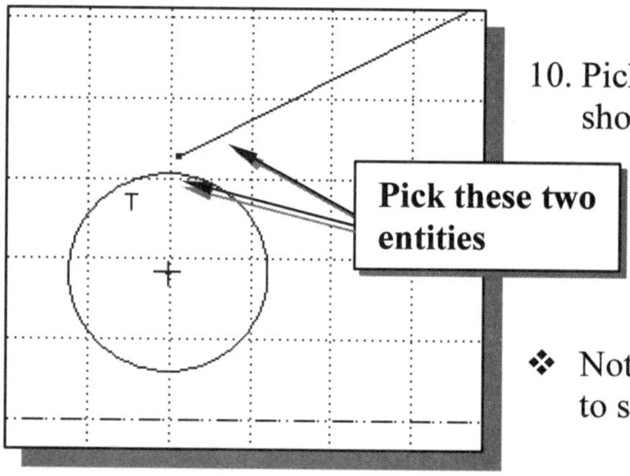

10. Pick the circle on the left and the line on top as shown.

❖ Note that *Pro/ENGINEER* adjusts the two entities to satisfy the applied tangent constraint.

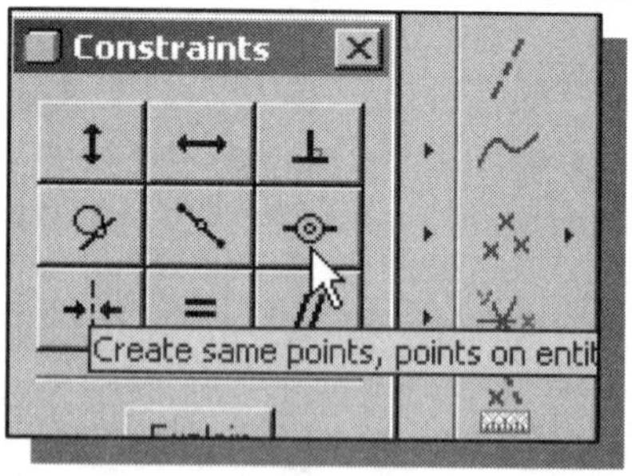

11. Pick **Point on Entities** in the **Constraints** submenu.

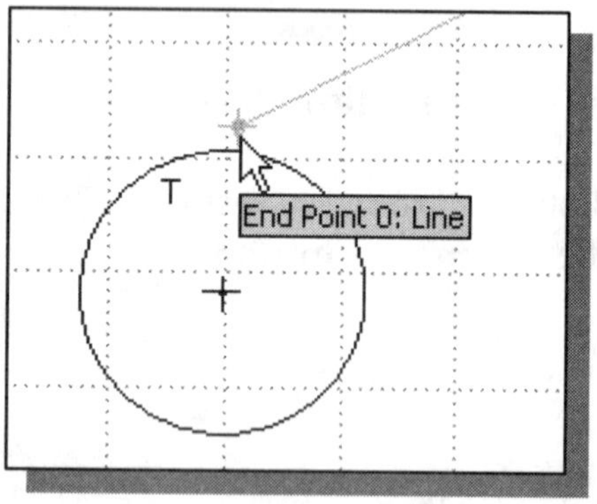

12. Pick the endpoint of the top line as shown.

13. Pick the circle on the left to assure the coincident constraint is applied.

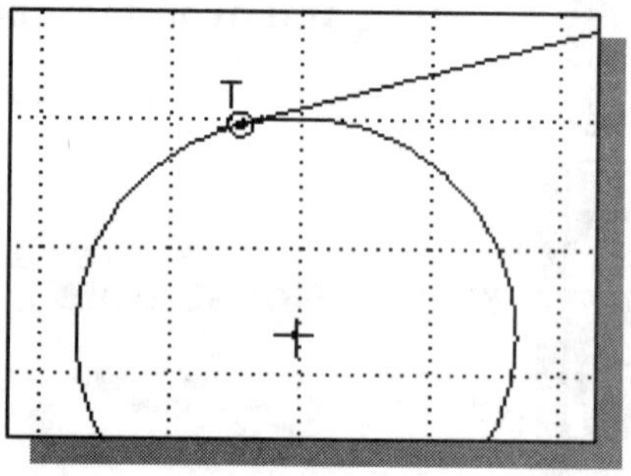

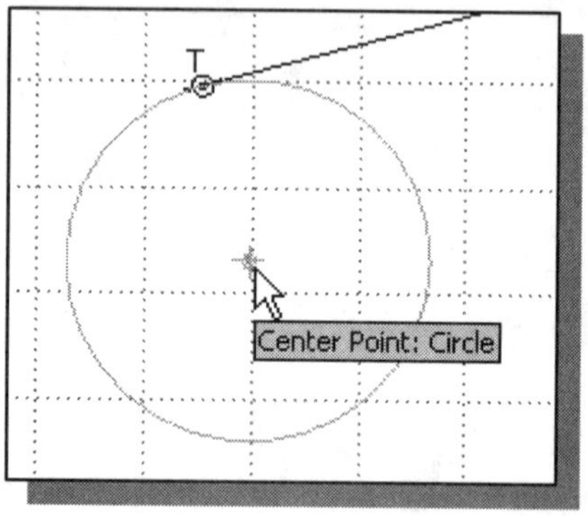

14. Pick the center point of the left circle as shown.

15. Pick the horizontal reference to align the center of the circle on the horizontal reference line.

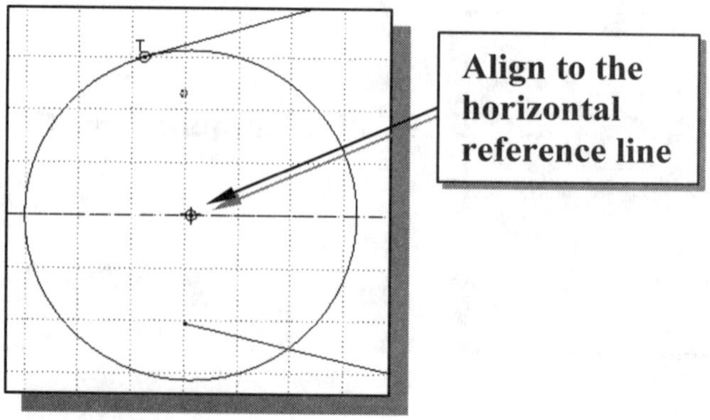

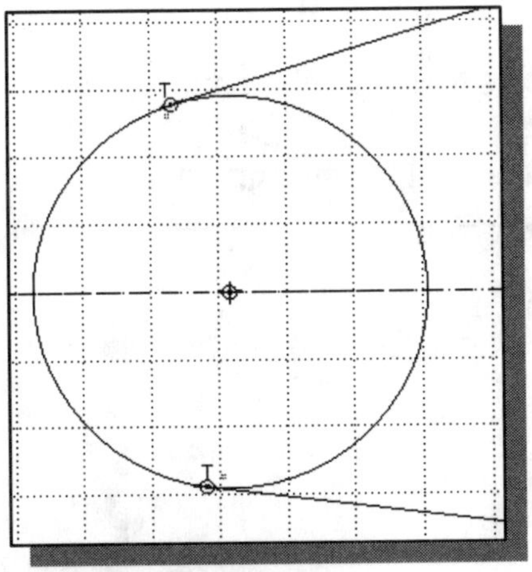

16. On your own, apply the proper constraints to align the lower line and the circle on the left as shown.

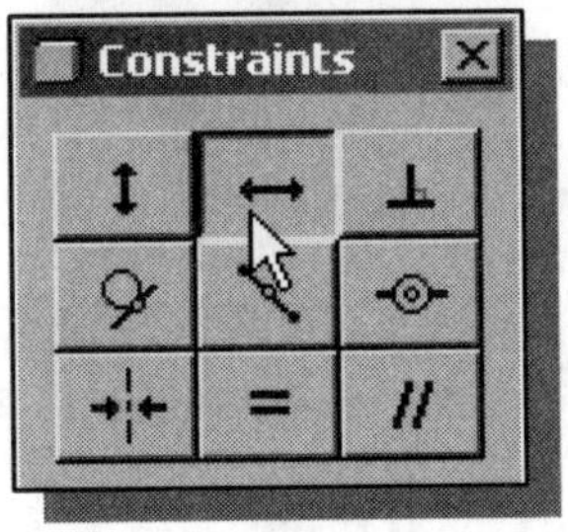

17. Pick **Align Horizontally** in the **Constraints** submenu.

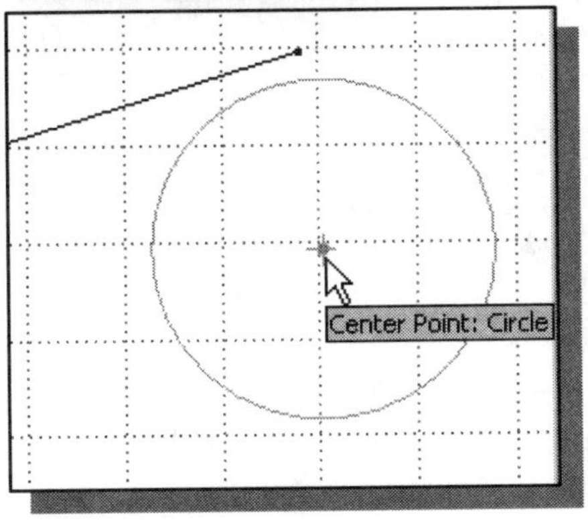

18. Pick the center point of the right circle as shown.

19. Pick the center point of the left circle to align the two center points horizontally.

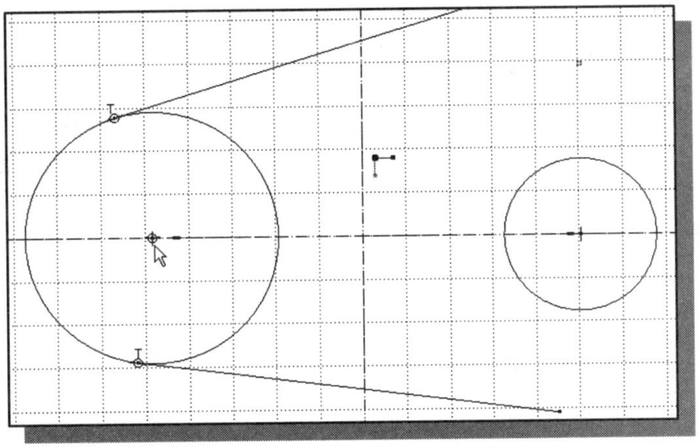

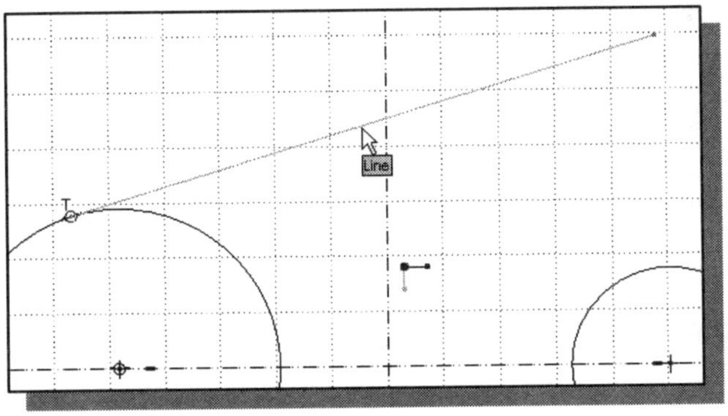

20. Pick the top line to make the line horizontal.

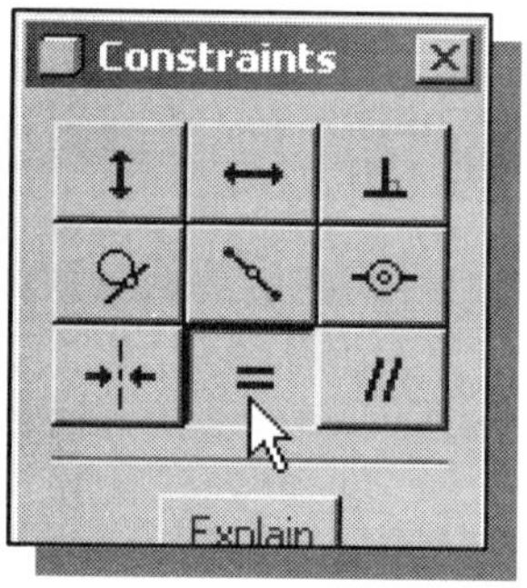

21. Pick **Equal Length** in the **Constraints** submenu.

22. Pick the two circles to apply the equal constraint.

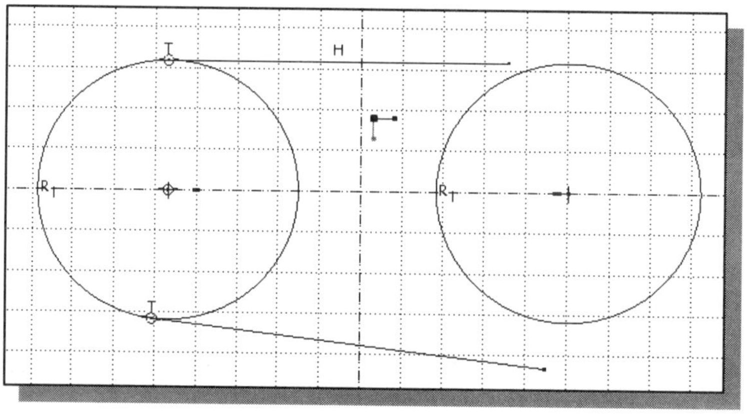

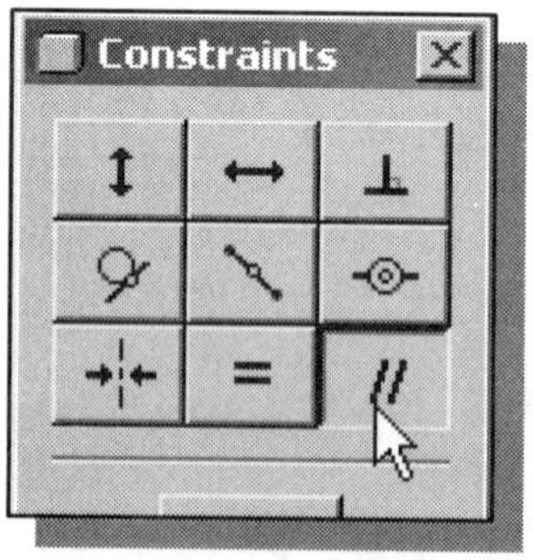

23. Pick **Parallel** in the **Constraints** submenu.

24. On your own, set the two lines parallel to each other.

25. Toggle **on** the **Dimension display** as shown.

❖ Examine the current sketch and compare it to the original sketch, it is clear that parametric modeling software can greatly enhance a designer's creativities.

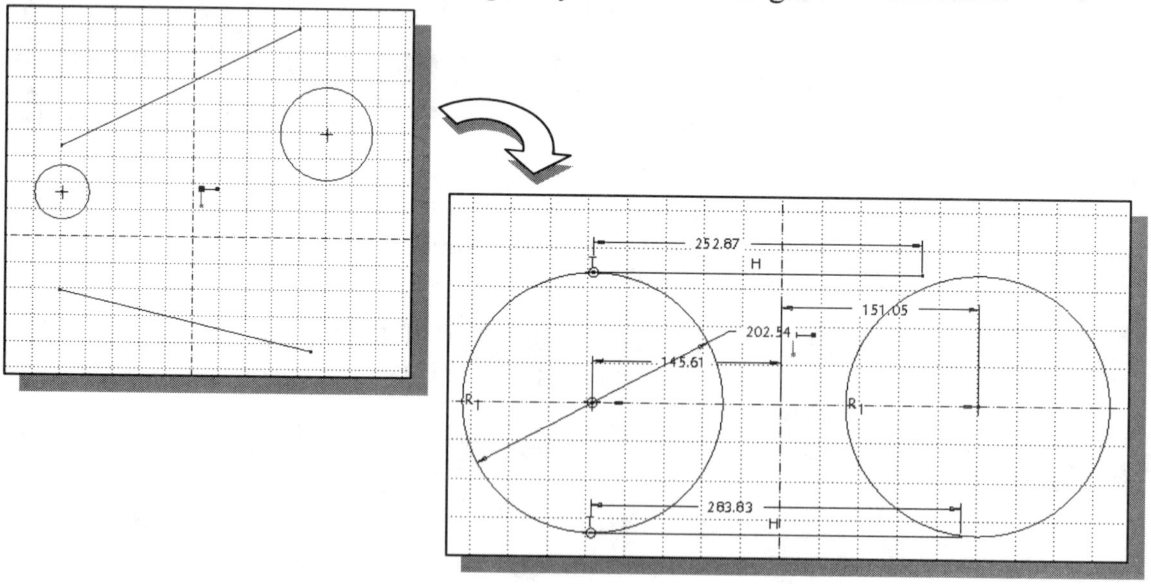

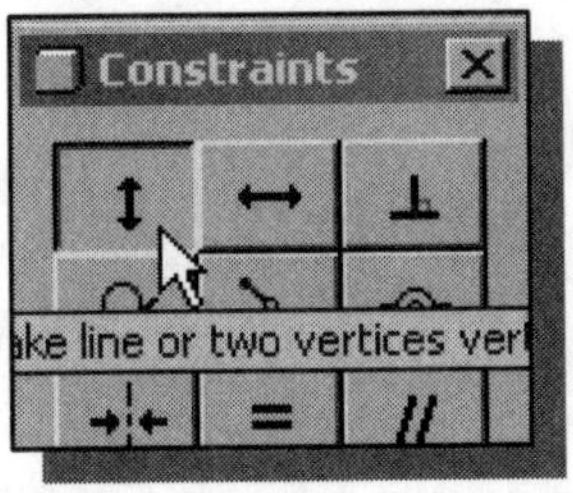

26. Pick **Align Vertically** in the **Constraints** submenu.

27. On your own, align the right endpoints of the two lines to the center point of the right circle as shown.

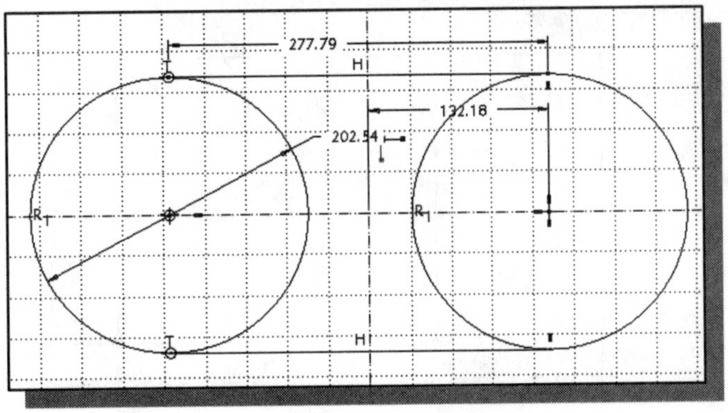

❖ Notice that some of the weak dimensions were removed as constraints are applied.

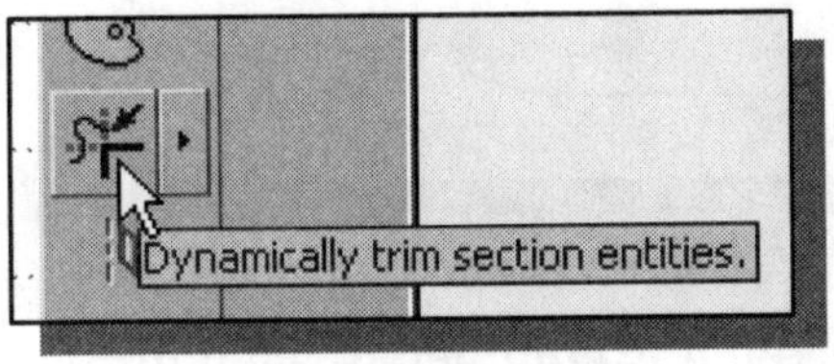

28. In the *Sketcher* toolbar, select **Dynamic Trim** as shown.

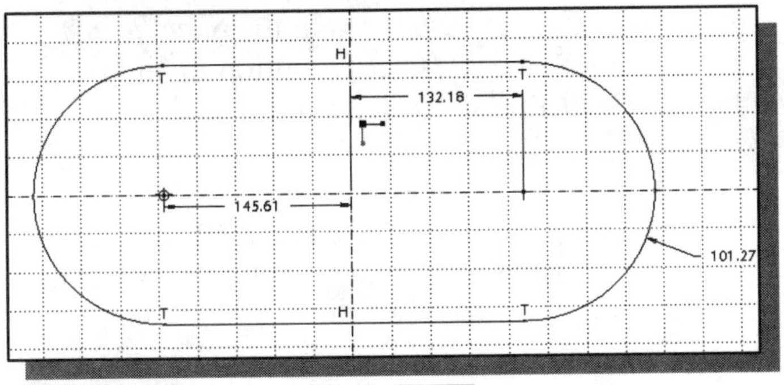

29. On your own, remove the inside halves of the two circles as shown.

❖ Note that some of the applied constraints have been removed and/or replaced, but the *design intent* is still maintained.

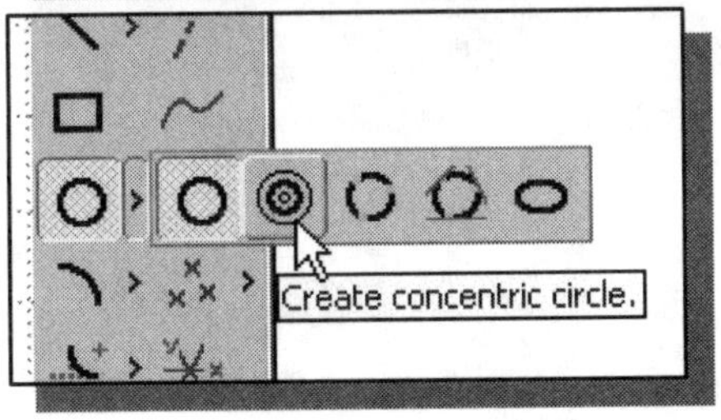

30. In the *Sketcher* toolbar, select the **Concentric Circle** option.

31. Create two circles that are of the same size and concentric to the two arcs as shown.

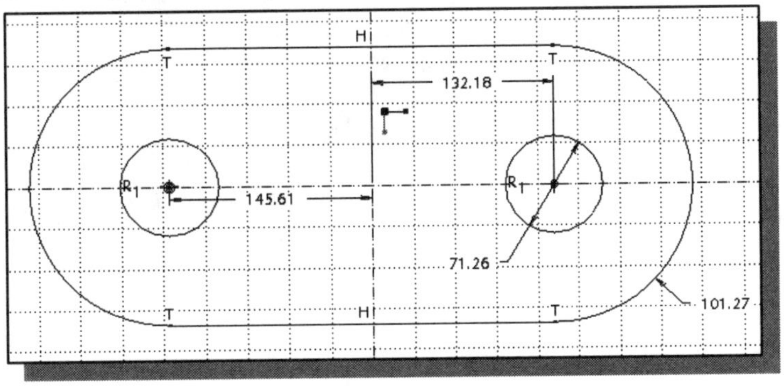

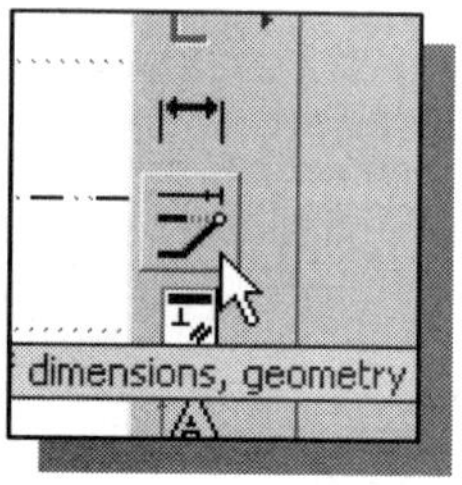

32. On your own, **modify** the dimensions as shown in the below figure.

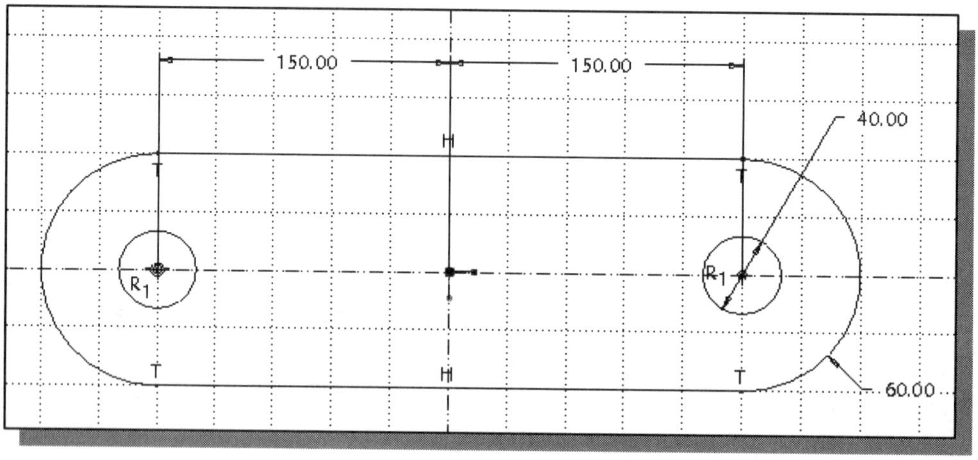

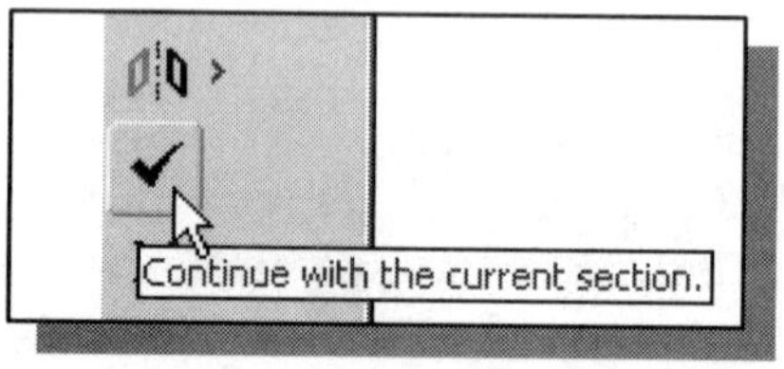

33. Now that the 2D sketch is completed, we will proceed to the next element. In the *Sketcher* toolbar, click on the **Accept** button to end the *Pro/ENGINEER* **Sketcher** command.

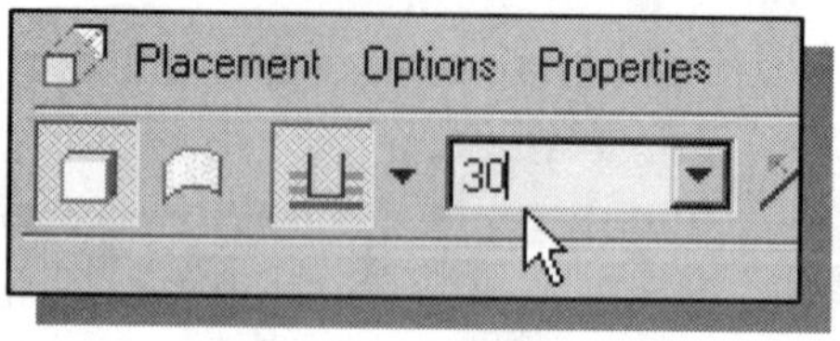

34. In the feature option area, enter **30** (mm) as the depth of extrusion and click **OK** to create the solid.

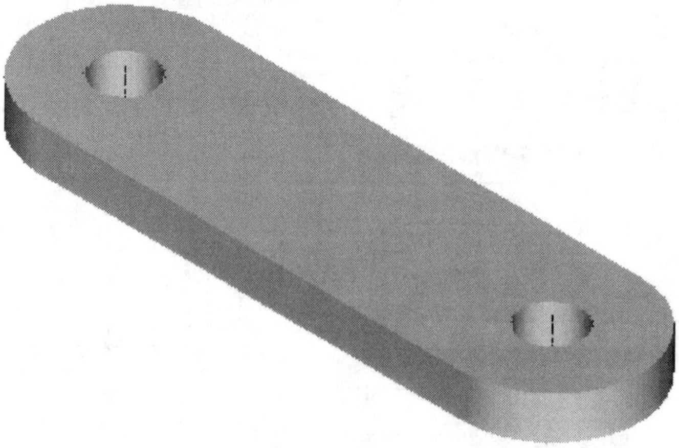

Creating the Second Solid Feature

- We will use the **Extrude Tool** to create the next solid feature.

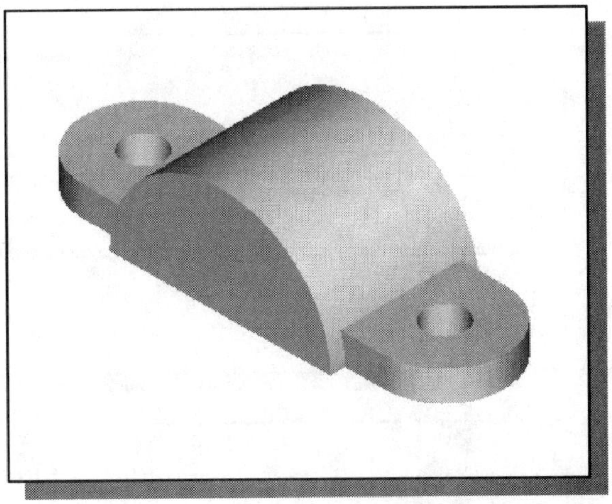

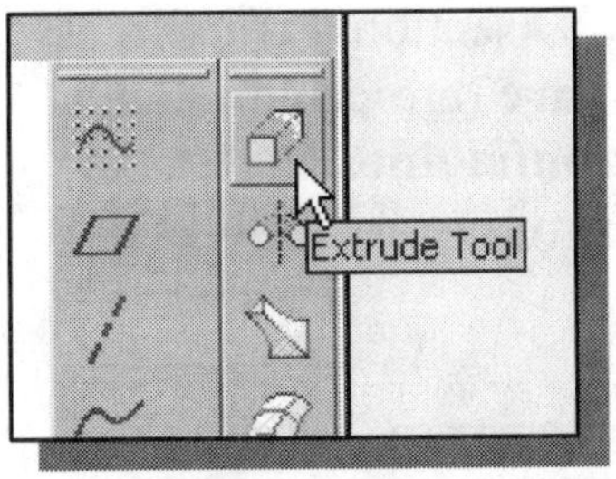

1. In the *Feature* toolbars (toolbars aligned to the right edge of the main window), select the **Extrude Tool** option as shown.

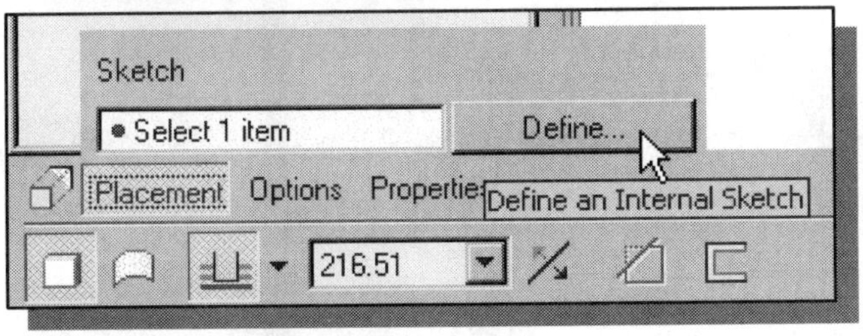

2. Click the **Placement** option and choose **Define** to begin creating a new *Internal sketch.*

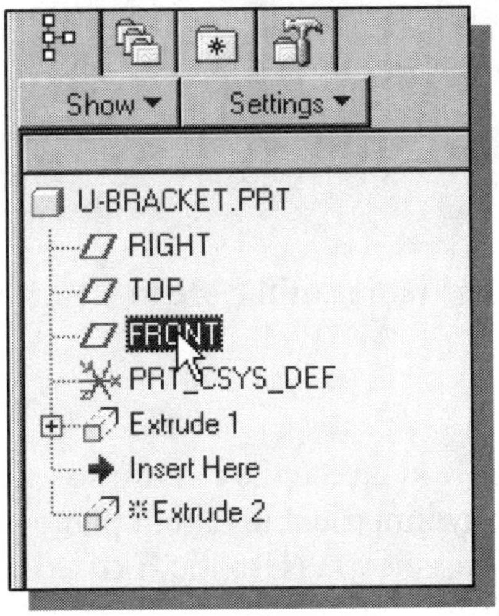

3. In the *Model Tree* window, select **Front** as the sketching plane.

4. Confirm the Sketch Orientation options is set to datum plane **Right** and **Right** orientation as shown in the figure below.

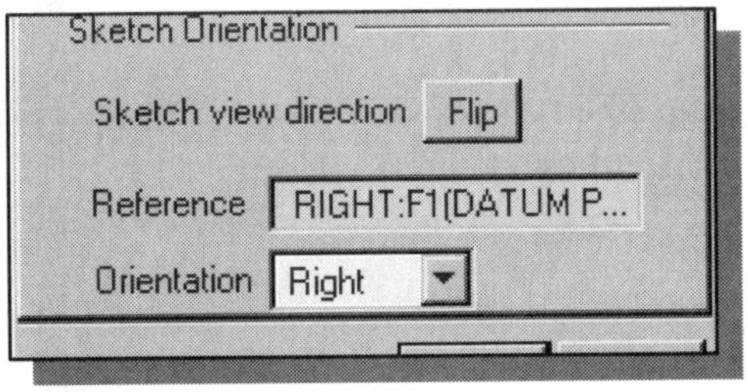

5. Pick **Sketch** to enter the *Pro/ENGINEER Sketcher* mode.

➢ Notice the use of **Right** and **Top** as the sketching references in the *Sketcher* mode.

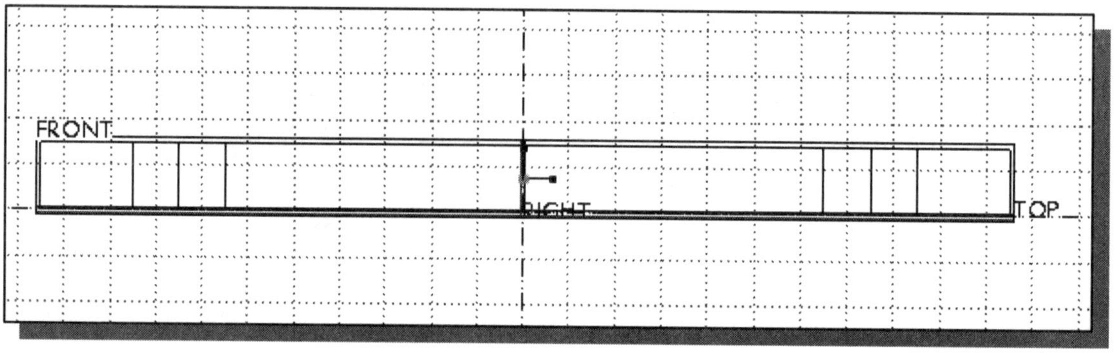

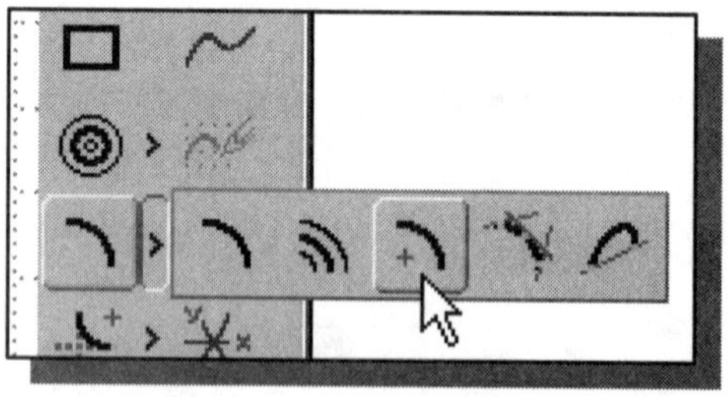

6. On your own, create the 2D section as shown. The sketch consists of an **arc** (arc-center aligned to the origin) and a **horizontal line**. The line is connected to the two ends of the arc to form a closed region.

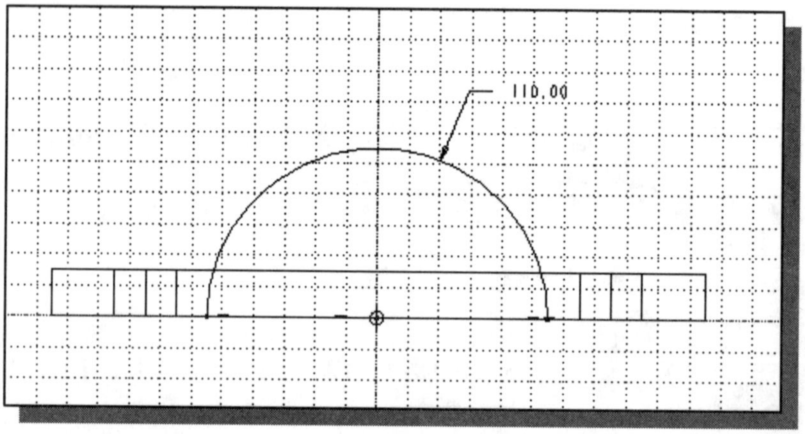

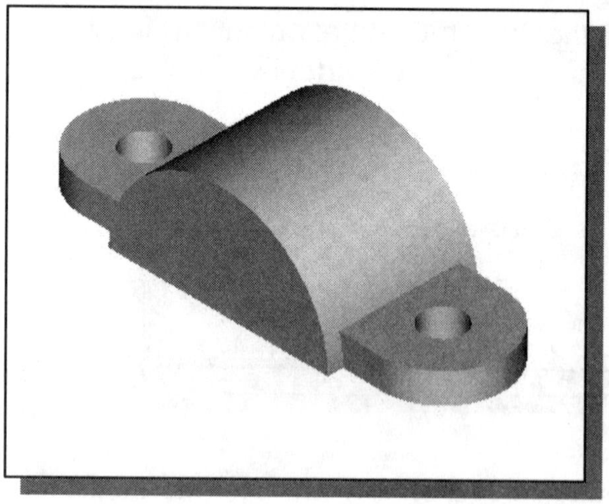

7. Modify the radius of the arc to **110 mm**.

8. On your own, create the feature so that it is symmetrical to datum plane FRONT as shown. (Use the **Extrude Both Sides** option and set the depth of extrusion to **150** mm.)

Creating a CUT Feature

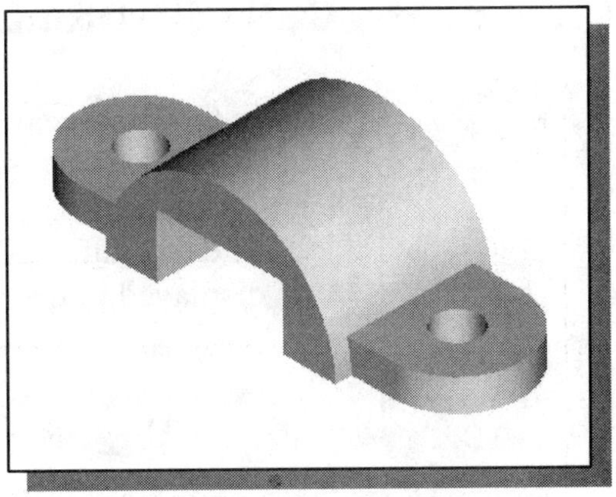

- We will create a rectangular cut as the next solid feature.

1. In the *Feature* toolbars (toolbars aligned to the right edge of the main window), select the **Extrude Tool** option as shown.

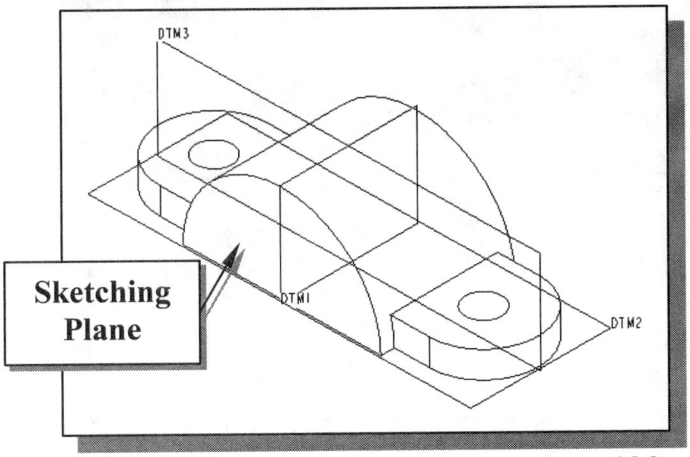

2. In the graphics window, select the **front face** of the model as the sketching plane.

3. On your own, create a rectangle (***120 x 60 mm***) positioned as shown in the figure.

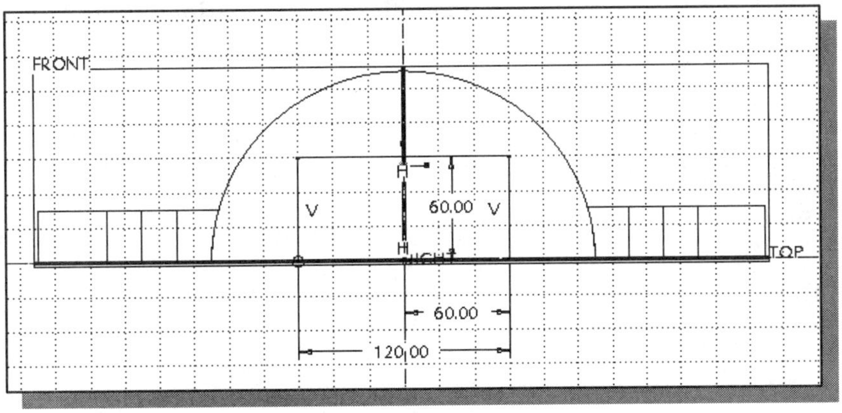

4. On your own, complete the cutout using the **Thru All** option.

Creating the CENTER_DRILL Feature

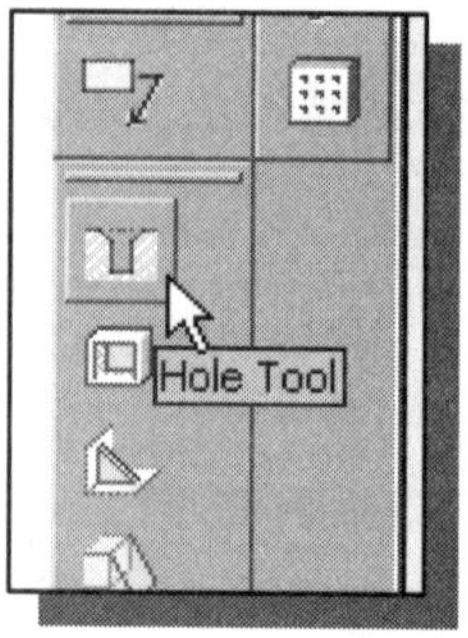

1. In the *Feature* toolbars (toolbars aligned to the right edge of the main window), select the **Hole Tool** option as shown.

❖ The message, "*Select a surface, axis or point to place hole*" is displayed in the message area.

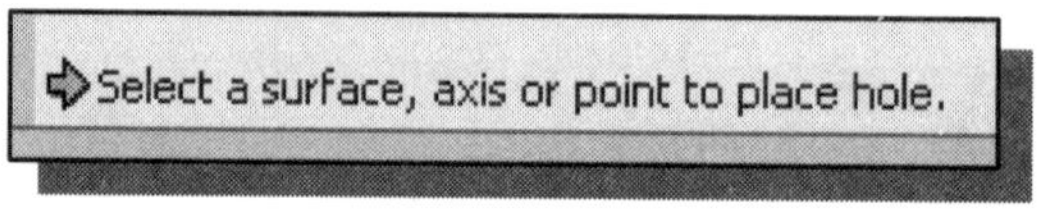

2. Select the horizontal face of the last cut feature as the **primary reference**, the *placement plane*.

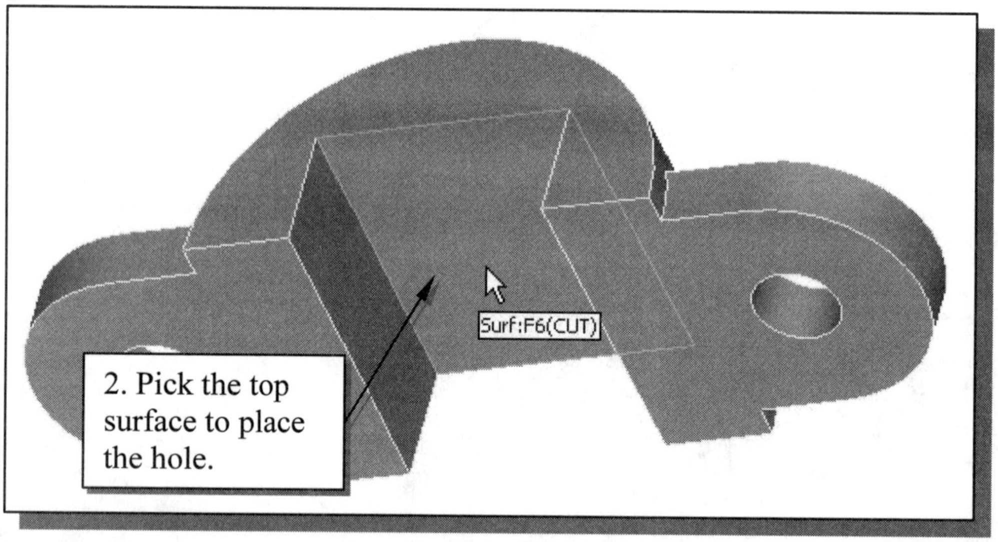

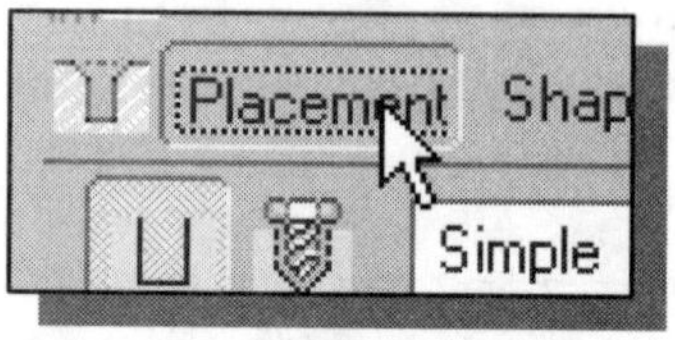

3. In the *Feature Option Dashboard*, select the **Placement** option to view the current settings.

4. Inside the graphics area, click once with the **right-mouse-button** to bring up the option menu.

5. Select the **Secondary Reference Collector** option as shown.

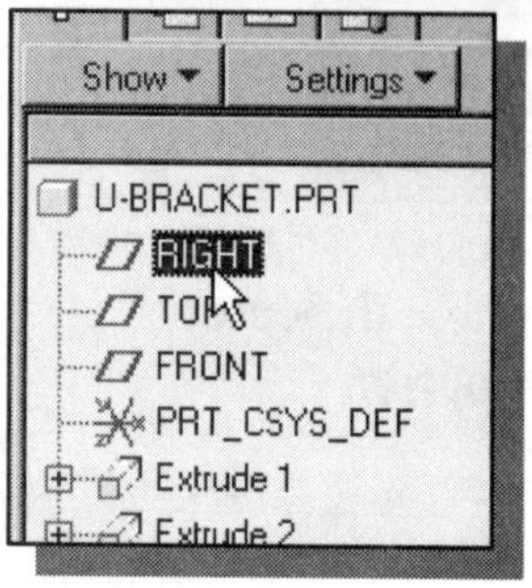

6. In the *Model Tree* window, pick **Right** as the first *secondary reference*.

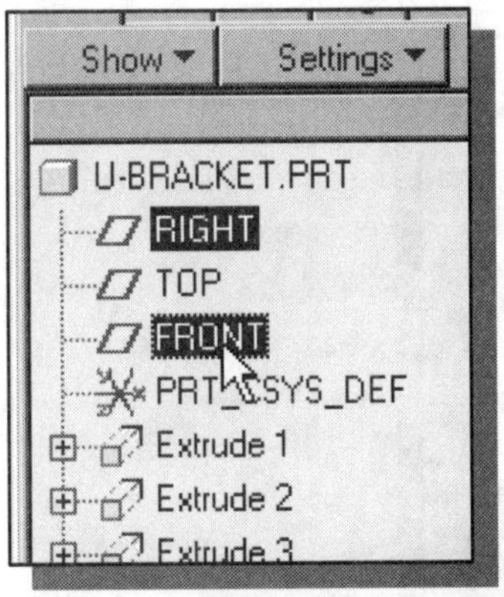

7. In the *Model Tree* window, hold down the **[CTRL]** key and pick **Front** as the second *secondary reference***.** Note that the selected references appear in the Placement option list.

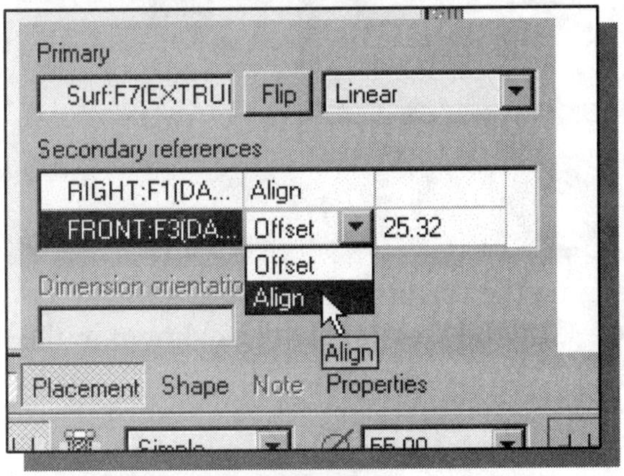

8. In the Placement option list, set the Secondary references option to **Align** as shown in the figure.

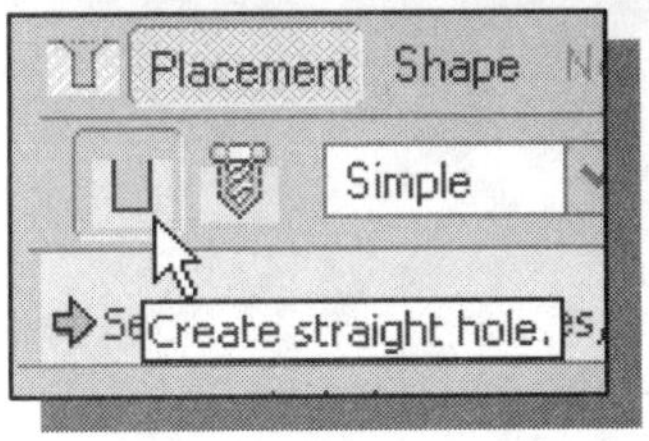

9. Confirm the hole type is set to **Straight hole** as shown.

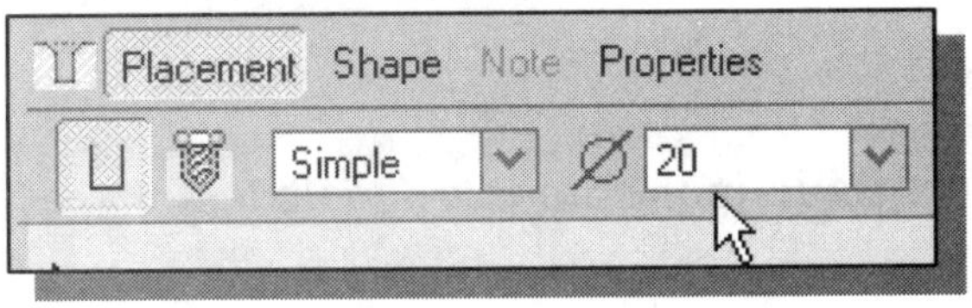

10. Set the hole diameter to **20 mm** as shown.

11. In the *Feature Option Dashboard*, set the extrusion *depth value* to **55mm** as shown.

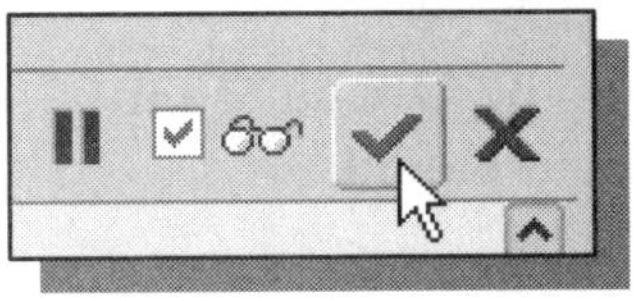

12. Click on the **Accept** icon and proceed to create the hole feature.

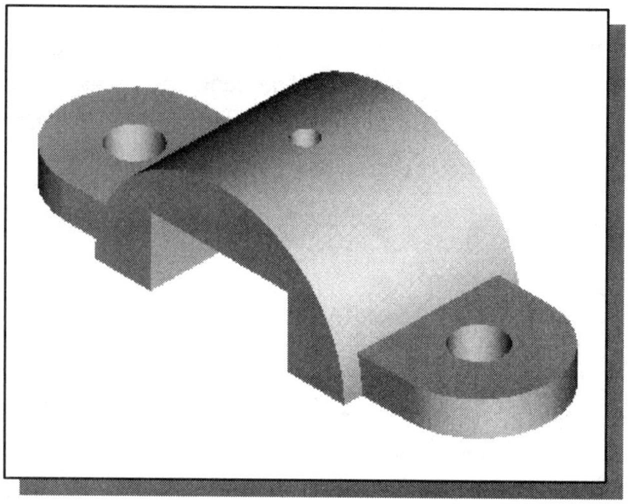

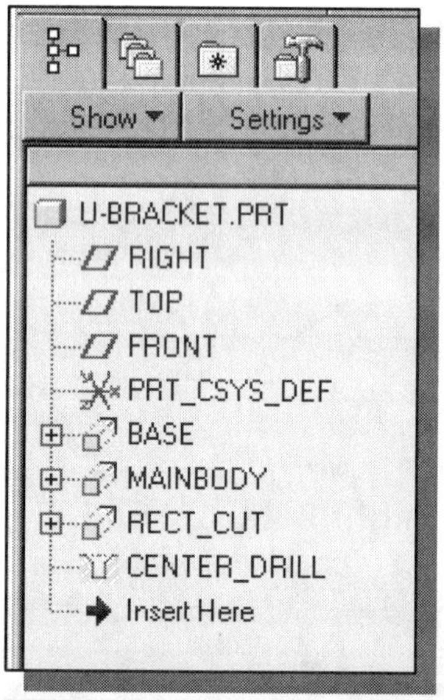

13. On your own, rename the feature names to: **Base**, **MainBody**, **Rect_Cut** and **Center_Drill** as shown in the figure.

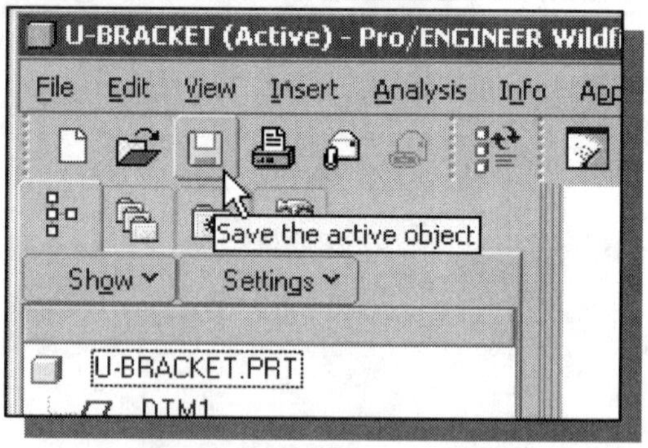

14. Select **Save** in the *Standard* toolbar.

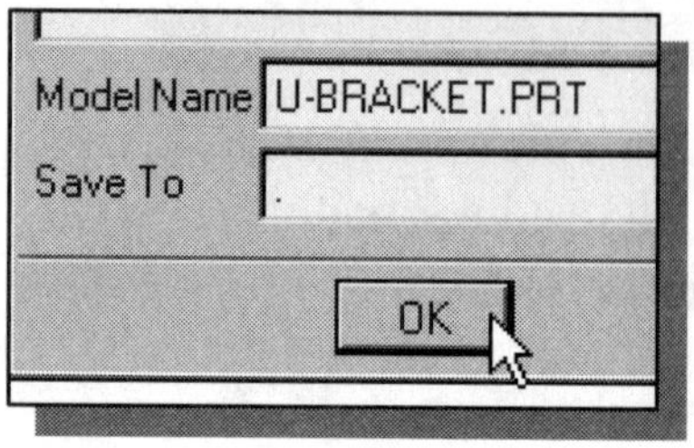

15. Press the **ENTER** key or click on the **OK** button to proceed with saving the completed model.

Examining the Parent/Child Relationships

The *Model Tree* window now contains seven items: three datum planes and four solid features. All of the parent/child relationships were established implicitly as we created the solid features. As more features are created, it becomes much more difficult to make a sketch showing all the parent/child relationships involved in the model. On the other hand, it is not really necessary to have a detailed picture showing all the relationships among the features. In using a feature-based modeler, the main emphasis is to consider the interactions that exist between the **immediate features**. Treat each feature as a unit by itself, and be clear on the parent/child relationships for each feature. Thinking in terms of features is what distinguishes *feature-based modeling* and the previous generation solid modeling techniques. Let us take a look at the last feature we created, the **CENTER_DRILL** feature. What are the parent/child relationships associated with this feature? (1) Since this is the last feature we created, it is not a parent feature to any other features. (2) Since we used one of the surfaces of the rectangular cutout as the sketching plane, the **RECT_CUT** feature is a parent feature to the **CENTER_DRILL** feature. (3) We also use datum planes Right and Front as placement references; therefore, Right and Front are parents to the **CENTER_DRILL** feature.

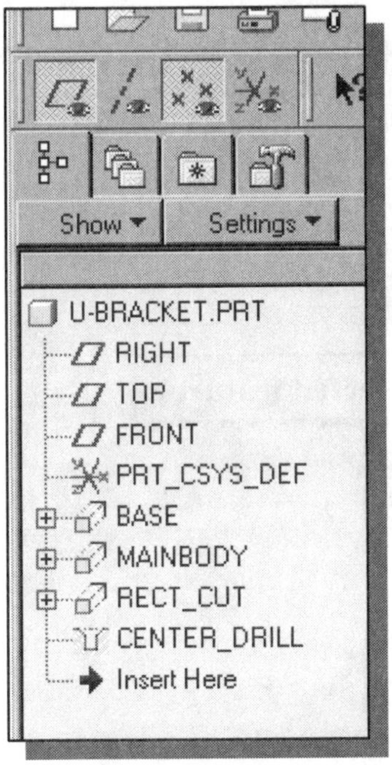

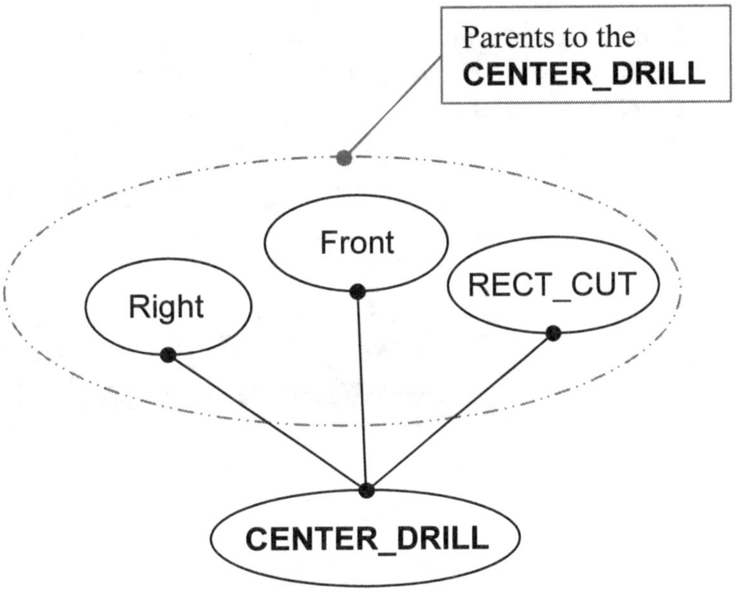

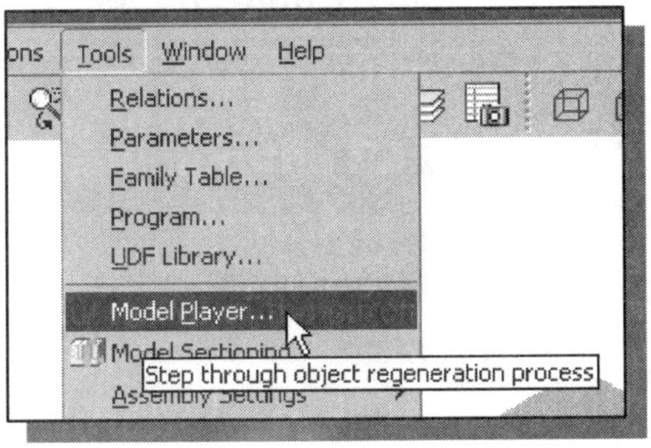

- On your own, use the **Model Player** command to review the *Model History* of the *U-Bracket* design. While replaying the construction of the features, consider the order of the features and the associated parent-child relationships.

Display Parent/Child Info

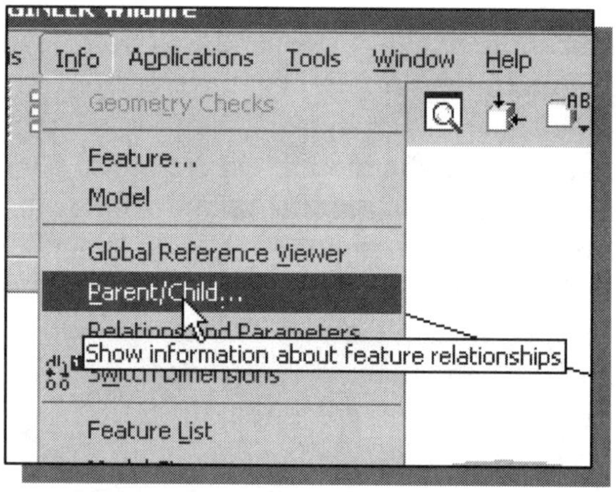

1. Select **Info** in the pull-down menus.
2. Pick the **Parent/Child** option. A selection appears to the right of the main *Pro/ENGINEER* window.

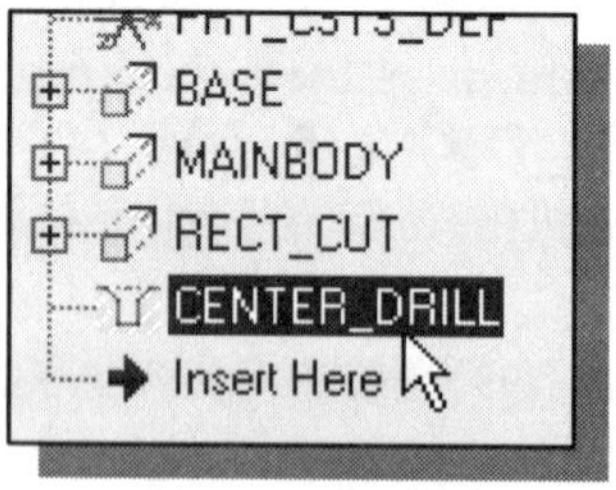

3. Pick **CENTER_DRILL** in the *Model Tree* window.

❖ In the reference information window, **Right, Front** and the **RECT_CUT** feature are listed in the **Parents Of Current Feature** area; they are parents to the **CENTER_DRILL**. Note that there is no child listed in the **Children** list area; this is what we have expected since this is the last feature of the model.

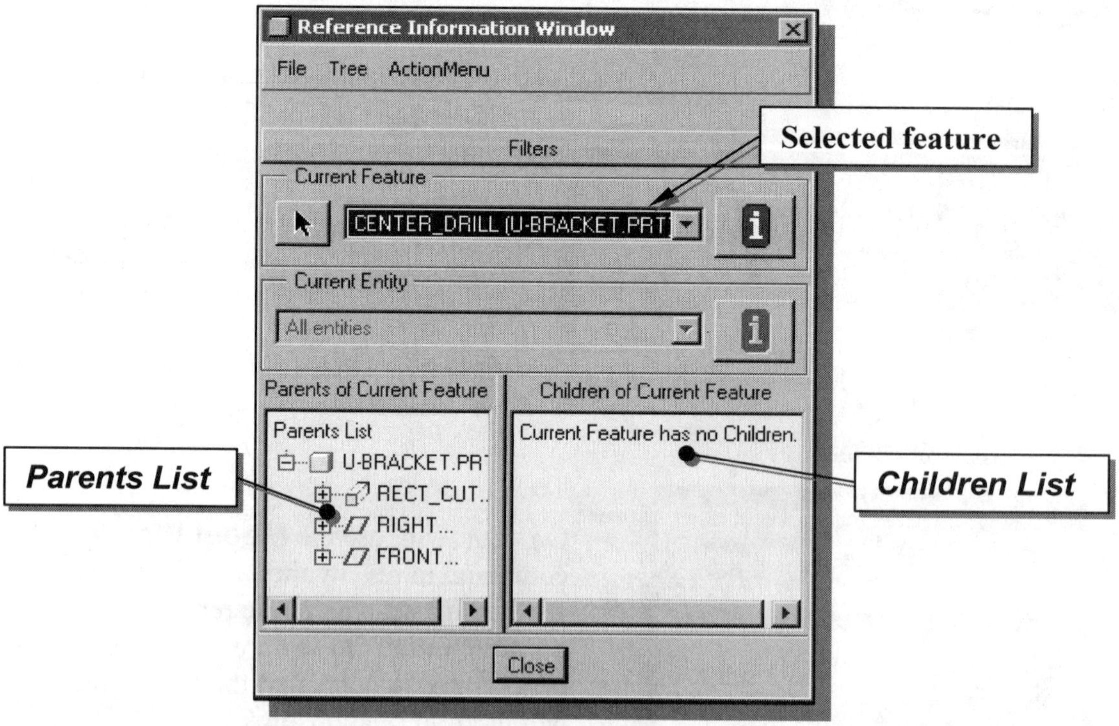

4. Click **Close** to end the *Reference Information Window.*

Modify a Parent Dimension

❖ Any changes to the parent features will affect the child feature. For example, if we modify the height of the **RECT_CUT** feature from 60 mm to 50 mm, the child feature (**CENTER_DRILL** feature) will be affected.

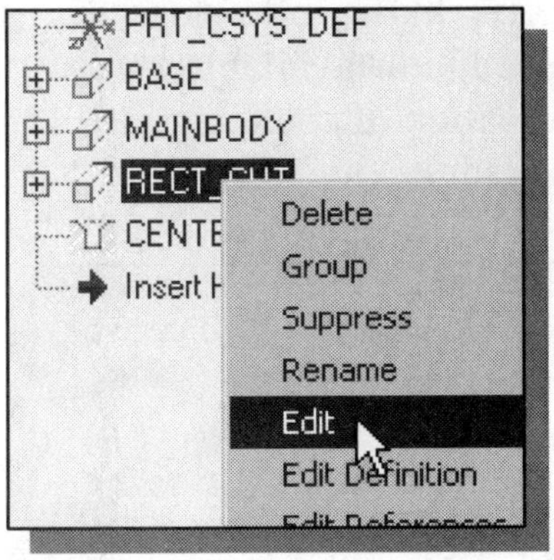

1. In the *Model Tree* window, move the cursor on top of RECT_CUT and click once with the right-mouse-button to bring up the option menu.

2. In the option menu, select **Edit.**

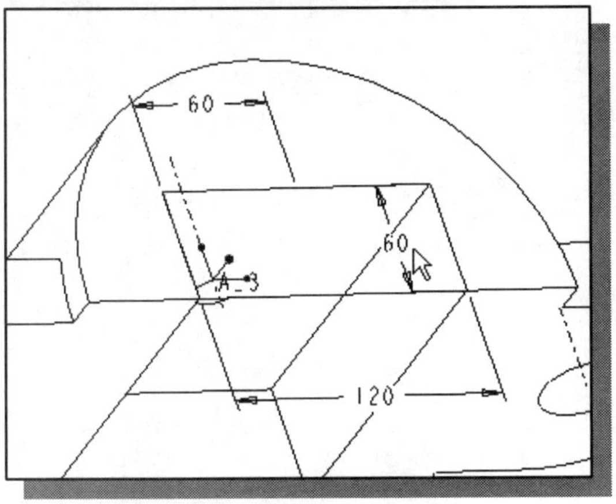

3. Double-click on the vertical size dimension text **60**.

4. Enter **50** as the new dimension value.

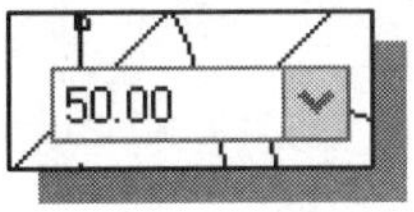

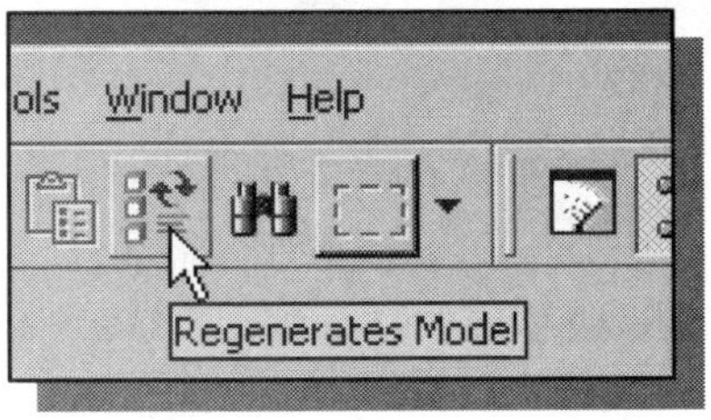

5. Click **Regenerate** in the *Standard* toolbar.

➢ The position of the **CENTER_DRILL** is moved downward by 10 mm since the placement plane is lowered by that amount; the drill does not go through the main body of the bracket anymore. If we are planning to make more dimensional changes to the rectangular cut, what would be your solution to make sure the drill will always go through the main body of the bracket?

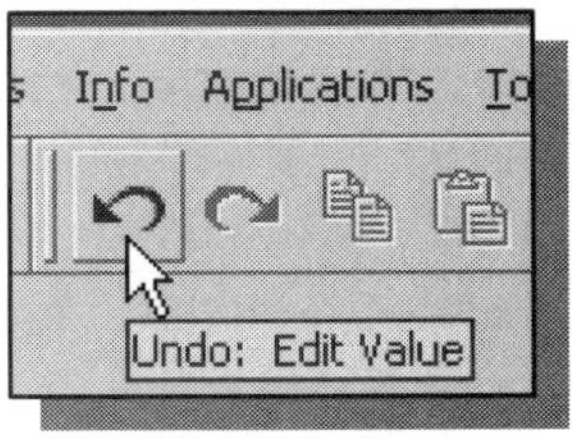

6. Click **Undo** once to the set the height of the **RECT_CUT** feature back to 60 mm.

7. On your own, adjust the depth of the **CENTER_DRILL** feature to **Thru All**.

A Design Change

➢ Engineering designs usually go through many revisions and changes. For example, a design change may call for a circular cutout instead of the current rectangular cutout feature in our model. *Pro/ENGINEER* provides an assortment of tools to handle design changes quickly and effectively. In the following sections, we will demonstrate some of the more advanced tools available in *Pro/ENGINEER*, which allow us to perform the modification of changing the rectangular cutout (120 x 60mm) to a circular cutout (radius: 80 mm).

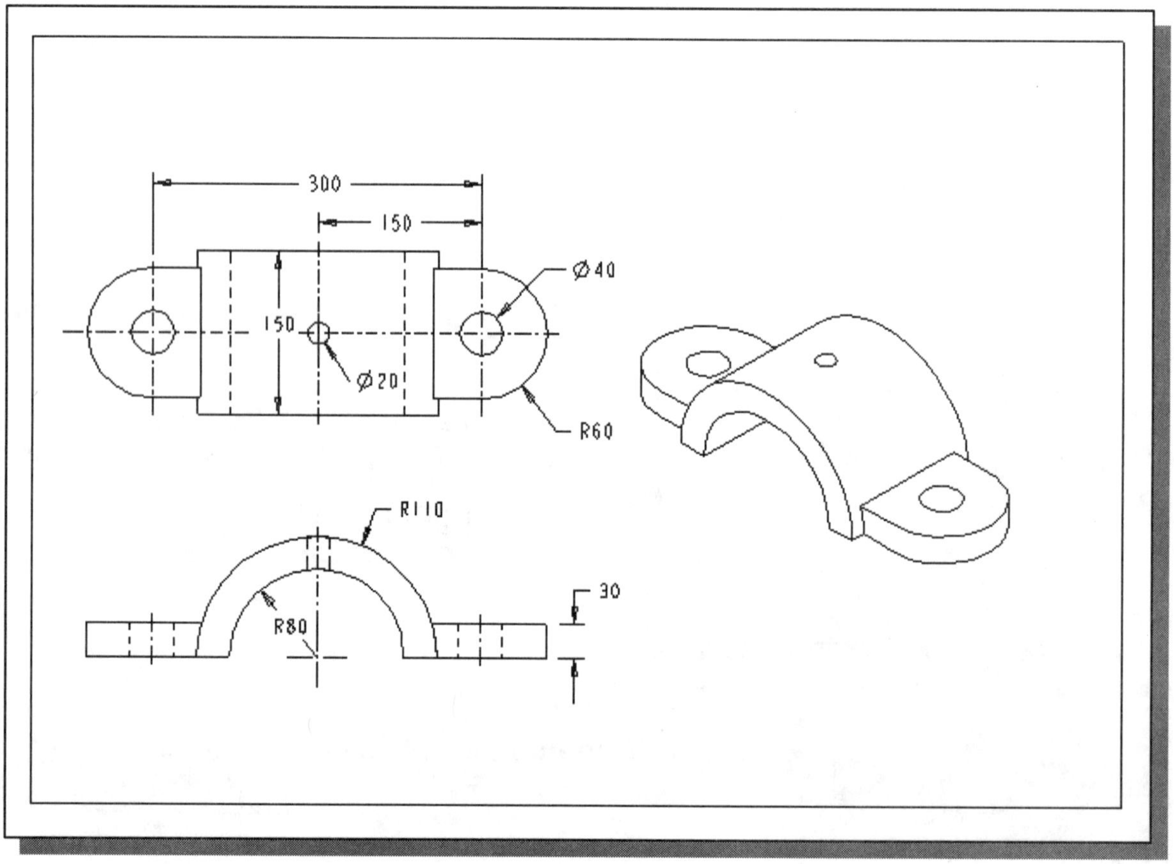

❖ Based on your knowledge of *Pro/ENGINEER* so far, how would you accomplish this modification? What other approaches can you think of that are also feasible? Of the approaches you came up with, which one is the easiest to do and which is the most flexible? If this design change were anticipated right at the beginning of the design process, what would be your choice in arranging the orders of the features? Take a few minutes to consider these questions and you are encouraged to perform the modifications prior to following through the rest of the tutorial.

➢ Before continuing to the next section, confirm the current ***U-Bracket*** model matches what is shown on page 4-3.

Feature Suppression

❖ With *Pro/ENGINEER*, we can take several different approaches to accomplish this modification. We could (1) create a new model, or (2) change the shape of the existing cut feature using the **Redefine** command, or (3) perform **feature suppression** on the rectangular cut feature and add a circular cut feature. The third approach offers the most flexibility and requires the least amount of editing to the existing geometry. **Feature suppression** is a method that enables us to disable a feature while retaining the complete feature information; the feature can be reactivated at any time. Prior to adding the new cut feature, we will first suppress the rectangular cut feature.

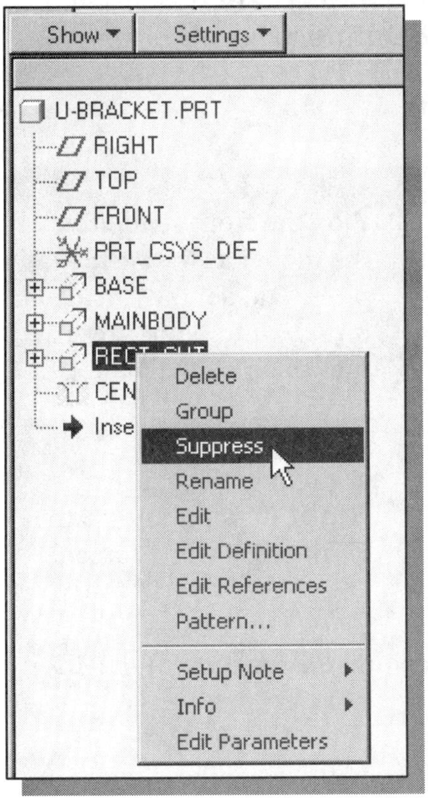

1. Move the cursor inside the *Model Tree* window. Click once with the right-mouse-button on top of RECT_CUT to bring up the option menu.

2. Pick **Suppress** in the pop-up menu. A pop-up window is displayed.

➢ In the display area and the *Model Tree* window, both the **RECT_CUT** feature and the **CENTER_DRILL** feature are highlighted. The child feature cannot exist without its parent(s), and any modification to the parent (RECT_CUT) influences the child (CENTER_DRILL).

3. Pick **OK** in the pop-up window to confirm the suppression of the RECT_CUT and CENTER_DRILL features.

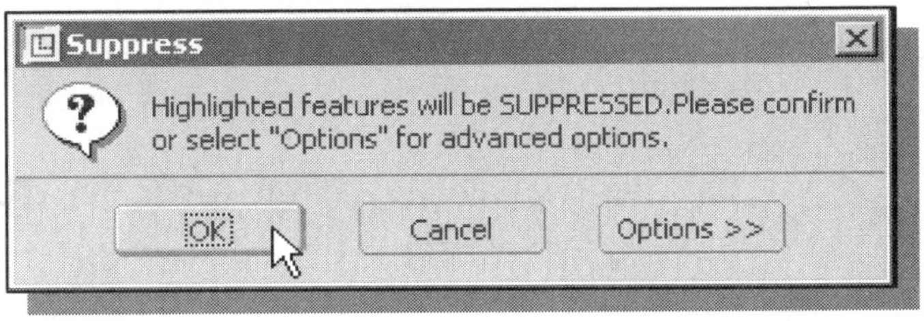

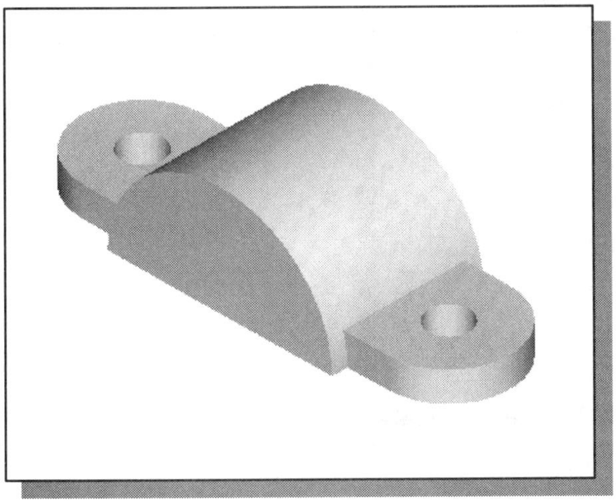

❖ We have literally ***gone back in time***. The last two features of the design, RECT_CUT and CENTER_DRILL, have disappeared in the *Model Tree* window and the display area.

Creating the Circular Cut Feature

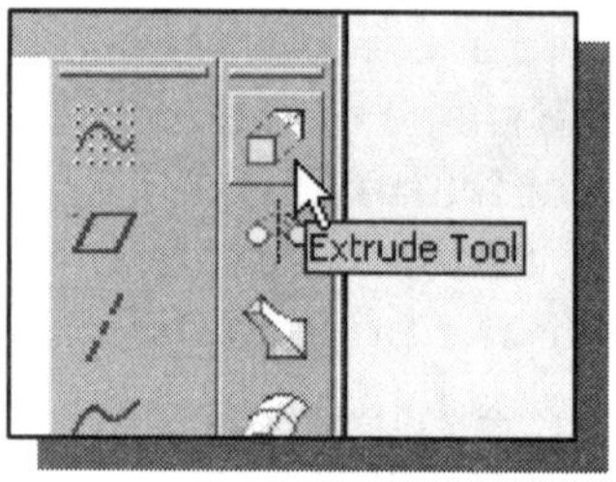

1. In the *Feature* toolbars (toolbars aligned to the right edge of the main window), select the **Extrude Tool** option as shown.

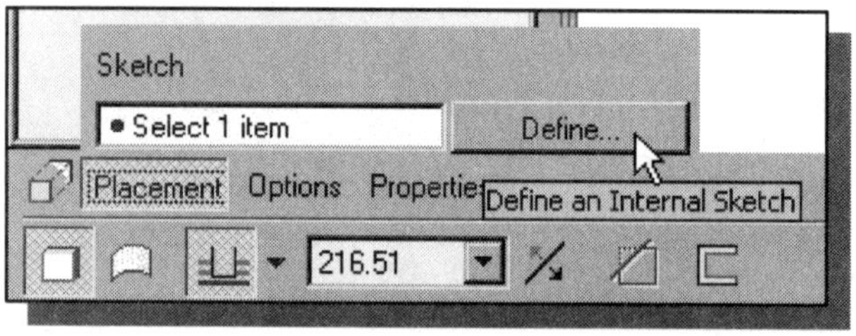

2. Click the **Placement** option and choose **Define** to begin creating a new *Internal sketch.*

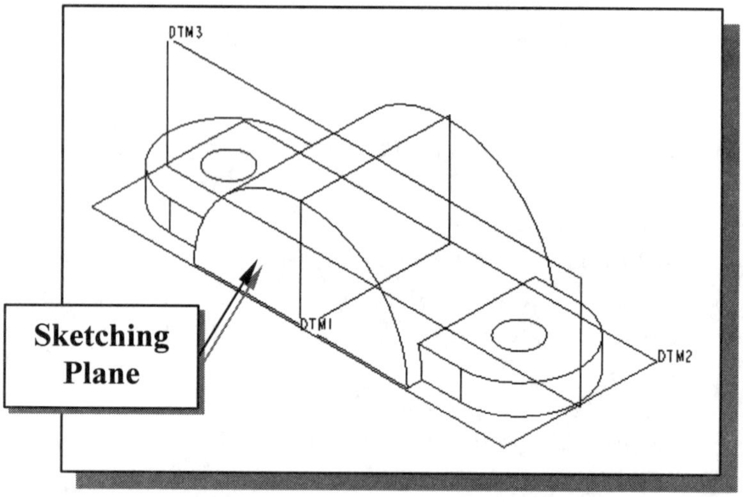

3. Pick the front semicircle of the model as the *sketching plane* and set the *extrusion direction* to point toward the center of the model.

4. For the orientation of the sketching plane, select **Right** to point toward the right edge of the screen.

5. Use **Right** and **Top** as the two references to position the location of the 2D sketch.

6. On your own, create a circle (diameter: **160 mm**), with the center aligned to the origin as shown in the figure below.

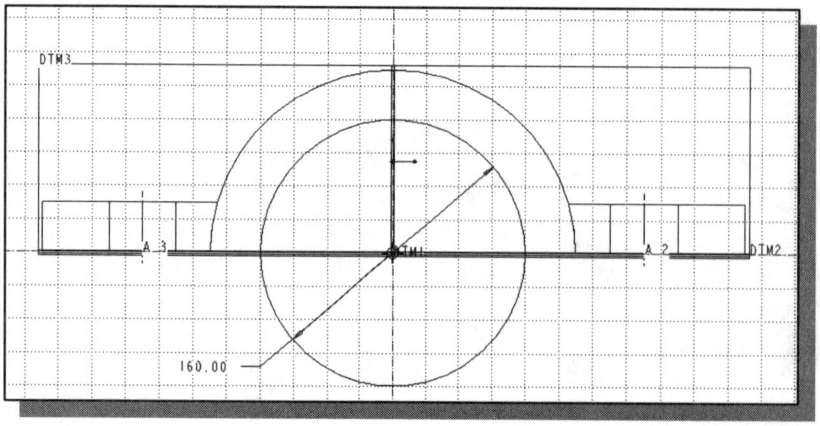

7. Complete the cutout using the **Thru All** option.

Listing Suppressed Features

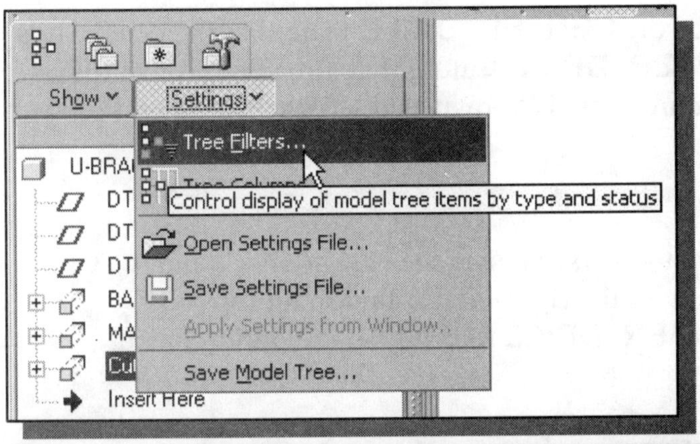

1. In the *Model Tree* window, click on the **Settings** icon to display the setting options.

2. Choose **Tree Filters** in the setting options as shown.

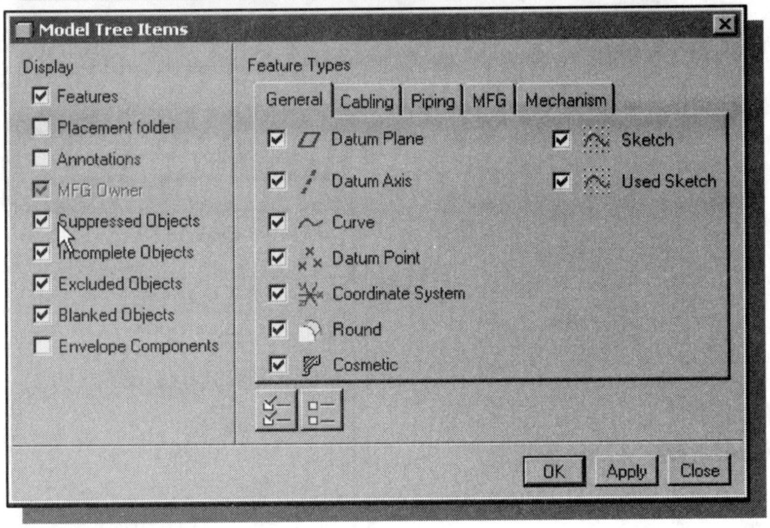

3. In the *Model Tree Items* window, click on the **Suppressed Objects** check-box to toggle **on** the display of suppressed features and components.

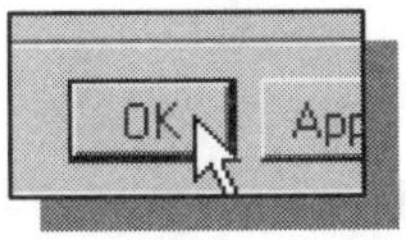

4. Click on the **OK** button to accept the settings.

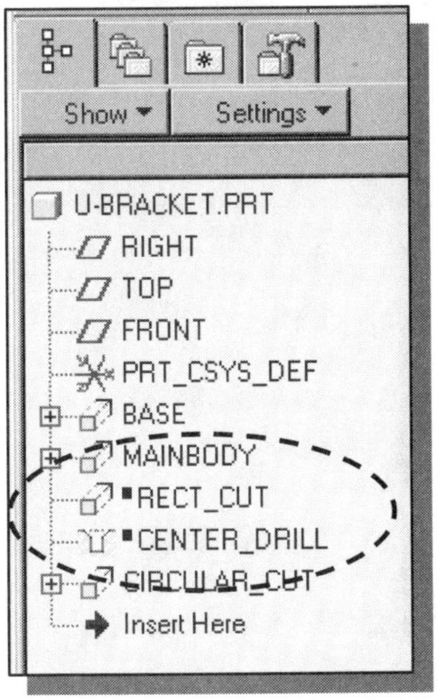

❖ Notice in the *Model Tree* menu window a complete list of features, including suppressed features, is displayed. A small rectangular marker next to the feature signifies the feature is suppressed.

Reactivate the CENTER_DRILL Feature

❖ To complete the model, we still need the **CENTER_DRILL** feature at the center. It is not necessary to duplicate the **CENTER_DRILL** feature that already exists in the database. We will reactivate the feature. This is known as the **Resume** operation in *Pro/ENGINEER*.

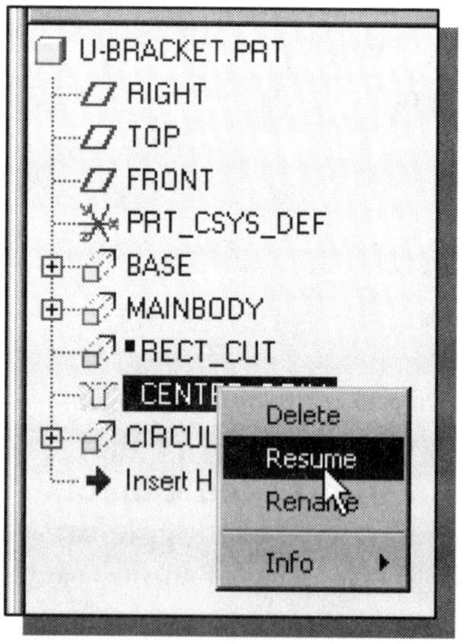

1. Move the cursor inside the *Model Tree* window. Click once with the right-mouse-button on top of **CENTER_DRILL** to bring up the option menu.

2. Pick **Resume** in the option menu.

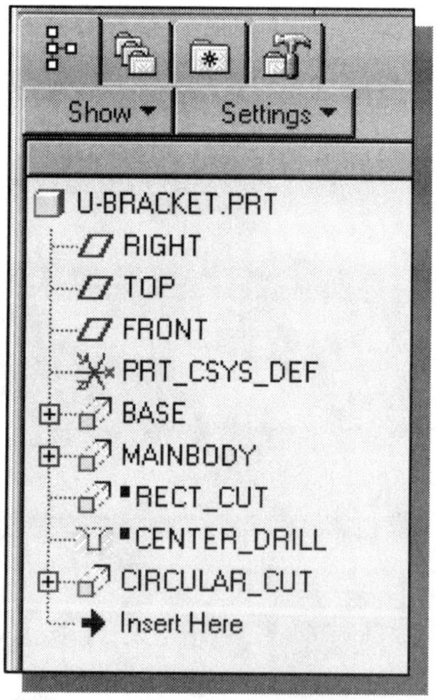

➢ Notice that both of the suppressed features (**RECT_CUT** and **CENTER_DRILL**) are reactivated, again. ***The child feature cannot exist without its parents***. All of the model's features are displayed in the display area and in the *Model Tree* window.

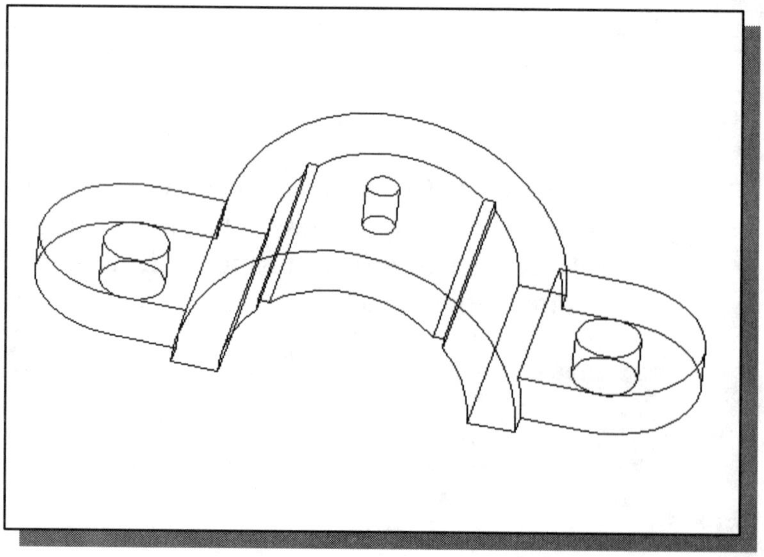

Reroute the CENTER_DRILL Feature

❖ In order to suppress only the RECT_CUT feature, we will need to remove the parent/child relationship that exists between RECT_CUT and CENTER_DRILL. The parent/child relationship was established when we used one of the RECT_CUT surfaces as the sketching plane. We will need to select a different sketching plane for the CENTER_DRILL feature. In *Pro/ENGINEER*, this is done by **Editing References** of the features.

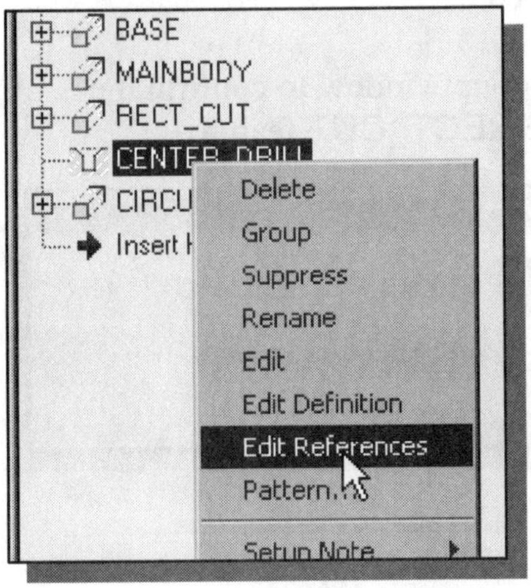

1. Move the cursor inside the *Model Tree* window. Click once with the right-mouse-button on top of **CENTER_DRILL** to bring up the option menu.

2. Pick **Edit References** in the option menu.

❖ The message "*Do you want to roll back the model?*" is displayed in the message area. The ***roll back*** option allows us to trace back in history and redefine the *sketching plane* and *reference planes* for the CENTER_DRILL feature.

3. Type **Yes** or pick the **Yes** icon in the message area.

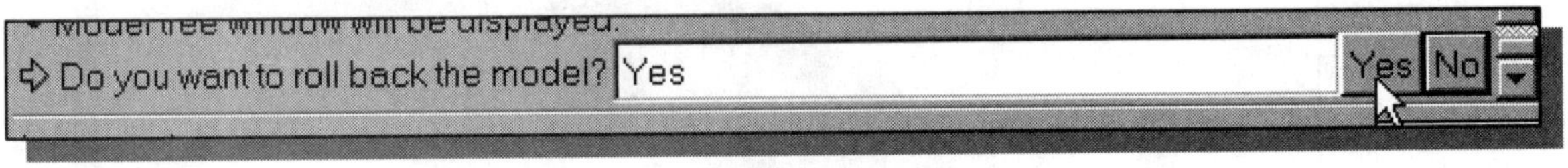

4. We are now asked to redefine the ***placement plane*** for the **HOLE** feature. Pick datum plane **Top** in the display area.

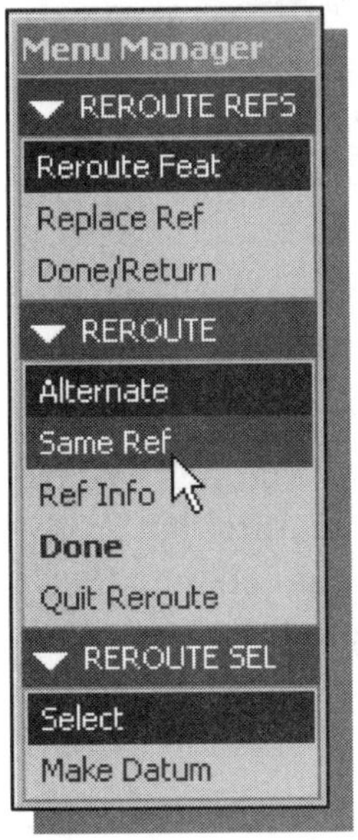

5. After selecting the placement plane, we will be asked to redefine the **two position reference planes**. Pick **Same Ref twice** in the *REROUTE* menu.

❖ *Pro/ENGINEER* will regenerate the modified features and display the regenerated model. Everything looks the same, but RECT_CUT is no longer a parent feature to the CENTER_DRILL feature.

Suppress the RECT_CUT Feature

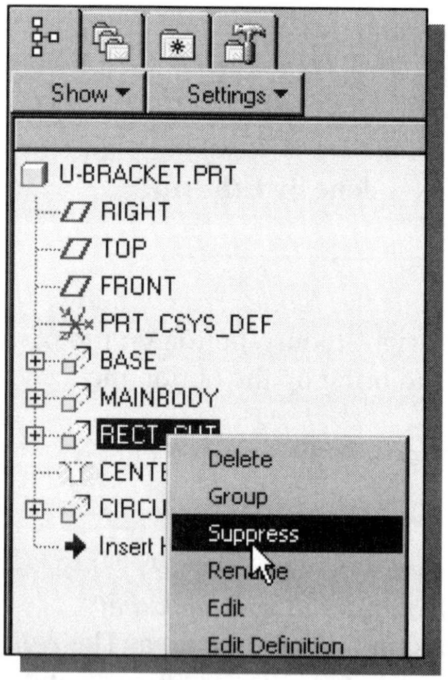

1. Move the cursor inside the *Model Tree* window. Click once with the right-mouse-button on top of RECT_CUT to bring up the option menu.

2. Pick **Suppress** in the option menu. A pop-up window is displayed.

3. Pick **OK** in the pop-up window to confirm the suppression of the RECT_CUT feature.

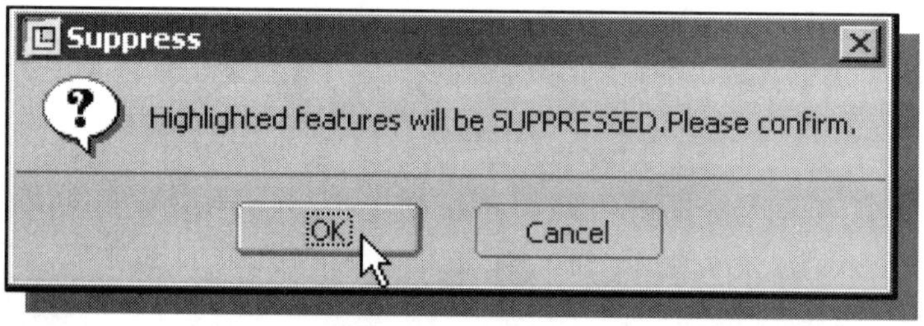

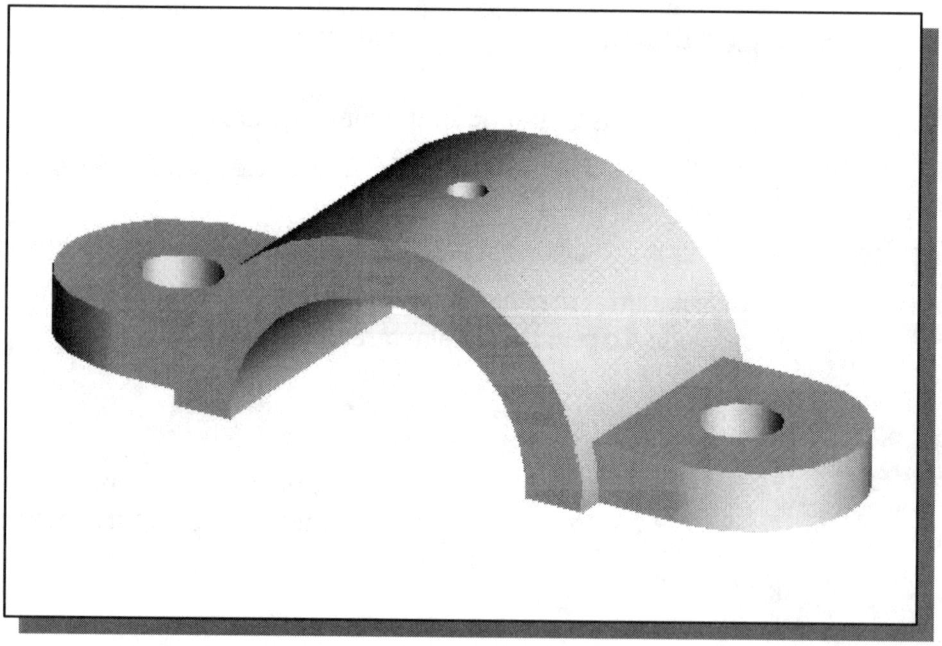

❖ Now, there is only one suppressed feature shown in the *Model Tree* window. We have successfully updated the model by switching off the RECT_CUT feature, but retained the complete feature information. On your own, do the design change that requires you to **suppress** the CIRCULAR_CUT feature and **resume** the RECT_CUT feature.

Questions:

1. How can we find out the order in which features are created in *Pro/ENGINEER*?
2. In *Pro/ENGINEER*, what does "rolling back the part" mean?
3. What is the difference between *Edit* and *Edit References*?
4. What determines how a feature reacts when other features in the model change?
5. Describe two methods in *Pro/ENGINEER* to access the *Redefine* the existing feature references.
6. Create sketches showing the steps you plan to use to create the two models shown on the next page:

Ex.1)

Ex.2)

Exercises: (All dimensions are in inches.)

1. Dimensions are in inches. Plate Thickness: 0.25

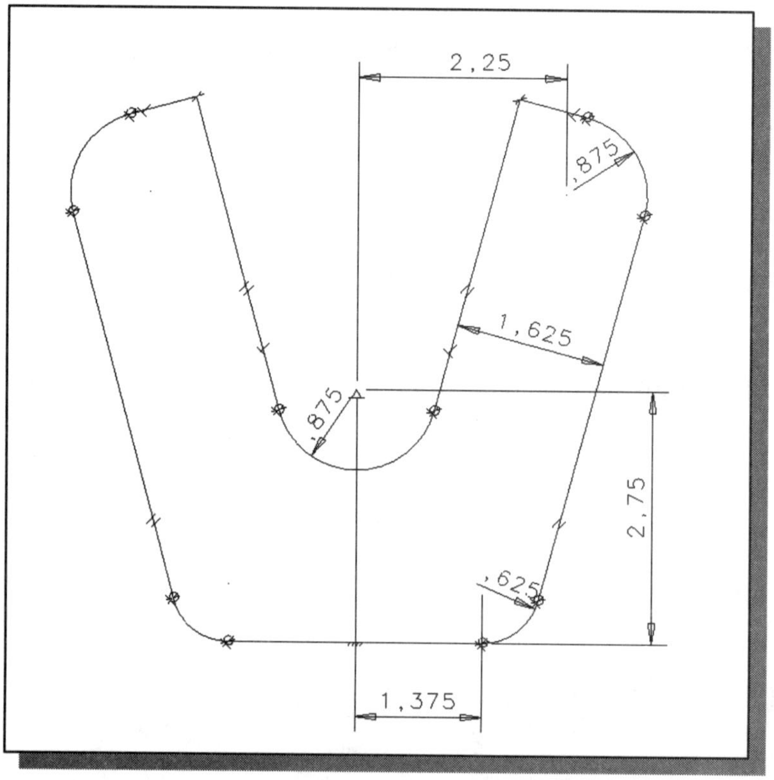

2. Dimensions are in millimeters.

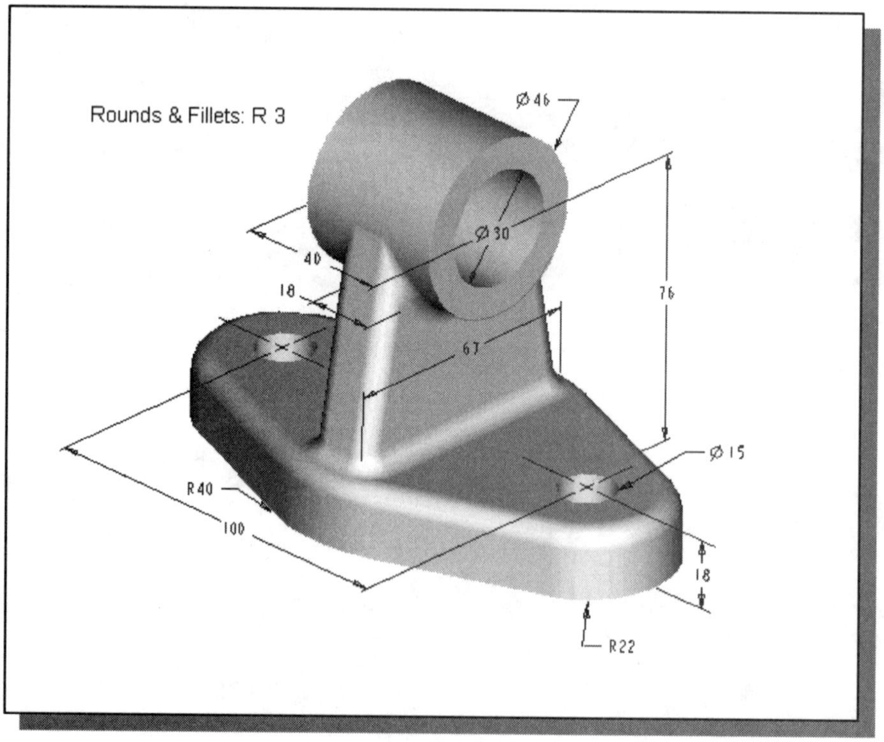

3. Dimensions are in inches.

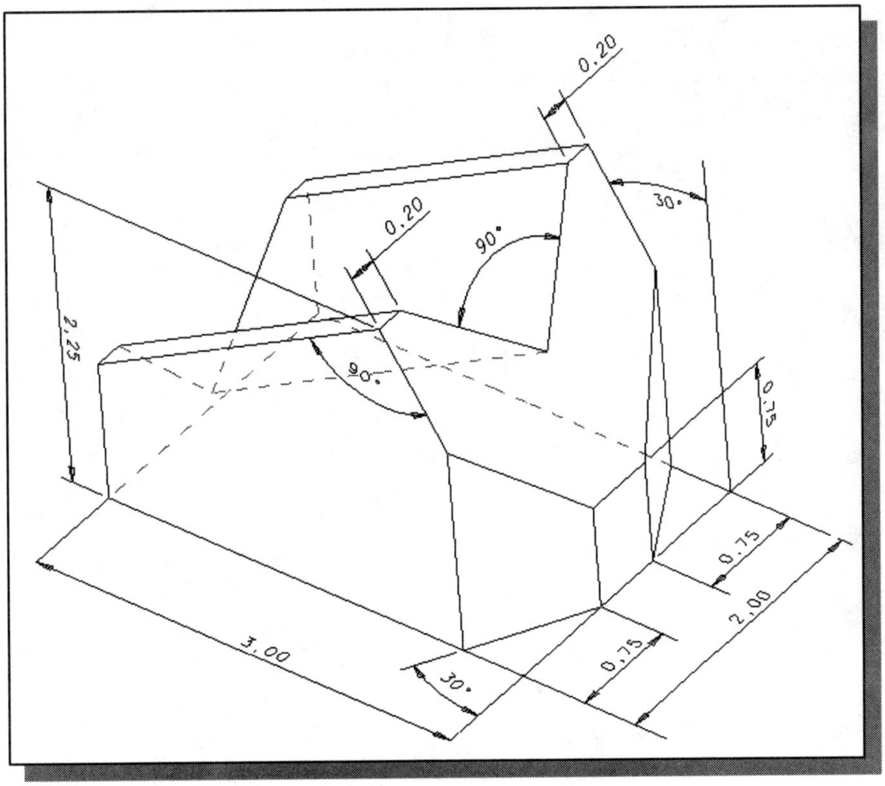

Notes:

Chapter 6
Datum Features, 3D Annotation and Part Drawings

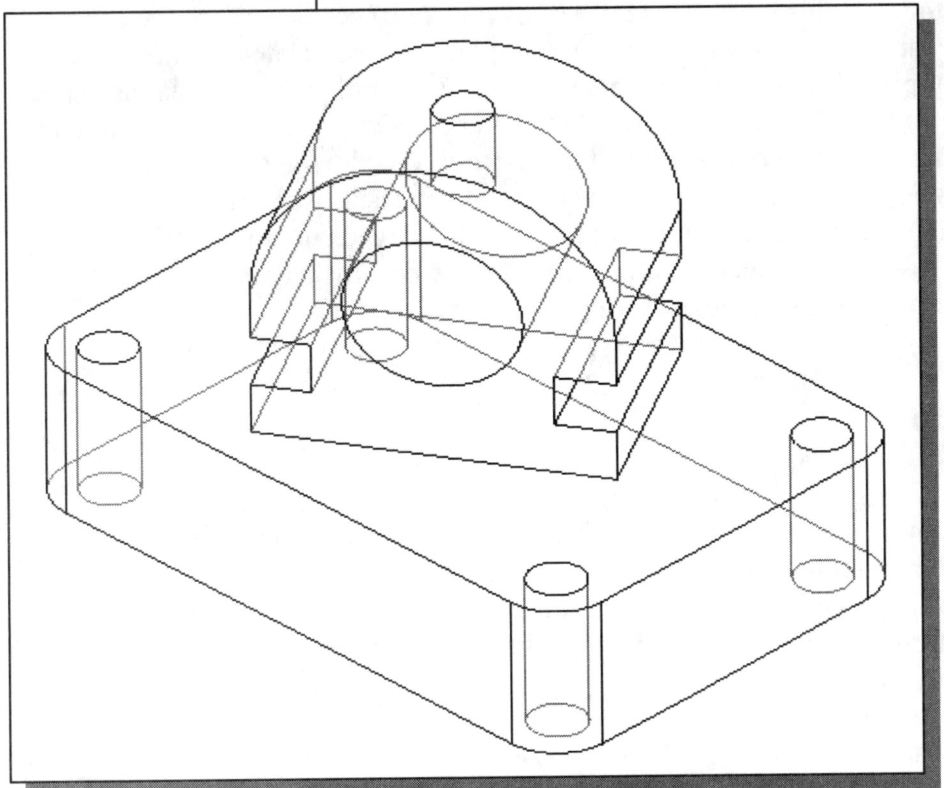

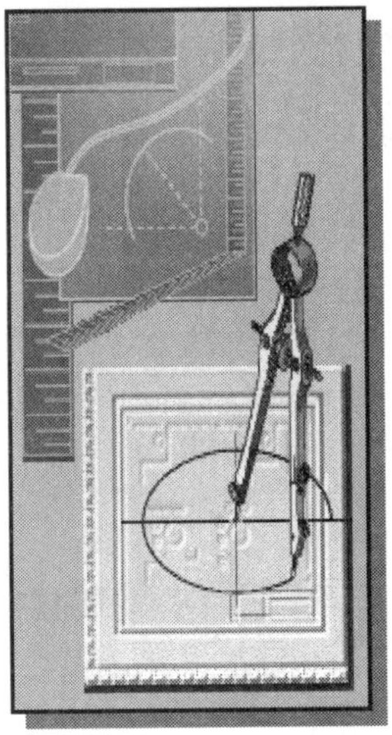

Learning Objectives

- **Understand the importance of Datum Features.**
- **Use the Selection Filter.**
- **Create Datum planes at an angle.**
- **Create 2D drawings from 3D parts.**
- **Using Pro/E Drawing Templates.**
- **Create Standard views and Auxiliary Views in 2D drawings.**
- **Using Feature/Driven Dimensions in 2D drawings.**

Datum Features

Feature-based parametric modeling is a cumulative process. A new feature can use previously defined features to define information such as size, shape, location and orientation. At times, it is necessary to create features at locations that are not easily referenced by the existing surfaces/features. Although several tools are available in *Pro/ENGINEER* to assist this process, the use of **datum features** is perhaps the most flexible and direct approach. For example, a datum plane can be created so that features can be sketched or placed on it. **Datum features**, such as datum points and datum planes, can be thought of as user-definable datum, which can be updated with the part geometry. Datum features can also be used to align features or to orient parts in an assembly. Note that in *Pro/ENGINEER*, datum features can also be parametrically linked to the references, which means associative functionality exists between datum features and solid features. By creating parametric datum features, the established feature interactions in the CAD database also assure the capturing of the design intent.

3D Annotation

Pro/Engineer provides users with the ability to access powerful digital product information for communication and in support of operations such as inspection, manufacturing, or marketing. With software/hardware improvements, it is now feasible to use the solid modeling software to document and communicate all production and manufacturing information in a three-dimensional (3D) environment. In *Pro/Engineer Wildfire 2.0*, exciting tools are now available for documenting and communicating product designs. We can apply 3D-based annotation, in conjunction with engineering drawing conventions, directly to the solid model or assembly. This new solids-based documentation approach provides the ability to create, manage, and deliver process-specific information without the need for paper-based documentation. With this new set of tools, the mental translation of 3D models to two-dimensional (2D) drawings for communicating information and then back to 3D models for manufacturing is replaced with a representative 3D prototype that provides all necessary information to communicate and manufacture the product. This approach means faster and more precise product communication, which can greatly improve the product development process. In *Pro/Engineer Wildfire 2.0*, the standard 2D drawing documentation symbols are also available in the 3D-based annotation mode. In this lesson, the general procedure of creating 3D-based annotation is illustrated.

Drawings from Parts

With the software/hardware improvements in solid modeling, the importance of two-dimensional drawings is decreasing. Drafting is considered one of the downstream applications of using solid models. In many production facilities, solid models are used to generate machine tool paths for *computer numerical control* (CNC) machines. Solid models are also used in *rapid prototyping* to create 3D physical models out of plastic resins, powdered metal, etc. Ideally, the solid model database should be used directly to generate the final product. However, the majority of applications in most production facilities still require the use of two-dimensional drawings. Using the solid model as the starting point for a design, solid modeling tools can easily create all the necessary two-dimensional views. In this sense, solid modeling tools are making the process of creating two-dimensional drawings more efficient and effective. In this lesson, the general procedure of creating multi-view drawings from solid models is illustrated.

The *Rod-Guide* Design

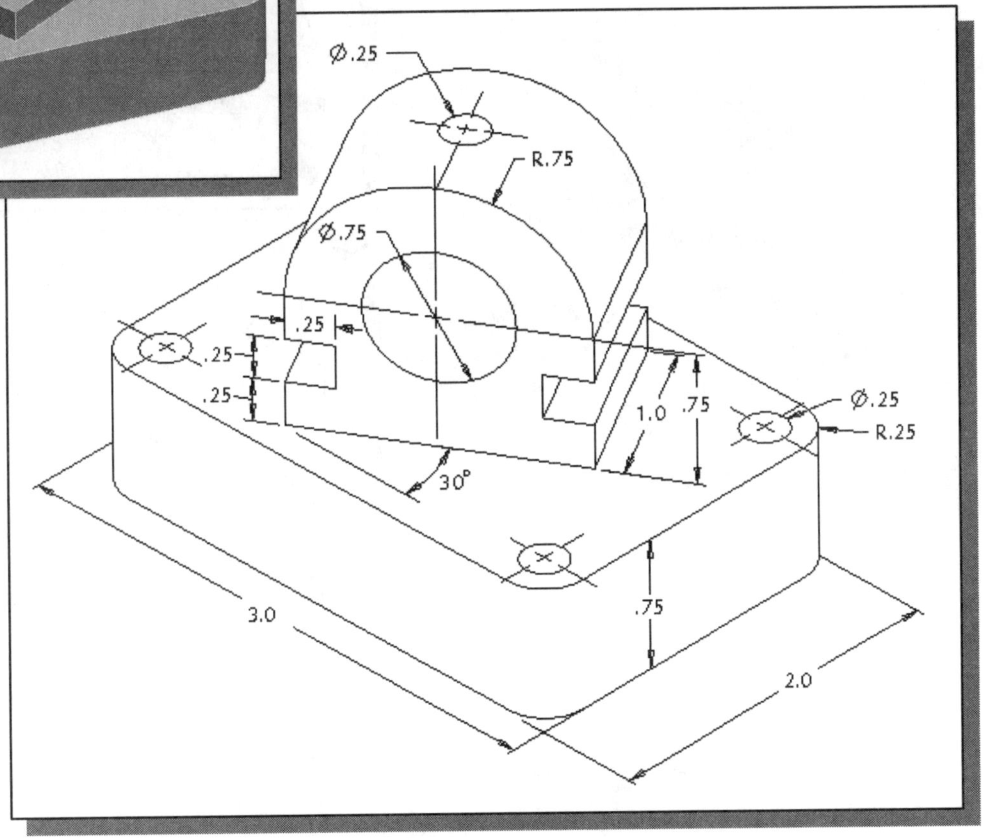

❖ Based on your knowledge of *Pro/ENGINEER* so far, how would you create this design? What are the more difficult features involved in the design? Take a few minutes to consider a modeling strategy and do preliminary planning by sketching on a piece of paper. You are also encouraged to create the design on your own prior to following through the tutorial.

Modeling Strategy

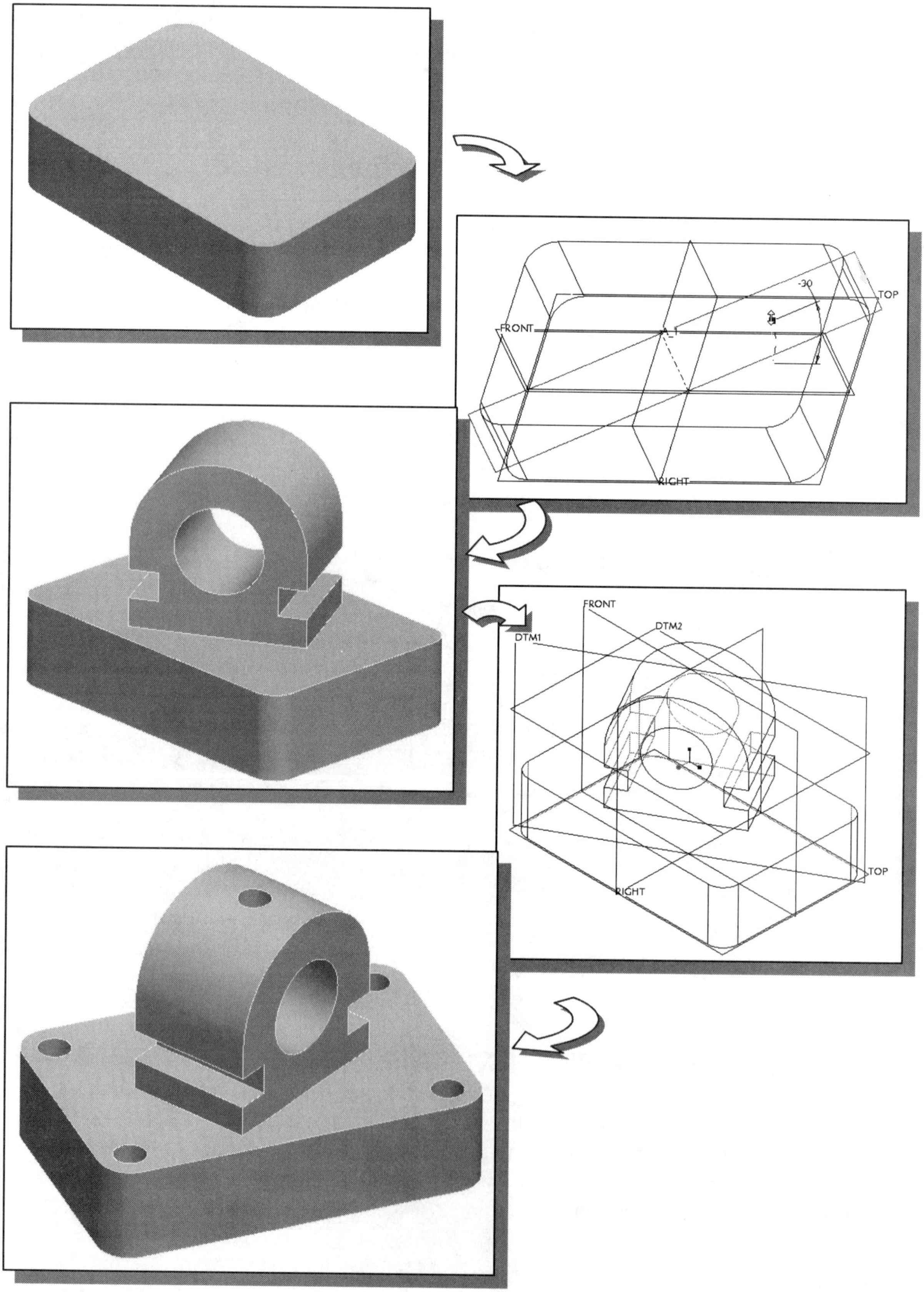

Starting Pro/ENGINEER

1. Select the **Pro/ENGINEER** option on the *Start* menu or select the **Pro/ENGINEER** icon on the desktop to start *Pro/ENGINEER*. The *Pro/ENGINEER* main window will appear on the screen.

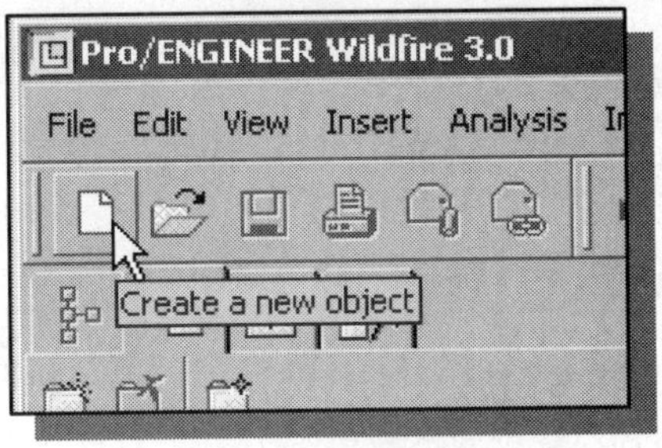

2. Click on the **New** icon, located in the *Standard* toolbar as shown.

3. On your own, start a new ***solid part*** file using ***Rod-Guide*** as the part **Name**.

4. Confirm the **Use default template** option is turned ***on*** so that the system units are set to the *Pro/ENGINEER* default settings (**Inch-lbm-Second**).

5. Click on the **OK** button to accept the settings.

Creating the Base Feature

1. In the *Feature* toolbars (toolbars aligned to the right edge of the main window), select the **Extrude Tool** option as shown.

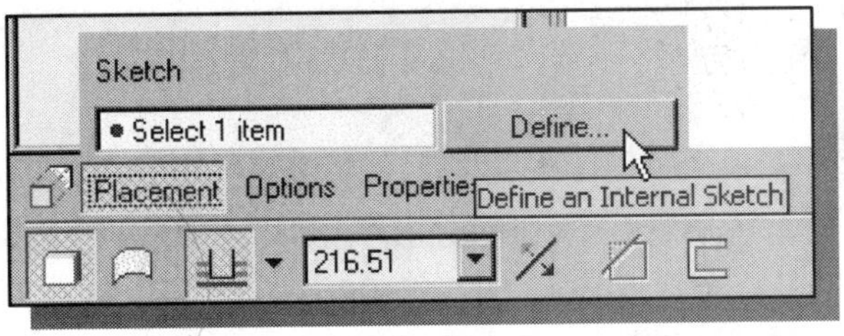

2. Click the **Placement** option and choose **Define** to begin creating a new *Internal sketch.*

3. On your own, set up the **Top** datum plane as the sketch plane with the **Right** datum plane facing the **right** edge of the computer screen and pick **Sketch** to enter the *Sketcher* mode.

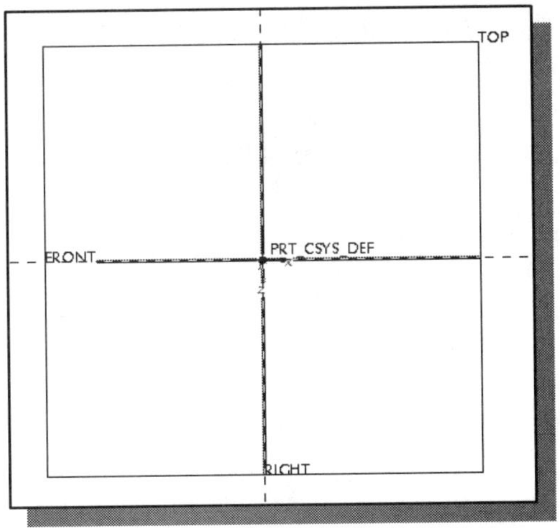

❖ Notice the default selection of the two datum planes, **Right** and **Front**, as the references for the 2D sketch.

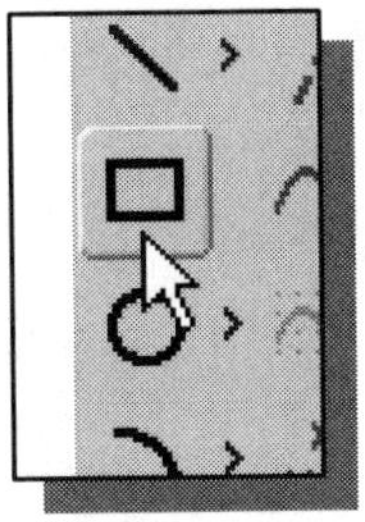

4. Choose the **Rectangle** command in the *Sketcher* toolbar.

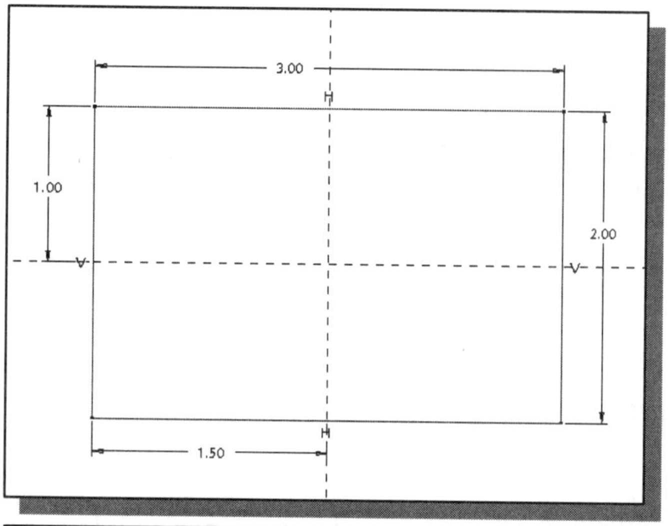

5. On your own, create a rectangle and modify the dimensions as shown in the figure.

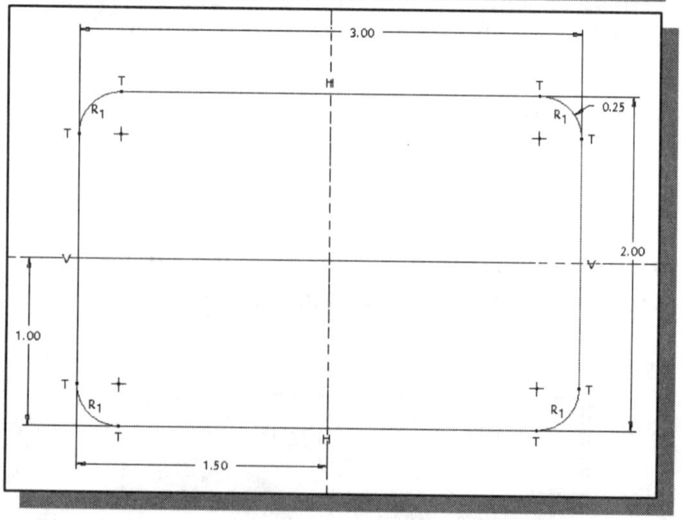

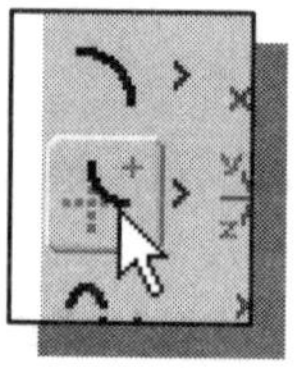

6. Choose the **Fillet** command in the *Sketcher* toolbar.

7. On your own, create the four rounded corners and modify the radius dimension to **0.25** as shown in the below figure. Note the applied **Equal radii** constraint (**R_1**).

8. On your own, establish the ***parametric relations*** so that the two referenced datum planes pass through the center of the 2D sketch.

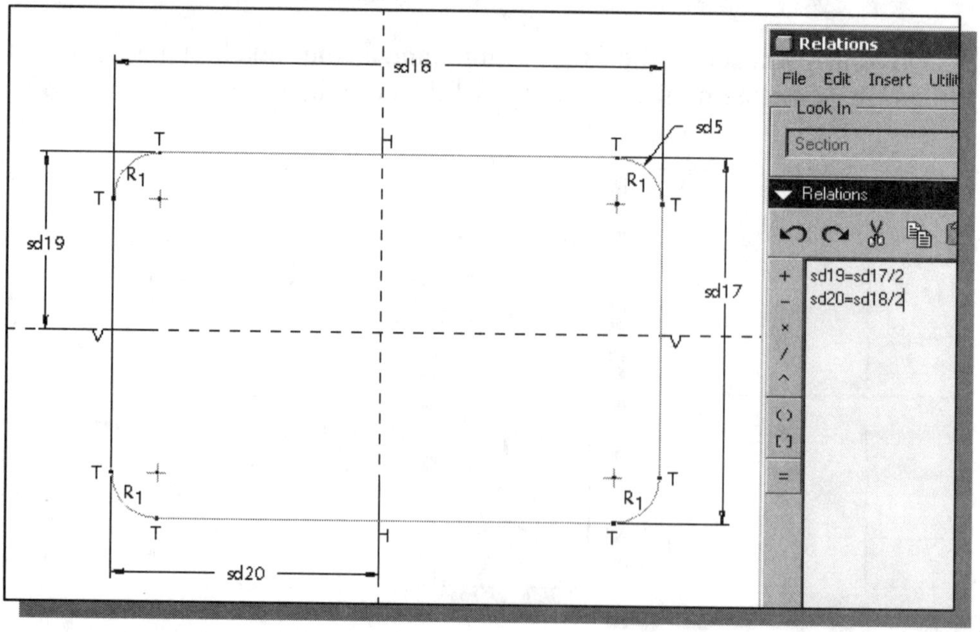

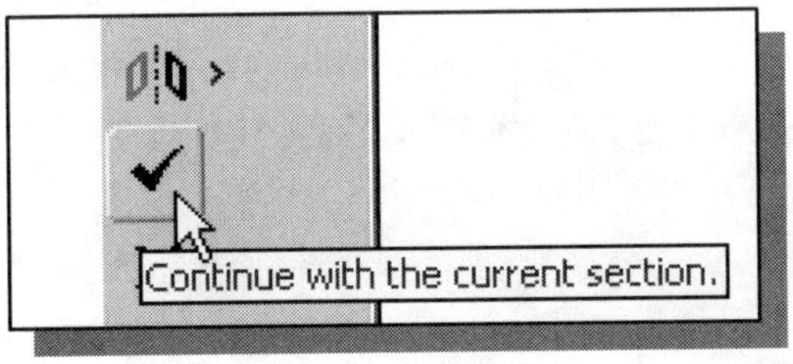

9. In the *Sketcher* toolbar, click on the **Accept** icon to end the *Pro/ENGINEER 2D Sketcher* and proceed to the next element of the feature definition.

10. Use the key combination **[Ctrl-D]** to change the view orientation in the display area.

11. In the *depth value* box, enter **0.75** as the extrusion depth.

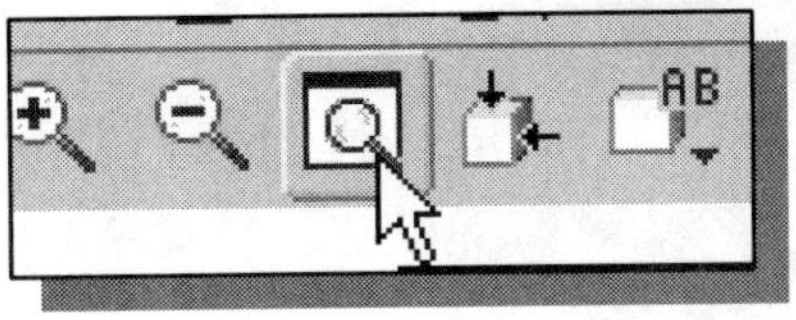

12. Click on the **Refit** icon to adjust the display so that all objects fit inside the display area. Also, use the *Dynamic Viewing* function to view the current model.

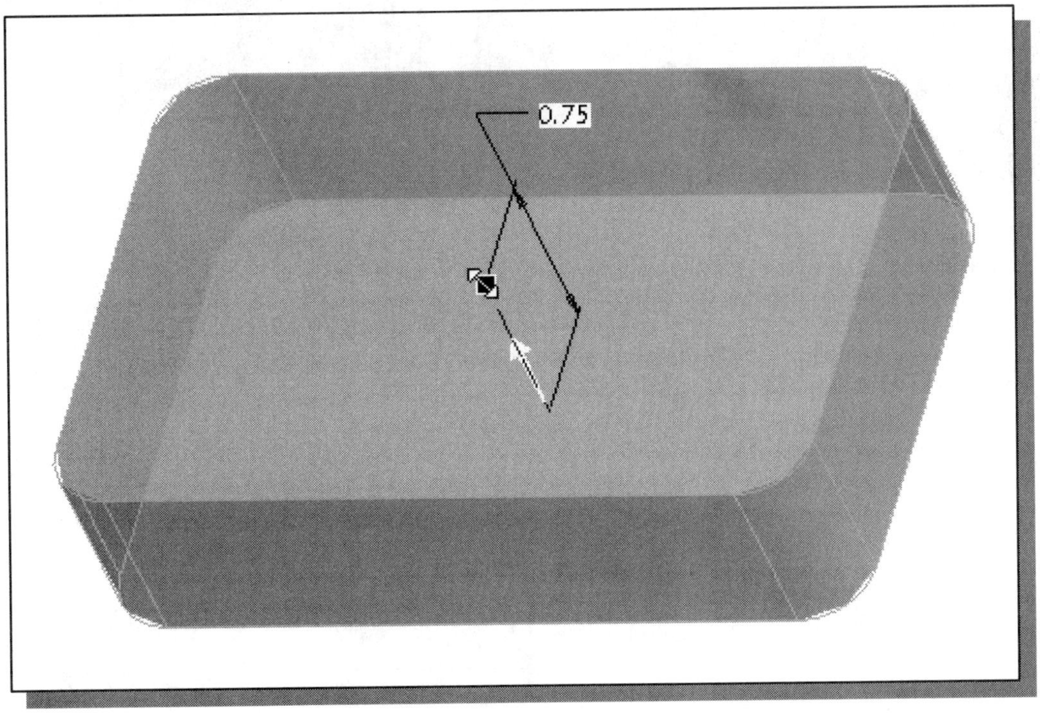

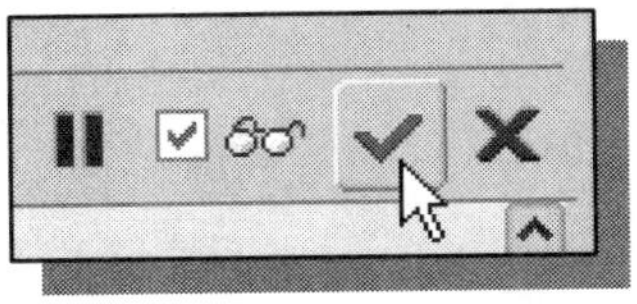

13. In the *Extrude Option Dashboard*, click on the **Accept** button to proceed with the feature creation.

Creating a Datum Axis

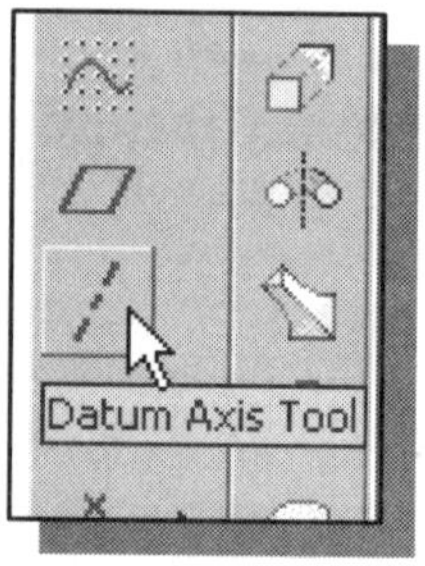

1. In the *Datum* toolbar, select **Datum Axis Tool** to start the creation of a new datum axis.

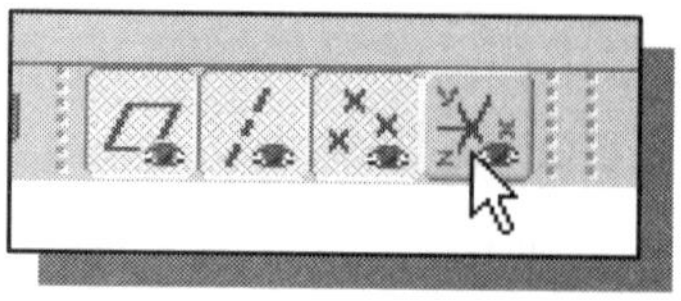

2. Click on the **Coordinate system display** icon once to toggle ***off*** the display of the coordinate system.

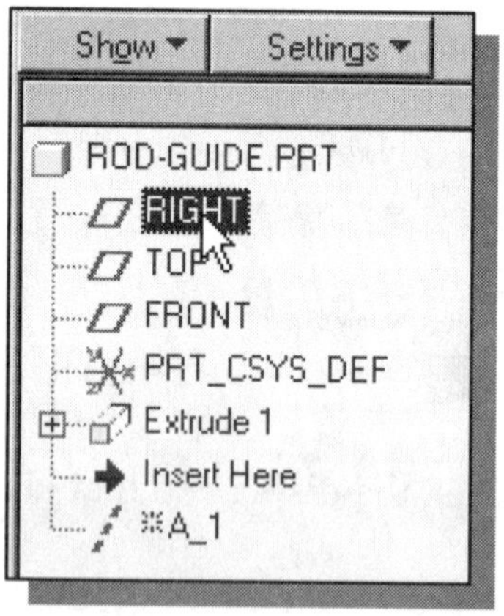

3. Inside the *Model Tree* area, pick the **RIGHT** datum plane as the first Placement Reference.

4. Inside the *Model Tree* area, hold down the **[CTRL]** key and select the **FRONT** datum plane as the second Placement Reference.

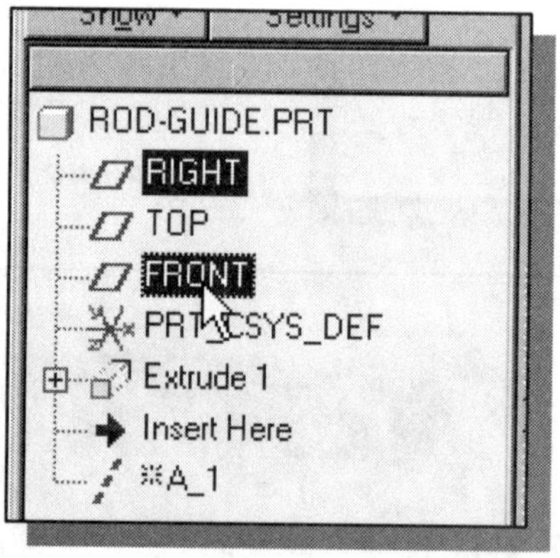

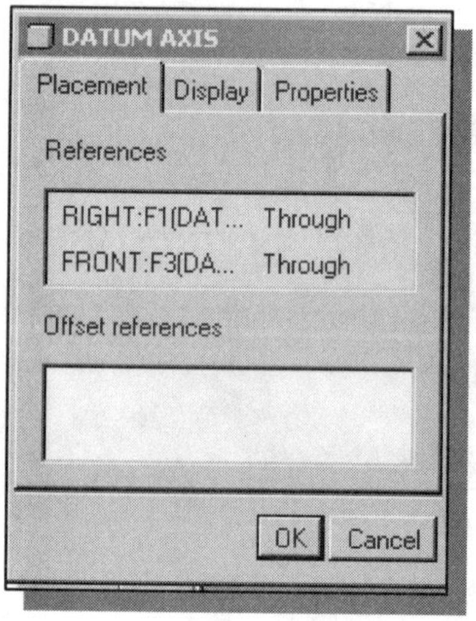

❖ Note that the selected placement references are displayed in the *Datum Axis* window and they are also highlighted in the *Model Tree* area.

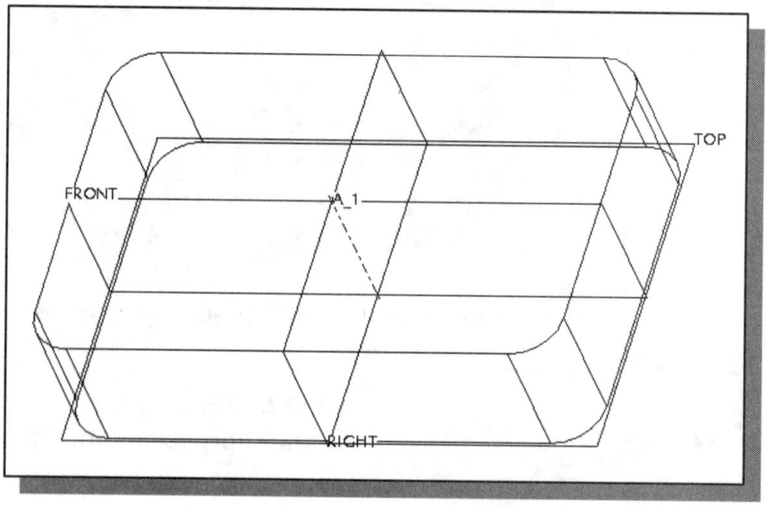

5. Click **OK** to accept the creation of the datum axis A_1, which passes through the intersection of the two selected datum planes.

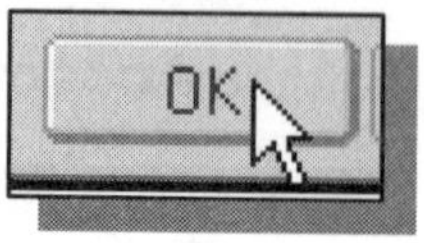

Creating a Datum Plane at an angle

1. In the *Datum Toolbar*, select **Datum Plane Tool** to start the creation of a new datum plane.

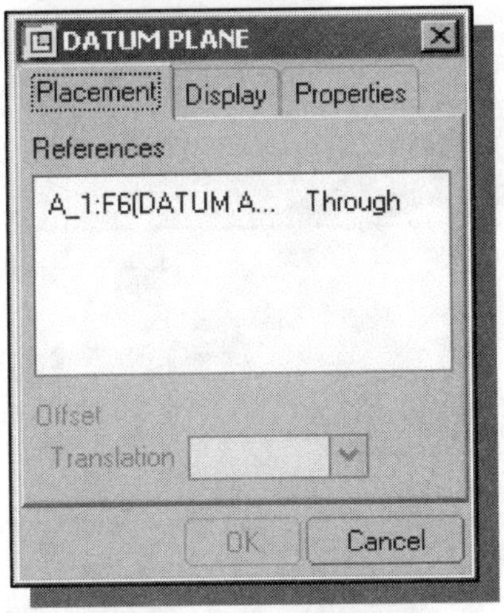

❖ Note that since datum axis A_1 was pre-selected, it is automatically placed in the *Datum Plane* window. *Pro/ENGINEER* will attempt to create a datum plane passing through the selected datum axis. But more definitions are required to fully place the desired datum plane, thus the **OK** button is greyed out.

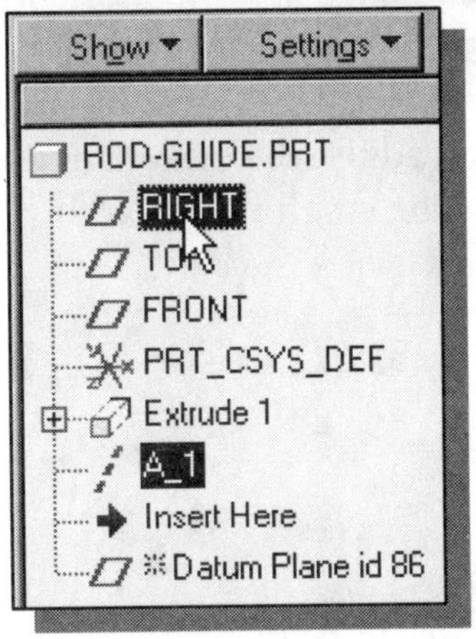

2. Inside the *Model Tree* area, hold down the [**CTRL**] key and pick the **RIGHT** datum plane as the second Placement Reference. Notice the selected reference plane is used, by default, as a rotation reference.

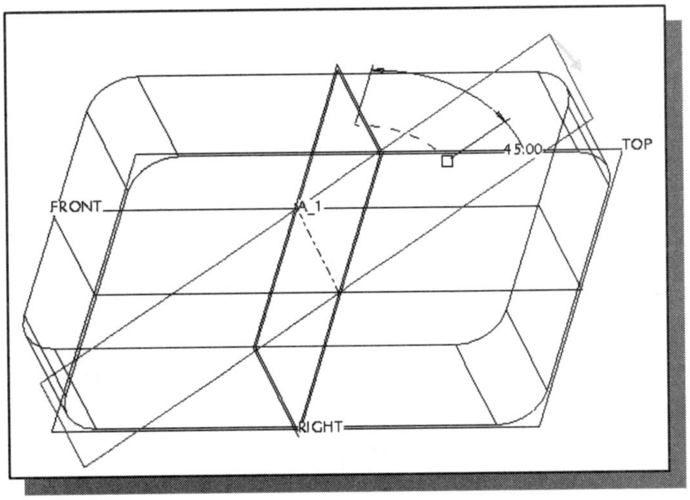

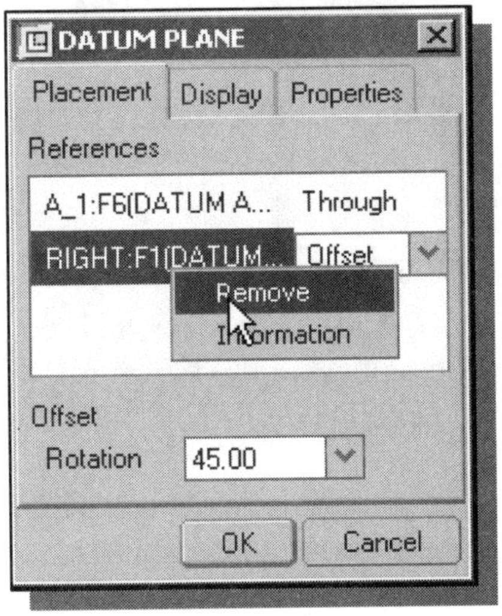

3. Inside the *Datum Plane* window, right-mouse-click once on the **RIGHT** datum plane and select the **Remove** option.

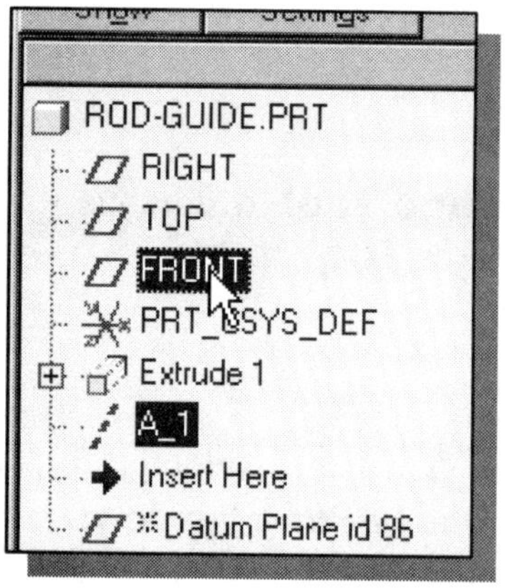

4. Inside the *Model Tree* area, hold down the [**CTRL**] key and pick the **FRONT** datum plane as the second Placement Reference.

5. Click once with the left-mouse-button on the **FRONT** datum plane inside the *Datum Plane* window as shown in the figure below.

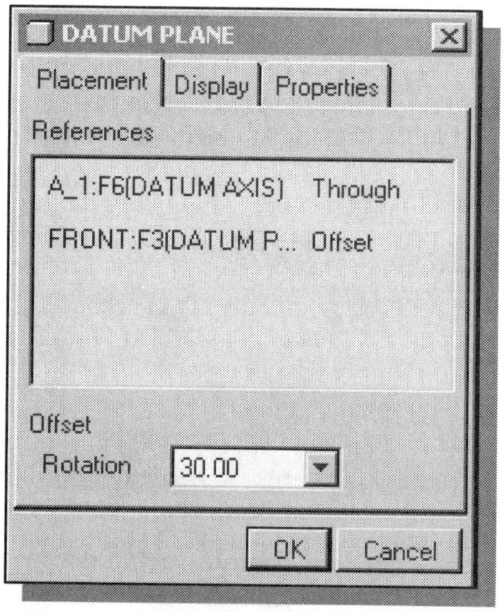

6. Click on the ***downward arrow*** to display the available placement options as shown.

7. On your own, click on the **Normal** option and observe the placement of the datum plane with this option.

8. Reset the placement option to **Offset** and enter **30** as the new Rotation angle as shown in the figure. (Note that we can also enter a negative value.)

9. Drag, with the left-mouse-button, on the ***adjust handle*** and adjust the offset angle dynamically on the screen as shown. (The adjust handle is the small box next to the dimension value)

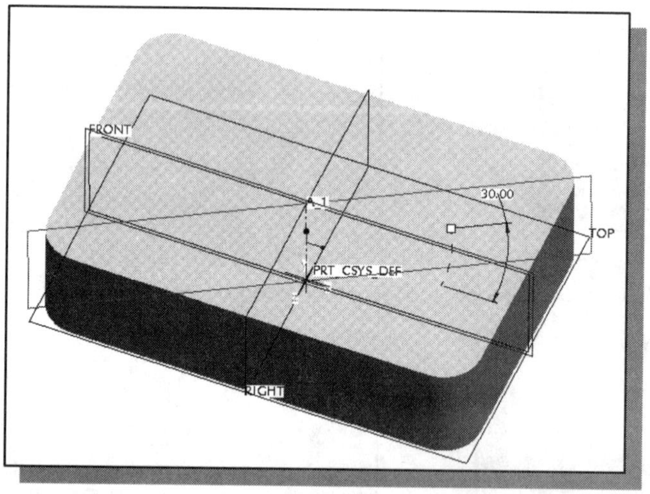

❖ Note that the new ***adjust handle*** option in *Pro/ENGINEER* allows the user to dynamically adjust dimension values without typing the values on the keyboard. The new handle box option is available in many of the *Pro/ENGINEER* commands.

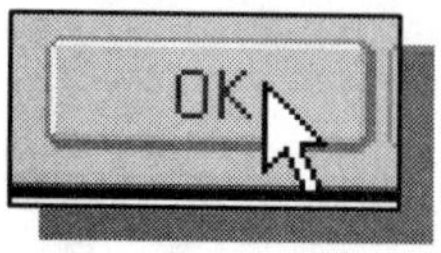

10. Click **OK** to accept the creation of the new datum plane **DTM1**.

Create the next Solid Feature

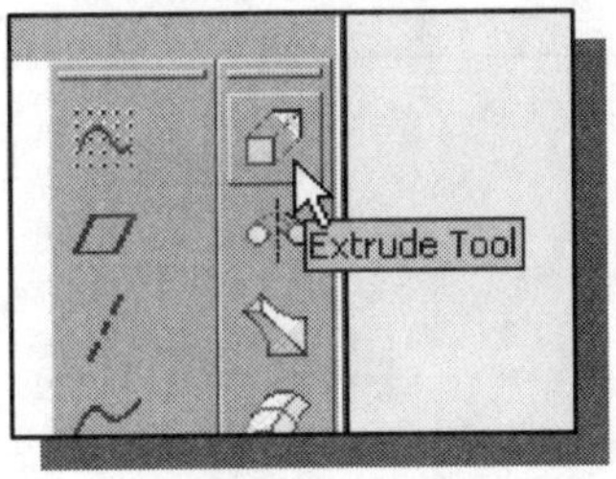

1. In the *Feature* toolbars (toolbars aligned to the right edge of the main window), select the **Extrude Tool** option as shown.

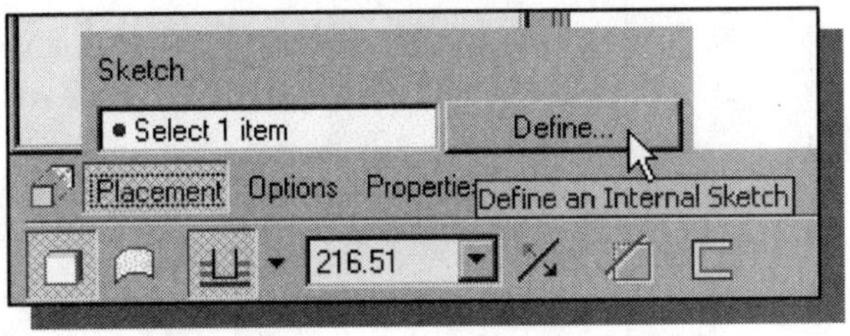

2. Click the **Placement** option and choose **Define** to begin creating a new *Internal sketch.*

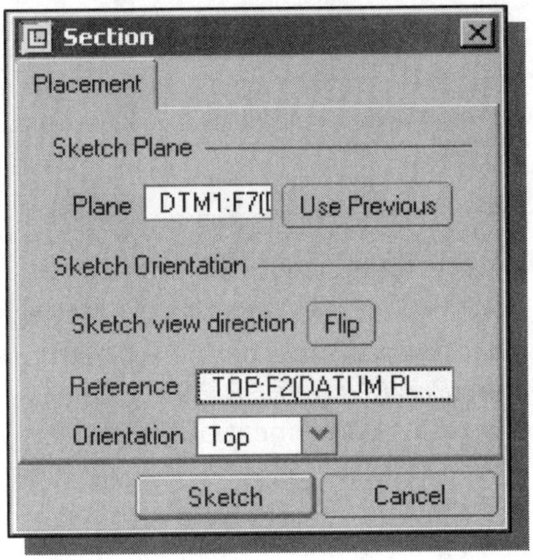

3. Pick **DTM1** as the sketching plane.

4. On your own, set the **Sketch Orientation** to use the **TOP** datum plane as the reference and the **Orientation** option to **Top** as shown in the figure.

5. Pick **Sketch** to exit the *Section Placement* window and proceed to enter the *Pro/ENGINEER Sketcher* mode.

6. Select datum axis **A_1** and the top edge of the solid model as the two primary sketch references as shown in the figure below.

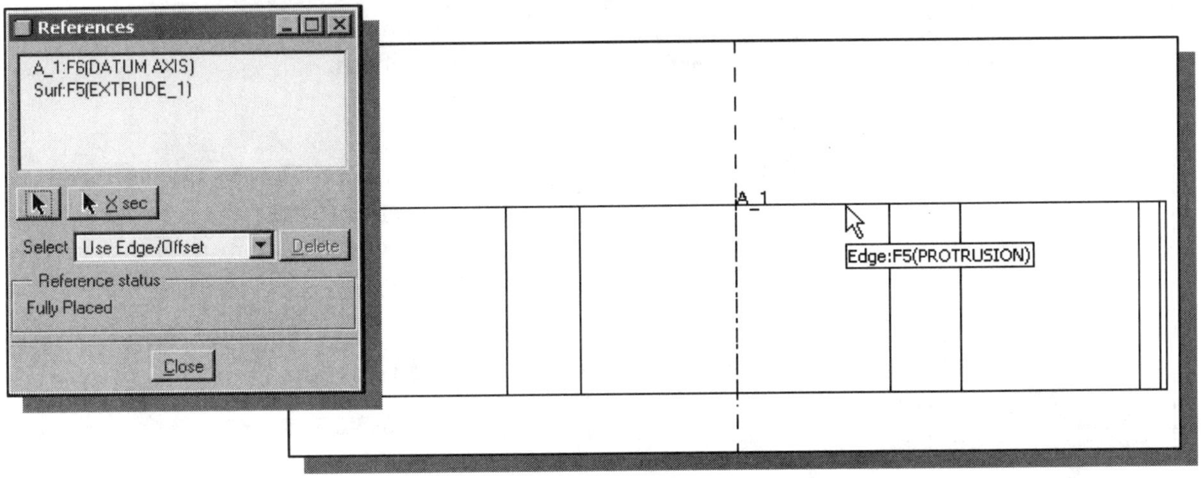

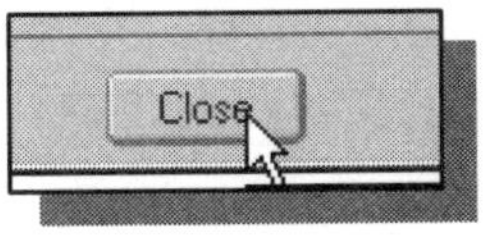

7. Click on the **Close** button to accept the settings.

8. On your own, create the sketch and modify the dimensions as shown in the figure below. Note the different constraints used in the sketch.

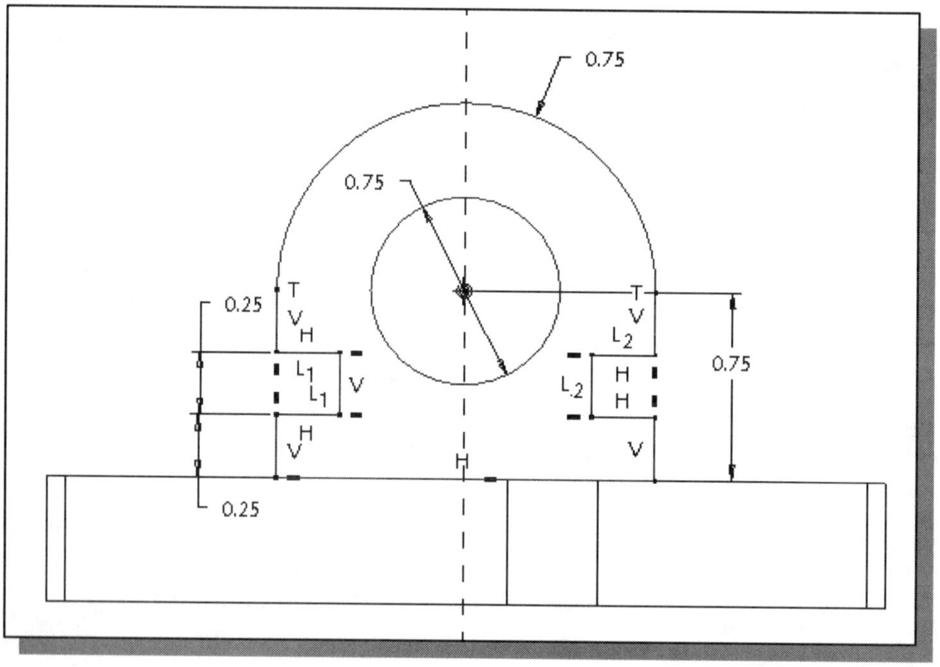

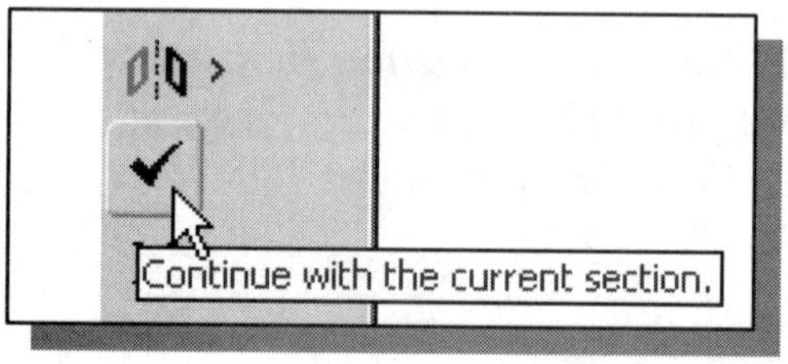

9. In the *Sketcher* toolbar, click on the **Accept** icon to end the *Pro/ENGINEER 2D Sketcher* and proceed to the next element of the feature definition.

10. In the *Feature Options Dashboard*, choose the **Both Sides** option as shown.

11. In the *depth value* box, enter **1.0** as the extrusion depth.

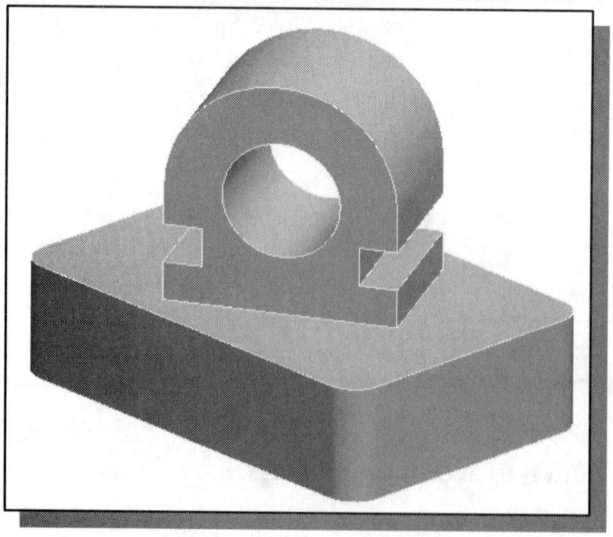

12. In the *Extrude Option Dashboard*, click on the **Accept** button to proceed with the feature creation.

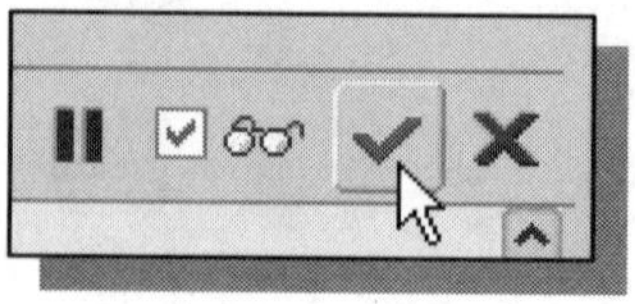

Creating a Datum Plane Using the Filter option

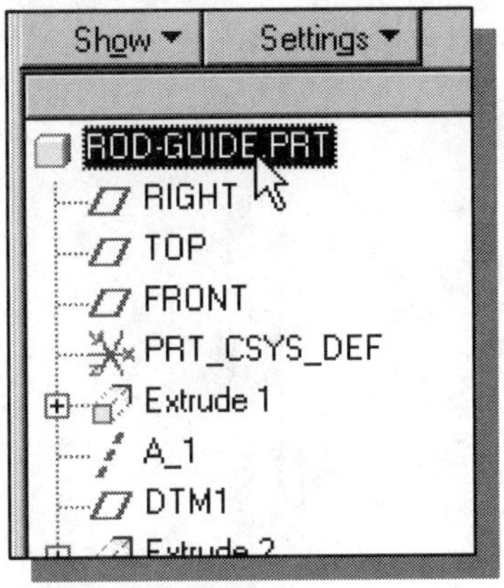

1. Inside the *Model Tree* area, select the **Rod-Guide** part name. By doing so, the pre-selected item (the solid feature we created in the previous section) is de-selected.

❖ Note that if we accept the default pre-selection to create the next datum plane, the sketching plane of the solid feature will be used as the first placement reference. To illustrate the use of the new **Selection Filter** option, we will first de-select any pre-selected item and.

2. In the *Datum* toolbar, select **Datum Plane Tool** to start the creation of a new datum plane. The message "*Select up to 3 references, such as plane, surface, edge or point to place plane.*" is displayed in the message area.

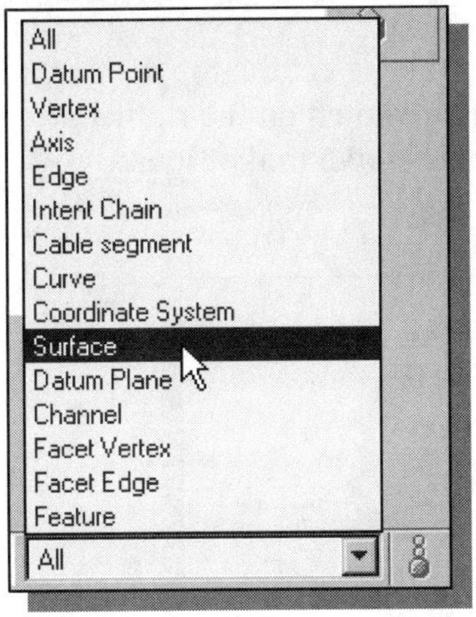

3. Click on the ***downward arrow*** of the *Selection Filter* option, which is located near the **bottom-right corner** of the main screen, to display the different selection options as shown. (Note that the default option is set to **All**, which simply means that all objects are selectable.)

4. Select **Surface** in the option list as shown.

5. Move the cursor on top of the solid model and notice only surfaces of the model can be selected. (*Pro/ENGINEER* automatically highlights selectable objects when cursor is on top of the object.)

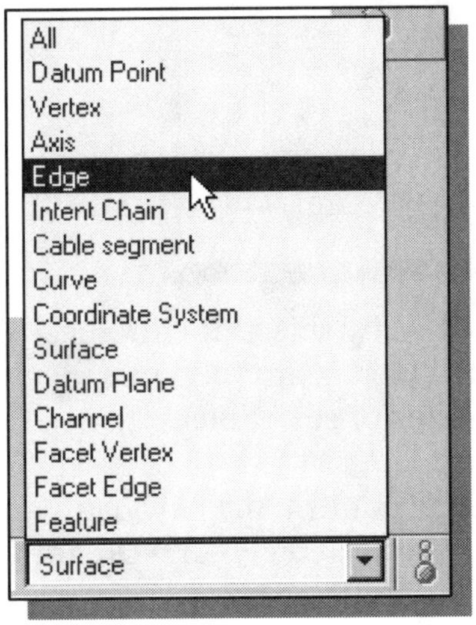

6. Reset the *Selection Filter* option to the **Edge** option as shown.

❖ The *Selection Filter* option allows us to quickly select a specified type of entity, which can be difficult to do when entities are on top of each others.

7. Inside the graphics area, select the ***front edge*** of the top surface of the base feature as shown.

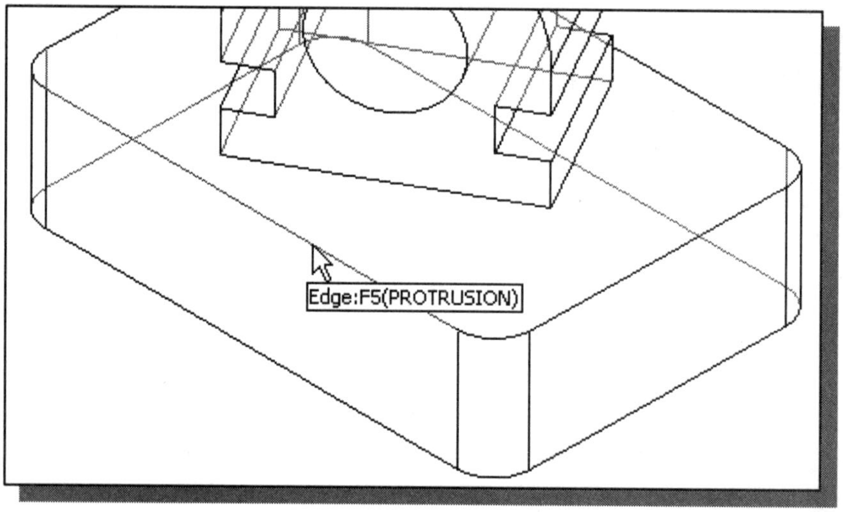

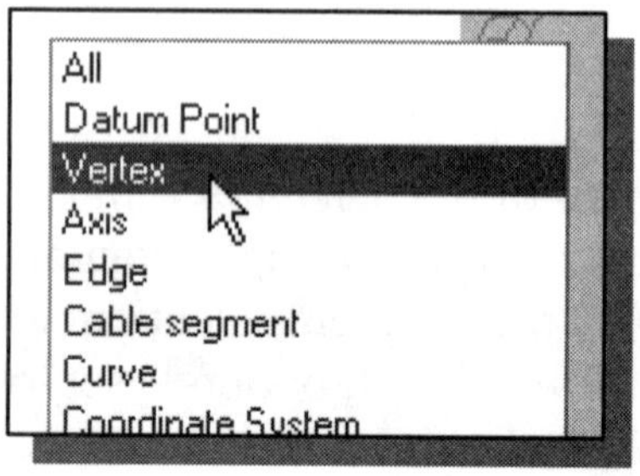

8. Reset the *Selection Filter* option to the **Vertex** option as shown.

9. Hold down the **[CTRL]** key and select the top inside front right corner of the square notch on the right side of the second solid feature as shown in the figure below.

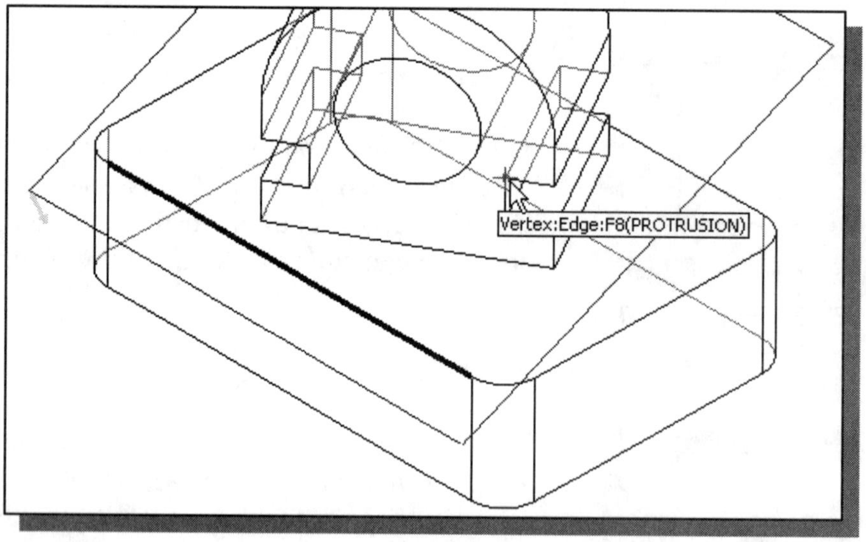

❖ The definition of a new datum plane can be accomplished by selecting different types of entities that belong to different features. Note that it is also possible to establish the same definitions with different combinations of selections.

10. On your own, use the *Dynamic Viewing* function and confirm the new datum plane passes through both the top front edge and the selected corner. (**Do not** end the *Datum Plane* window; more options are illustrated next.)

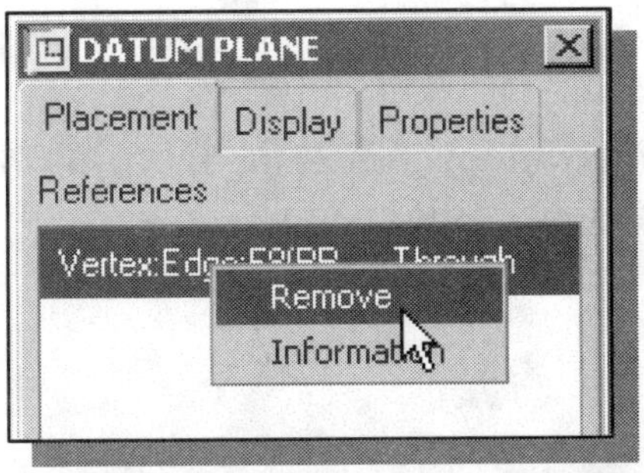

11. On your own, **Remove** both of the references in the *Datum Plane* window. (Hint: Use the right-mouse-button to bring up the option menu.)

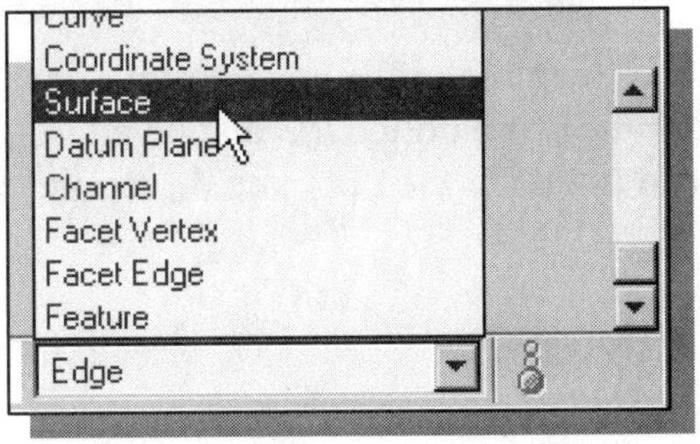

12. Reset the *Selection Filter* option to the **Surface** option as shown.

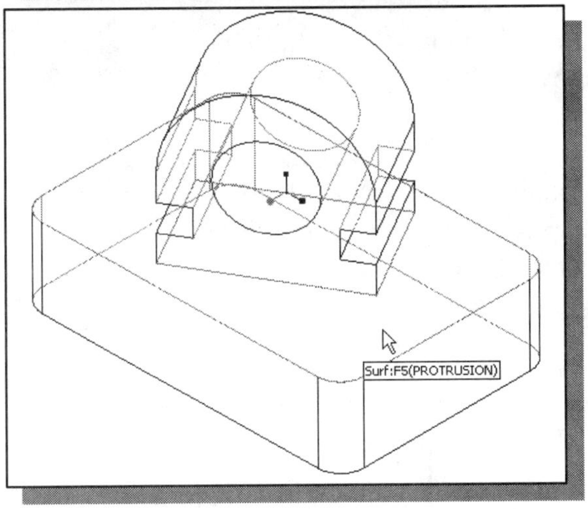

13. Select the ***top surface*** of the base feature when the surface is highlighted as shown.

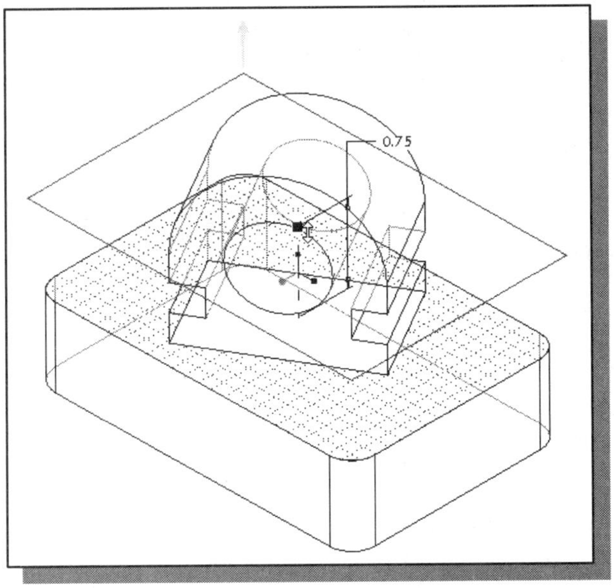

14. On your own, use the adjust handle and set the new datum plane to be **0.75** inches above the selected reference as shown.

❖ Note that negative values can also be used for the offset distance. A negative value simply means that the measurement is opposite to the positive side of the datum plane.

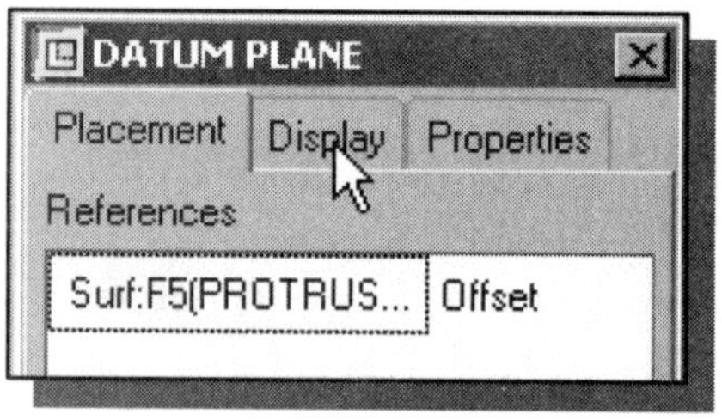

15. Inside the *Datum Plane* window, click the **Display** tab to switch to the display option page.

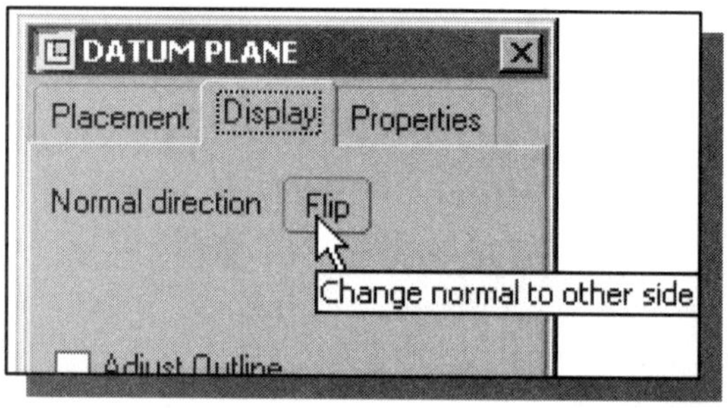

16. Click once, with the left-mouse-button, on the **Flip** button and notice the displayed arrowhead points in the opposite direction.

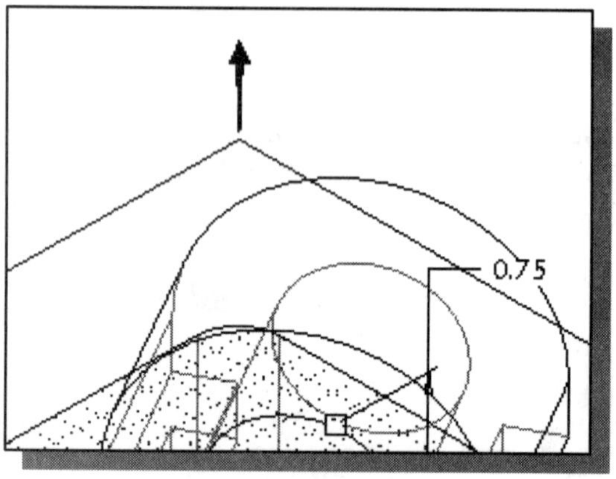

17. On your own, set the Arrow direction as shown.

❖ In *Pro/ENGINEER*, each plane/surface has a positive side and a negative side. The displayed arrowhead identifies the direction vector of the plane, which is defined as the positive side of the plane.

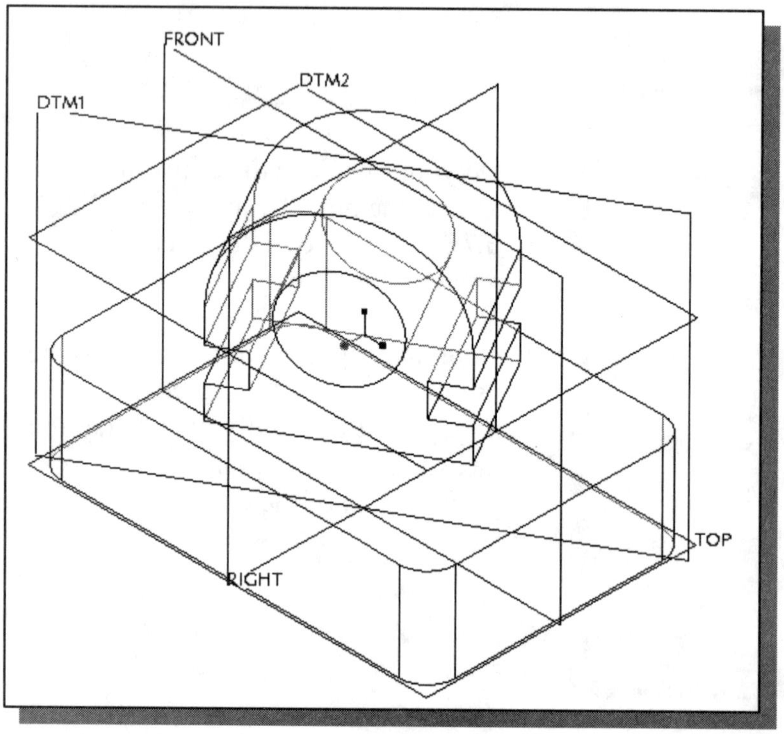

18. Click **OK** to accept the creation of the new datum plane **DTM2**.

Creating a Placed Feature

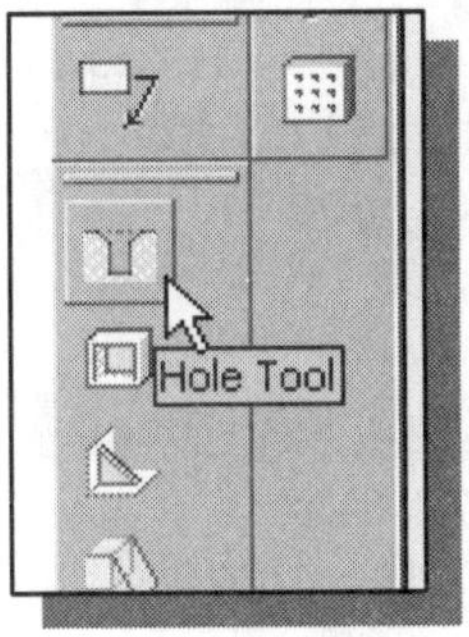

1. In the *Feature* toolbars (toolbars aligned to the right edge of the main window), select the **Hole Tool** option as shown.

❖ The message, "*Select a surface, axis or point to place hole*" is displayed in the message area.

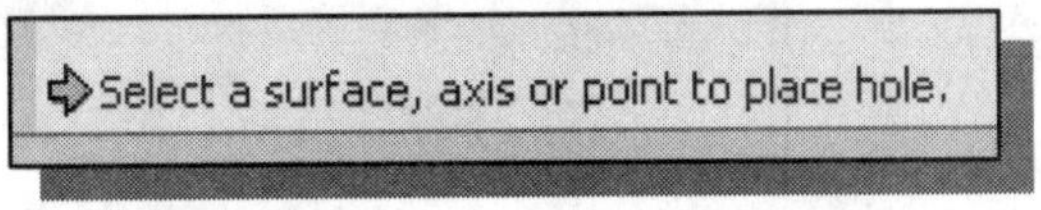

2. Select the datum axis **A_1** as the **primary reference**.

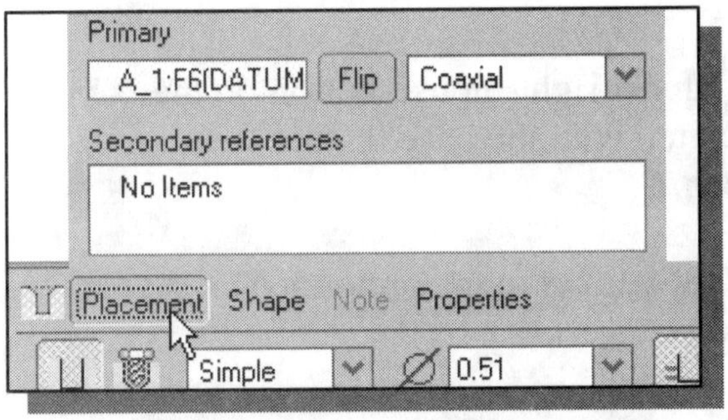

3. In the *Feature Options Dashboard*, select the **Placement** option to view the current setting. Note that since a datum axis has been selected as the primary reference, the **Coaxial** placement option is set. The center axis of new hole feature is aligned to the selected datum axis.

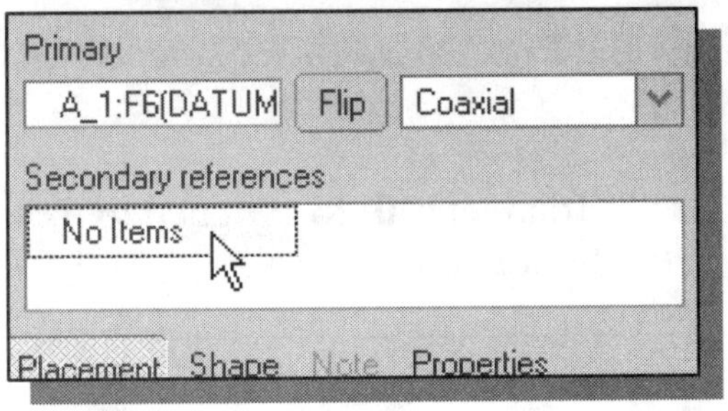

4. Click once with the left-mouse-button inside the list area of the secondary references area as shown.

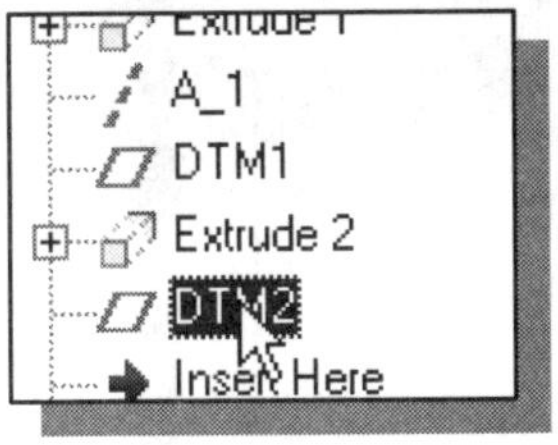

5. Select datum plane **DTM2** as the secondary placement reference.

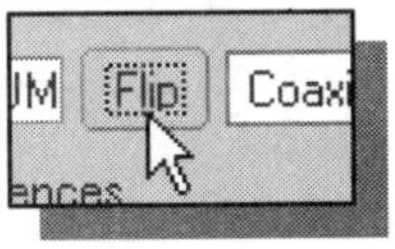

6. Click on the **Flip** button once to switch the placement direction.

7. Set the *hole diameter* to **0.25** and the *depth option* to **Thru All** as shown.

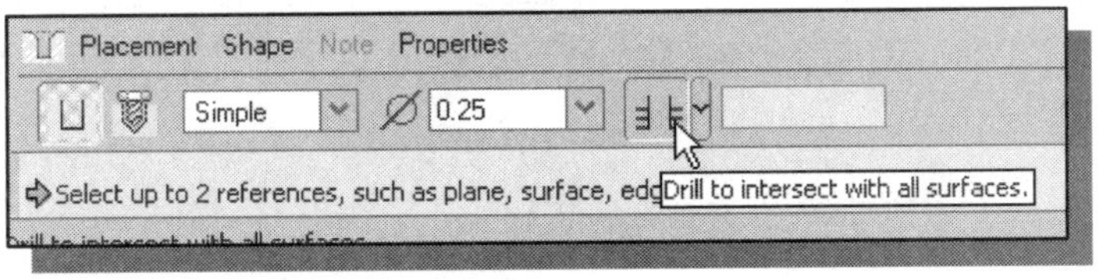

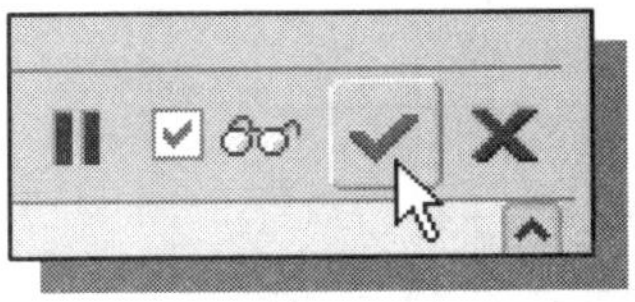

8. Click on the **Accept** icon and proceed to create the hole feature.

Add four holes to the Base Feature

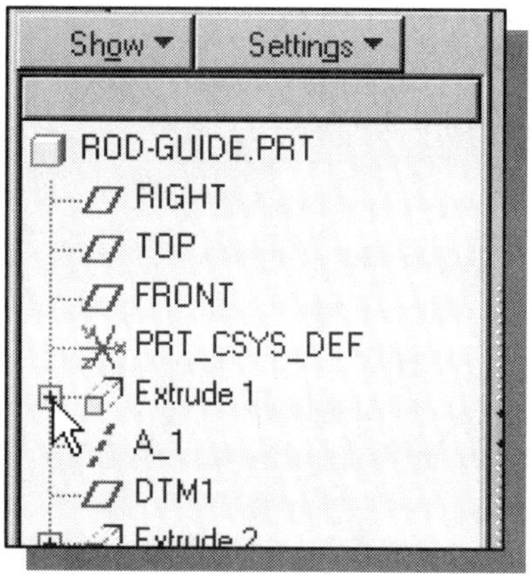

1. Inside the *Model Tree* area, click on the [**+**] marker in front of the first protrusion feature to expand the display of the feature list.

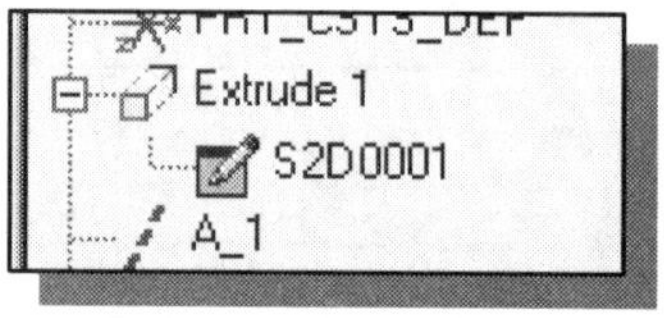

❖ The 2D section associated with the feature is displayed in the *Model Tree* area.

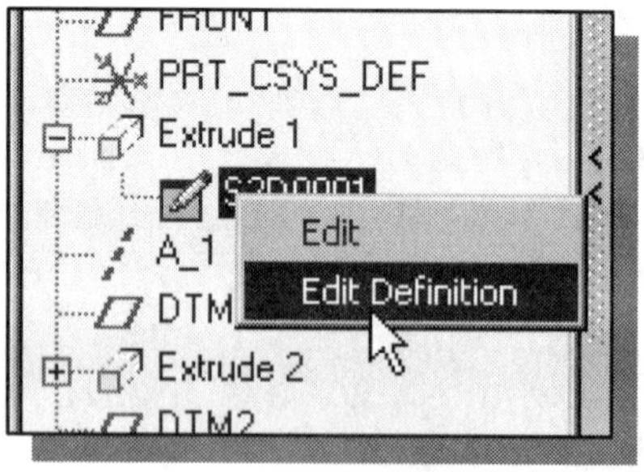

2. Click once with the right-mouse-button on the 2D section associated with the base feature to bring up the ***option menu***.

3. Select **Edit Definition** in the option list as shown.

❖ Note that we are switched automatically into the *Pro/ENGINEER Sketcher* mode.

4. On your own, create the four equal radii circles (diameter **0.25**) aligned to the centers of the rounded corners as shown in the figure below.

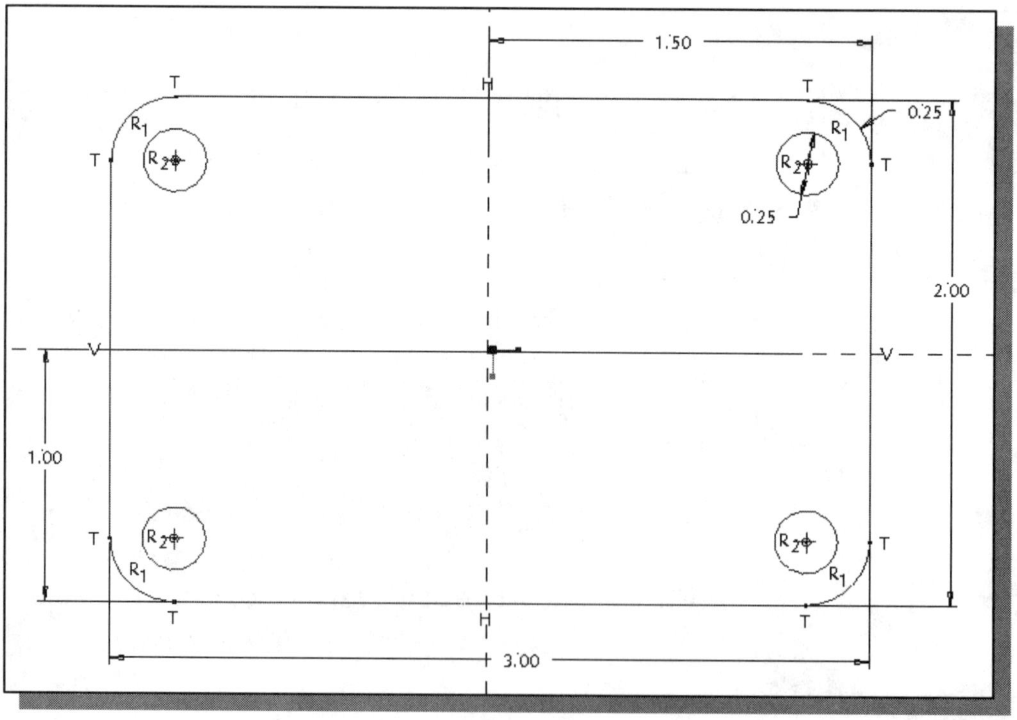

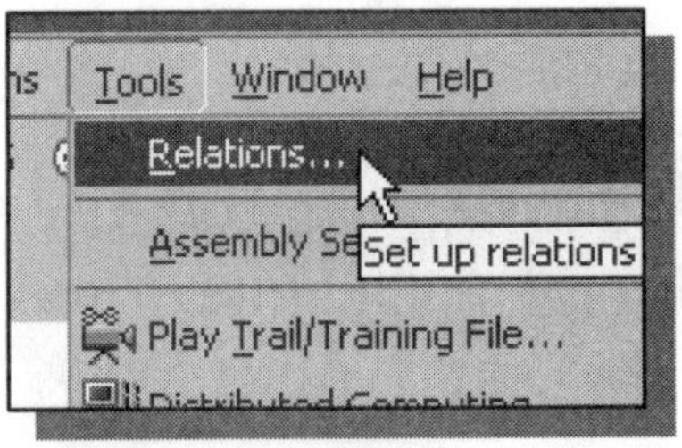

5. Pick **Relations** in the **Tools** pull-down menu.

6. Confirm the relations established earlier are still maintained.

7. Click **OK** to close the *Relations* window.

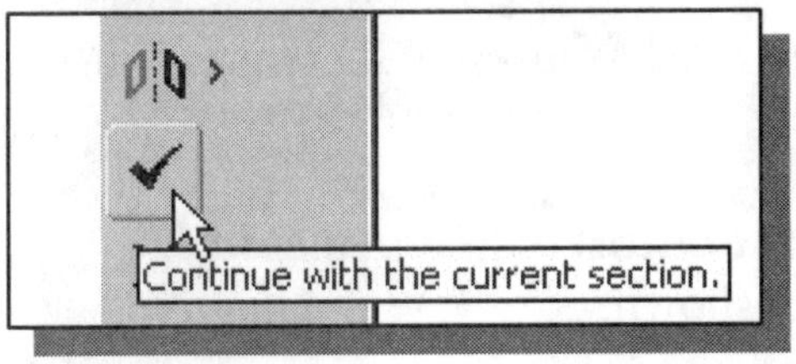

8. In the *Sketcher* toolbar, click on the **Accept** button to accept the modification and exit *Pro/ENGINEER Sketcher*.

9. Pick **OK** in the section dialog box.

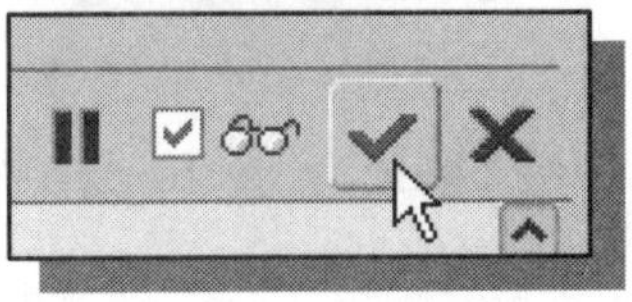

10. Pick **Accept** to complete the modification.

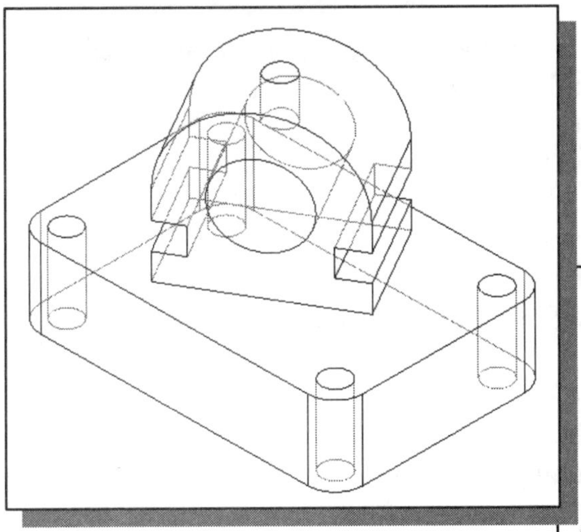

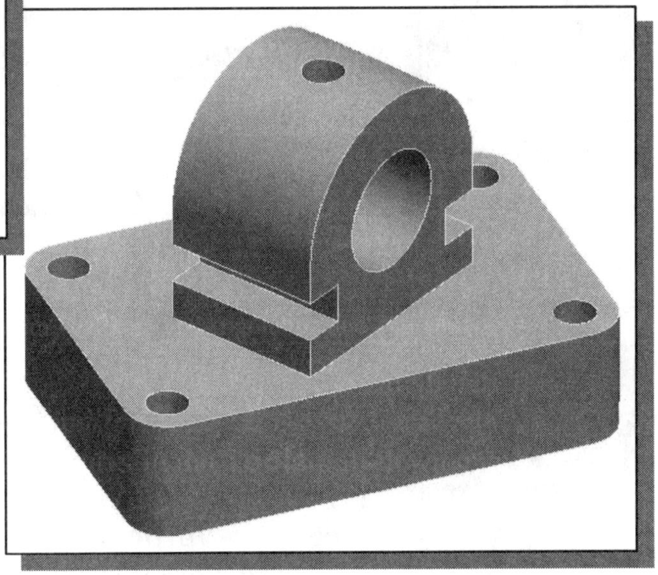

3D Annotation

1. In the *Feature* toolbar, select **Annotation** to start the creation of a 3D annotation feature. The message "*Insert an annotation feature.*" is displayed in the message area.

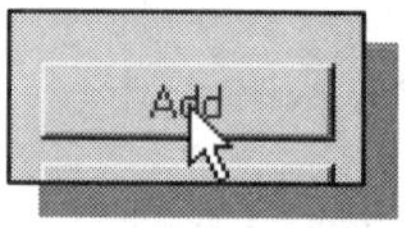

2. In the *Annotation* dialog box, click **Add** to insert a 3D annotation.

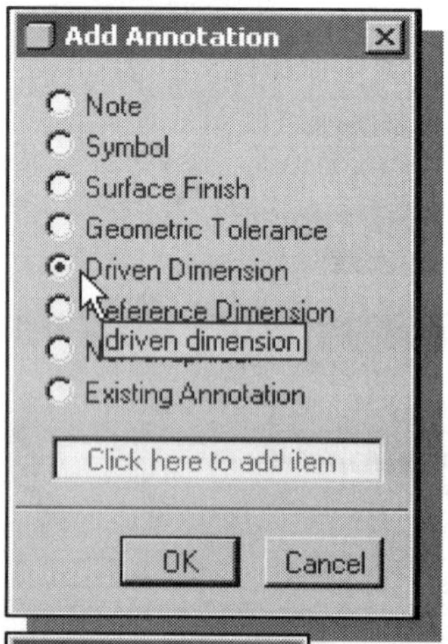

3. Activate the **Driven Dimension** option in the *Add Annotation* dialog box as shown.

4. Click **OK** to proceed with the option. The message "*Select a datum plane or symbol to which the new symbol will be parallel.*" is displayed in the message area.

5. In the display area, pick the **TOP** datum plane with the left-mouse-button.

6. In the *Menu Manager*, click **Okay** to accept the default setting in setting the orientation of the annotation.

7. Click **Done** to accept the 0 degree rotation.

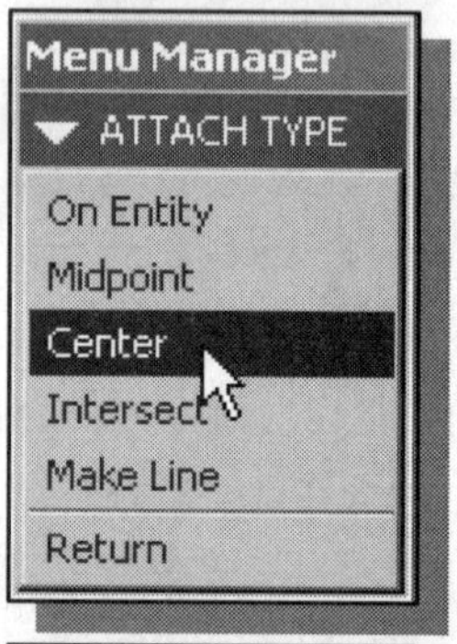

8. In the *Menu Manager*, set the *Attach Type* to **Center** as shown.

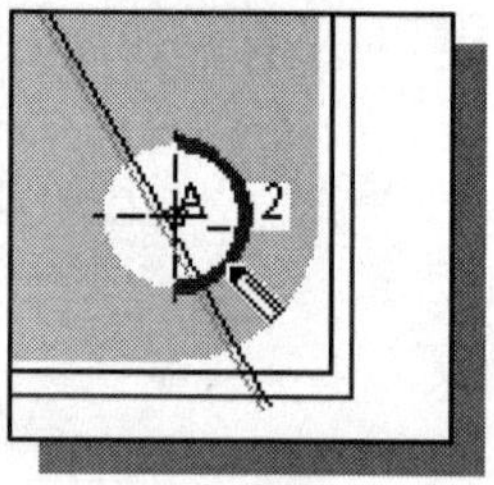

9. Select the two circles on the right side by clicking once on the entities.

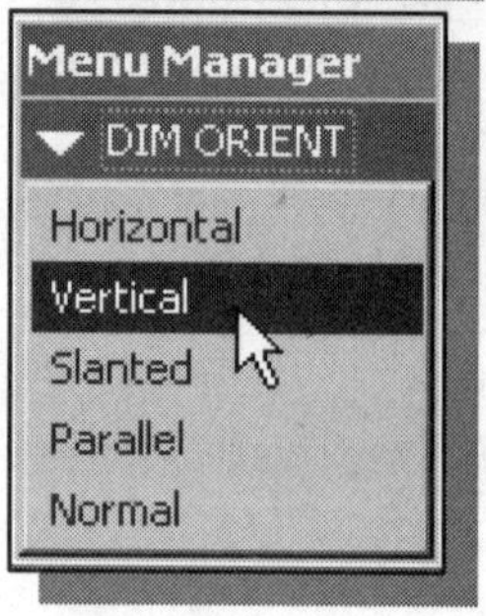

10. Accept the selections by clicking once with the **middle-mouse-button** at the desired location.

11. In the *Menu Manager* dialog, set the *Dimension Orientation* option to **Vertical** as shown. A driven dimension is created to document the 3D model.

12. Use the quick-key option **[Ctrl-D]** to change the view orientation in the display area.

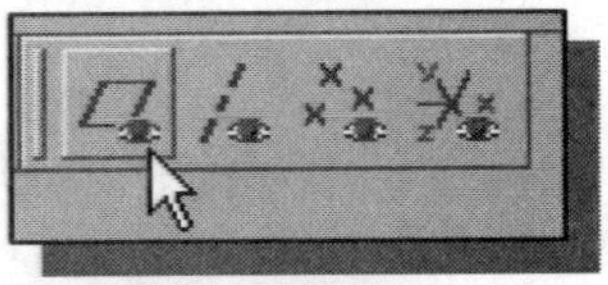

13. Turn ***off*** the display of the *datum features* by clicking on the associated icons in the *Standard* toolbar area.

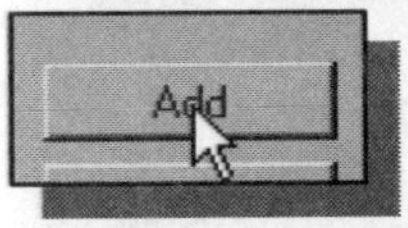

14. In the *Annotation* dialog box, click **Add** to insert a 3D annotation.

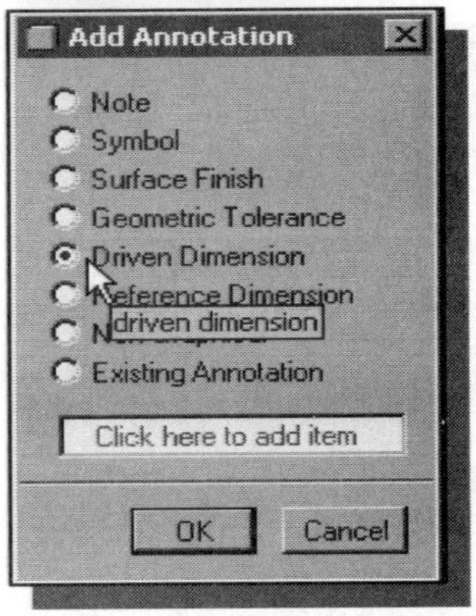

15. Activate the **Driven Dimension** option in the *Add Annotation* dialog box as shown.

16. Click **OK** to proceed with the option. The message "*Select a datum plane or symbol to which the new symbol will be parallel.*" is displayed in the message area.

17. In the display area, pick the **driven dimension** that we just created.

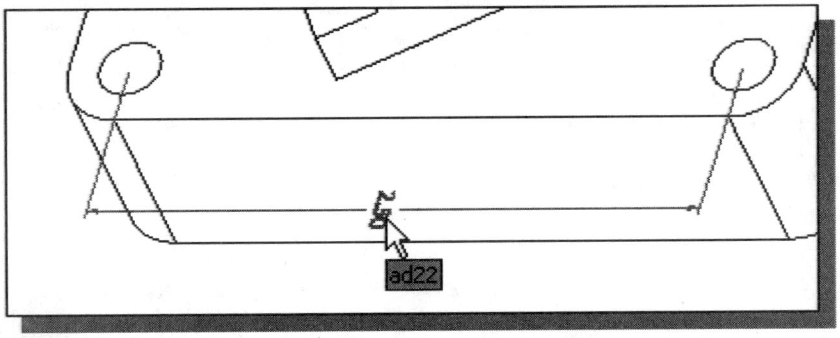

18. In the *Menu Manager*, click **Okay** to accept the default setting in setting the orientation of the annotation.

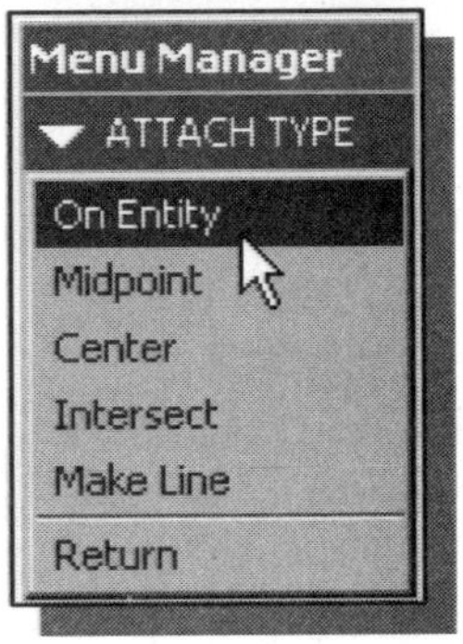

19. In the *Menu Manager*, set the *Attach Type* to **On Entity** as shown.

20. Use the quick-key option **[Ctrl-D]** to switch to the default 3D-view orientation in the display area.

21. Select the inside edge of the vertical section by clicking once on the entity as shown in the figure.

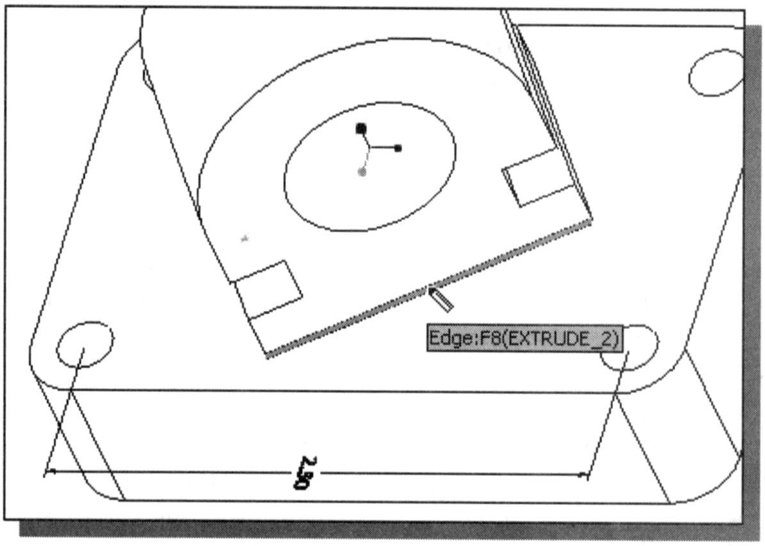

22. Pick the front top edge of the base feature as shown.

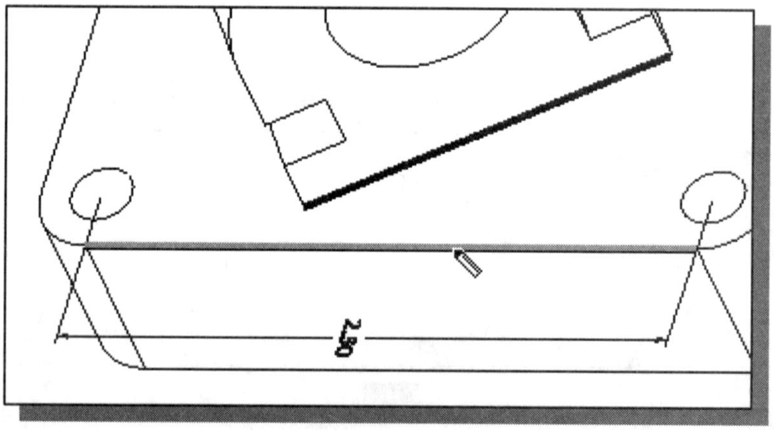

23. Click once with the middle-mouse-button inside the small region, toward the right side, between the two selected lines to create an **Angle Dimension**.

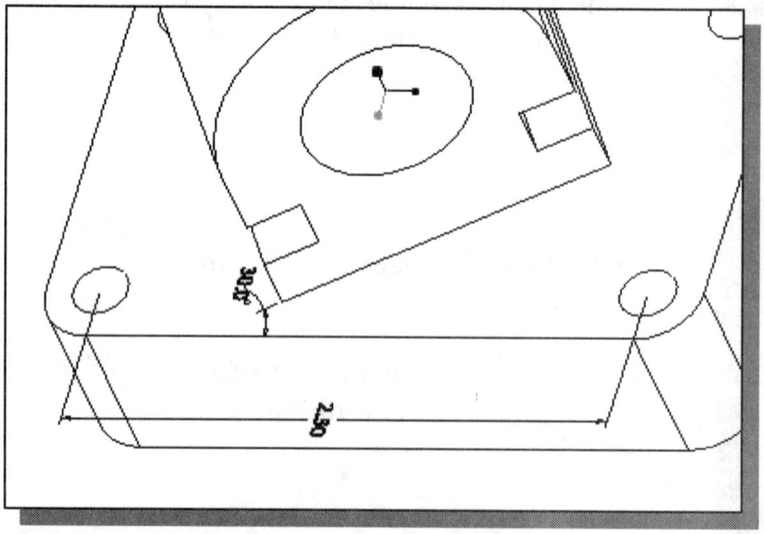

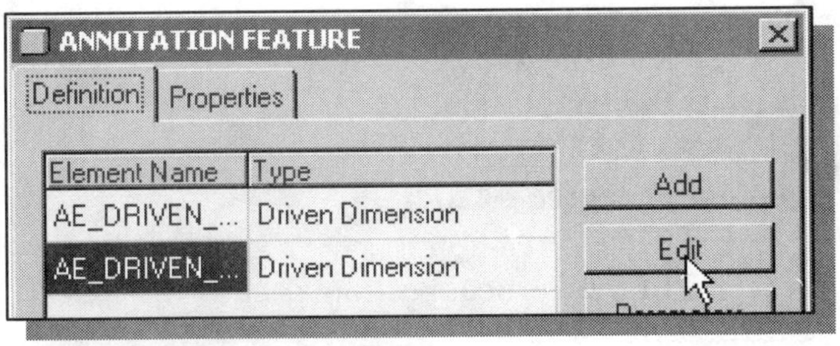

24. In the *Annotation Feature* dialog box, click **Edit** to modify the currently highlighted annotation.

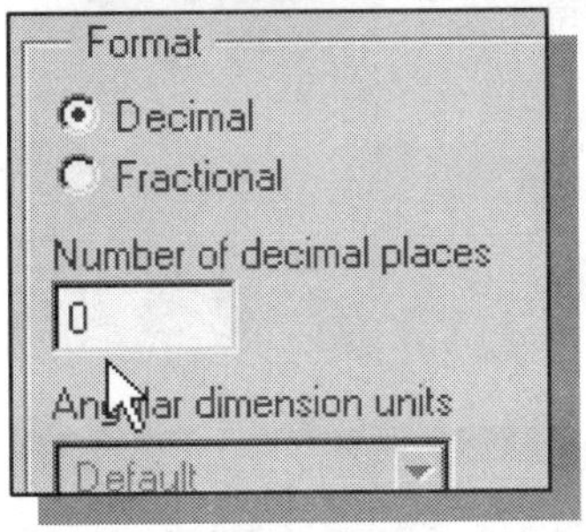

25. In the *Dimension Properties* dialog box, set the display of ***Number of decimal places*** to **0** as shown.

26. Click **Move** to reposition the highlighted dimension.

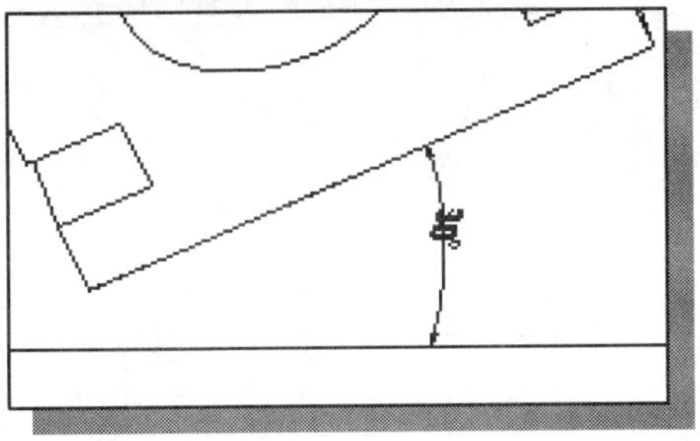

27. Inside the display area, pick a location to move the dimension to the new location.

28. Click **OK** to end the *Edit annotation* option.

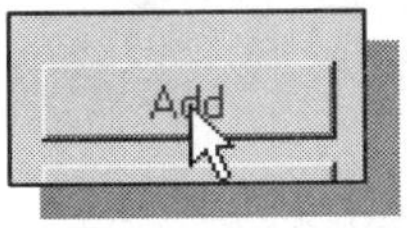

29. In the *Annotation* dialog box, click **Add** to insert a 3D annotation.

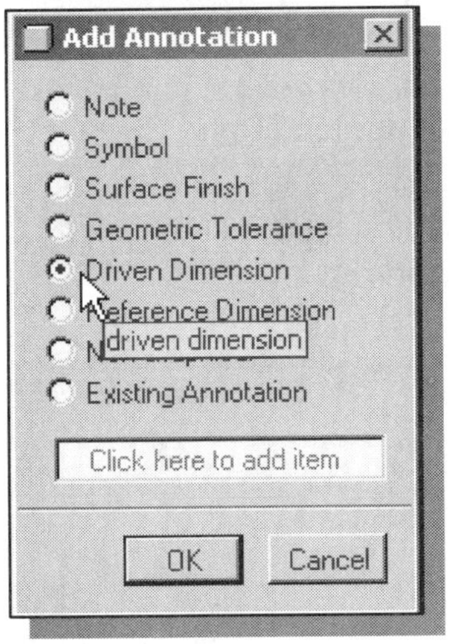

30. Activate the **Driven Dimension** option in the *Add Annotation* dialog box as shown.

31. Click **OK** to proceed with the option. The message "*Select a datum plane or symbol to which the new symbol will be parallel.*" is displayed in the message area.

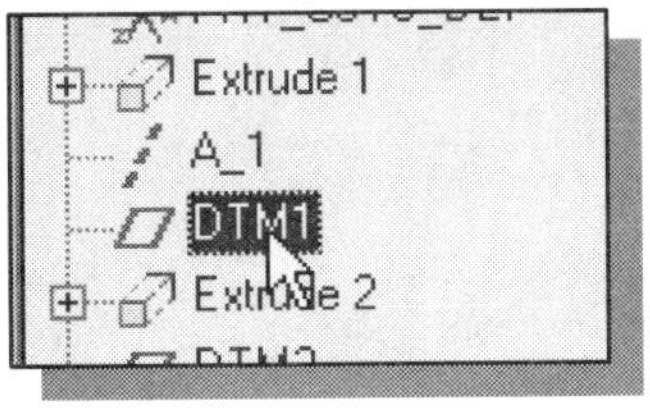

32. In the *Model Tree area*, pick the **DTM1** datum plane with the left-mouse-button.

33. In the *Menu Manager*, click **Okay** to accept the default setting in setting the orientation of the annotation.

34. Click **Done** to accept the 0 degree rotation.

35. Select the front large arc of the vertical section as shown.

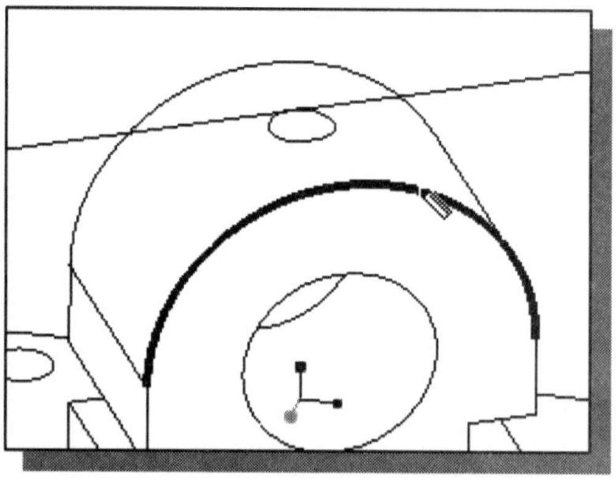

36. Accept the selection by clicking once with the middle-mouse-button at the desired location.

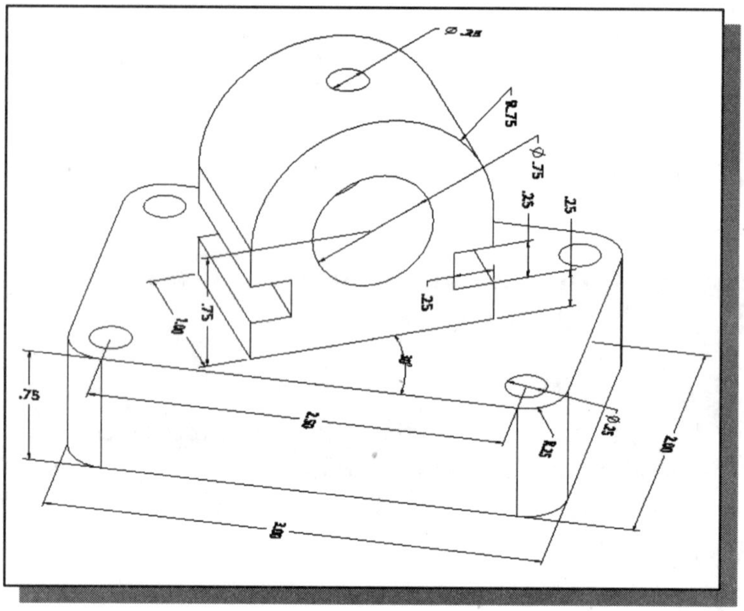

37. On your own, repeat the above steps and complete the 3D annotation of the model as shown.

❖ Note that the dimensions created with the *3D annotation* command cannot be used as control variables; they are used to document the completed solid model.

38. Inside the *Model Tree* window, **right-mouse-click** once on the Annotation feature to display the option menu. Note the different options available. Hide the 3D annotation by picking **Hide** before continuing to the next section.

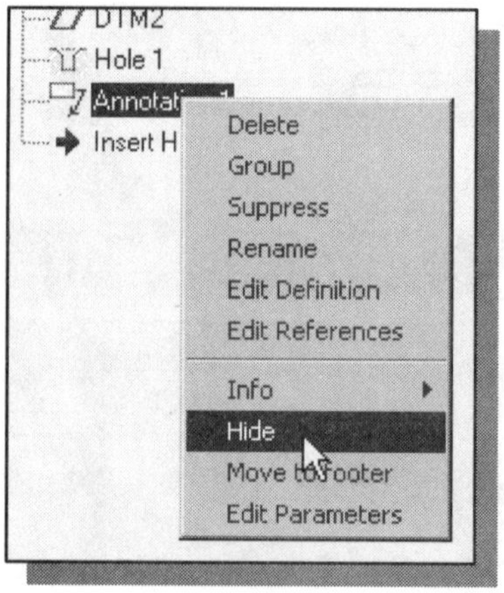

Creating a Multi-View Drawing using a Drawing Template

➤ *Pro/ENGINEER* allows us to generate 2D engineering drawings directly from the 3D solid models. An engineering drawing is a tool that can be used to communicate engineering ideas/designs to manufacturing, purchasing, service, and other departments. Until now we have been working in *Part Modeling* mode to create our design in ***full size***. We can arrange our design on a two-dimensional sheet of paper so that the plotted hardcopy is exactly what we want. This two-dimensional sheet of paper is known as the *Drawing* mode in *Pro/ENGINEER*. In general, each company uses a set of standards for drawing content, based on the type of product and also on established internal processes. The appearance of an engineering drawing varies depending on when, where, and for what it is produced. However, the general procedure for creating an engineering drawing from a solid model is fairly well defined. In *Pro/ENGINEER*, creation of 2D engineering drawings from solid models consists of four basic steps: drawing sheet formatting, creating/positioning views, annotations, and printing/plotting.

1. Pick the **New Object** icon in the toolbar. (We can also use the key combination **Ctrl-N** to start a new object.)

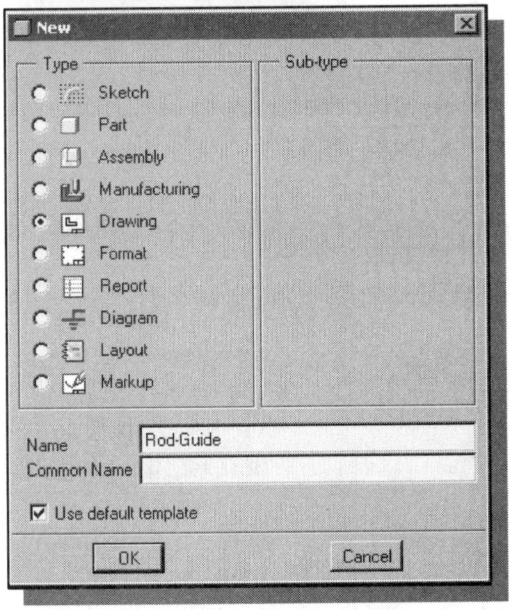

2. In the *New* form, select **Drawing** under the **Type** list.

3. Enter ***Rod-Guide*** as the drawing file **Name**.

4. Confirm the **Use default template** option is set.

5. Pick **OK** to start a new drawing file.

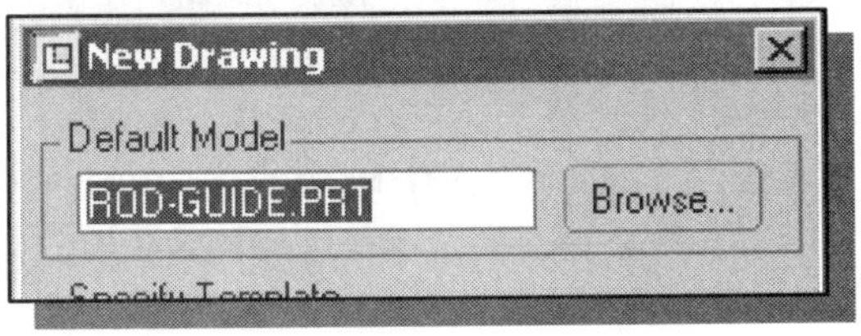

6. In the *New Drawing* form, the currently active part is automatically selected as the drawing model.

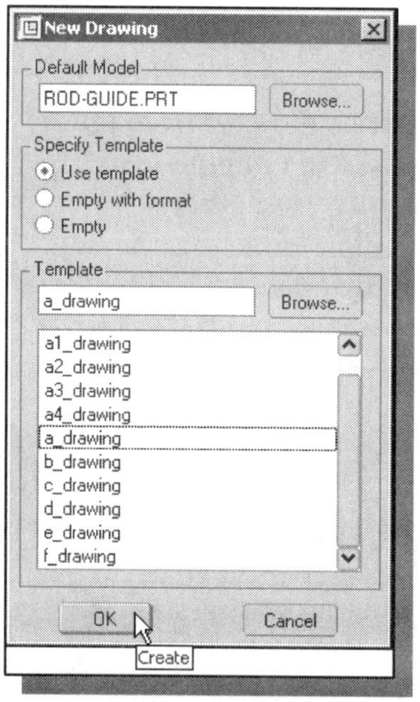

7. Confirm the **Specify Template** option is set to **Use template**.

8. Pick **a_drawing** size (8.5″ × 11″) in the **Template** list.

9. Pick **OK** to proceed with the drawing layout.

❖ Drawing templates can be used when creating a new drawing. The *Pro/ENGINEER* drawing templates, such as the **a_drawing** template used in this example, automatically create standard views and set the view display options.

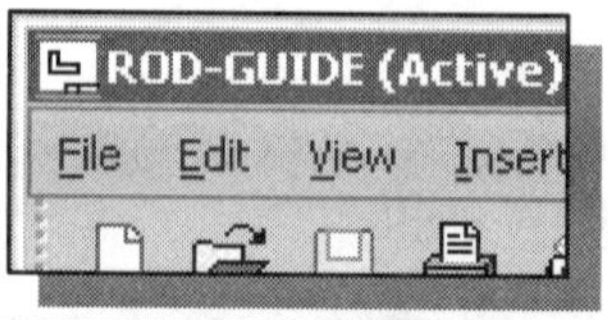

❖ A new window will appear with the title "***Rod-Guide***." Notice the icon at the top left corner of the window indicating the file type. Note that the part window is pushed into the background and can be made active at any time.

10. On your own, toggle ***off*** the display of the datum planes, datum axis, datum points and coordinate systems as shown in the figure.

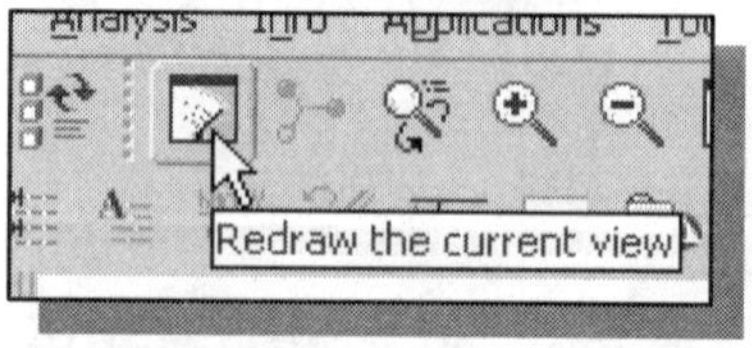

11. In the *Standard* toolbar, click **Redraw** to refresh the display in the graphics area.

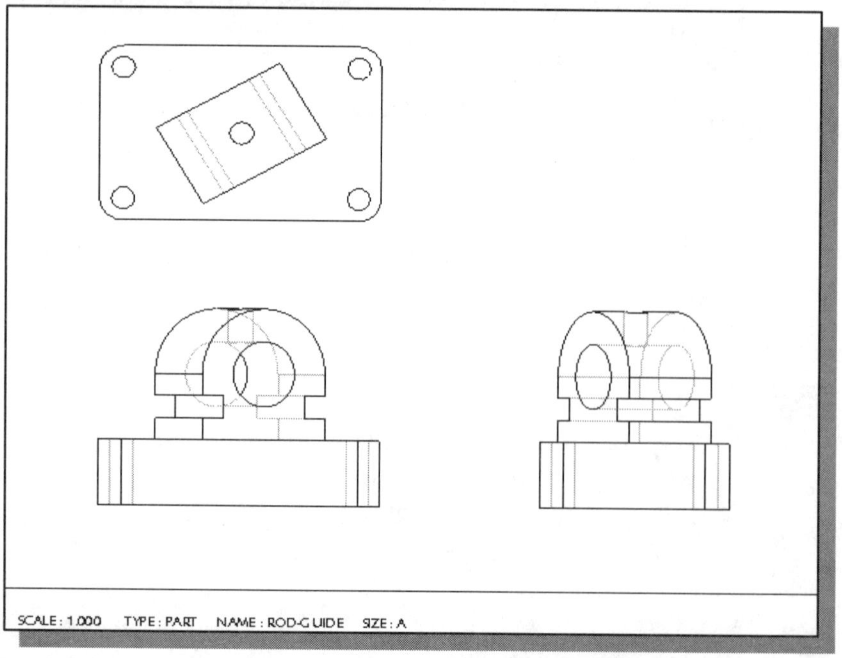

❖ The **a_drawing** template is set to the three standard views (**Top, Front** and **Right**) with the *view display option* set to the **Hidden line** option as shown in the figure. Note the ***Model Tree*** of the **Rod-guide** design is also displayed in the *Navigator* area.

Creating Another Multi-View Drawing

➢ For the ***Rod-Guide*** design, a drawing with an auxiliary view is needed for the angled feature of the design. In this section, the procedure required to create a drawing from scratch, along with the creation of specific views, is illustrated.

1. Pick the **New Object** icon in the toolbar. (We can also use the key combination **Ctrl-N** to start a new object.)

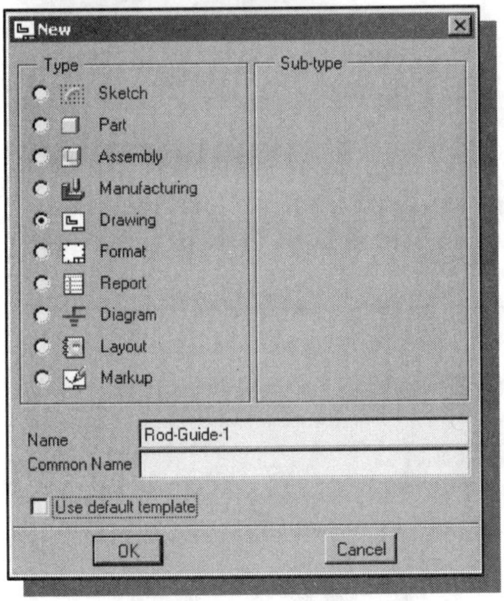

2. In the *New* form, select **Drawing** under the **Type** list.

3. Enter ***Rod-Guide-1*** as the drawing file **Name**.

4. Toggle *off* the **Use default template** option.

5. Pick **OK** to start a new drawing file.

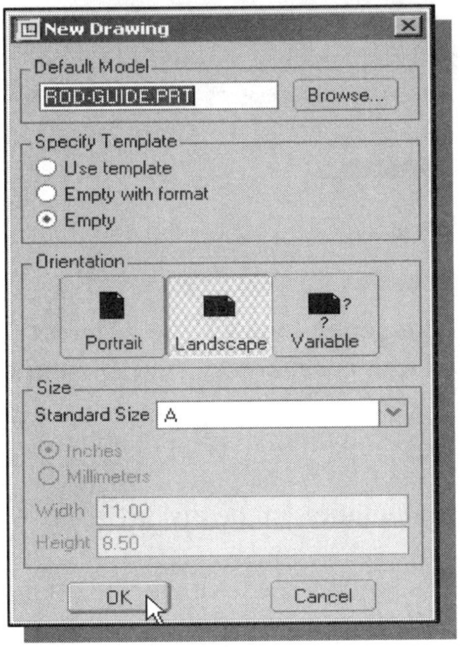

6. In the *New Drawing* form, the currently active part (Rod-Guide) is automatically selected as the drawing model.

7. Confirm the **Specify Template** option is set to **Empty** and the **Orientation** option is set to **Landscape**.

8. Pick **A** size (8.5″ × 11″) in the **Standard Size** list.

9. Pick **OK** to proceed with the drawing layout.

➢ A new window will appear with the title "***Rod-Guide-1***." Note that we currently have three files opened: The Rod-Guide part, the Rod-Guide drawing and the Rod-Guide-1 drawing.

Adding a Primary Main View

❖ In *Pro/ENGINEER Drawing* mode, the first drawing view we create is called a **main view**. A *main view* is the primary view in the drawing; other views can be derived from this view. When creating a main view, *Pro/ENGINEER* allows us to specify the view to be shown. We can select any view to be used as the base view. Note that there can be more than one main view in a drawing.

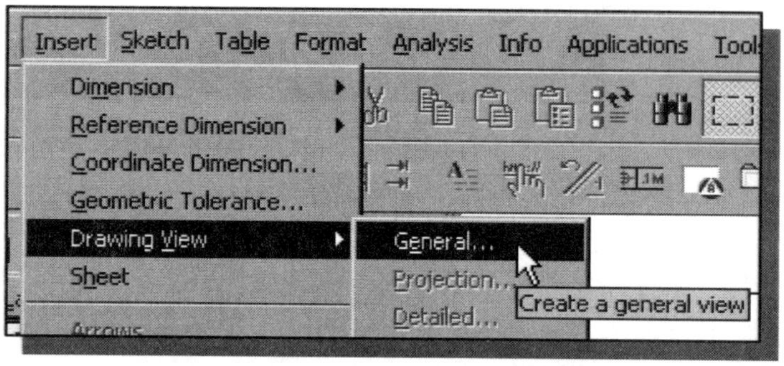

1. In the **Insert** pull-down menu, select **Drawing view → General** as shown.

❖ The first view in a drawing is a main view and it is created through the **General** view option.

2. The message "*Select CENTER POINT for drawing view*" is displayed in the message area. Pick a location that is about one-third from the left side of the display window as shown in the figure below. The default view of the model is displayed at the selected position.

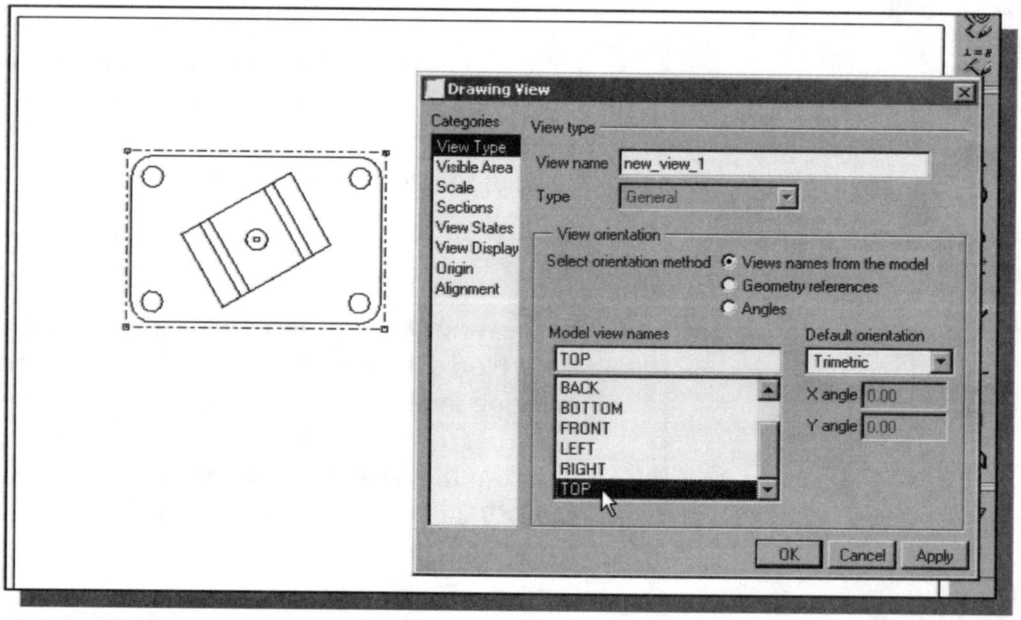

❖ Different options are available for the set up of a general view. The different options are organized into categories, such as *View Type*, *Scale*, *View Display*.

3. Choose **TOP** in the **Model View Names** list as shown in the figure above.

4. Click on the **Apply** button to adjust the view orientation in the display area.

5. Click on the **Close** button to accept the settings and proceed with the creation of the main view.

Adding a Projected Auxiliary View

❖ To create an auxiliary view, the definition of a view projection direction is required. In *Pro/ENGINEER*, the **view projection direction** can be defined by selecting an edge, an axis or a datum plane of the primary view.

1. Click once with the left-mouse-button in an empty area to deselect the main view.

2. In the **Insert** pull-down menu, select **Drawing view → Auxiliary** as shown.

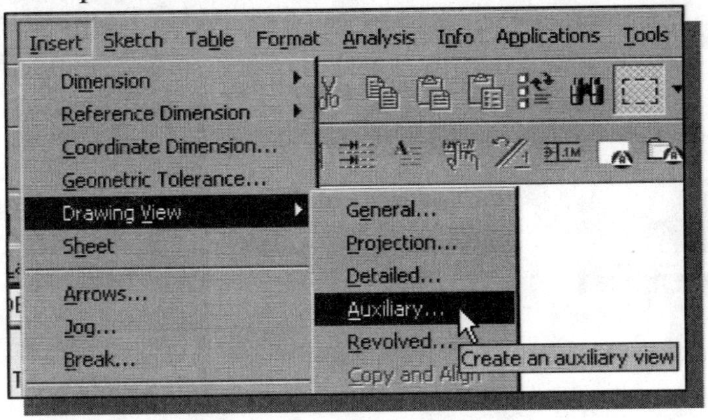

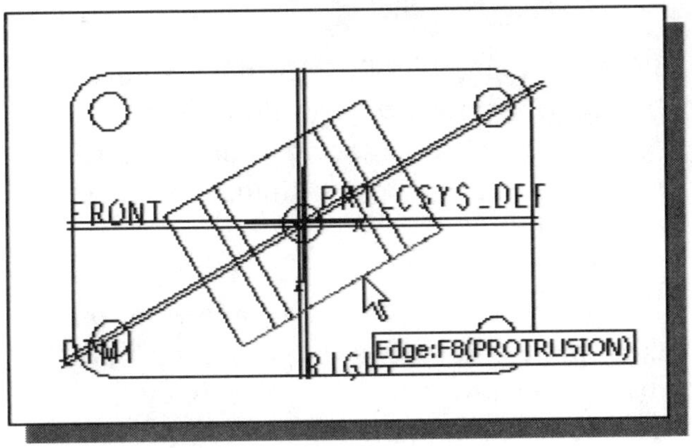

3. The message "*Select edge of or axis through, or datum plane as, front surface on main view*" is displayed in the message area. Pick the **front edge** of the angled feature as shown.

4. The message "*Select CENTER POINT for drawing view*" is displayed in the message area. Pick a location that is to below and toward the right of the main view.

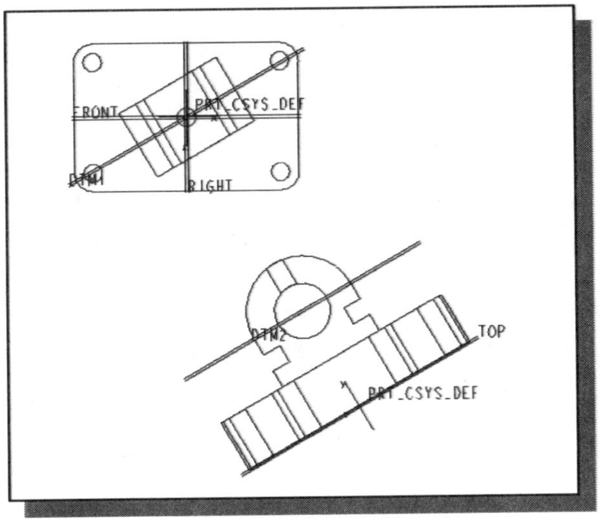

❖ An auxiliary view of the model, projected from the main view, is placed in the display window. The alignment of the auxiliary view to the main view is set automatically and maintained by the system.

Adjust the Overall Drawing Display

1. Click on the **Display Datum Planes On/Off** icon in the toolbar to toggle ***on/off*** the display of the datum planes. On your own, toggle **off** the display of datum planes, datum axes and coordinate systems.

2. Press and hold down the [**CTRL**] key and press the [**R**] key once. This is the quick-key combination to **Repaint** the display area.

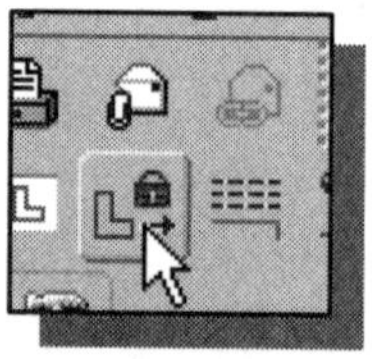

3. In the *Standard* toolbar area, click on the **Lock Views** icon to toggle ***off*** this option. (This option is set to ***on*** by default, which does not allow the moving of the views.)

4. Set the *Selection Filter* option to **Drawing View** as shown.

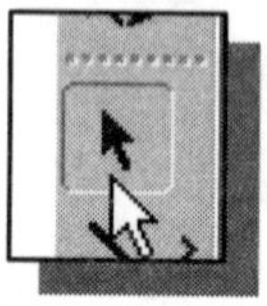

5. Click on the **Select** icon to activate the Select command.

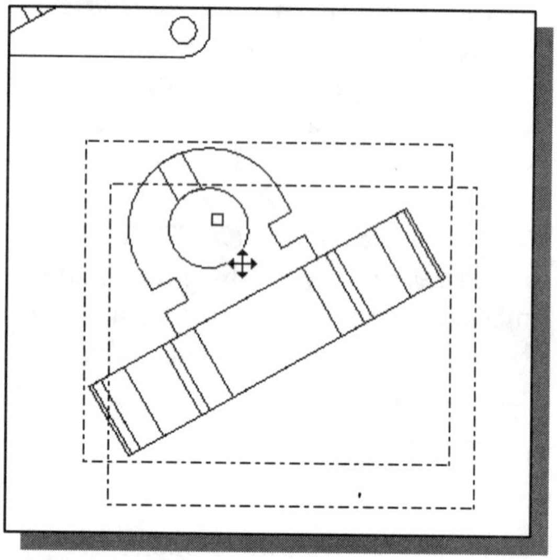

6. Pick and drag, with the left-mouse-button, the ***auxiliary view*** to reposition the view. Note the position of the auxiliary view is restricted to alignment with the top view, the primary view.

7. Pick and drag, with the left-mouse-button, the ***top view*** to reposition the views. Note that the position of the auxiliary view is also adjusted to maintain its alignment to the top view.

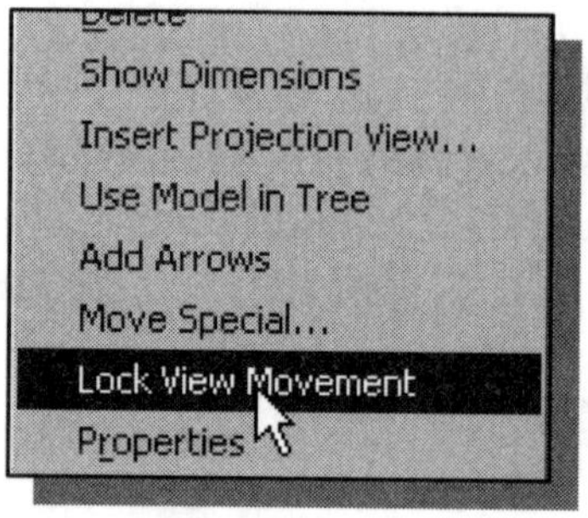

8. Inside the display area, press down the right-mouse-button and switch ***on*** the **Lock View Movement** option.

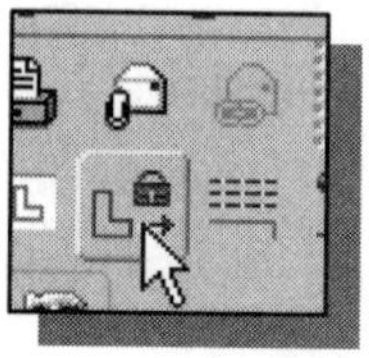

❖ Note that this option can also be set by clicking on the **Lock View** icon in the *Standard* toolbar area.

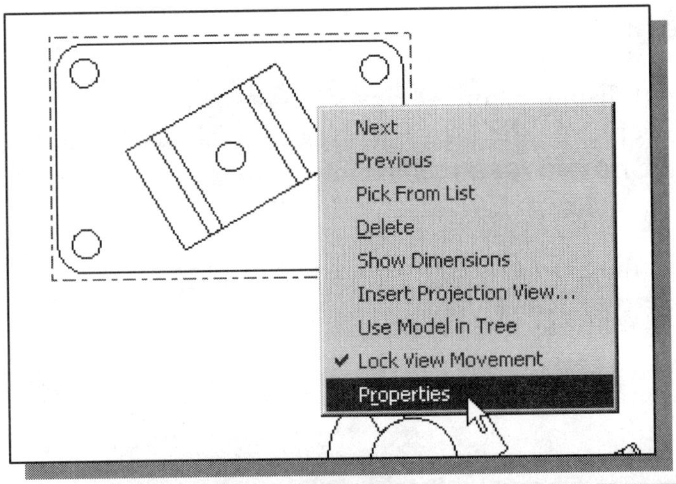

9. Pick the top view by clicking once inside the view with the left-mouse-button.

10. Press down the right-mouse-button and select **Properties** as shown.

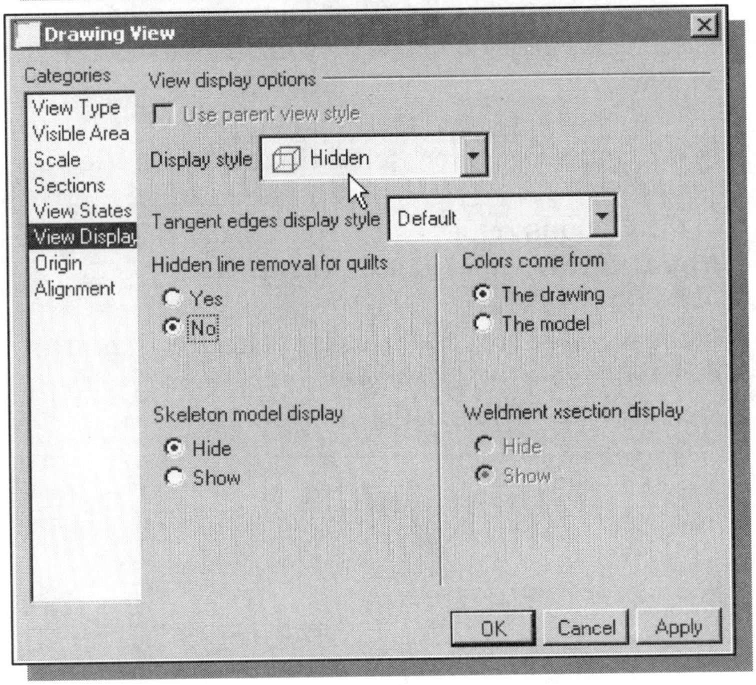

11. In the ***Drawing View*** window, select **View Display** to change the view display settings.

12. In the **Display style** list, select **Hidden** to display hidden edges as dashed lines. Note that the default view display setting is **Wireframe**.

13. Click **OK** to accept the settings and exit the ***Drawing View*** window.

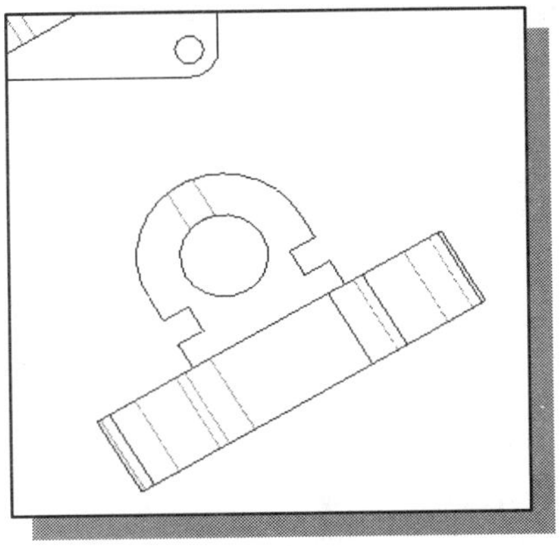

14. On your own, repeat the above procedure and adjust the display option of the *auxiliary view* to **Hidden Line**.

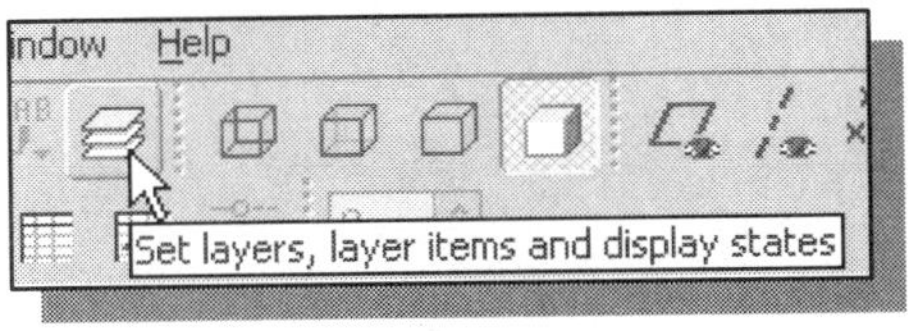

15. In the *Standard* toolbar, toggle ***off*** the **Layer Control Panel**. Note the *Model Tree* of *Rod-Guide* is displayed in the *Navigator* area.

Displaying Feature Dimensions

- By default, feature dimensions are not displayed in 2D views in *Pro/ENGINEER*. We can change the default settings while creating the views or switch on the display of the parametric dimensions using the **Show/Erase** command.

1. In the *Standard* toolbar area, select the **Show/Erase** command.

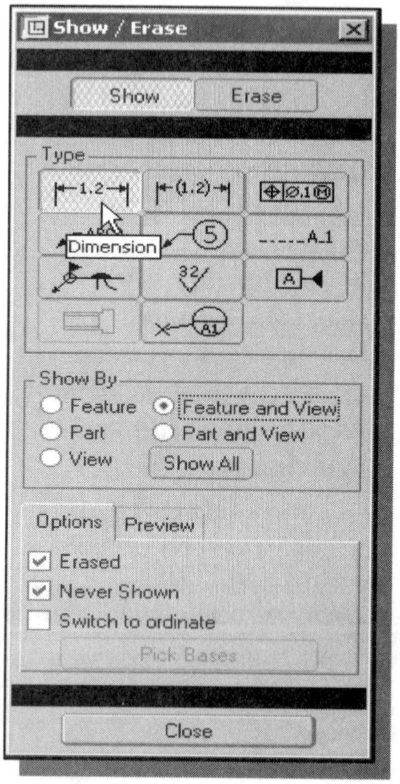

2. Choose to Show the **Regular Dimension** by toggling ***on*** the icon in the *Show/Erase* window as shown.

3. In the Show By option, select **Feature and View**.

4. In the Options settings area, confirm the **Erased** and **Never Shown** options are switched ***on***.

5. In the message area, the message "*Select a feature in the picked view. Middle button to finish*" is displayed. Pick the base feature in the top view as shown.

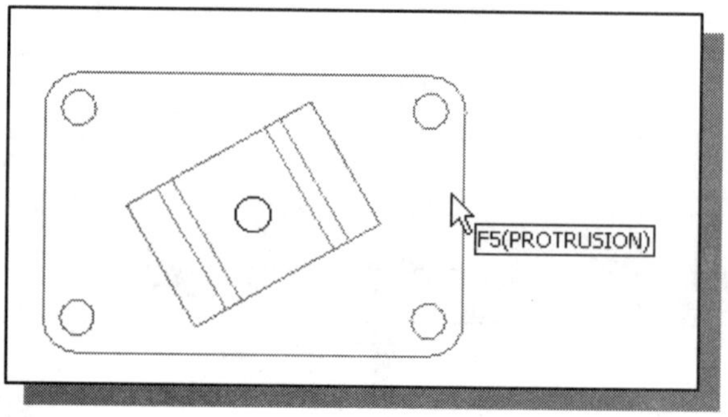

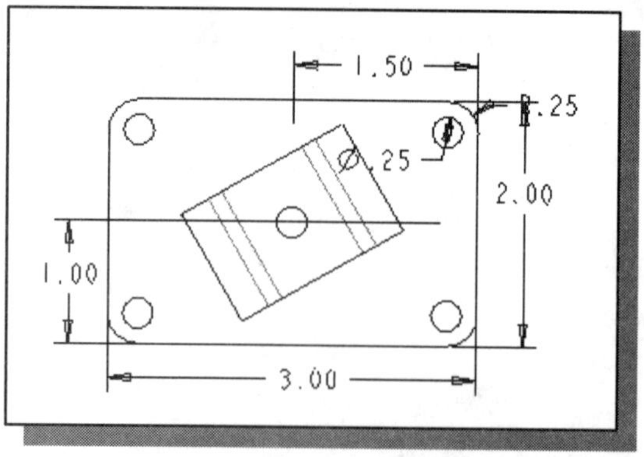

6. Inside the display area, click **twice** with the middle-mouse-button to accept the displayed dimensions.

7. Click **Close** to end the command.

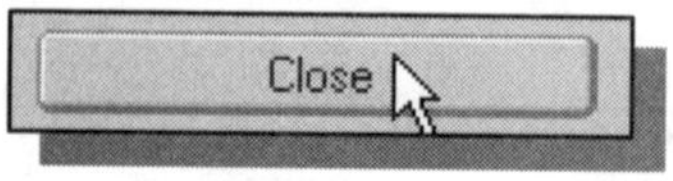

❖ Note that the **Feature and View** option is used to display feature dimensions of the selected feature that are viewable in the view.

Dimension Appearances

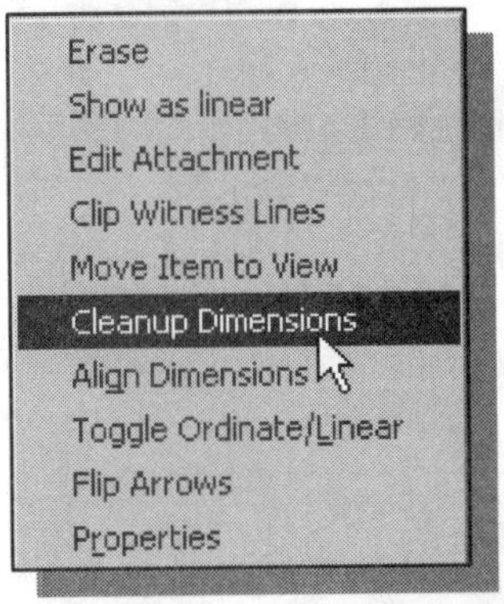

1. Inside the display area, press down the right-mouse-button and select the **Cleanup Dimensions** option.

❖ Note that six dimensions of the base feature are pre-selected. The displayed option menu lists only options that are applicable to the selected drawing dimensions.

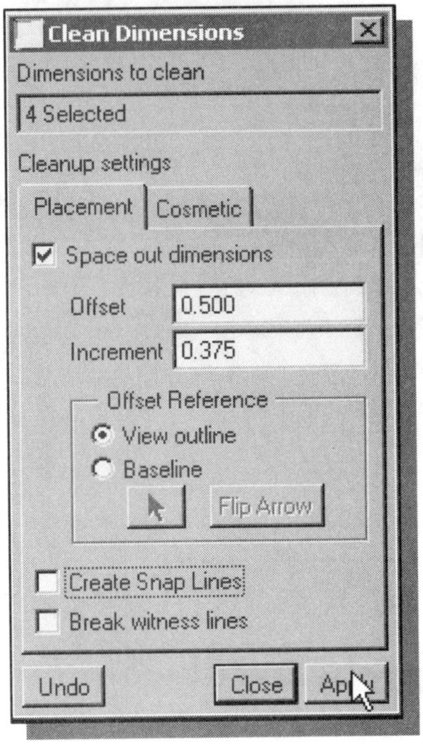

2. In the *Clean Dimensions* window, turn ***off*** the **Create Snap Lines** option.

❖ Note that only four dimensions are selected. The **Clean Dimensions** option is applicable only to the linear dimensions.

3. Click **Apply** to accept the settings and adjust the spacing of dimensions.

❖ **Clean Dimensions** offers a quick and easy way to **Space out dimensions**, which improves the legibility of the drawing. On your own, examine the different settings available in the *Clean Dimensions* dialog box.

4. Click **Close** to end the **Clean Dimensions** command.

5. Inside the graphics area, click once with the left-mouse-button in an empty area. This will ***deselect*** any pre-selected items. Note that the deselect command can also be found in the pull-down menu: **Edit → Select → Deselect All.**

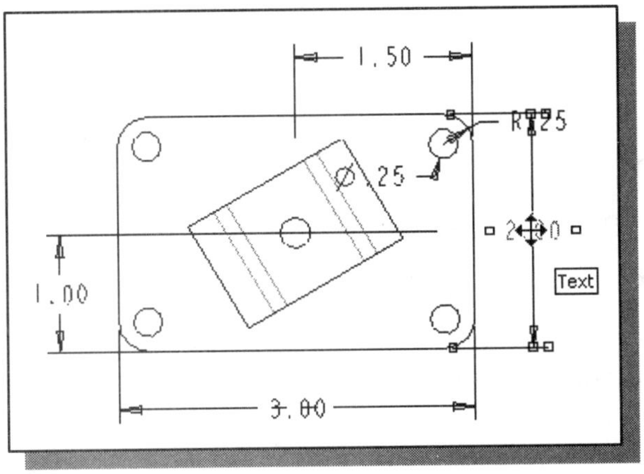

6. Pick one of the dimensions to move. Several small markers appear on the selected dimension.

7. Move the cursor on top of the small markers and note the arrow icon indicates the direction in which the dimension can be moved.

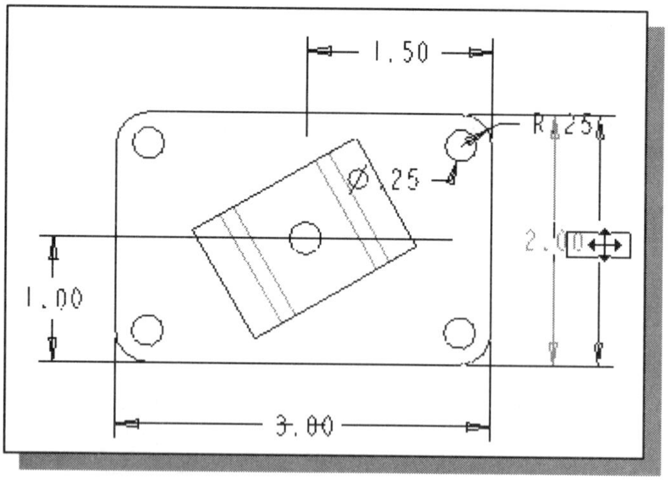

8. Use the left-mouse-button to drag one of the markers to reposition the text and/or dimension lines, as well as the arrowheads.

9. Pick a new location to place the dimension.

10. On your own, repeat the above steps and reposition the displayed dimensions.

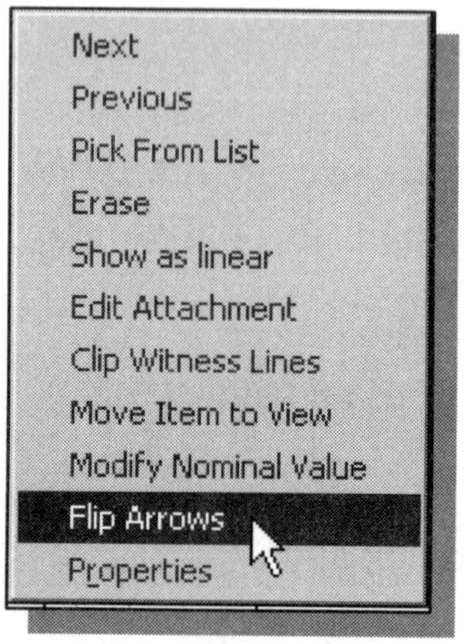

11. Select the **radius** dimension, **R.25**, by clicking once with the left-mouse-button on the dimension text.

12. Press down the right-mouse-button on the selected dimension text to display the option menu and select **Flip Arrows** as shown.

13. On your own, adjust the dimensions so that they appear as shown in the figure below.

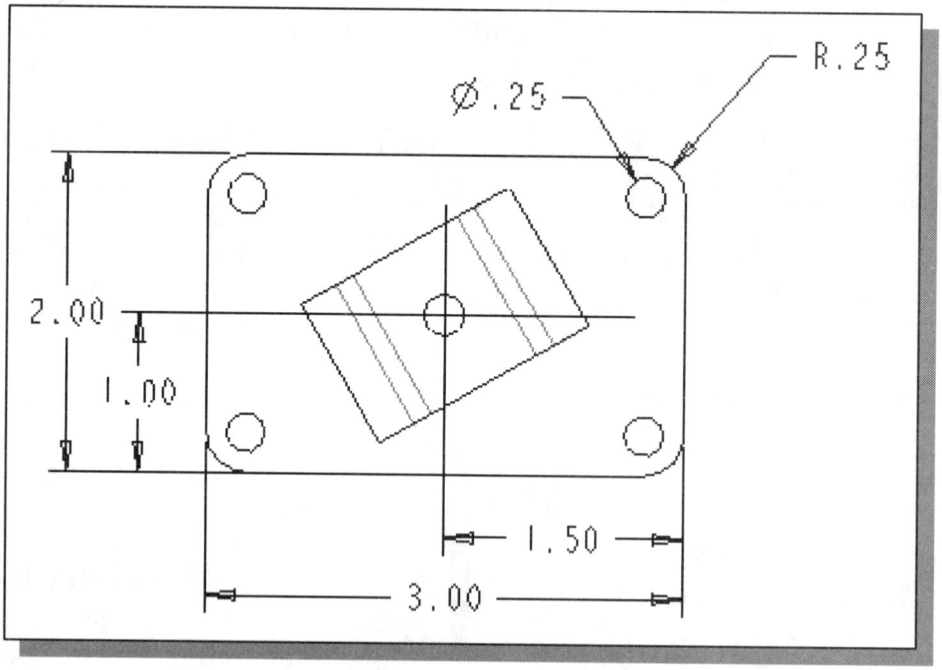

❖ Note that with modern parametric modeling software, the feature dimensions used in the creation of the solid model are easily displayed in the drawing mode.

Displaying Centerlines

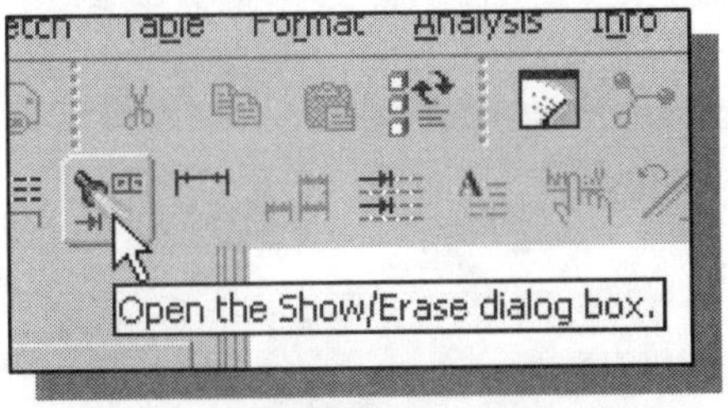

1. In the *Standard* toolbar area, select the **Show/Erase** command.

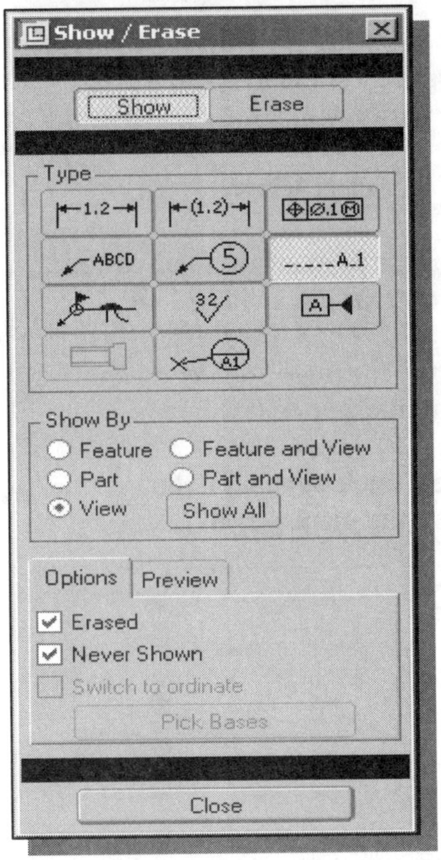

2. Choose to Show the **Centerlines** by toggling ***on*** the icon in the *Show/Erase* window as shown.

3. In the Show By option, select **View**.

4. In the Options settings area, confirm the **Erased** and **Never Shown** options are switched ***on***.

5. In the message area, the message "*Select model view or window.*" is displayed. Pick the ***top view*** as shown.

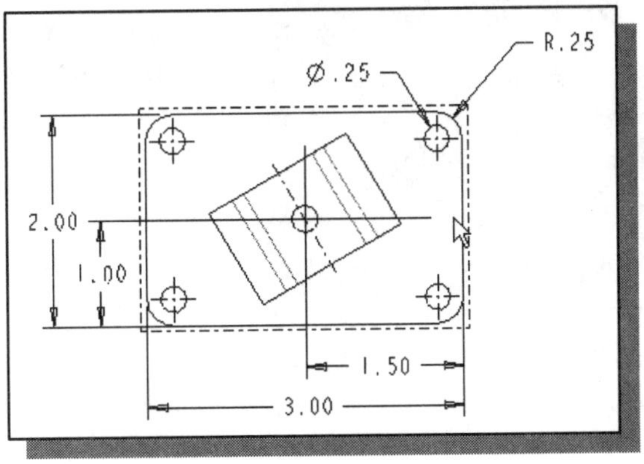

6. In the *Show/Erase* form, click on the **Accept All** button to keep all of the displayed centerlines.

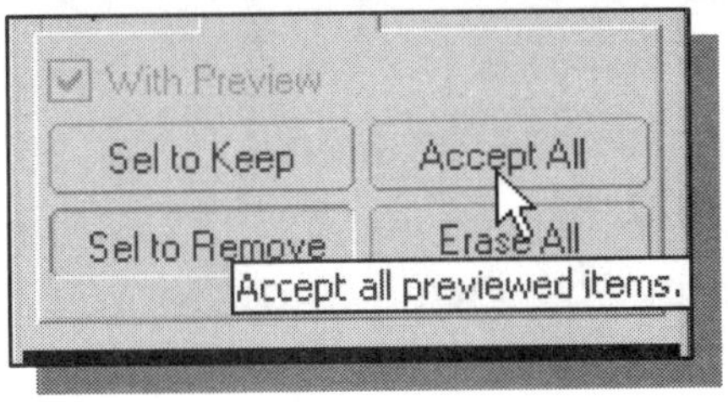

7. Pick the **Close** button in the *Show/Erase* form to exit the Show/Erase command.

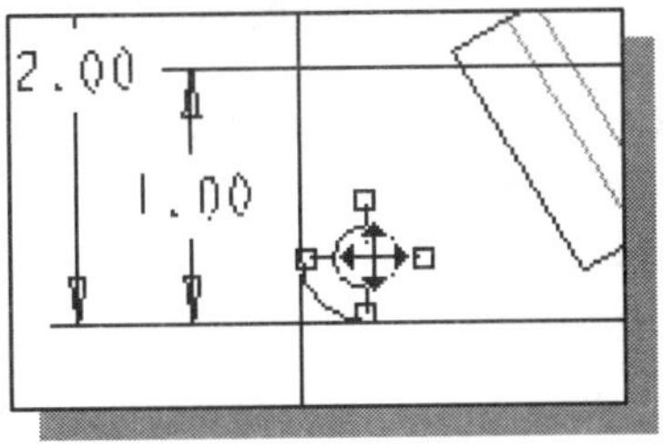

8. On your own, adjust the length of the displayed centerlines by selecting and dragging with the left-mouse-button as shown.

Displaying Remaining Annotations

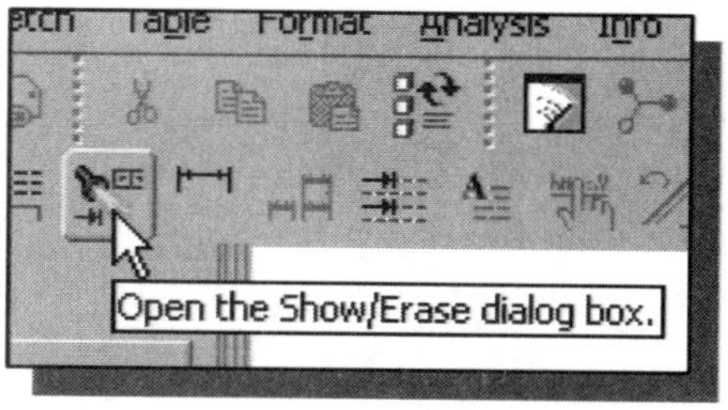

1. In the *Standard* toolbar area, select the **Show/Erase** command.

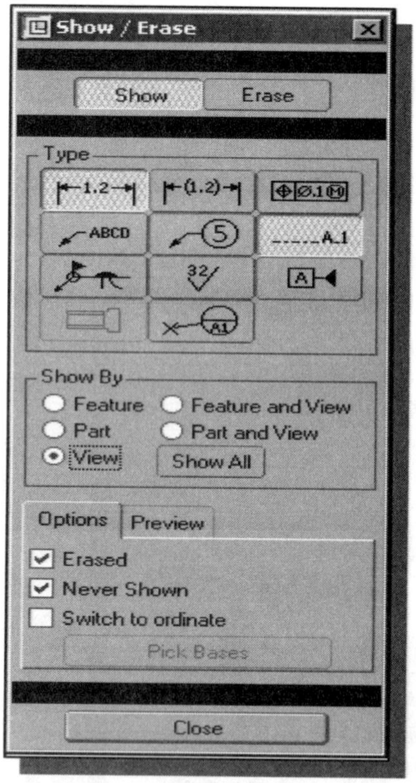

2. Choose to **Show** the **Regular Dimension** and the **Centerlines** by toggling ***on*** the icons in the *Show/Erase* window as shown.
3. In the **Show By** option, select **View**.
4. In the **Options** settings area, confirm the **Erased** and **Never Shown** options are switched ***on***.
5. In the message area, the message "*Select a view.*" is displayed. Pick the auxiliary view as shown.
6. Inside the display area, click **once** with the **middle-mouse-button** to keep the displayed dimensions.

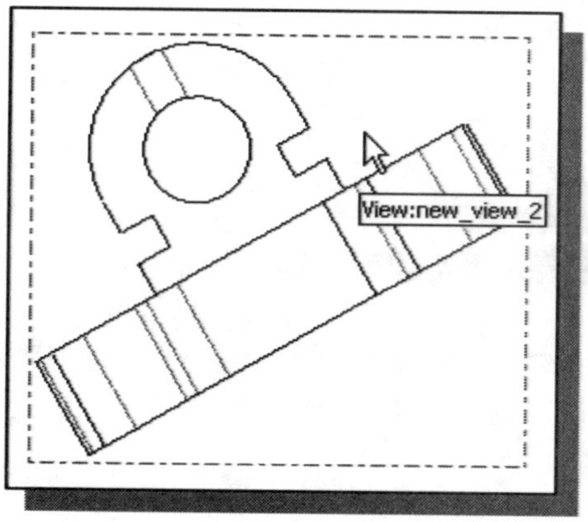

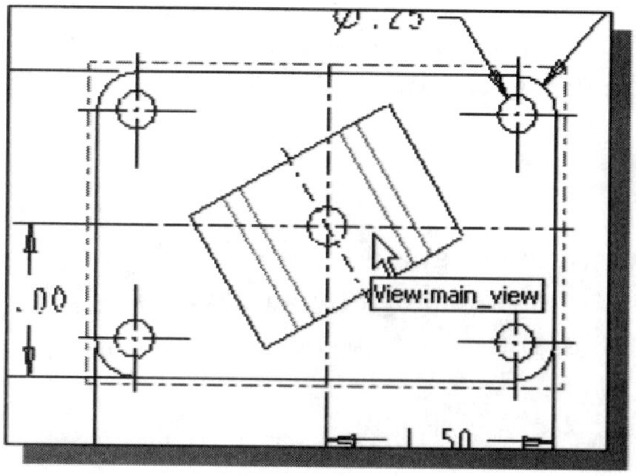

7. In the message area, the message "*Select a view.*" is displayed. Pick the main view to display the dimensions that are not displayed.
8. Inside the display area, click **once** with the middle-mouse-button to keep the displayed dimensions.
9. Click **Close** to end the command.

Annotation Appearances

1. Inside the graphics area, click once with the left-mouse-button in an empty area. This will ***deselect*** any pre-selected items.

2. On your own, adjust the locations of the dimensions so that they appear as shown in the figure below.

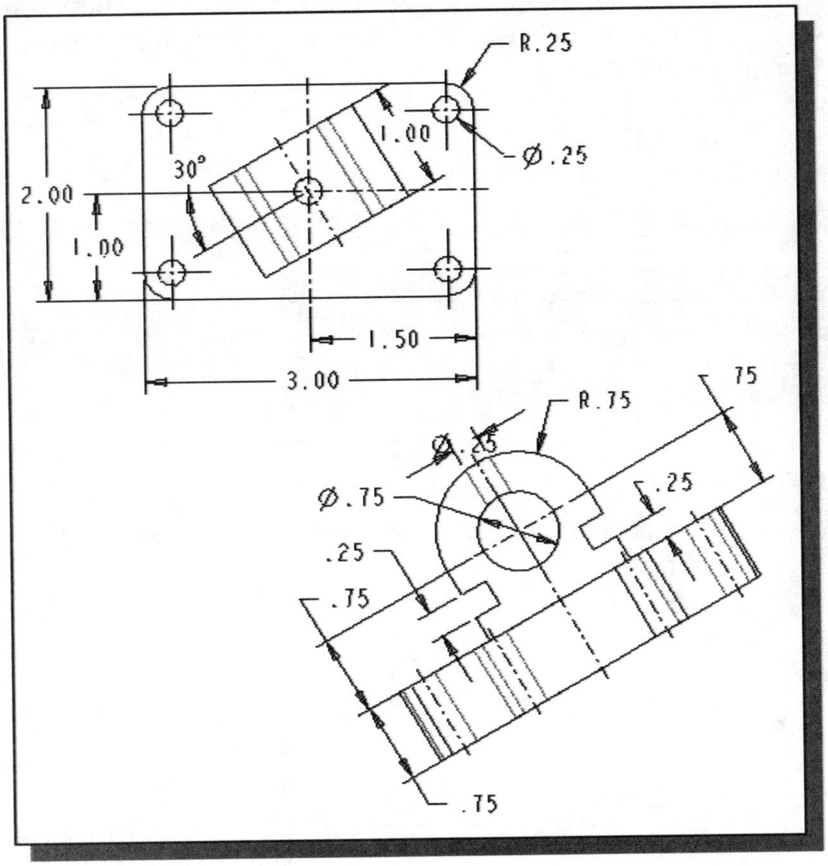

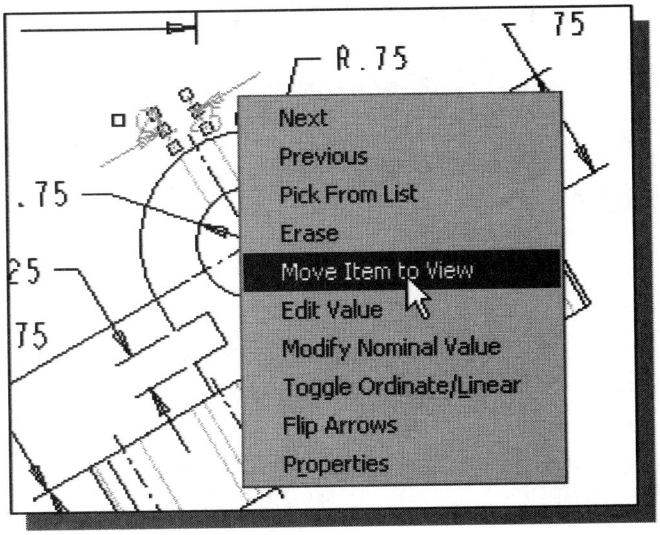

3. Pick the dimension text of the center hole (diameter **0.25**) in the auxiliary view as shown.

4. Inside the display area, click once with the right-mouse-button to bring up the option menu.

5. Select **Move Item to View** in the option list.

6. Pick the main view, the ***top*** view, in the display area.

Adding/Deleting Dimensions – Feature/Driven Dimensions

❖ Besides displaying the **feature dimensions**, dimensions used to create the features, we can also add additional **driven dimensions** to the drawing. *Feature dimensions* are used to control the geometry, where *driven dimensions* are controlled by the existing geometry. In the drawing layout, we can therefore **delete** the *driven dimensions* but we can only **hide** the *feature dimensions*. One should try to use as many *feature dimensions* as possible and add *driven dimensions* only if necessary. It is also more effective to use *feature dimensions* in the drawing layout since they were created when the model was built.

1. In the **Insert** pull-down menu, select **Dimension → New References**

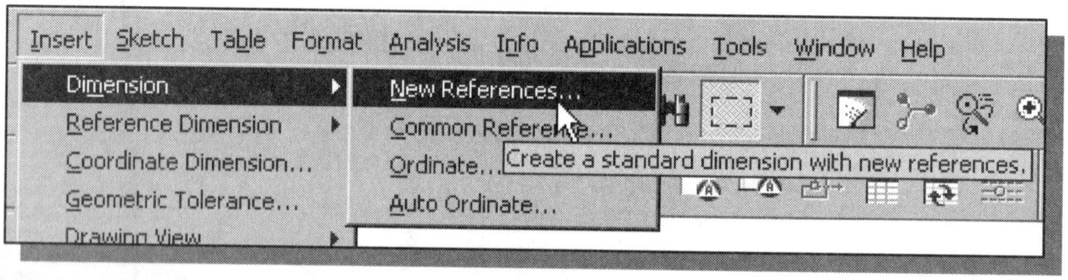

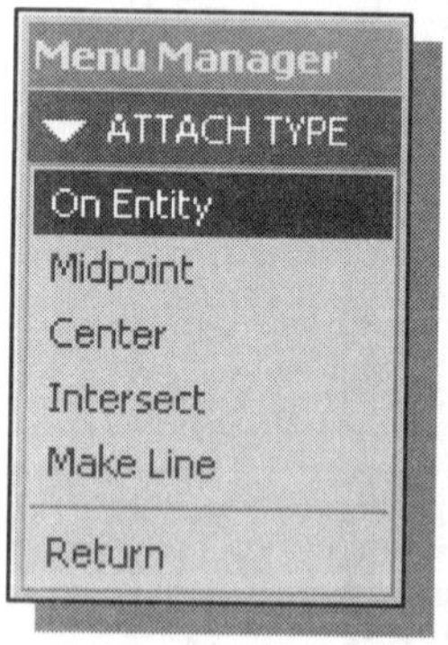

2. In the **Attach Type** submenu, confirm the selection of **On Entity**.

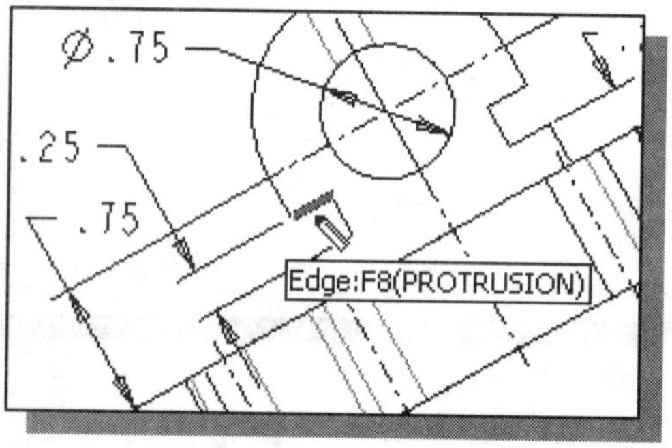

3. Pick the top inside edge of the notch as shown.

4. Click once with the middle-mouse-button at a location inside the display area to create the dimension.

❖ Note that we add dimensions the same way we did in the *2D Sketcher*.

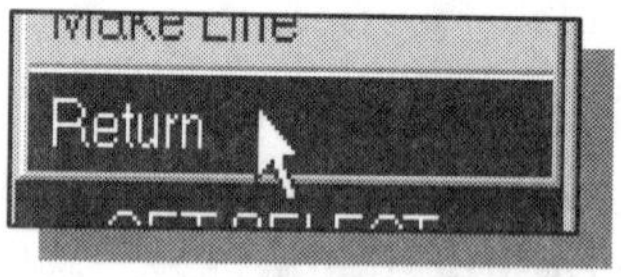

5. Pick **Return** in the **Attach Type** submenu to exit the **Insert Dimension** command.

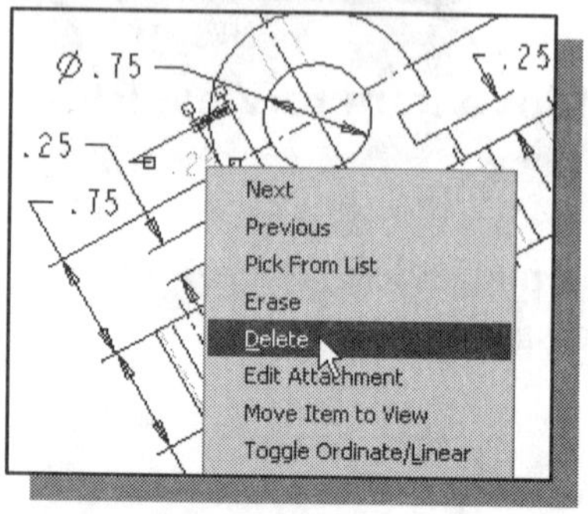

6. Pick the driven dimension we added in the above steps.

7. Press down the right-mouse-button to bring up the option menu. Note that **Delete** appears in the option list.

8. Press the [**ESC**] key once to abort the option.

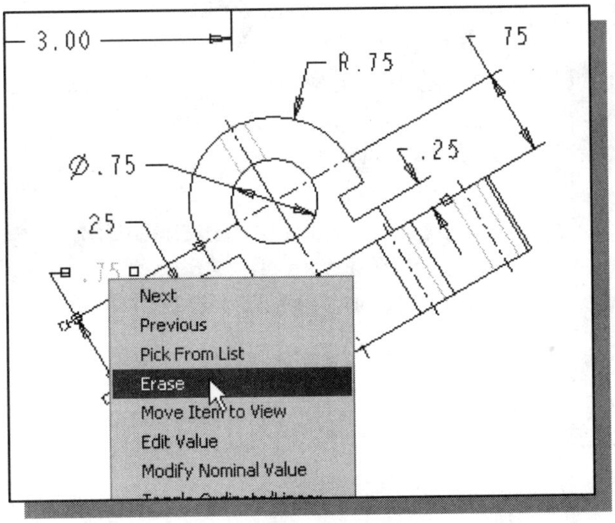

9. Select one of the duplicated dimensions in the center of the vertical feature.

10. Choose **Erase** in the option list. Note that there is no **Delete** option.

❖ The two dimensions are for two different features. The first one was added in the 2D section of the vertical feature and the second one was the location dimension of the horizontal datum plane **DTM2**.

❖ In the *Pro/ENGINEER Drawing* mode, the **Delete** option is used to remove entities from the database. The **Erase** option is used to temporarily remove the highlighted entity from the display. *Feature dimensions* can only be ***erased***, not ***deleted*** from the drawing. Driven dimensions exist only in the drawing mode and can therefore be deleted.

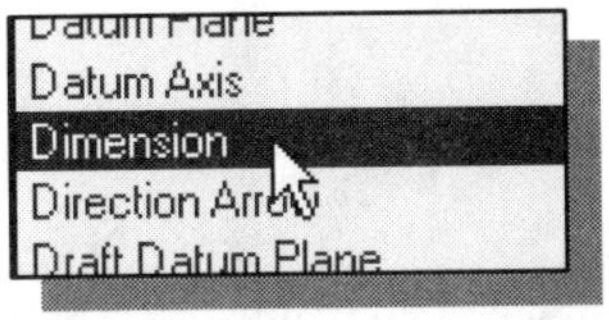

11. Set the **Selection Filter** option to **Dimension**.

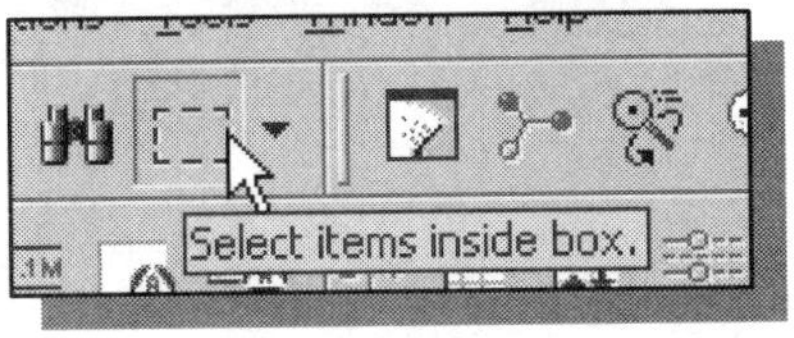

12. In the *Standard* toolbar area, confirm the box selection option is set to **Select items inside box** as shown.

13. Inside the display area, select all of the dimensions by enclosing all dimensions inside a selection box with the left-mouse-button.

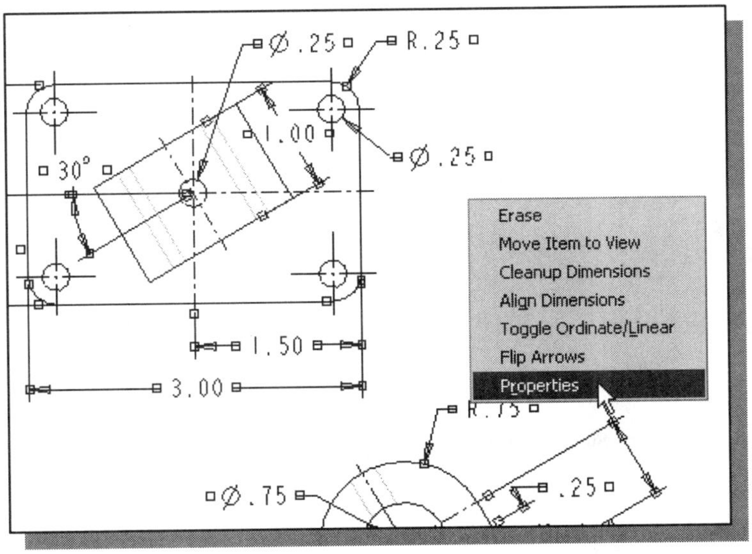

14. Inside the display area, press down the right-mouse-button to bring up the option menu.

15. Select **Properties** in the option list.

❖ We will modify the dimension settings for all of the selected dimensions.

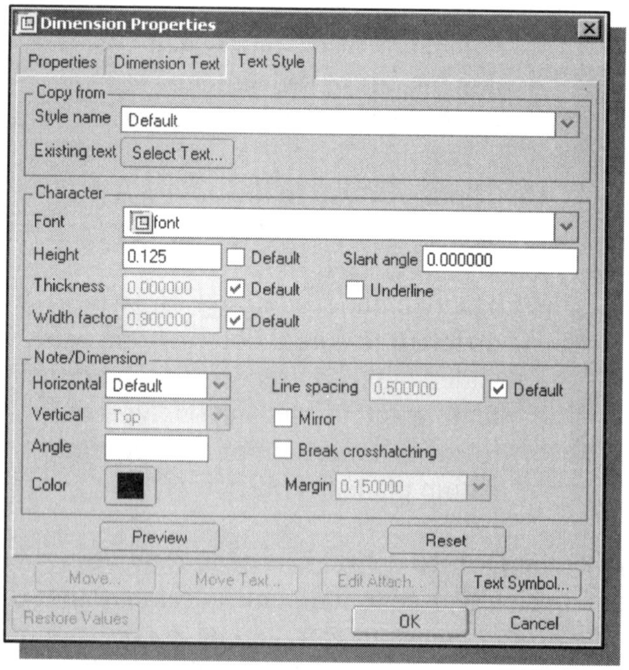

16. In the *Dimension Properties* window, click on the **Text Style** tab.

17. Toggle ***off*** the **Default Height** option and enter **0.125** as the character height as shown.

18. Click **OK** to accept the settings.

19. On your own, adjust the dimensions and centerlines so that the drawing appears as shown in the figure below.

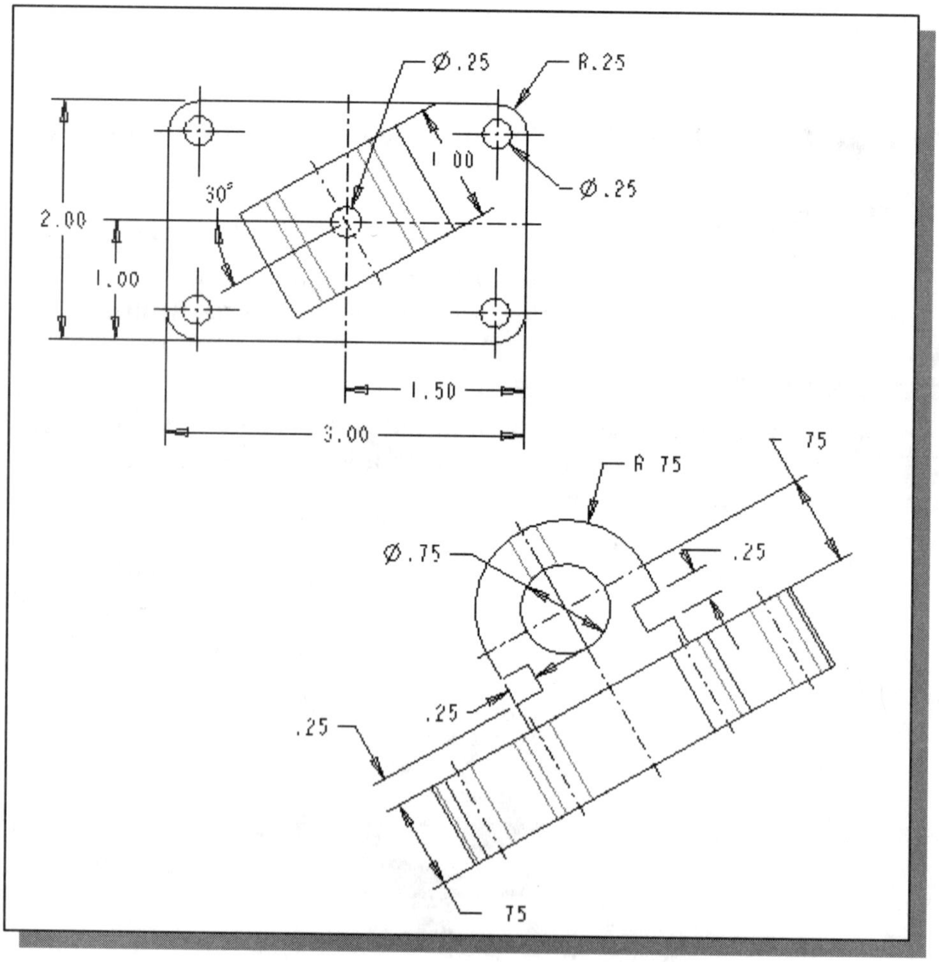

Adding an Isometric View

❖ In *Pro/ENGINEER*, the default 3D orientation is set to display the model in the ***trimetric*** orientation. In order to add an ***isometric*** view in the *Drawing* mode, the easiest approach is to reset the default orientation to *isometric*, in the *Part* mode, before creating the view.

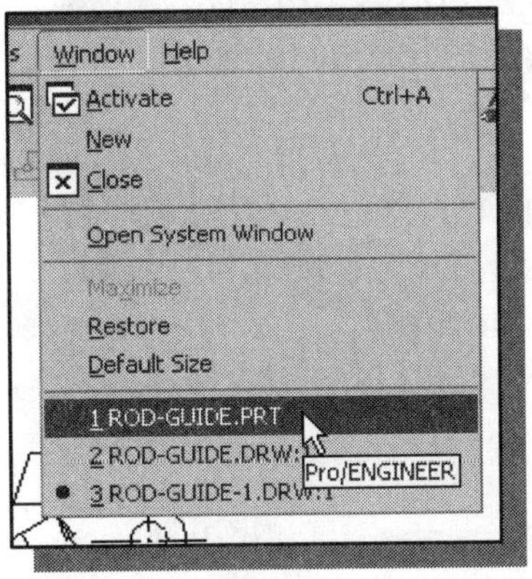

1. In the **Window** pull-down menu, select the ***Rod-Guide*** part (file name: Rod-Guide.prt) as shown.

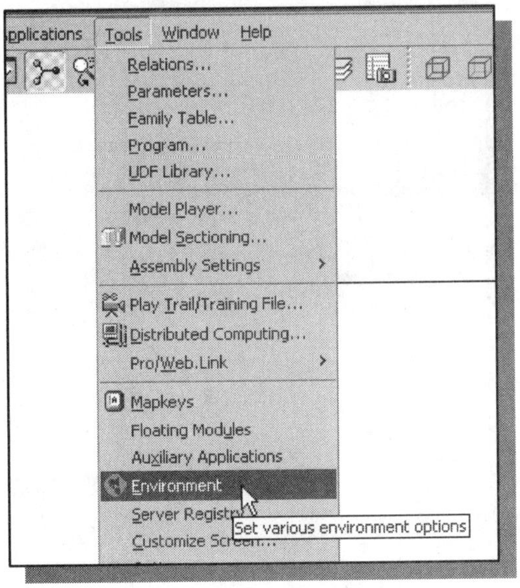

2. Select **Tools** in the pull-down menus. Pick the **Environment** option. The environment option allows us to adjust various environment settings.

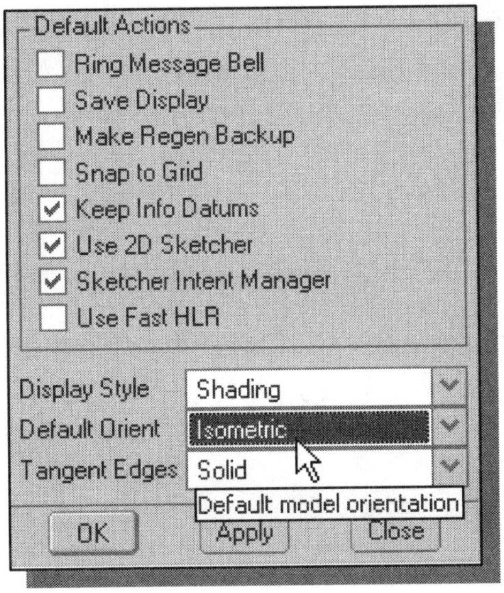

3. At the bottom of the *Environment* form, three groups of display settings are available. Pick **Isometric** as the *default view orientation.*

4. Pick **OK** to exit the form.

5. Use the quick key combination **[Ctrl-D]** to change the view orientation in the display area.

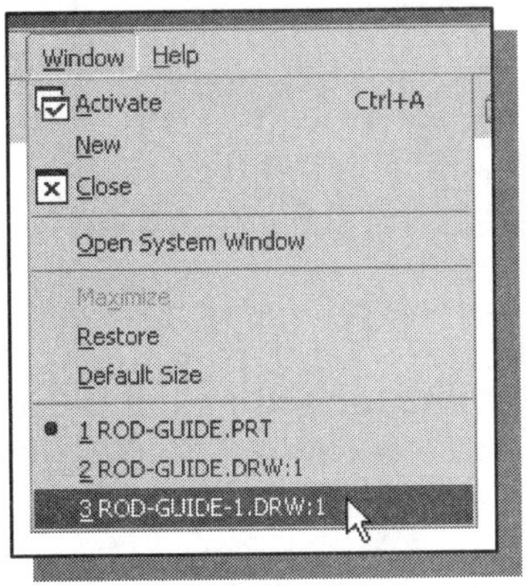

6. In the **Window** pull-down menu, select the ***Rod-Guide-1*** drawing (file name: Rod-Guide-1.DRW) to switch back to the 2D drawing *Rod-Guide-1*.

7. In the *Standard* toolbar area, click on the **Insert view** button as shown.

8. The message "*Select CENTER POINT for drawing view*" is displayed in the message area. Pick a location that is to the upper right corner of the display window. The default view (*isometric* view) of the model is placed at the selected location in the display window.

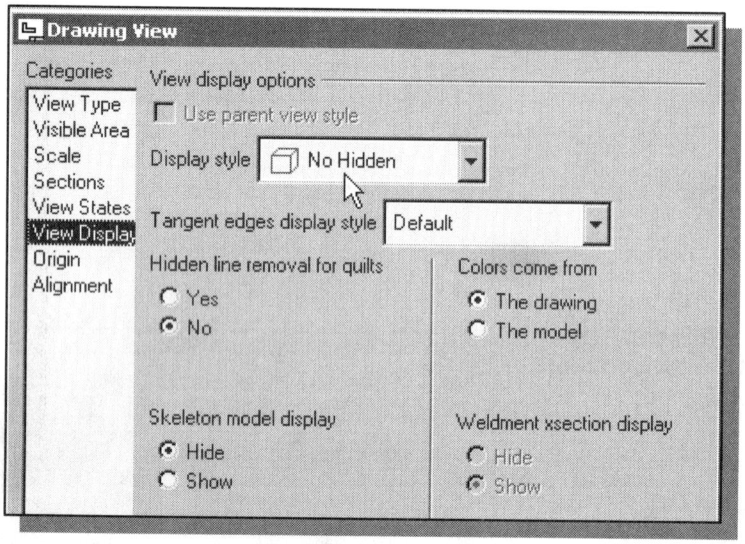

9. In the ***Drawing View*** window, adjust the **View Display** of the *isometric* view to **No Hidden** as shown.

❖ Note the default display option, **Follow Environment,** can be used to create shaded image.

10. Click **OK** to accept the settings and end the Drawing View command.

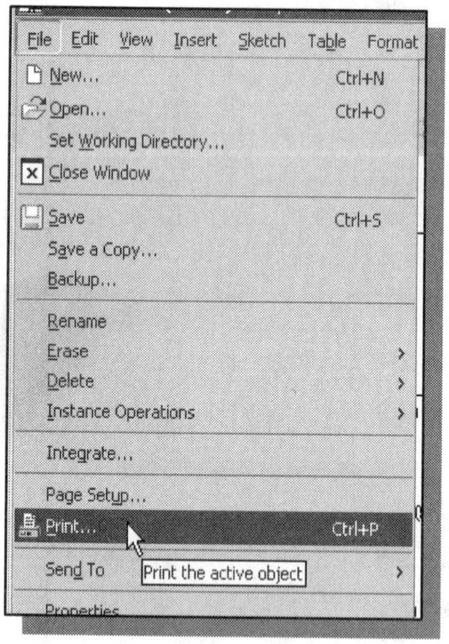

11. In the **File** pull-down menu, select **Print** as shown.

12. Click **Config** to view/adjust the printer configuration.

13. Confirm the **Page Size** is set to **A** size as shown.

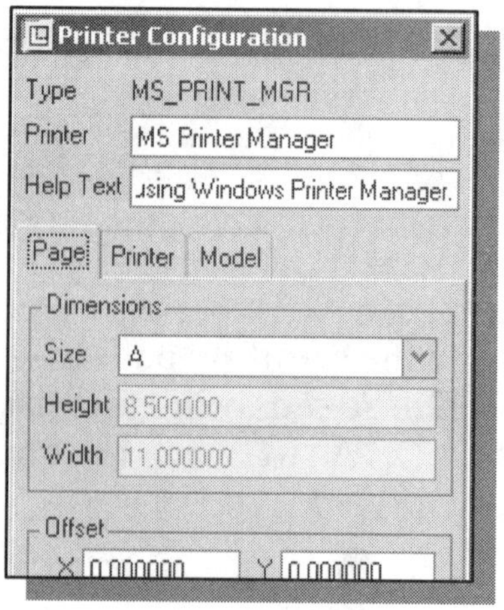

14. Click **OK** to proceed with the settings and enter the *MS Printer Manager*.

15. Examine the settings in the *MS Printer Manager* and print a copy of the *Rod-Guide* drawing.

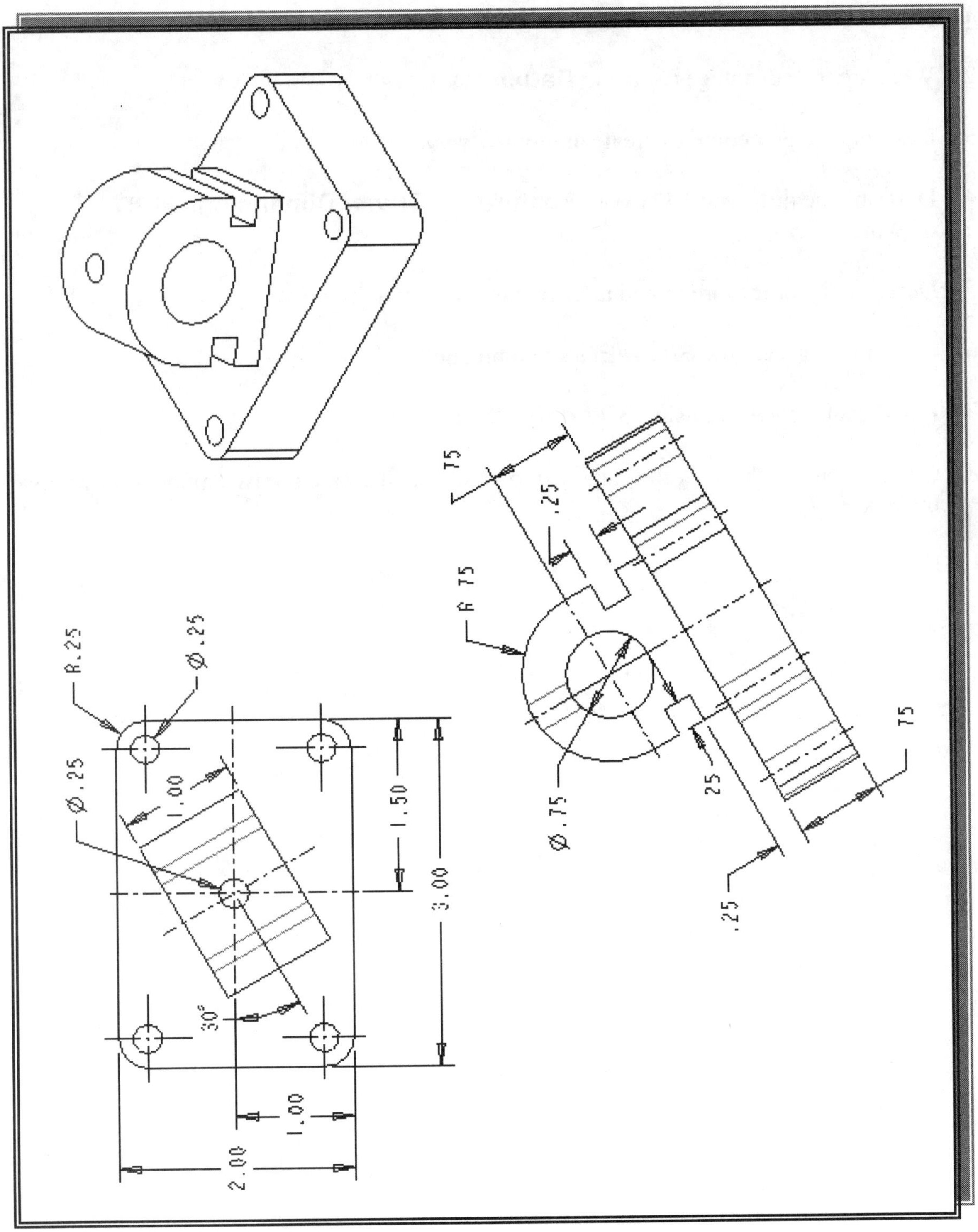
R.25
Ø.25
Ø.25
1.00
1.50
3.00
30°
1.00
2.00
.75
.25
R.75
Ø.75
.25
.75
.25

Questions:

1. How do we create geometry on a surface that does not already exist?

2. What are the advantages of using **Datum features**?

3. Describe the procedure to create an *angled datum plane*.

4. Describe the differences between **Feature** and **Driven Dimensions** in 2D drawings.

5. Describe the procedure to add an *Auxiliary view* in a drawing

6. How do we access the **Show/Erase** command?

7. List the advantages of using **3D annotation**.

8. Create sketches showing the steps you plan to use to create the two models shown on the next page:

Ex.1)

Ex.2)

Exercises:

1. Dimensions are in inches.

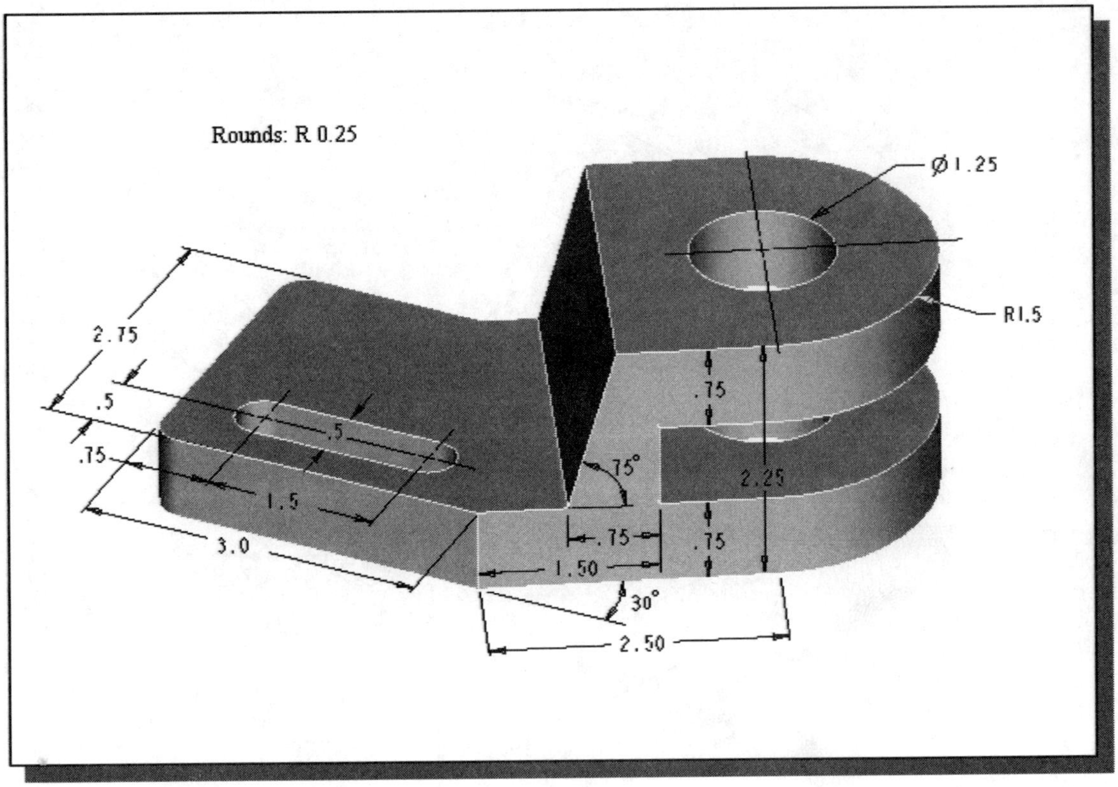

2. Dimensions are in inches.

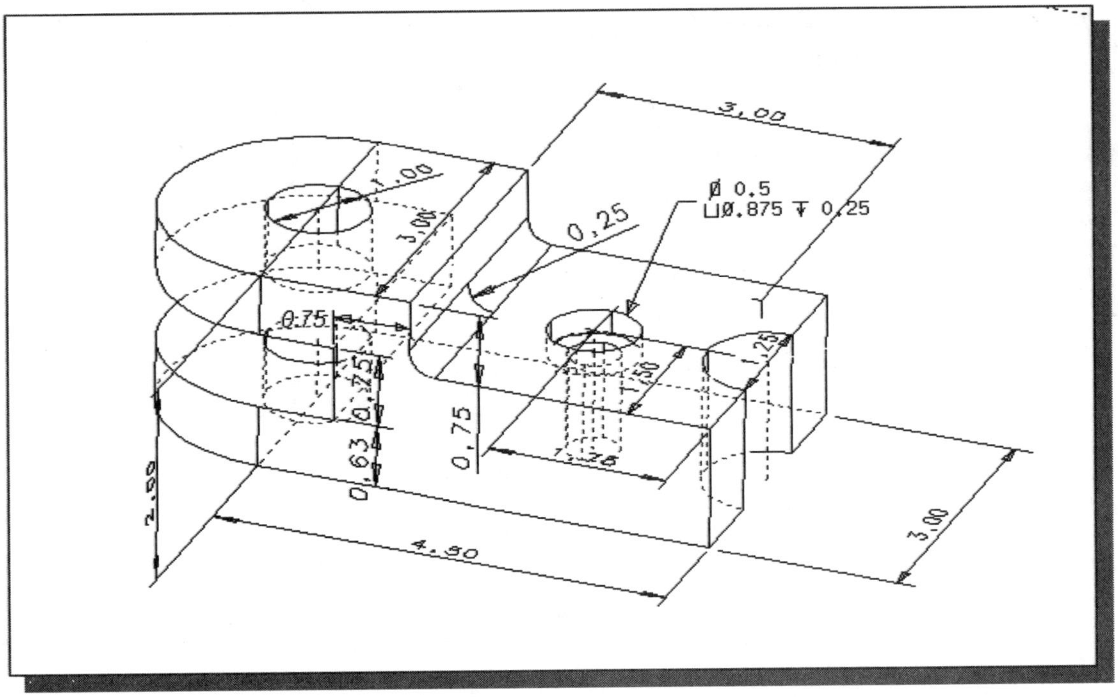

3. Dimensions are in Millimeters.

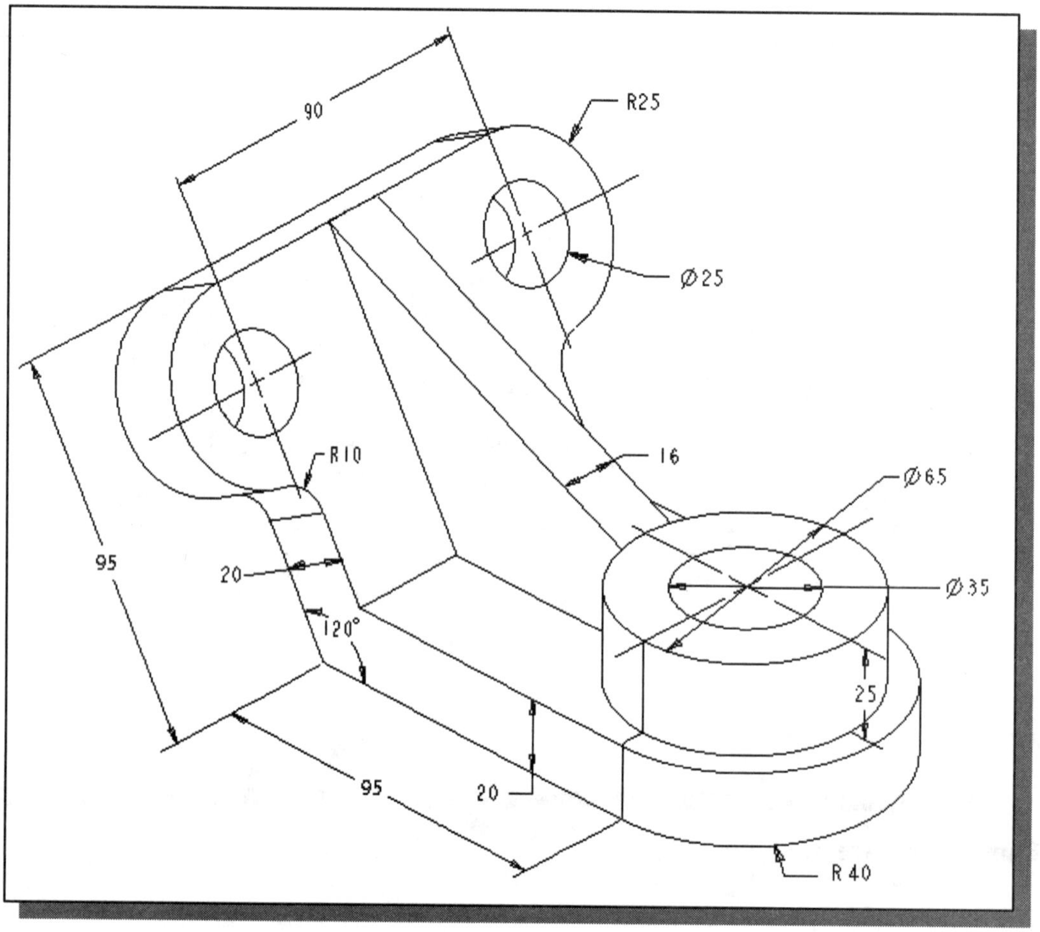

Chapter 7
Symmetrical Features in Designs

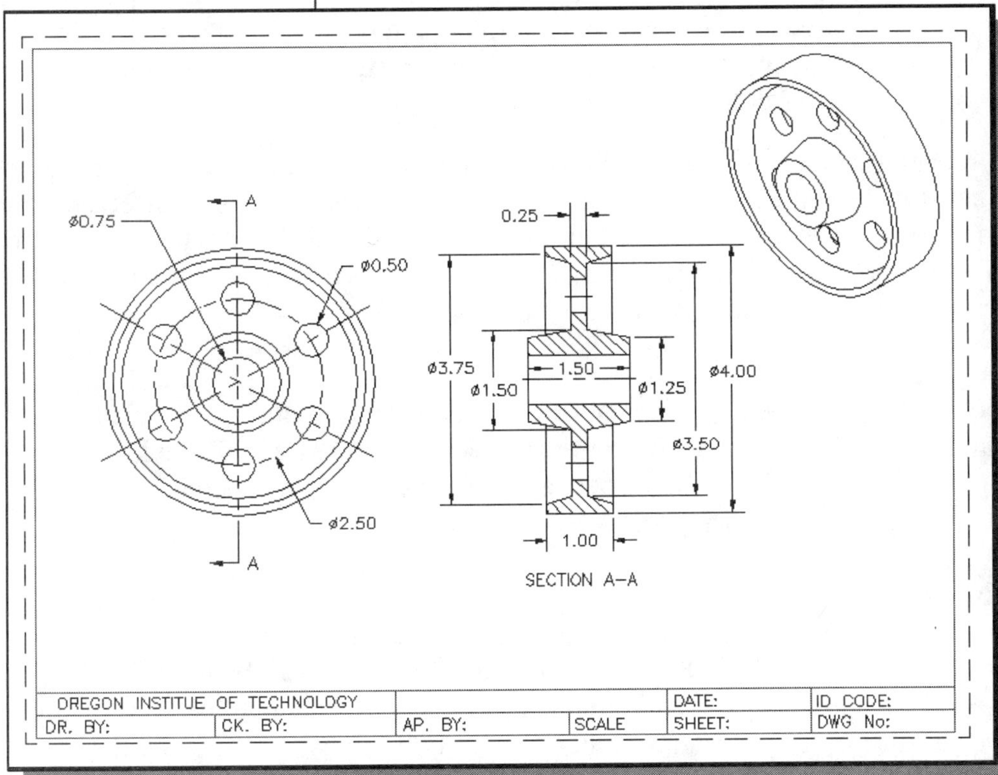

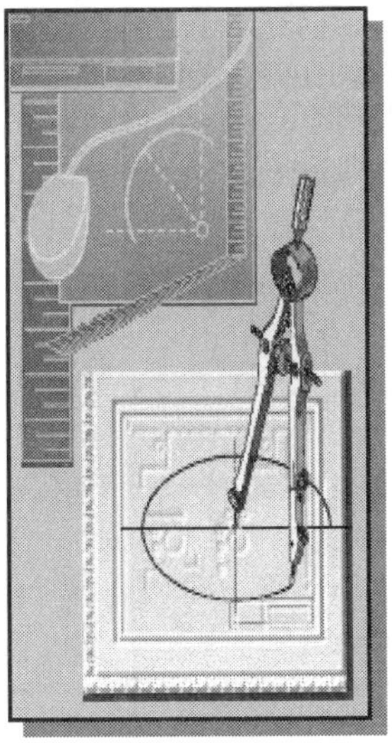

Learning Objectives

- **Create Drawing Layouts from Solid Models.**
- **Understand Associative Functionality.**
- **Add Borders and Title Block in the Layout Mode.**
- **Arrange and Manage 2-D Views in Drawing Mode.**
- **Create Sectioned Orthogonal Views.**
- **Display and Hide Feature Dimensions.**
- **Create Reference Dimensions.**

Symmetrical Features in Designs

In parametric modeling, it is important to identify and determine the features that exist in the design. *Feature-based parametric modeling* enables us to build complex designs by working on smaller and simpler units. This approach simplifies the modeling process and allows us to concentrate on the characteristics of the design. Symmetry is an important characteristic that is often seen in designs. Symmetrical features can be easily accomplished by the assortment of tools that are available in feature-based modeling systems, such as *Pro/ENGINEER*.

The modeling technique of extruding two-dimensional sketches along a straight line to form three-dimensional features, as illustrated in the previous chapters, is an effective way to construct solid models. For designs that involve cylindrical shapes, shapes that are symmetrical about an axis, revolving two-dimensional sketches about an axis can form the needed three-dimensional features. In solid modeling, this type of feature is called a **revolved feature**.

In *Pro/ENGINEER*, besides using the **Revolve** command, several options are also available to handle symmetrical features. For example, we can create multiple identical copies of symmetrical features with the **Feature Pattern** command, or create mirrored images of models or 2D sections using the **Mirror** command.

Pro/ENGINEER provides full associative functionality in all *Pro/ENGINEER* software modules. This functionality allows us to change the design at any level, and the system reflects it at all levels automatically. For example, the solid model can be modified in the standard *Part Modeler* and the system automatically reflects that change in the *Drawing* layout. And we can also modify a feature dimension in the *Drawing* layout, and the system automatically updates the solid model in all applications.

A Revolved Design: *Wheel*

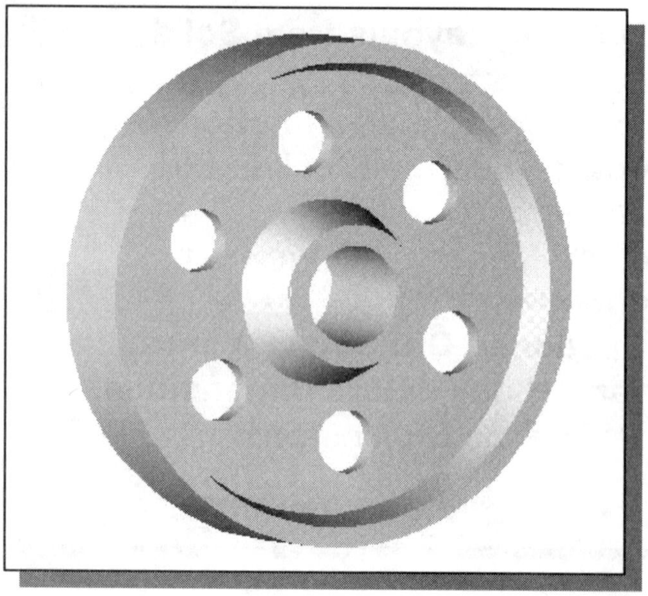

- Based on your knowledge of *Pro/ENGINEER*, how many features would you use to create the design? Which feature would you choose as the **base feature** of the model? Identify the symmetrical features in the design and consider other possibilities in creating the design. You are encouraged to create the model on your own prior to following through the tutorial.

Modeling Strategy - A Revolved Design

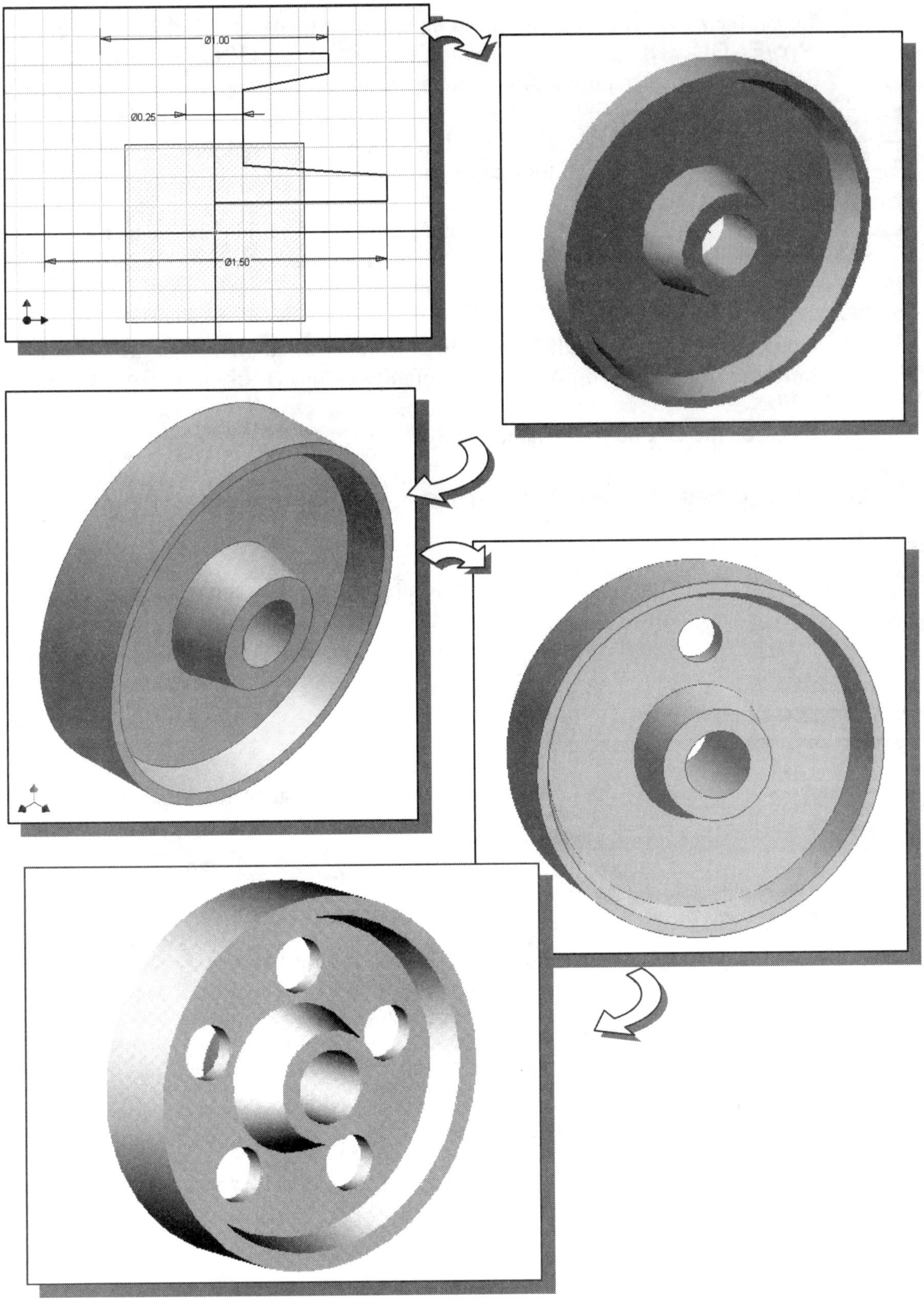

Starting Pro/ENGINEER

1. Select the **Pro/ENGINEER** option on the *Start* menu or select the **Pro/ENGINEER** icon on the desktop to start *Pro/ENGINEER*. The *Pro/ENGINEER* main window will appear on the screen.

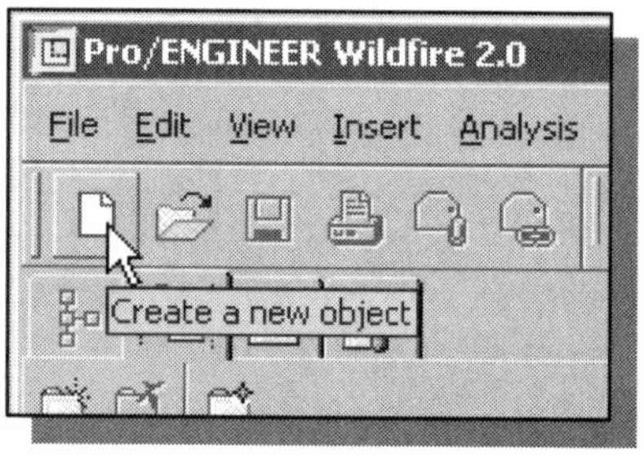

2. Click on the **New** icon, located in the *Standard* toolbar as shown.

3. On your own, start a new solid part file using **Wheel** as the part Name.

4. Confirm the **Use default template** option is turned ***on*** so that the system units are set to the *Pro/ENGINEER* default settings (**Inch-lbm-Second**).

5. Click on the **OK** button to accept the settings.

Creating the Base Revolved Feature

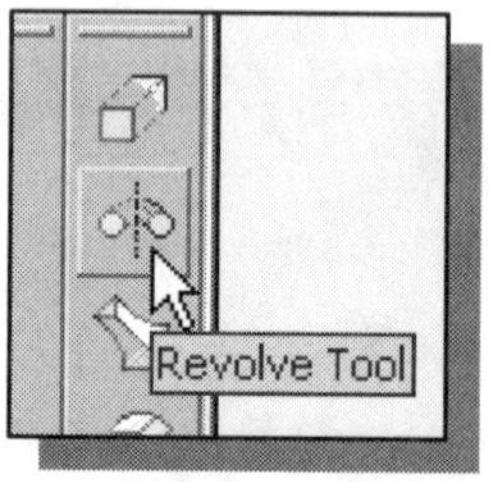

1. In the *Feature* toolbars (toolbars aligned to the right edge of the main window), select the **Revolve Tool** option as shown.

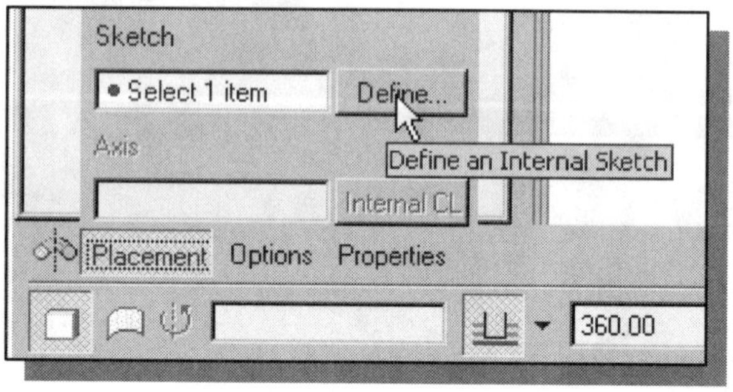

2. Click the **Placement** option and choose **Define** to create a new *Internal sketch.*

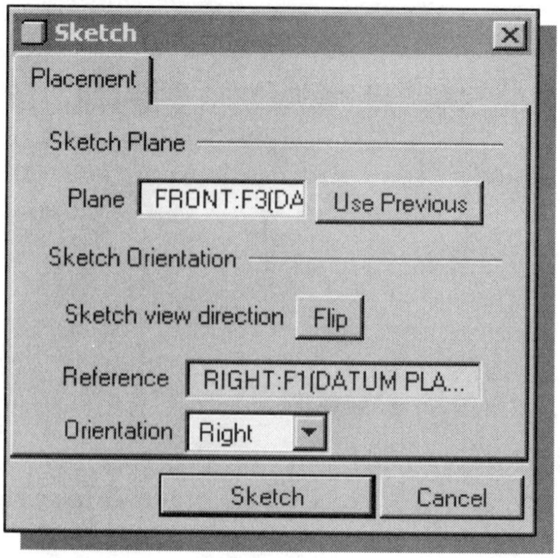

3. On your own, set up the **Front** datum plane as the sketch plane with the **Right** datum plane facing the **right** edge of the computer screen and pick **Sketch** to enter the *Sketcher* mode.

❖ Note the default selection of the two datum planes, **Right** and **Top**, as the references for the 2D sketch.

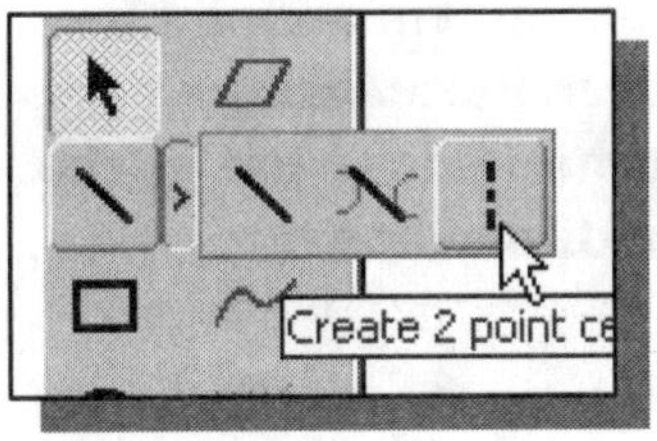

4. In the *Sketcher* toolbar, select **Centerline**.

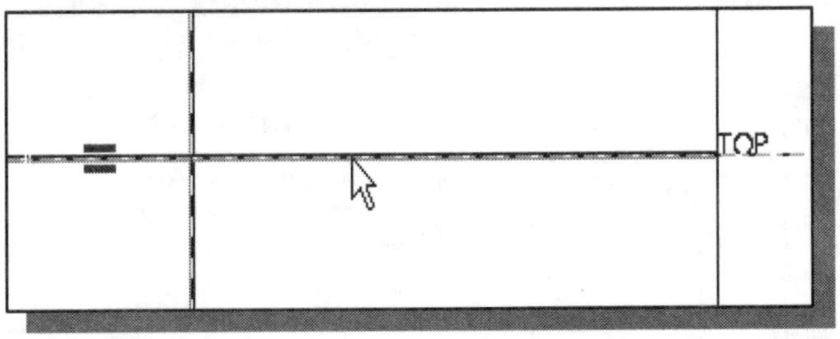

5. First, select a point on the horizontal axis (**TOP**).

6. Select a second point aligned to datum plane **TOP** to create a horizontal centerline.

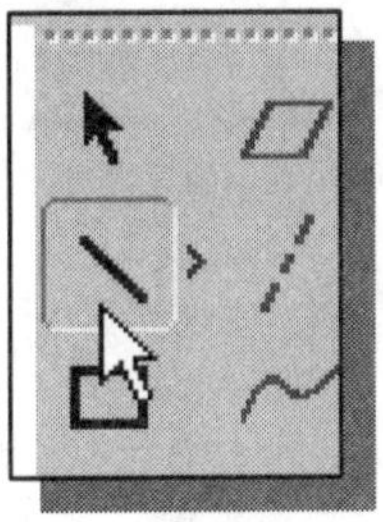

7. Next, switch to sketching regular line entities. In the *Sketcher* toolbar, select **Line**.

8. On your own, create a closed region sketch approximate in shape as shown below. At this point, concentrate on the **shape** and **form** of the design. The dimension values may appear differently on your screen. Note there is no *Equal Length* constraint in the sketch.

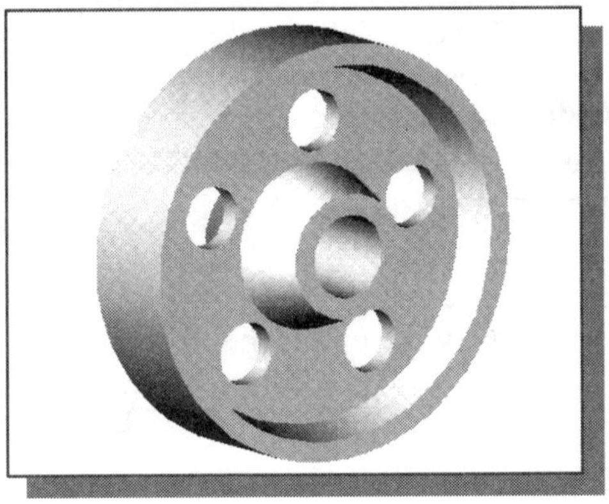

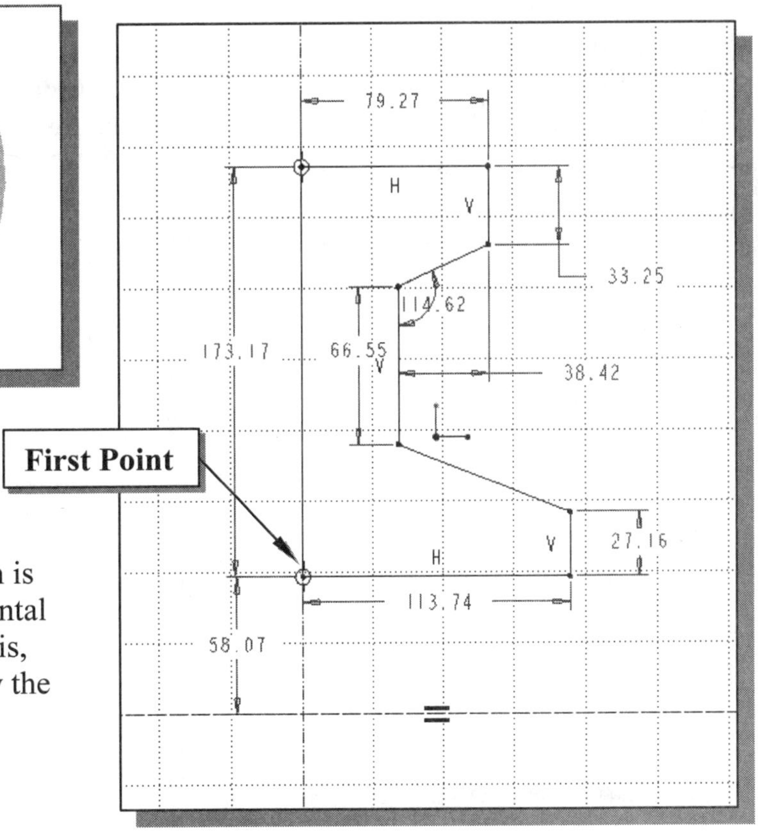

❖ Note that the *Wheel* design is symmetrical about a horizontal axis as well as a vertical axis, which allows us to simplify the 2D sketch as shown.

Creating Diameter Dimensions

➢ *Pro/ENGINEER* provides a special option to create dimensions that will account for the symmetrical nature of the design; this is known as the ***diameter dimension***.
To create a diameter dimension: pick the entity, pick the centerline to use as the axis of revolution, and pick the entity again; then place the dimension.

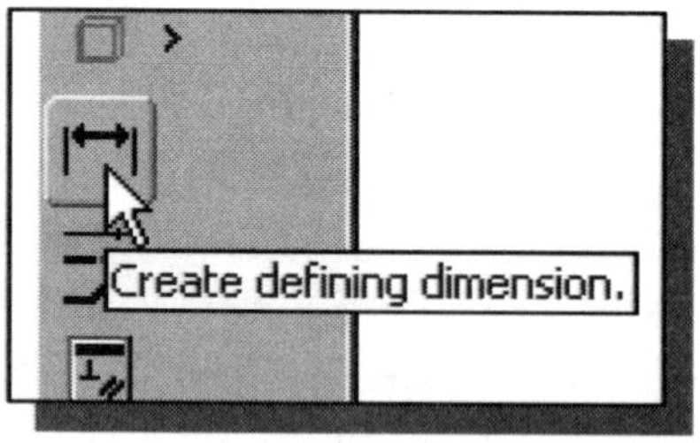

1. Select **Dimension** in the *Sketcher* toolbar.

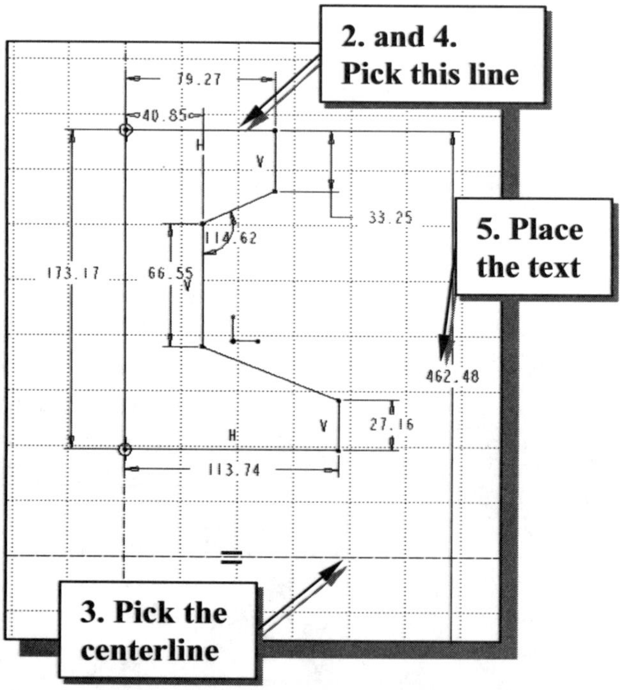

2. Pick the **top horizontal line** as the first entity to dimension.

3. Pick the horizontal **centerline.**

4. Pick the **top horizontal line** again.

5. Pick a location, with the middle-mouse-button, toward the right side of the sketch as shown.

❖ The diameter dimension uses the centerline as a reference to calculate the dimension value.

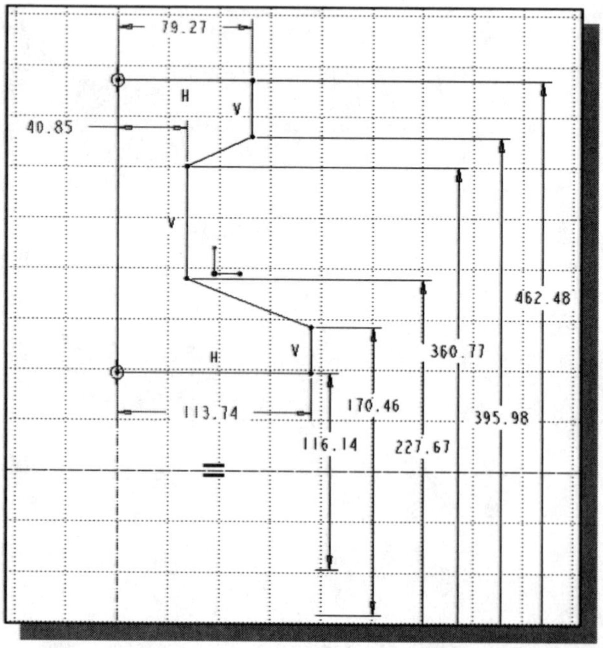

6. On your own, repeat the above steps and create the desired dimensions as shown.

❖ Notice many of the *weak* dimensions are removed by the system as the desired dimensions are added.

Modify the Values of Dimensions

1. Choose **Select** in the *Sketcher* toolbar.

2. Select dimensions by enclosing all of the dimension text inside a *selection window*. (Use the left-mouse-button and drag inside the graphics area.)

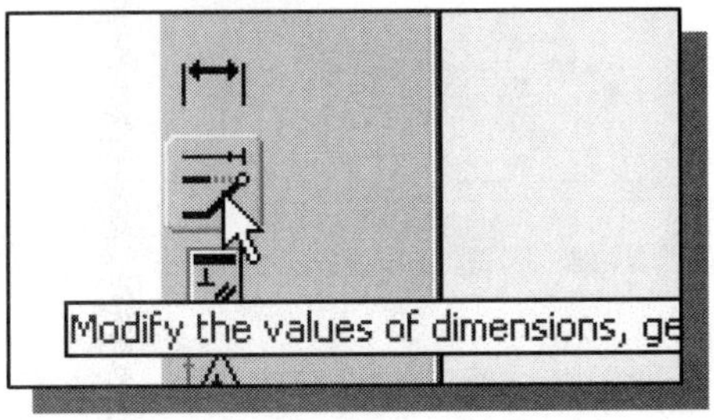

3. Choose **Modify Dimension** in the *Sketcher* toolbar.

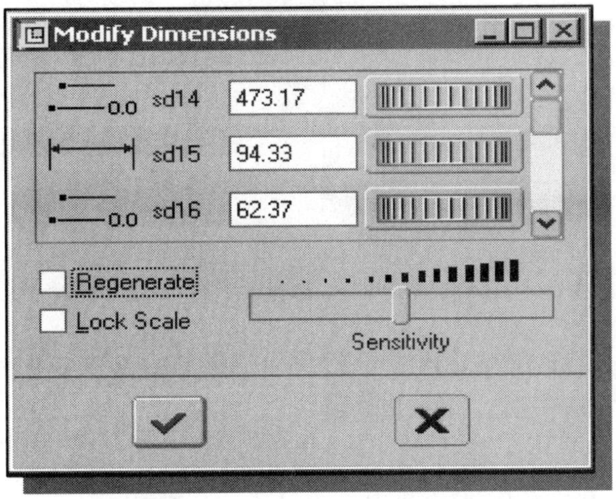

4. In the *Modify Dimensions* window, toggle ***off*** the **Regenerate** option by left-clicking once on the option.

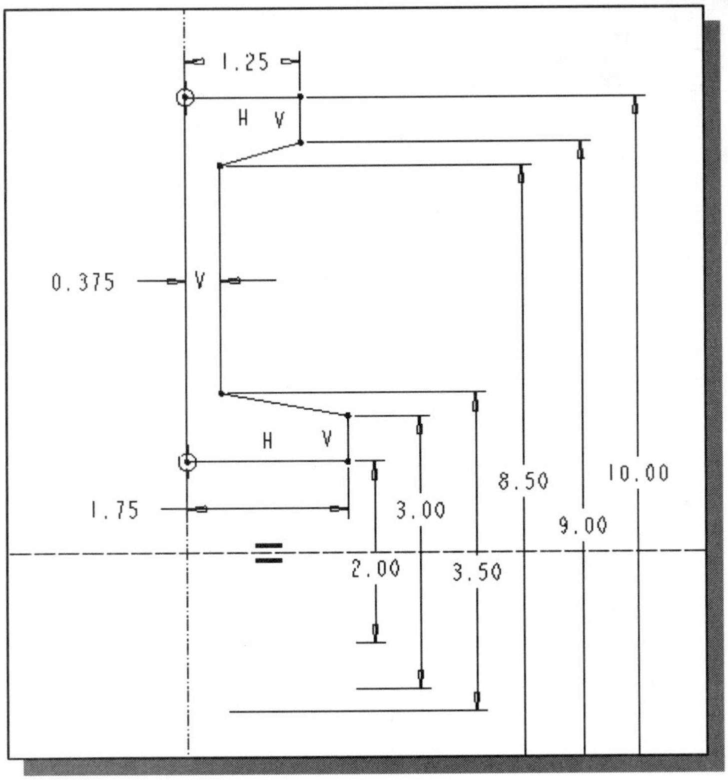

5. On your own, modify the dimension values so that they appear as shown in the figure.

6. Click the **Accept** button to regenerate the sketch and exit the *Modify Dimensions* dialog box.

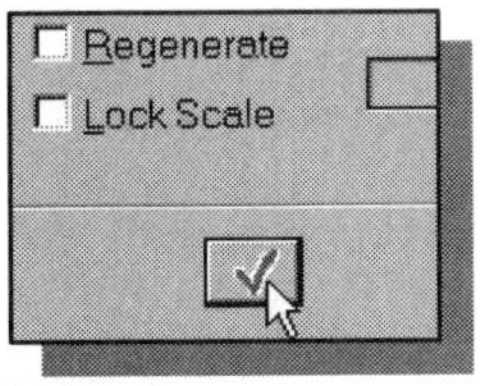

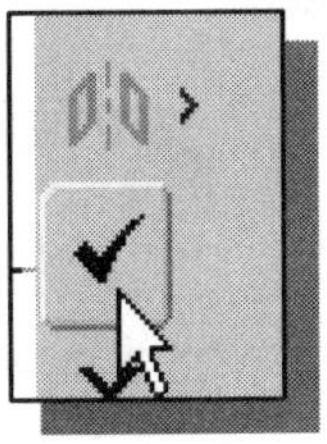

7. Click **Accept** to accept the completed section.

Completing the Revolved Feature

❖ For the **Revolve** operation, the 2D section needs to be a valid *section* (no self-intersecting edges), and the axis selected cannot cause the resultant model to intersect with itself. The axis of rotation can be one of the edges, but the axis cannot go through the *middle* of the section.

1. Confirm the revolve angle is set to **360**.

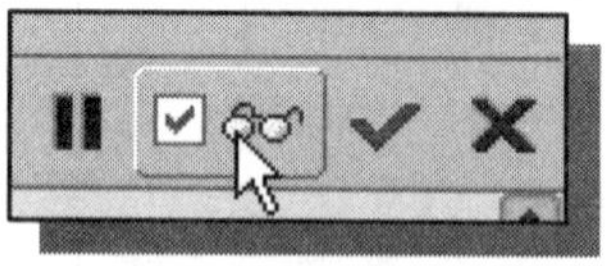

2. Pick **Preview** in the *Feature Option Dashboard* to examine the solid feature.

3. On your own, use the *Dynamic Viewing* functions to view the 3D model.

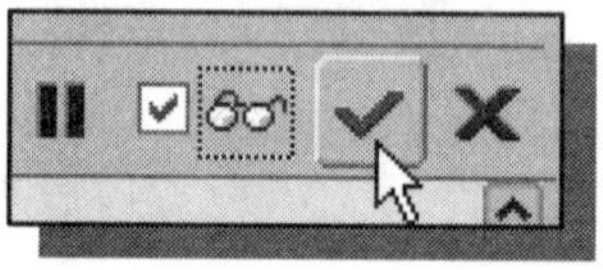

4. Pick **Accept** in the *Feature Option Dashboard* to create the solid feature.

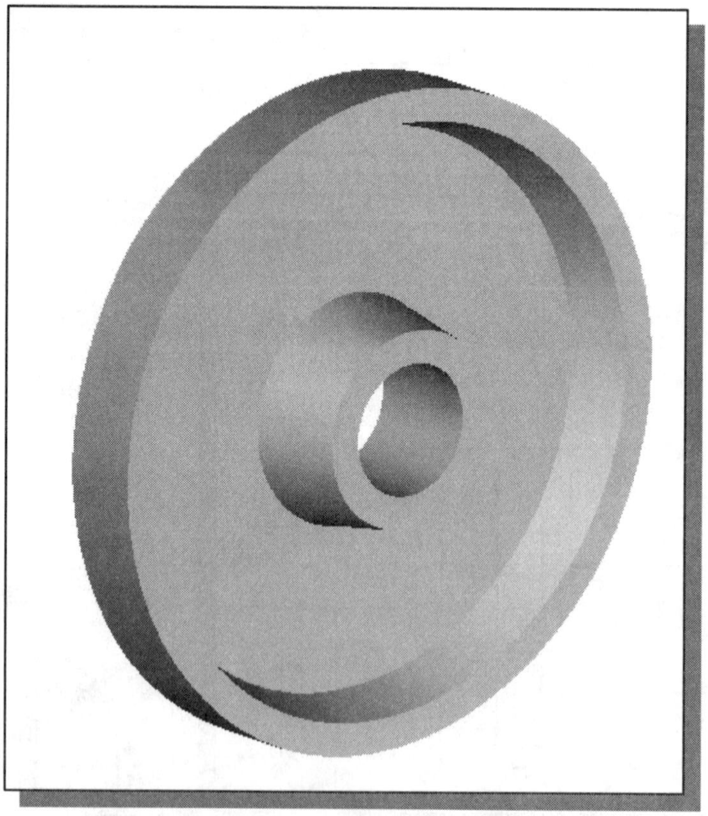

Using the Mirror Option

- In *Pro/ENGINEER*, features can be mirrored to create and maintain complex symmetrical features. We can mirror a feature about a datum plane or a specified surface. We can create a mirrored feature while maintaining the original parametric definitions, which can be quite useful in creating symmetrical features. For example, we can create one quadrant of a feature, then mirror it twice to create a solid with four identical quadrants.

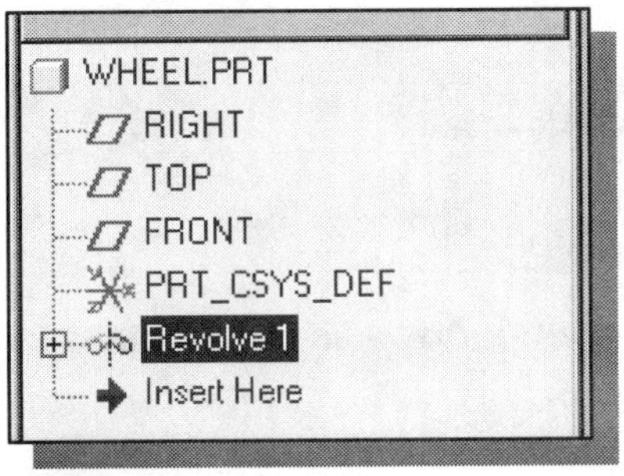

1. In the *Model Tree* window, confirm that the **Revolve 1** feature is preselected.

2. Choose **Mirror Tool** in the *Feature* toolbar.

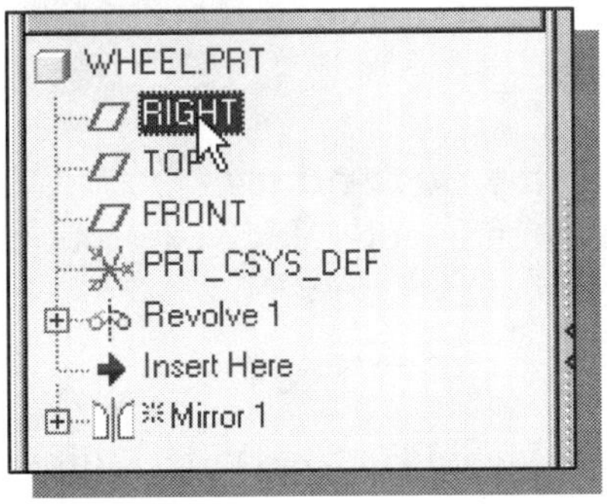

3. In the message area, the message "*Select a plane or create a datum to mirror about*" is displayed. Pick datum plane **RIGHT**, in the *Model Tree* window, as the plane about which to mirror.

4. Pick **Accept** in the *Feature Option Dashboard* to create the solid feature.

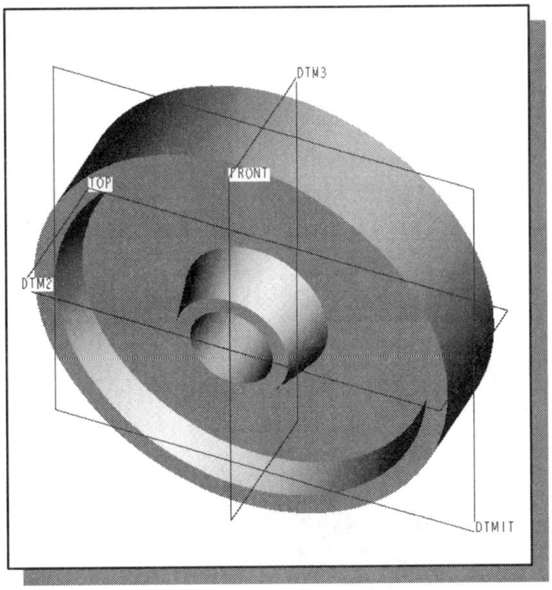

➢ Now is a good time to save the model (Quick key: [**Ctrl**] + [**S**]). It is a good habit to save your model periodically, just in case something might go wrong while you are working on the model. You should also save the model after you have completed any major constructions.

Adding a Radial Hole Feature

❖ We will create a circular hole as the next solid feature, which will be used as a ***pattern leader*** for the subsequent pattern feature.

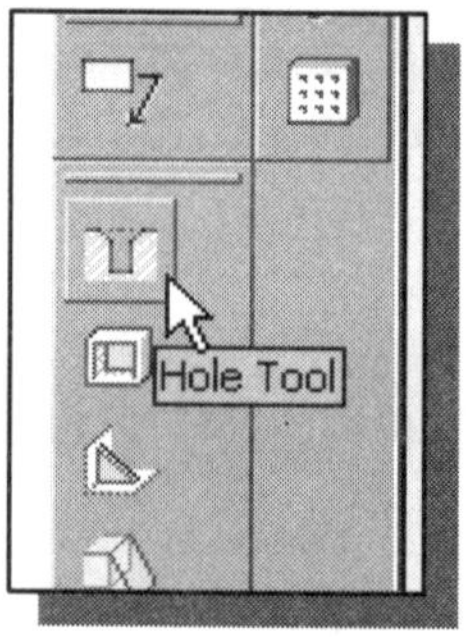

1. In the *Feature* toolbars (toolbars aligned to the right edge of the main window), select the **Hole Tool** option as shown.

❖ The message, "*Select a surface, axis or point to place hole*" is displayed in the message area.

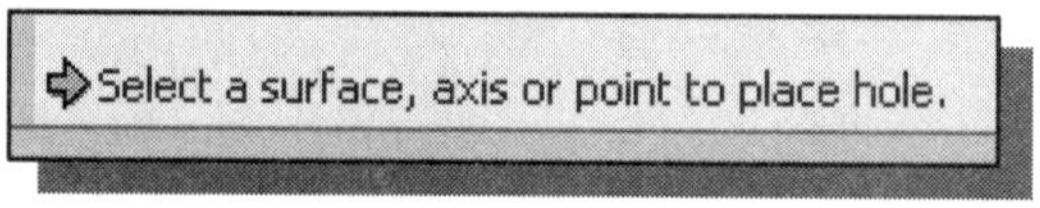

2. Select datum plane **RIGHT** (pick in the *Model Tree* window) as the **Primary** placement reference.

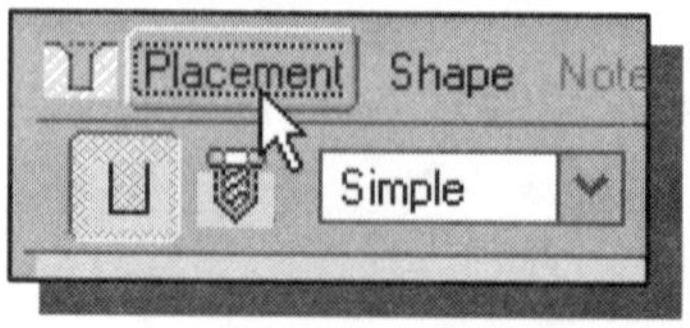

3. In the *Feature Option Dashboard*, click the **Placement** option icon and examine the placement settings.

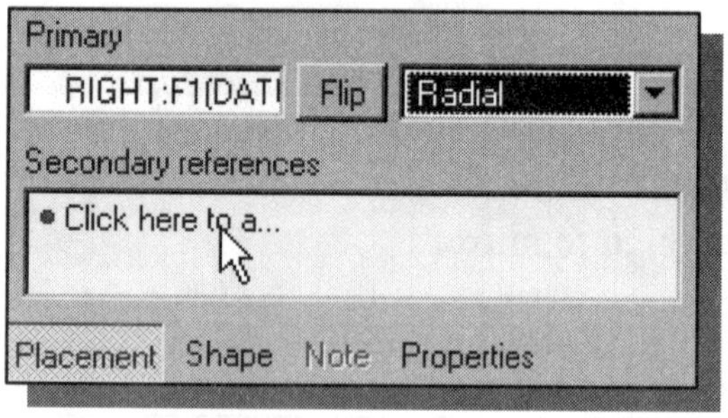

4. Set the hole placement option to **Radial** as shown.

5. Click once inside the **Secondary references** list area to activate the selection of secondary references.

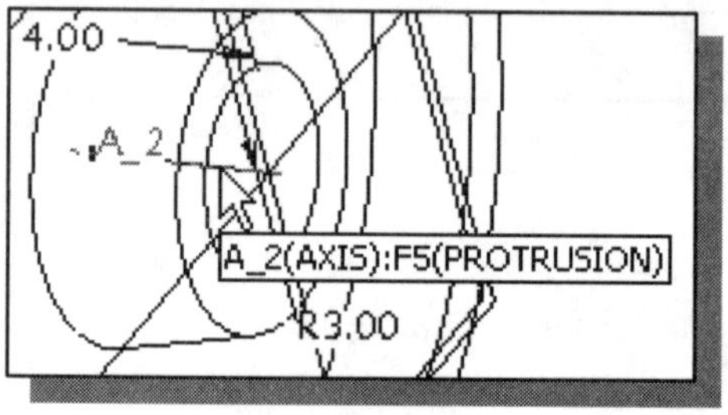

6. Select the ***datum axis*** located at the center of the model.

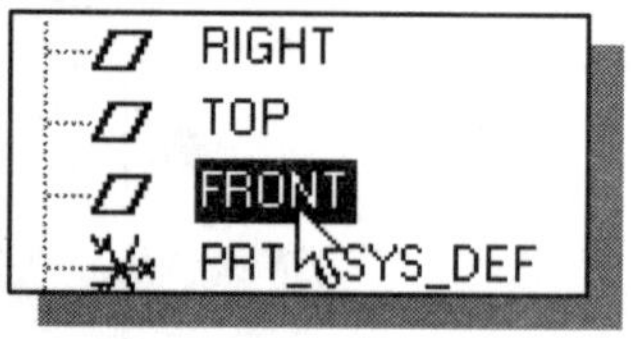

7. In the *Model Tree* area, hold down the [**CTRL**] key and select **Front** as a secondary reference.

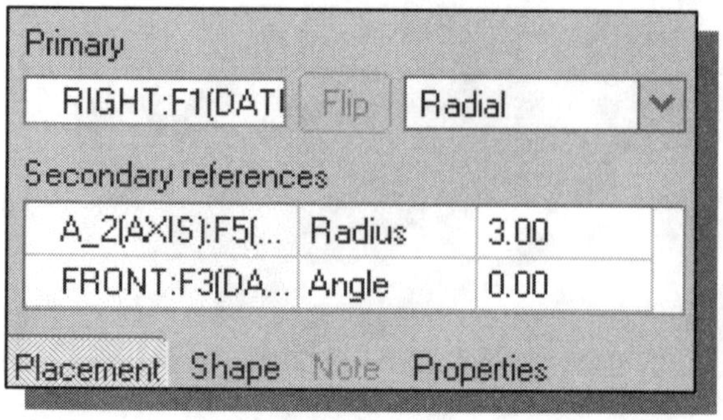

8. In the **Secondary references** list, adjust the **Radius** to **3.0** and **Angle** to **0.0** (or **180.0**).

❖ Note that the **Radial Hole** option uses a *Polar coordinate system* to position the center of the hole feature.

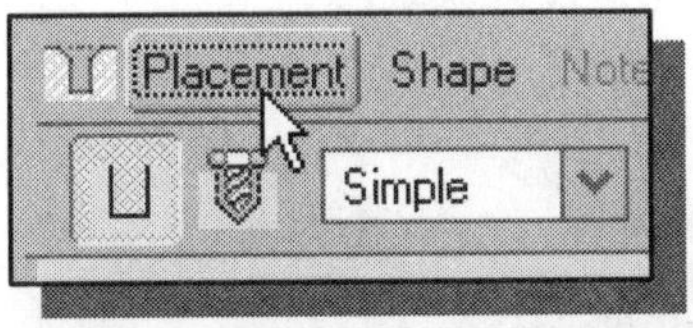

9. In the *Feature Option Dashboard*, click the **Placement** icon to close the **Placement** option list .

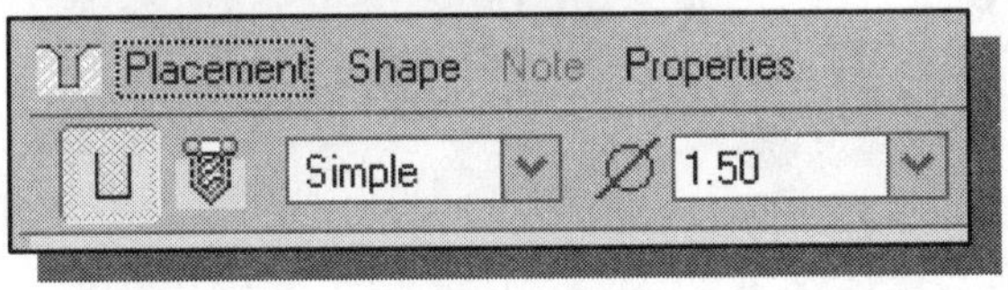

10. In the *Feature Option Dashboard*, set the *feature diameter* to **1.5**.

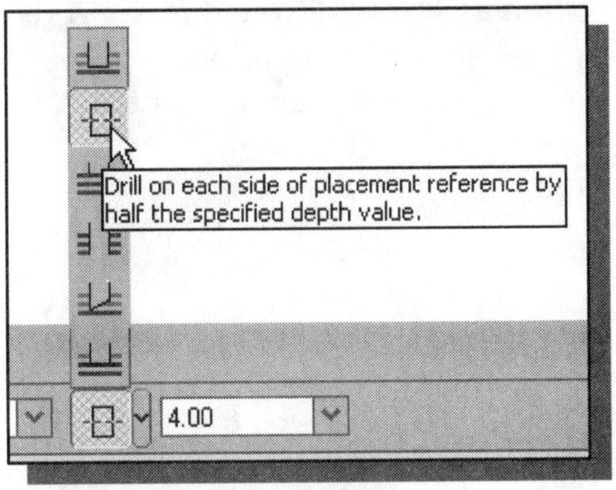

11. Set the *extrusion option* to **Both Sides**.

12. Set the *depth value* to **4.0**.

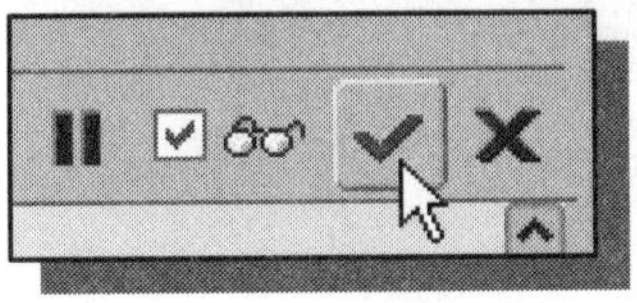

13. Click on the **Accept** icon and proceed to create the hole feature.

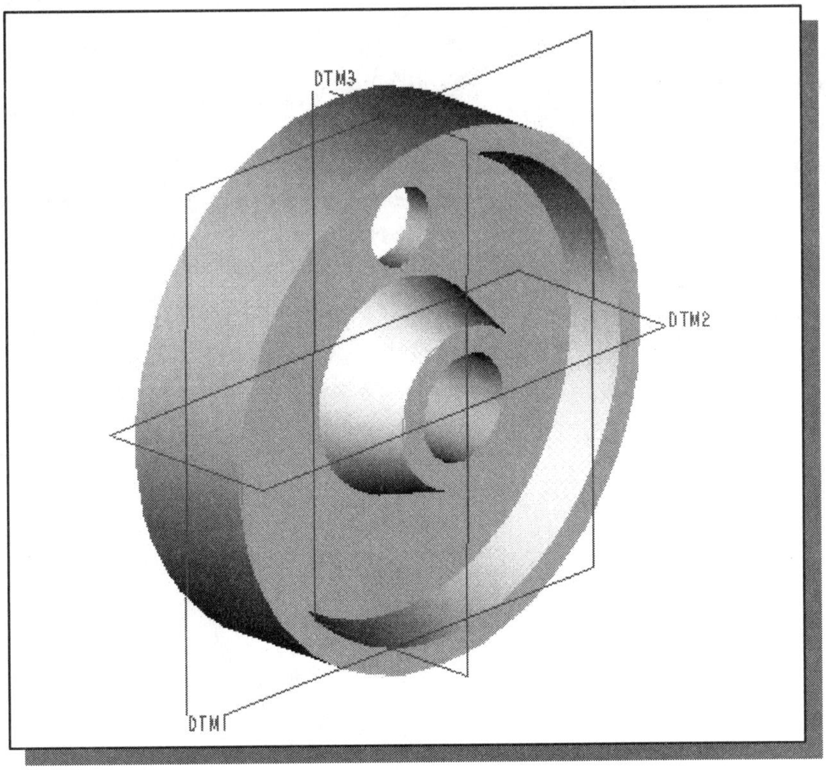

Create a Circular Pattern

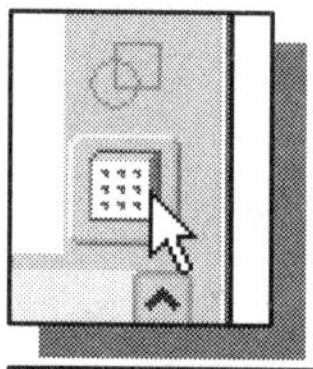

1. In the *Feature* toolbar, select **Pattern**. (Note that the **Hole** feature is currently highlighted in the *Model Tree* area, which means it is pre-selected.)

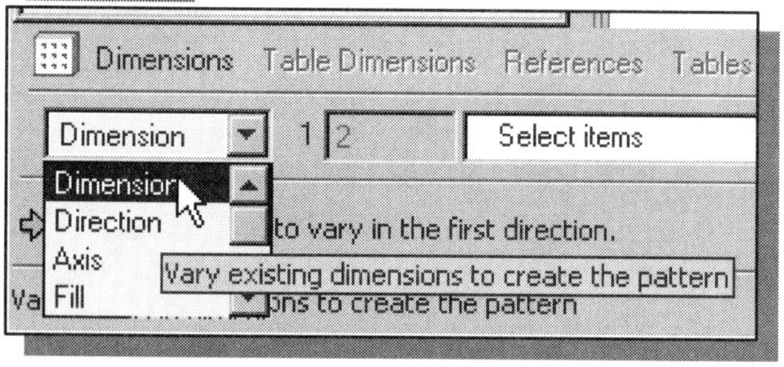

2. Select **Dimension** as the pattern option as shown. (Note that the **Axis** option is a feasible option for most designs, you are encouraged to repeat this section using the *Axis* option.)

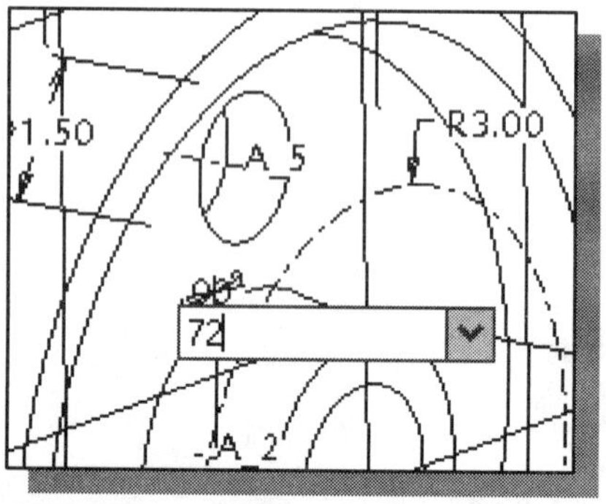

3. Pick the angle dimension (**0.0**).

4. Enter **72** in the value box as the *increment dimension.*

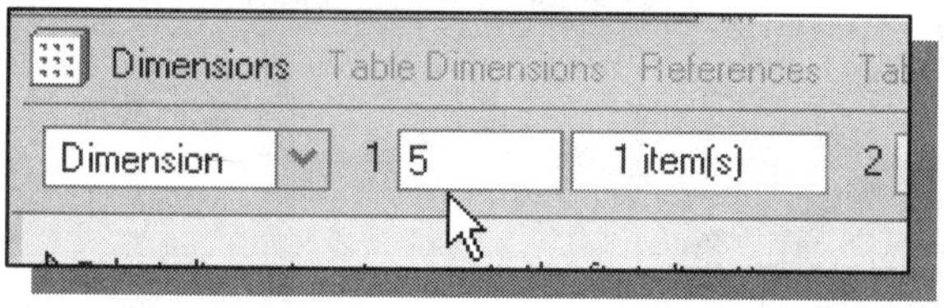

5. Enter **5** in the pattern number box for the total number of copies.

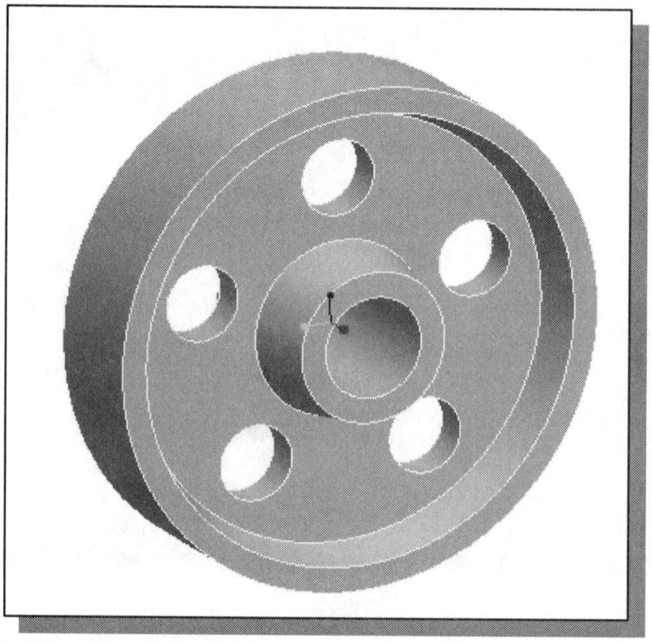

6. Click **Accept** to create the pattern.

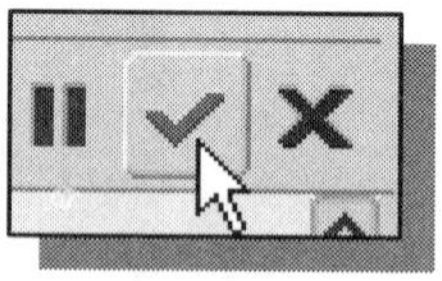

Parametric Relations

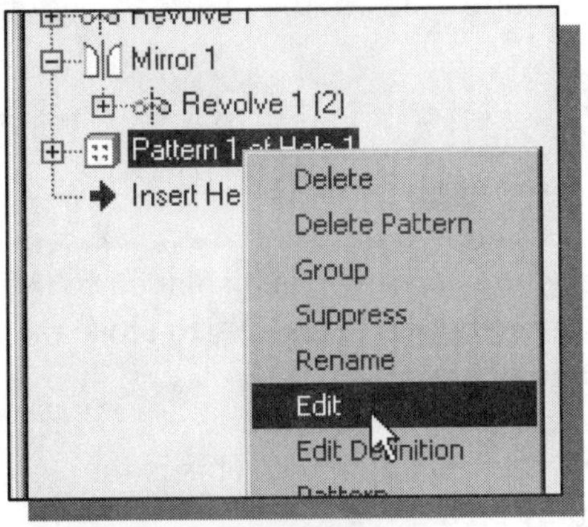

1. Inside the *Model Tree* window, right-mouse-click on the *Pattern* feature and select **Edit** to display the dimensions related to the pattern.

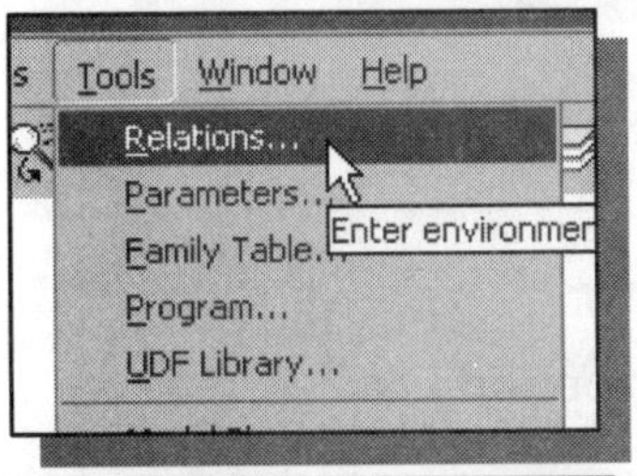

2. Pick **Relations** in the **Tools** pull-down menu.

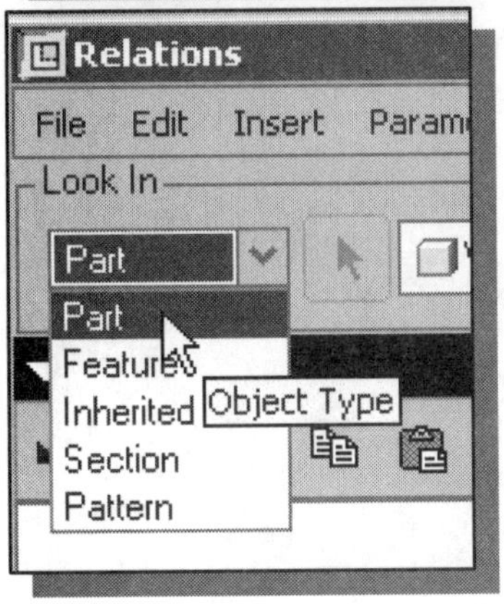

In part files, four types of **parametric relations** can be established:

- **Part Relations** relate different feature parameters to each other in a single model.
- **Feature Relations** relate specific parameters within one feature in the model.
- **Section Relations** relate specific parameters within one 2D sketch in the model.
- **Pattern Relations** relate specific parameters within a pattern in the model.

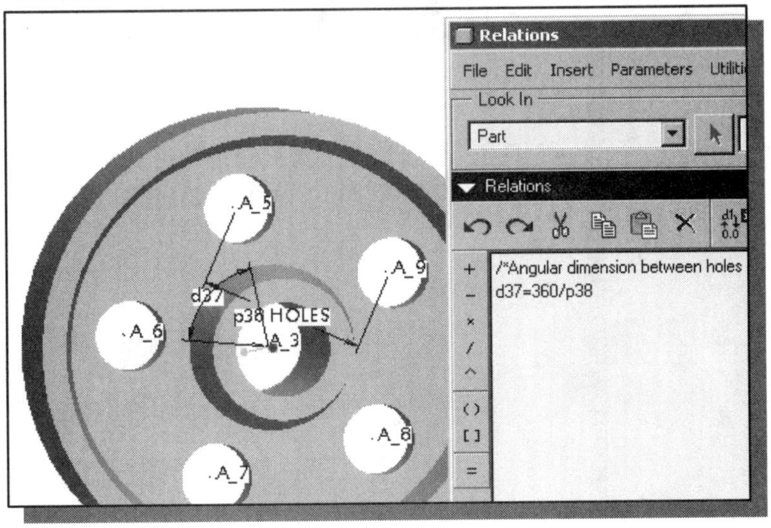

3. Enter **d37=360/p38** in the *Relations* window as the first relation. Use the corresponding variable names that appeared on your screen.

❖ The line above the relation is a ***note statement.*** Any line that begins with the characters **/*** is treated as a *note statement*.

4. Click on the **Execute/Verify** icon to confirm the entered relation is a valid relation.

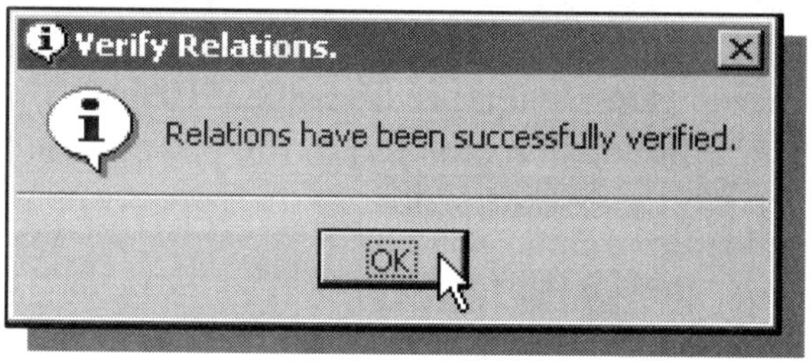

5. The message "*Relations have been successfully verified.*" is displayed in the message window. Click **OK** to close the message window.

6. Press and hold down the [**Ctrl**] key and press the [**R**] key once. This is the quick-key combination to **Repaint** the display area.

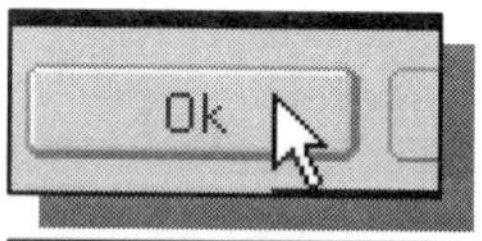

7. Click **OK** to accept the relation settings and exit the relation window.

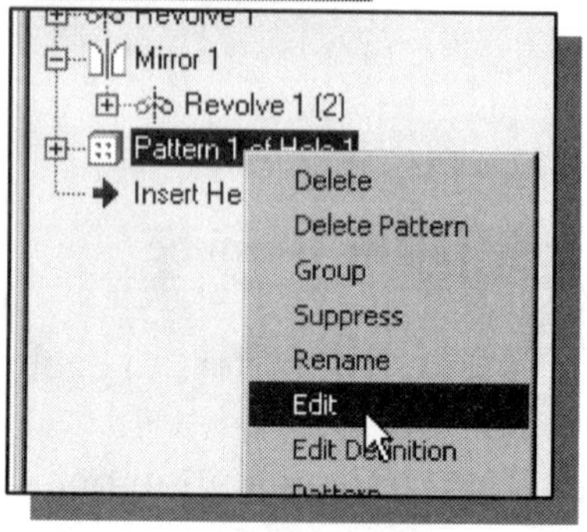

8. In the *Model Tree* area, right-mouse-click the pattern feature and choose **Edit** in the option list.

9. On your own, change the ***number of holes*** to **6**.

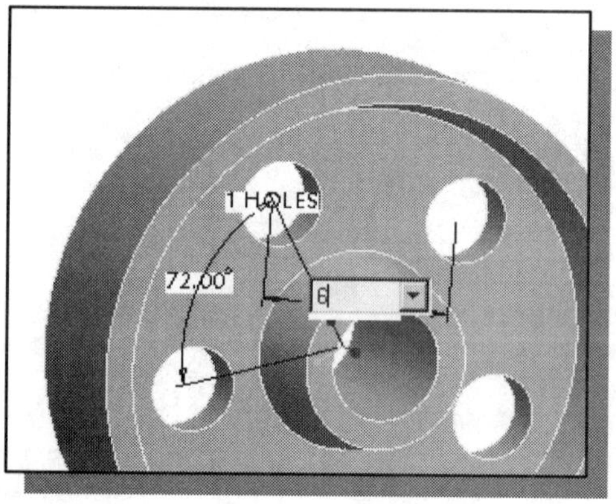

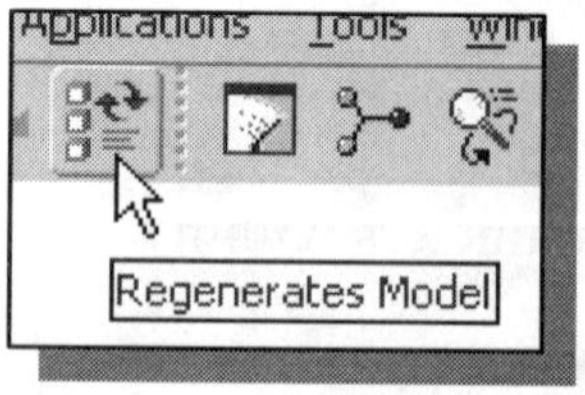

10. Pick **Regenerate** in the *Standard* toolbar.

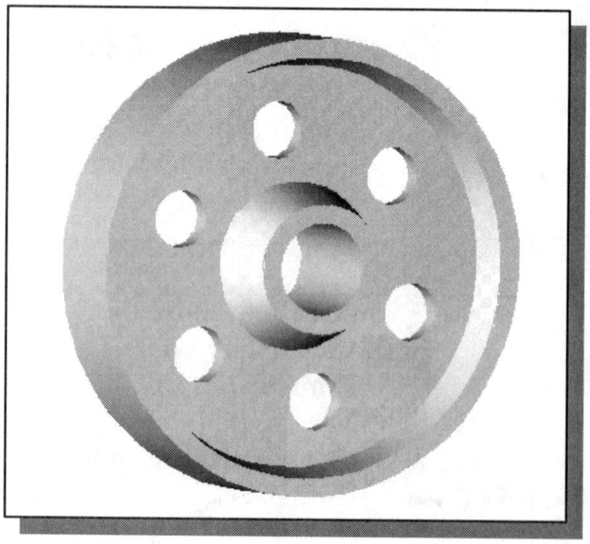

- The design is updated while the established relation is maintained.

Creating a Multi-View Part Drawing

1. Pick the **New Object** icon in the toolbar. (We can also use the key combination **Ctrl-N** to start a new object.)

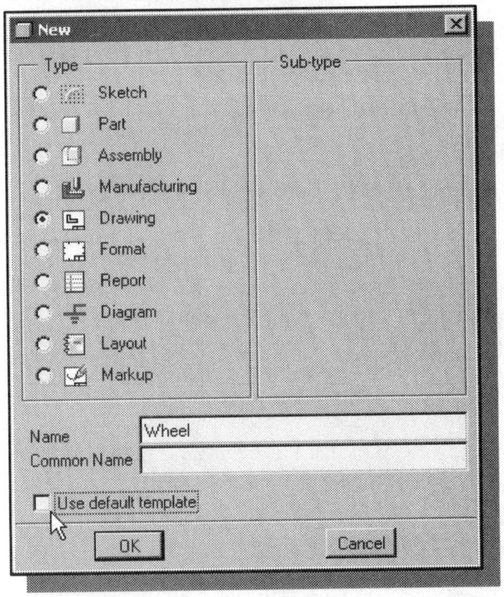

2. In the *New* form, select **Drawing** under the **Type** list.

3. Enter ***Wheel*** as the drawing file Name.

4. Toggle ***off*** the **Use default template** option.

5. Pick **OK** to start a new drawing file.

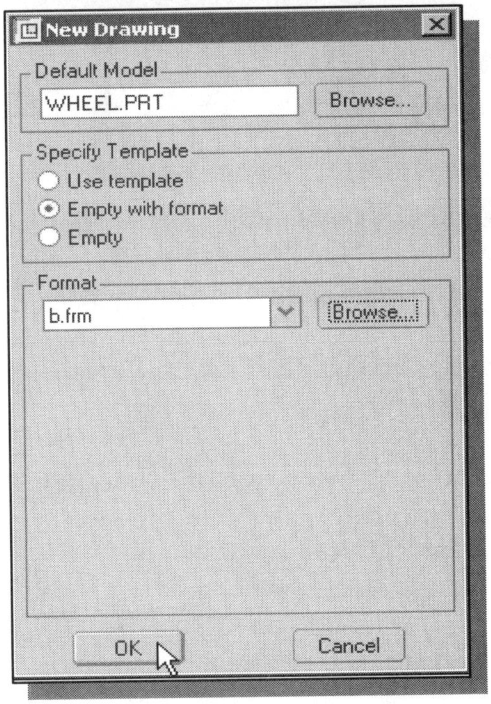

❖ In the *New Drawing* form, the currently active part is automatically selected as the drawing model.

6. In the **Specify Template** option, choose **Empty with Format**.

7. Pick **b.frm** in the **Format** list. (Use the **Browse** button to locate the *Pro/ENGINEER format files* if the list is blank.)

8. Pick **OK** to proceed with the drawing layout.

❖ A new window will appear with the title "***WHEEL***." Note the b.frm format file sets the paper size to *B size*, orientation to *Landscape* and a *title block* is placed in the drawing area.

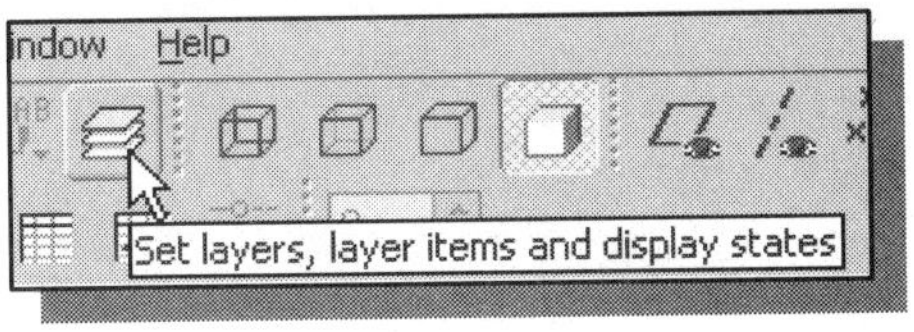

9. In the *Standard* toolbar, toggle ***off*** the **Layer Control Panel**.

Adding a Primary Main View

1. In the *Standard* toolbar area, click on the **Create a general view** button as shown. Note that this is a shortcut to the same **Drawing view → General** command in the pull-down menu.

2. The message "*Select CENTER POINT for drawing view*" is displayed in the message area. Pick a location that is about one-third from the left side of the display window as shown in the figure below.

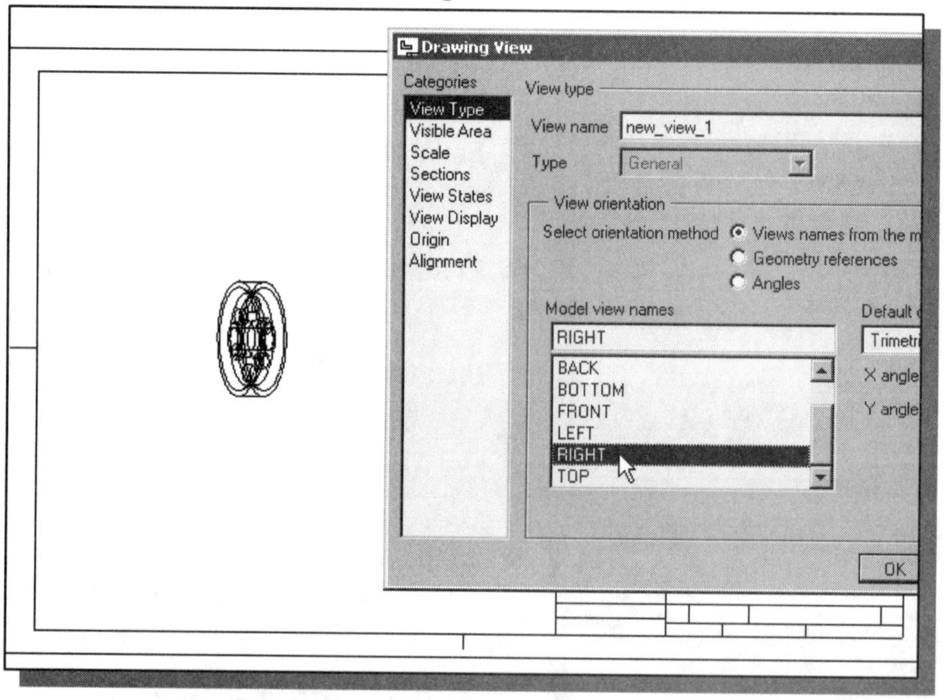

3. Choose **RIGHT** in the **Model view names** list as shown in the figure above.

4. Click on the **Apply** button to adjust the view orientation in the display area.

5. Click on the **Close** button to accept the settings and proceed with the creation of the main view.

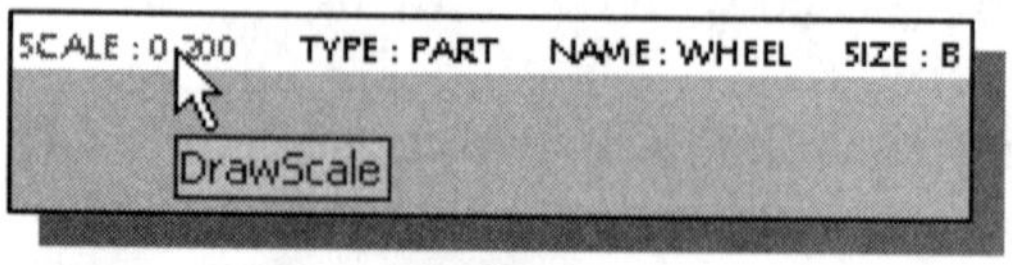

6. Move the cursor to the bottom of the display area and **double click** with the left-mouse-button on **DrawScale**.

7. In the message area, enter **0.5** as the new scale factor.

Adding a Projected Sectional View

❖ To create a section view, a view projection line is required. The **view projection line**, also known as the **cutting plane line**, defines a cutting line for the section view. Depending upon how a view projection line is drawn, the line can be used to define the type of section view or define a boundary for a partial view. If the view projection line is created outside the base view, it defines a plane from which to project an auxiliary view.

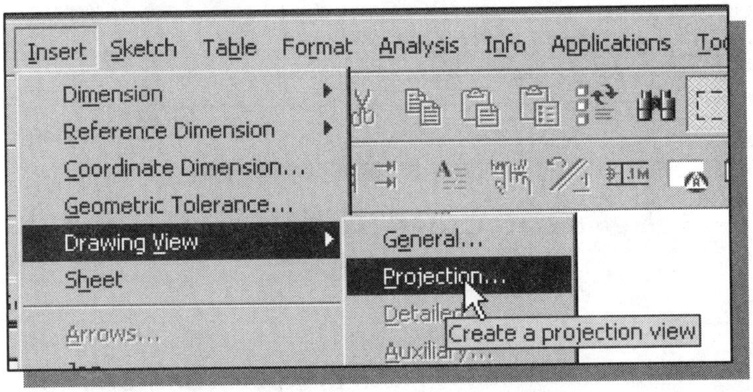

1. Click the primary view in the display area to **pre-select** the view.

2. In the **Insert** pull-down menu, select **Drawing view → Projection** as shown.

3. The message "*Select CENTER POINT for drawing view*" is displayed in the message area. Pick a location that is toward the right side of the primary main view in the display area.

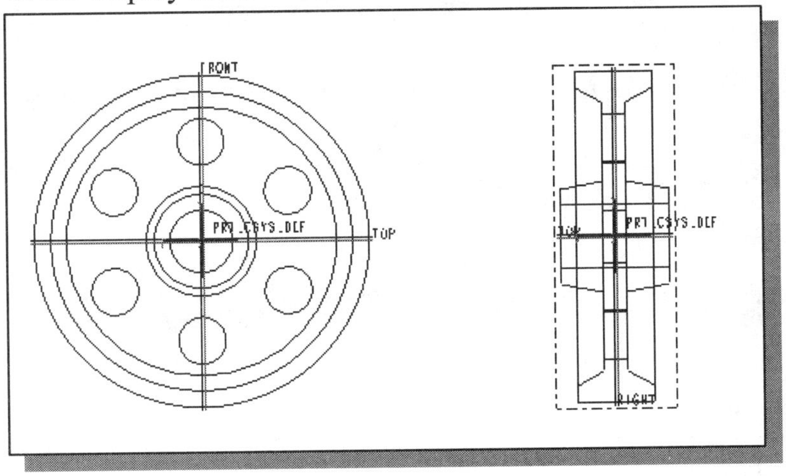

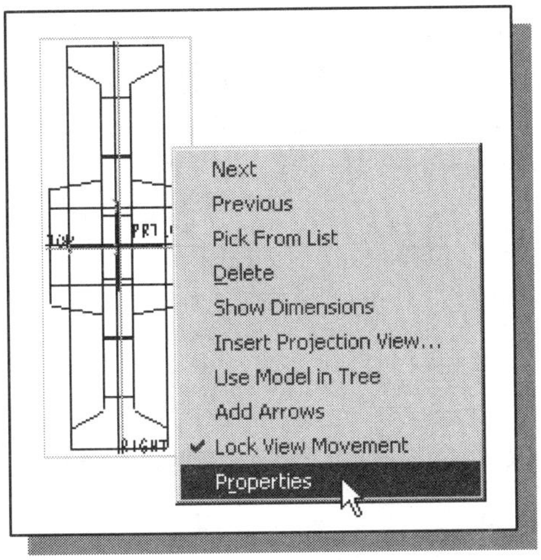

4. Inside the display area, click once with the right-mouse-button on the projection view and select **Properties** in the option list as shown.

5. In the *Drawing View* window, select **Sections** in the **Categories** list.

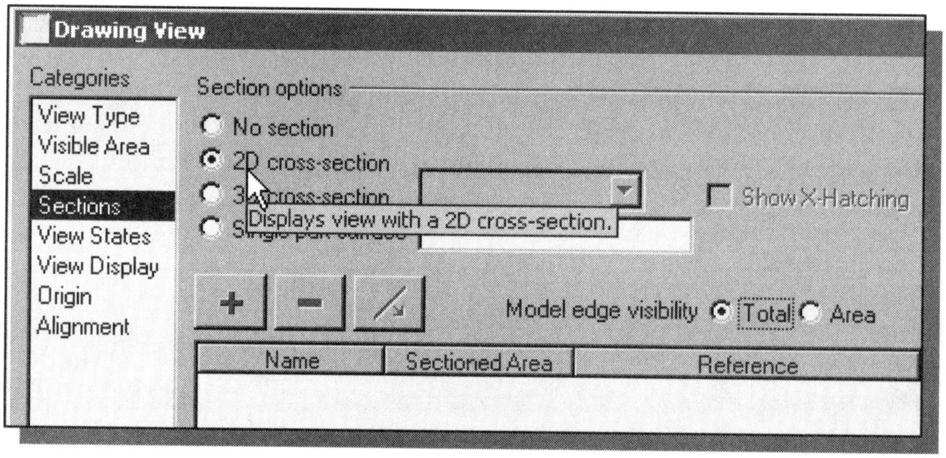

6. In the **Sections** option, switch ***on*** the **2D Cross-Section** as shown in the figure above.

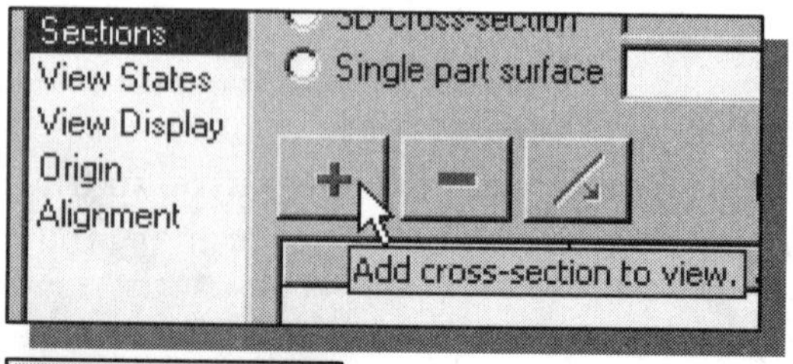

7. Click the [**+**] icon to **Add cross-section to view**.

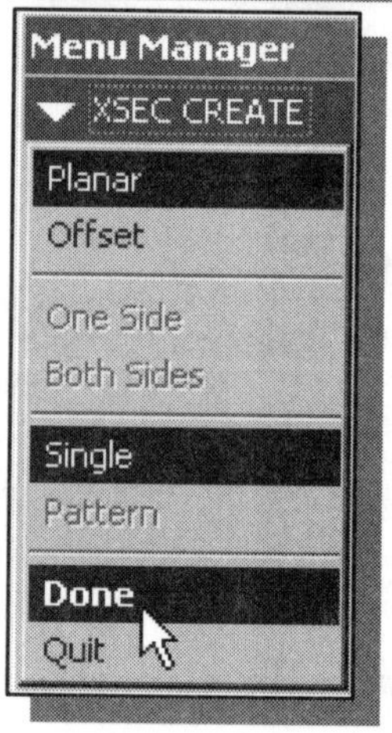

8. In the *Menu Manger*, select **Planar → Single→ Done**.

9. In the prompt area, type **A** as the name for the new cross-section.

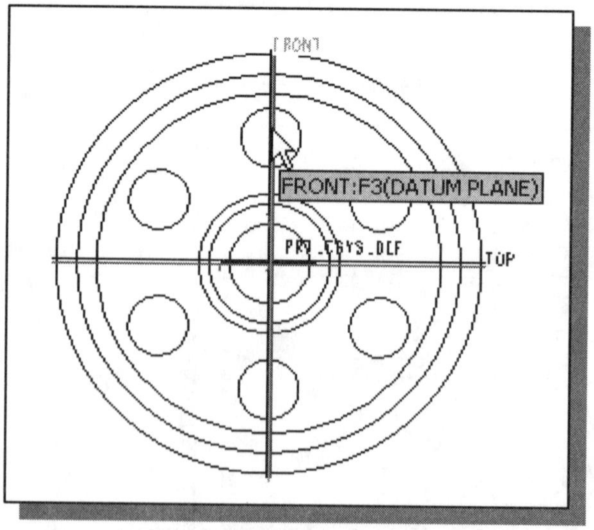

10. Pick the **Front Datum Plane** in the ***primary main view*** (front view) to use as the cutting plane location.

11. Click **Apply** to accept the settings.

12. Select **Sections** in the **Categories** list.

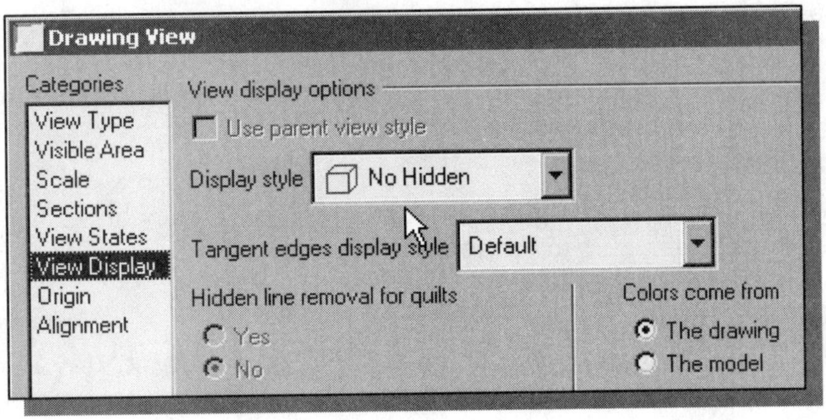

13. Set the **Display style** to **No Hidden** as shown in the above figure.

14. Click **Apply** to accept the settings.

15. Click **Close** to end the **Drawing View Properties** command.

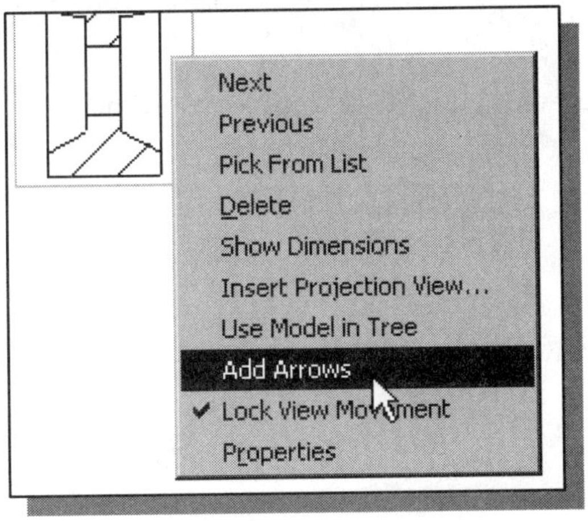

16. Inside the display area, click once with the right-mouse-button on the projection view and select **Add Arrows** in the option list as shown.

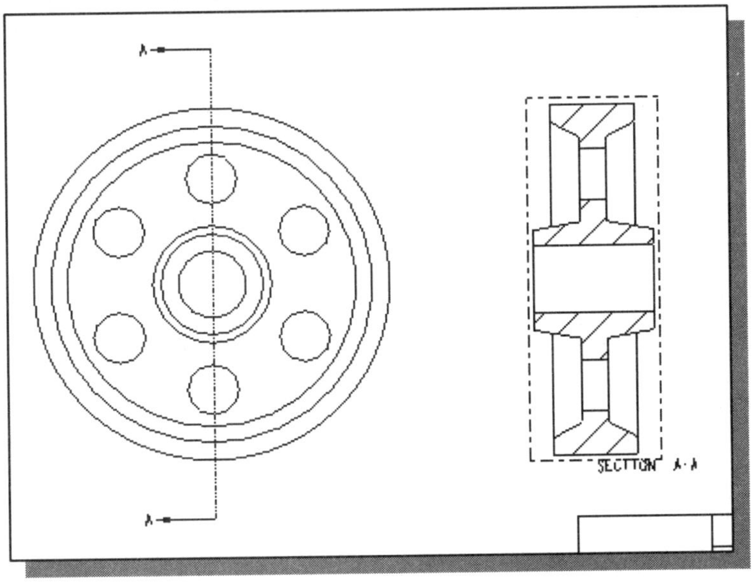

17. Pick the primary view by clicking on the view to add the cutting plane line and the two arrows.

Modify the Overall Drawing Display

1. On your own, toggle ***off*** the display of the datum planes, datum axis, datum points and coordinate systems as shown in the figure.

2. Press and hold down the [**Ctrl**] key and press the [**R**] key once. This is the quick key combination to **Repaint** the display area.

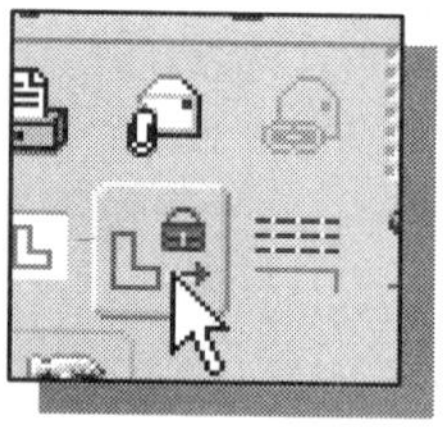

3. In the *Standard* toolbar area, click on the **Lock Views** icon to toggle ***off*** this option. (This option is set to ***on*** by default, which does not allow the moving of the views.)

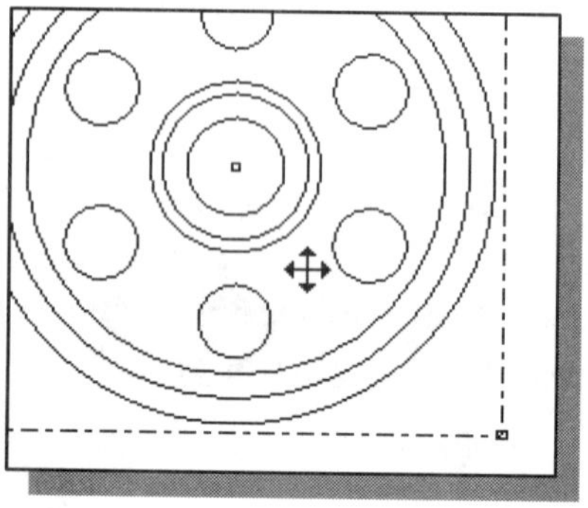

4. Pick the primary view by clicking once on the view. Note the cursor is changed to a four-arrow display indicating the moveable directions.

5. Drag with the left-mouse-button to reposition the view.

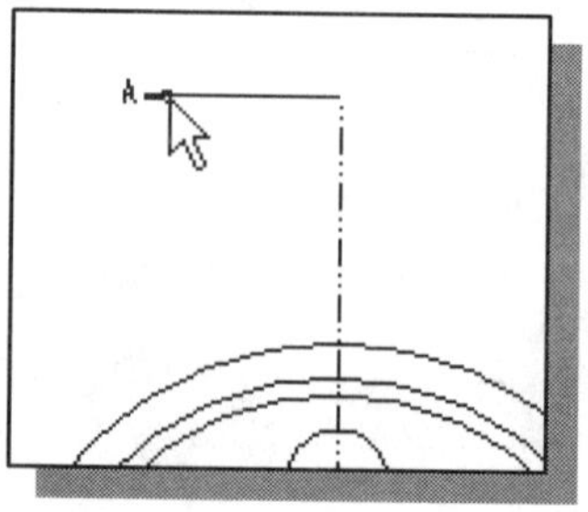

6. Click on the upper arrow to select the entity.

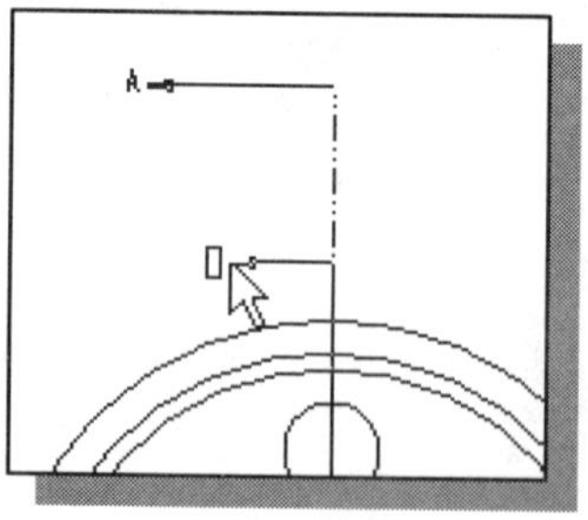

7. Reposition the location of the selected arrow by dragging the arrow with the left-mouse-button.

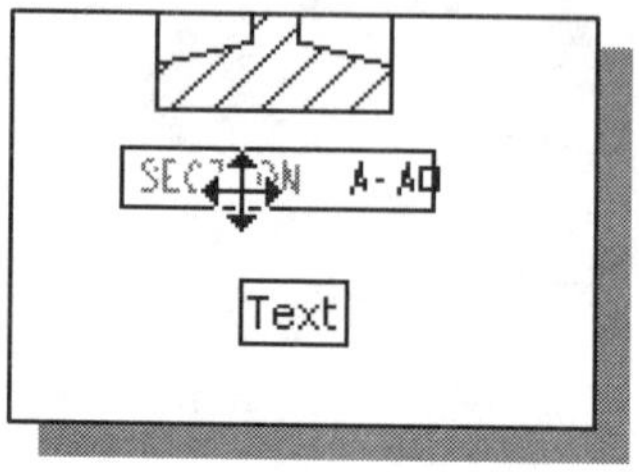

8. Pick the section view label, **SECTION A-A** next to the section view, with the left-mouse-button.

9. Drag and drop with the left-mouse-button to reposition the label.

10. On your own, repeat the above steps to adjust the views, and reposition the arrows and text as shown in the below figure.

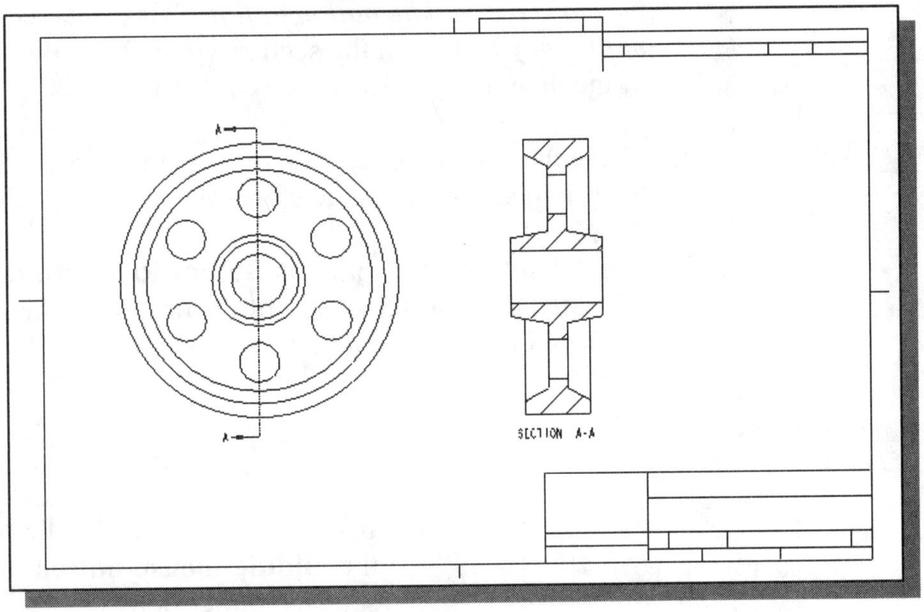

Displaying Feature Dimensions

- By default, feature dimensions are not displayed in 2D views in *Pro/ENGINEER*. We can change the default settings while creating the views or switch on the display of the parametric dimensions using the **Show/Erase** option menu.

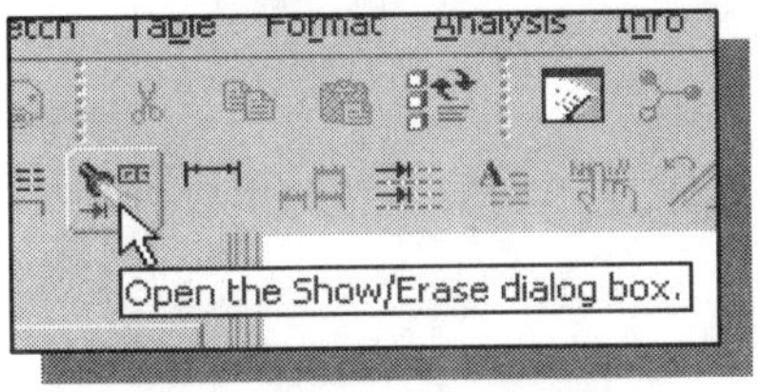

1. In the *Standard* toolbar area, select the **Show/Erase** command.

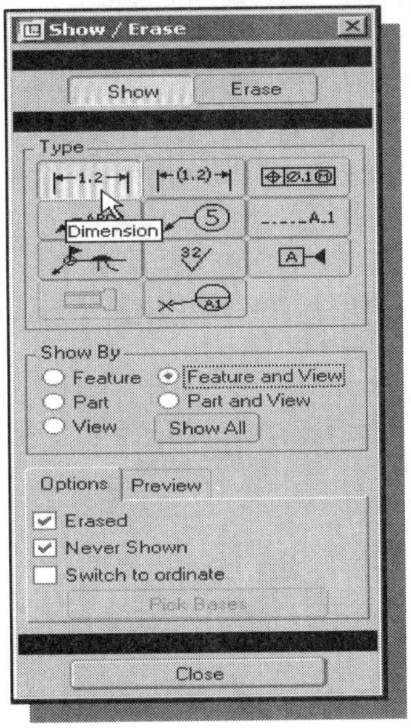

2. Choose to Show the **Regular Dimension** by toggling ***on*** the icon in the *Show/Erase* window as shown.

3. In the Show By option, select **Feature and View**.

4. In the Options settings area, confirm the **Erased** and **Never Shown** options are switched ***on***.

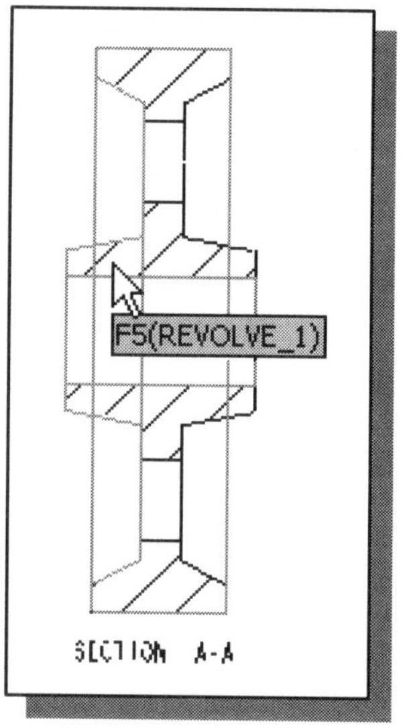

5. In the message area, the message "*Select a feature in the picked view. Middle button to finish*" is displayed. Pick the ***main body*** feature in the section view. (Note the feature label of the main body is Revolve_1.)

6. Inside the display area, click **once** with the **middle-mouse-button** to accept the selected feature.

7. The default remove option is **Select to Remove**, hold down the [**Ctrl**] key and select any of the unwanted dimensions displayed in the section view.

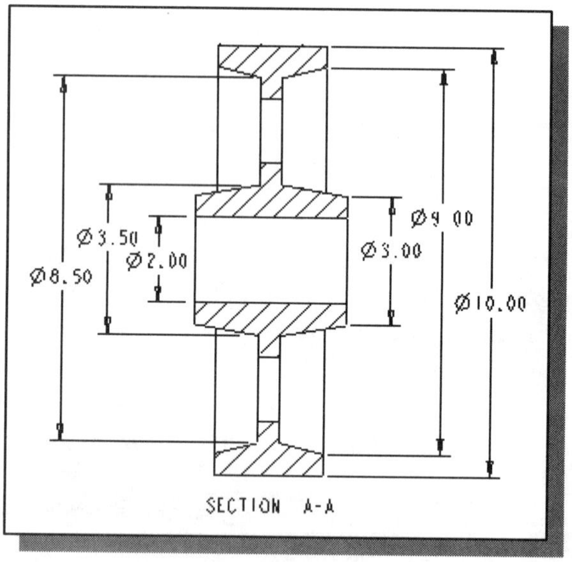

8. Inside the display area, click **once** with the **middle-mouse-button** to remove the selected dimensions.

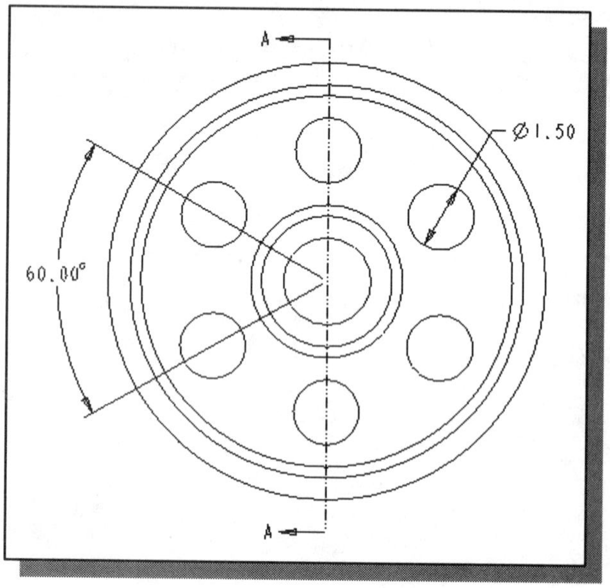

9. Pick the 3rd hole of the **hole pattern** in the *front view* and, on your own, remove any unwanted dimensions.

10. Pick the **Close** button in the *Show/Erase* form to exit the Show and Erase command.

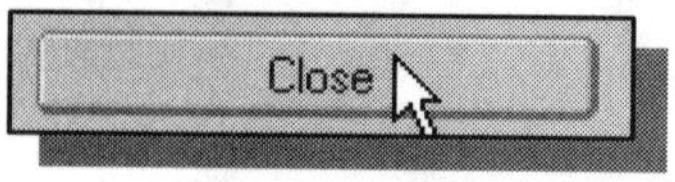

Adding Additional Driven Dimensions

1. In the *Standard* toolbar area, select **Create Dimension**.

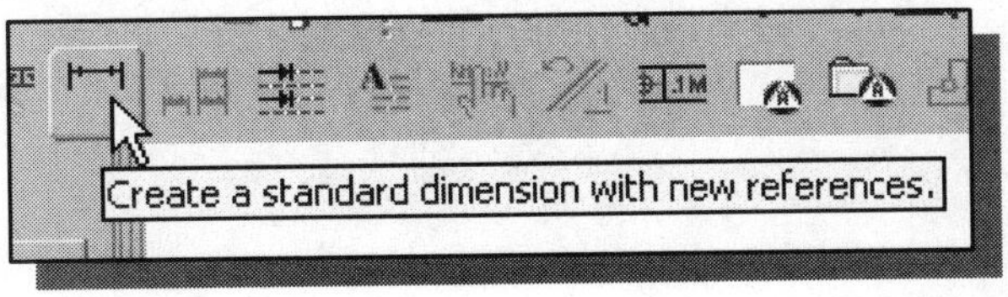

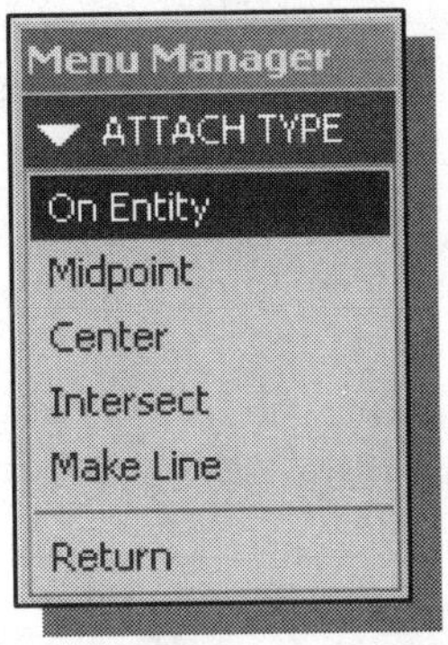

2. In the **ATTACH TYPE** submenu, confirm the selection of **On Entity**.

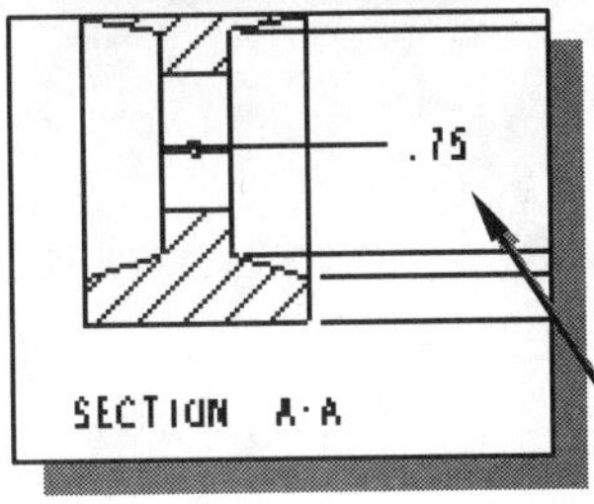

❖ Note that we add dimensions the same way we did in the *2D Sketcher*.

3. Pick the two inside vertical edges and create the dimension as shown.

Create this dimension.

4. Inside the display area, click once with the **middle-mouse-button** to end the Create Dimension command.

5. On your own, repeat the above steps and create the desired dimensions in the 2D views.

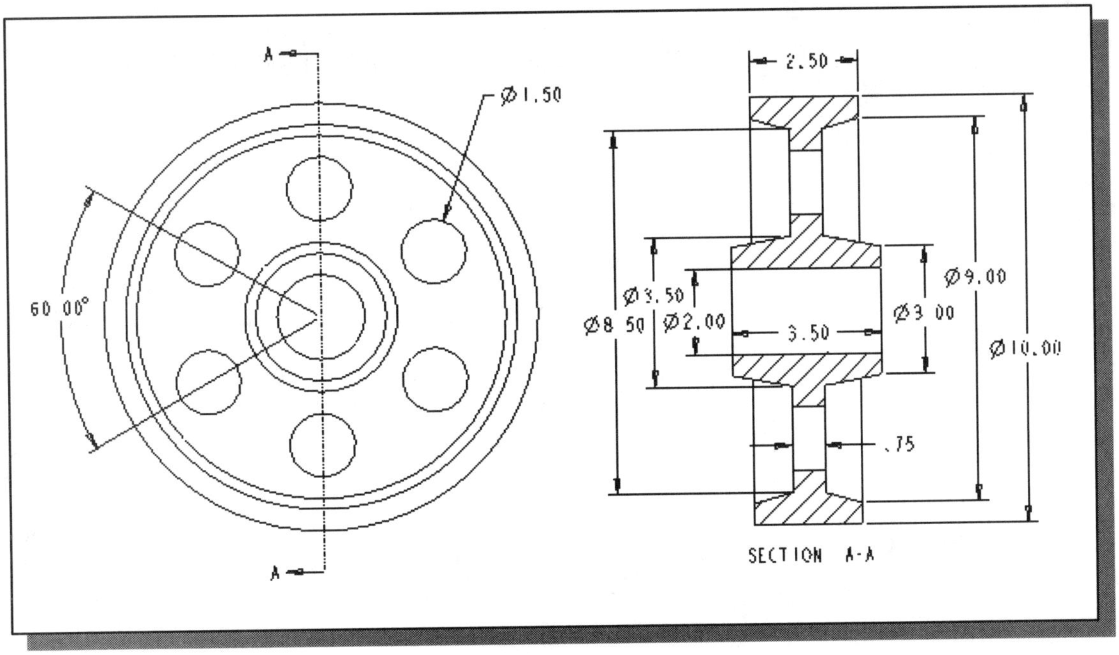

Dimension Appearances

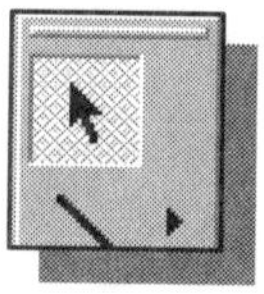

1. Exit any command and confirm the **Select** option is active.
2. On your own, reposition the displayed dimensions.

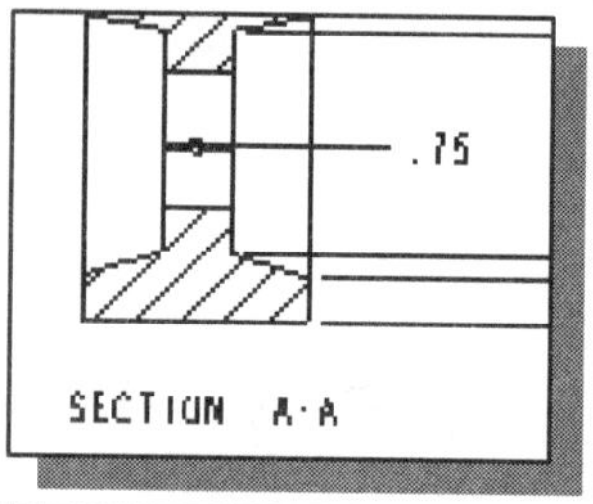

3. Pick the driven dimension we added in the above steps.
4. Inside the display area, click once with the right-mouse-button to bring up the option menu.

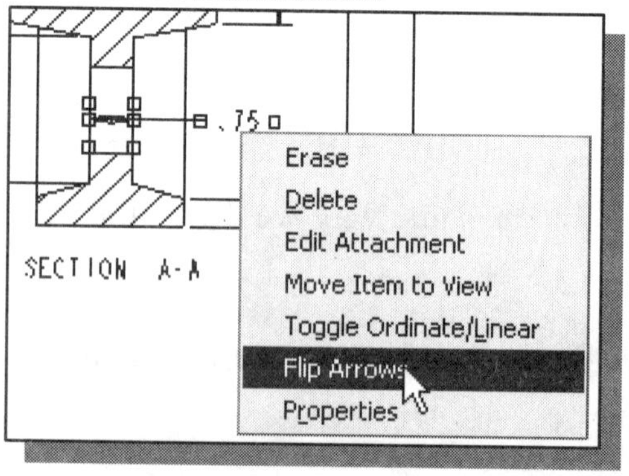

5. Select **Flip Arrows** in the option list

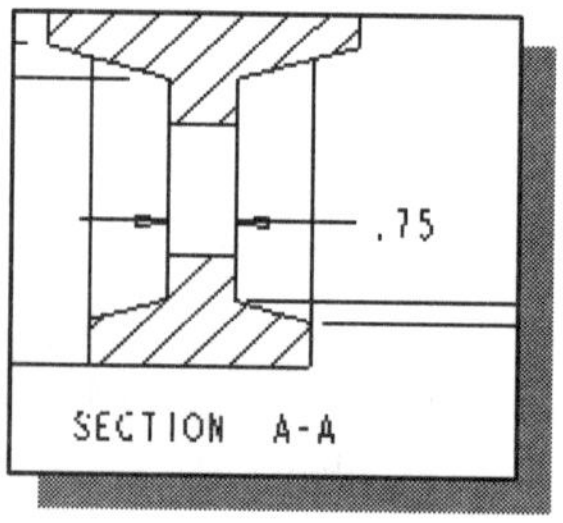

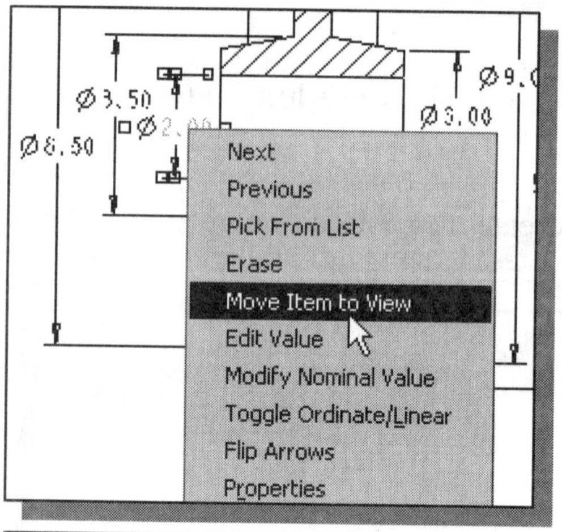

6. Pick the dimension text of the center hole (diameter 2.0) in the section view as shown.
7. Inside the display area, click once with the right-mouse-button to bring up the option menu.
8. Select **Move Item to View** in the option list.

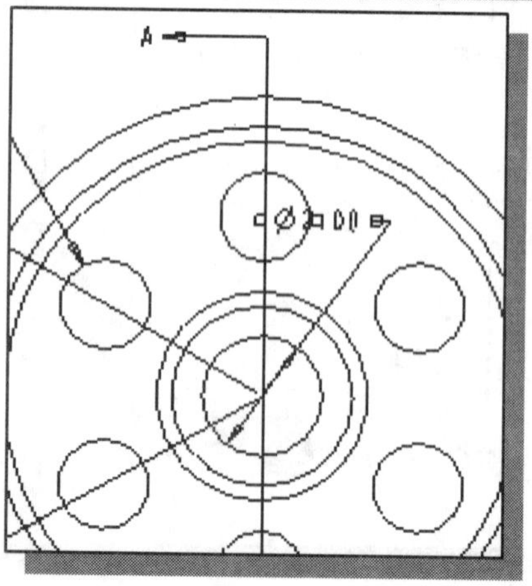

9. The message "*Select model view or window*" is displayed in the message area. Pick the primary view, the ***front*** view, in the display area.

- The dimension is moved to the front view.

10. On your own, adjust the dimensions so that they appear as shown in the figure below.

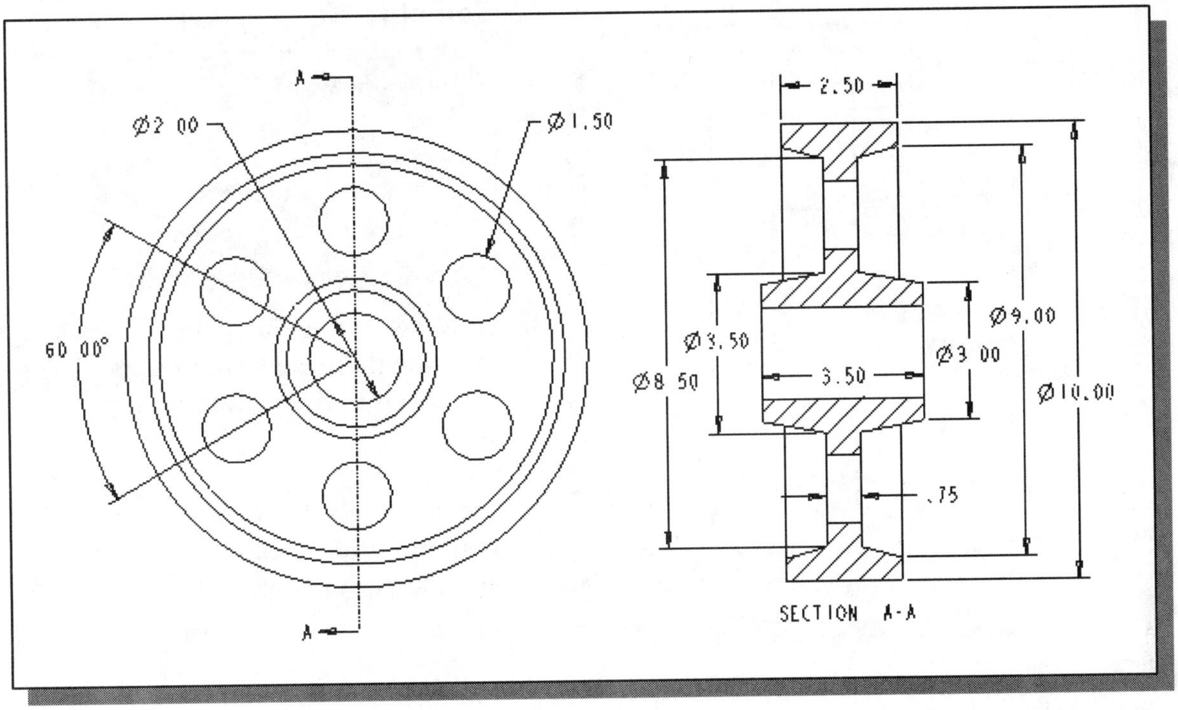

Displaying Centerlines

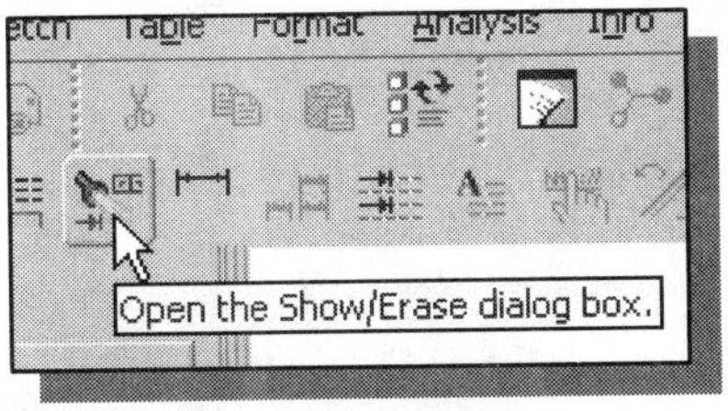

1. In the *Standard* toolbar area, select the **Show/Erase** command.

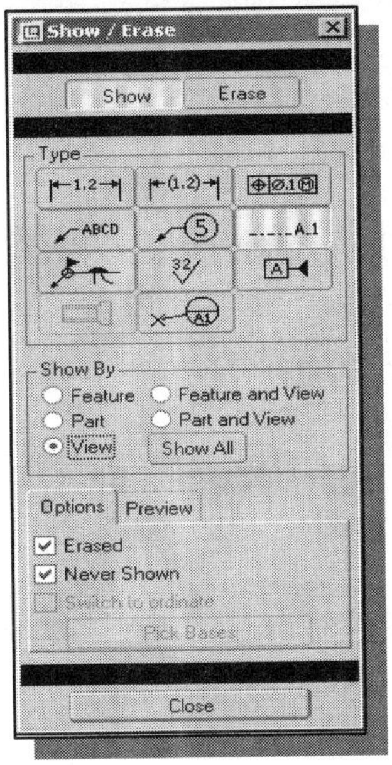

2. Pick the **Axis** icon in the *Show/Erase* form.
3. On your own, toggle ***off*** any other types of dimension details under the **Type** list.
4. In the **Show By** list, select **View**.
5. In the **Options** list, confirm the **Erased** and **Never shown** options are switched ***on***.
6. In the message area, the message "*Select model view or window.*" is displayed. Pick the section view.

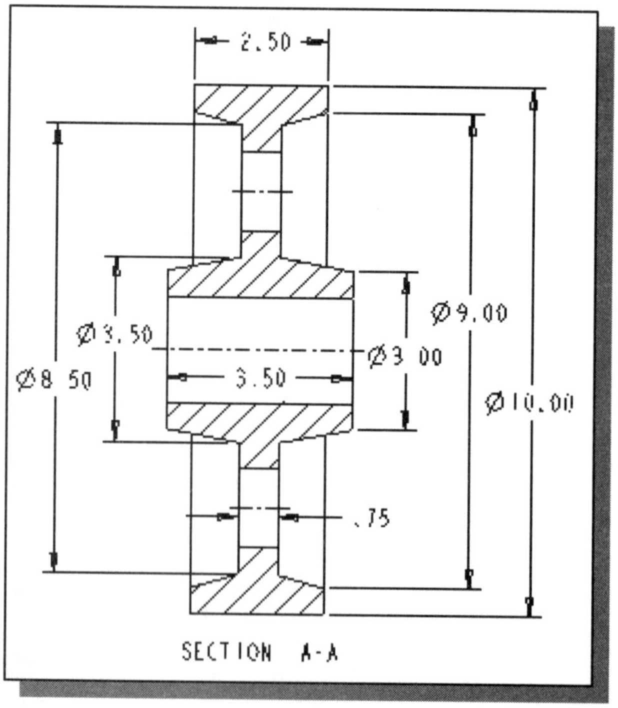

7. In the *Show/Erase* form, set the **Select to Keep** option.

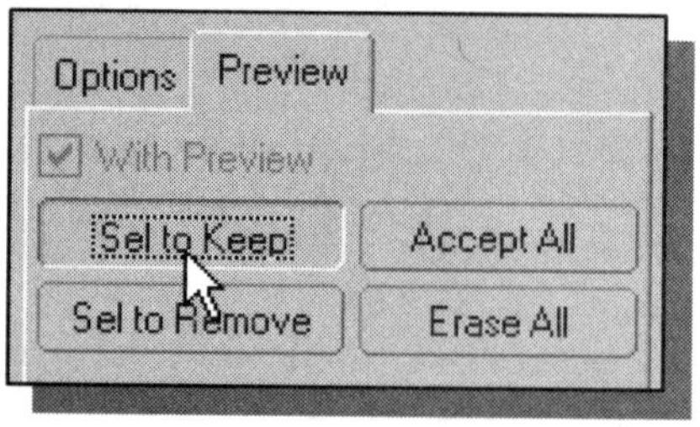

8. Hold down the [**Ctrl**] key and pick only the three centerlines to be displayed in the section view as shown.

9. Inside the display area, click once with the **middle-mouse-button** to end the current command.

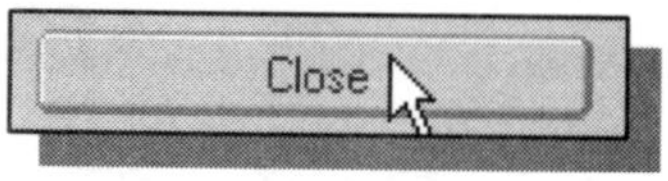

10. Pick the **Close** button in the *Show/Erase* form to exit the Show/Erase command.

Adjusting the Length of Centerlines

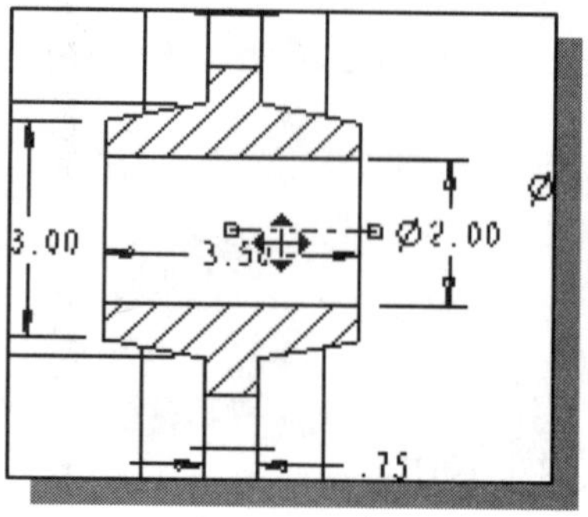

1. Pick the axis that is through the center of the section view as shown.

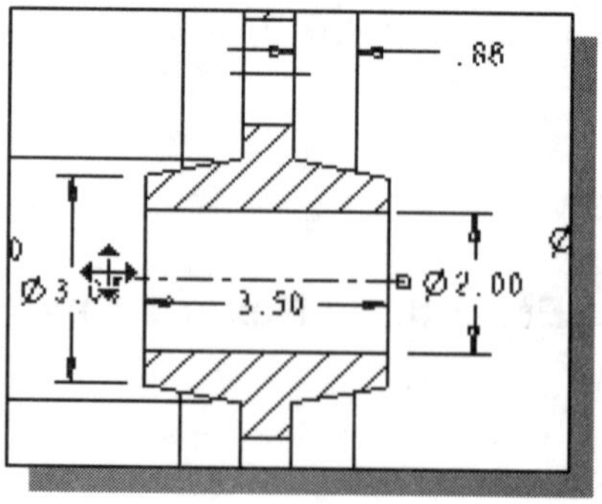

2. Two small markers appear on the centerline. Use the left-mouse-button to drag one of the markers to lengthen/ shorten the centerline.

3. On your own, repeat the above steps and adjust the length of the displayed centerlines.

❖ Note that the same procedure is used to adjust the location of any dimensions as well as the size of dimension related entities.

Creating Additional Centerlines

1. In the *Sketch* toolbar, select the **Circle** command.

- Note that the *Sketch* option allows us to create 2D entities that exist only in *Drawing* mode.

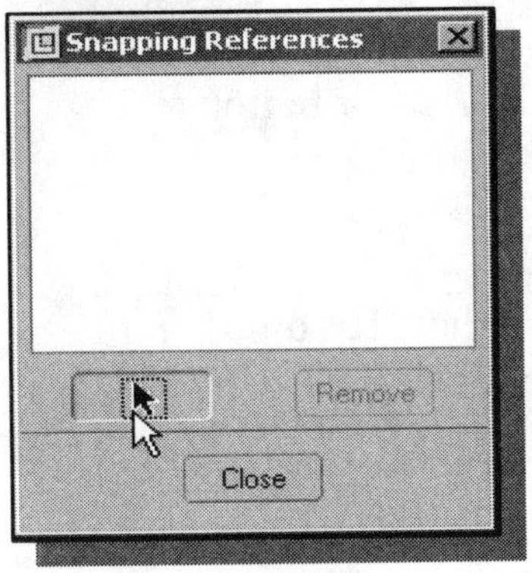

2. Click on the **Arrow** icon to activate the selection of *Snapping References* as shown.

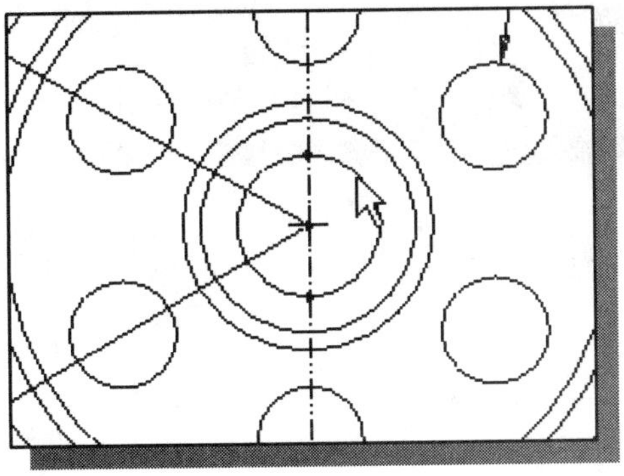

3. Select the center circle as a *snapping reference* as shown.

4. Inside the display area, click once with the **middle-mouse-button** to end the snapping references selection.

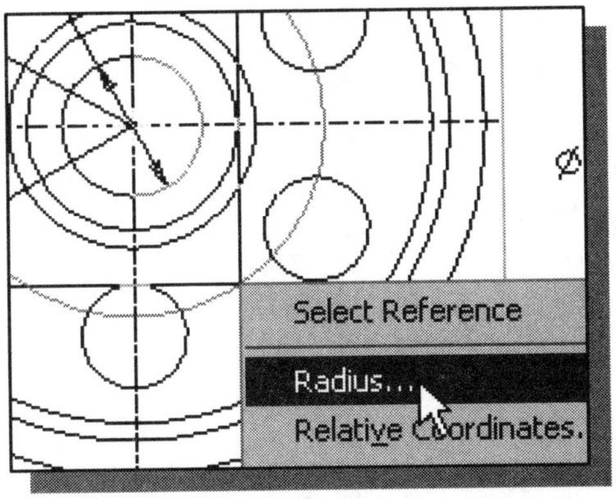

5. Pick the center point of the snapping reference we just identified.

6. Press down the right-mouse-button to bring up the option menu.

7. Select **Radius** as shown.

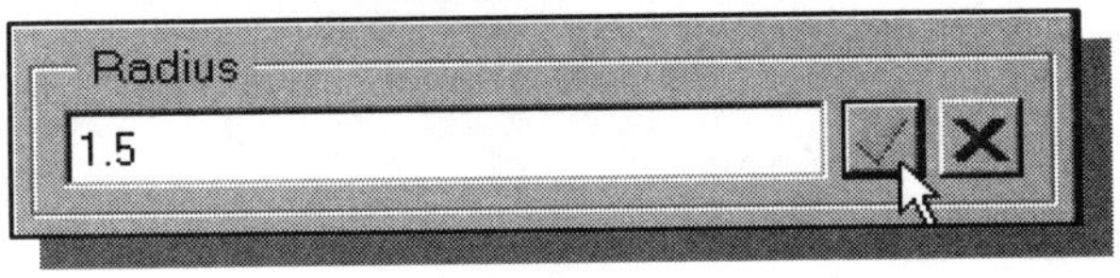

8. Enter **1.5** as the radius of the circle as shown. (Why 1.5?)

9. Inside the display area, click once with the **middle-mouse-button** to end the Sketch Circle command.

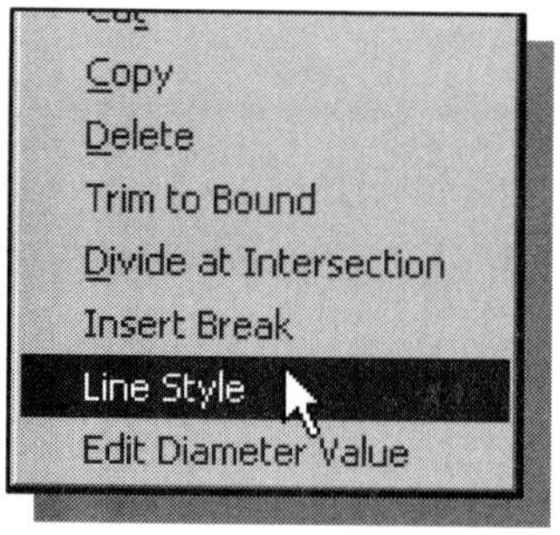

10. Confirm the circle we just created is highlighted.

11. Inside the display area, click once with the right-mouse-button to bring up the option menu

12. Select **Line Style** in the option list as shown.

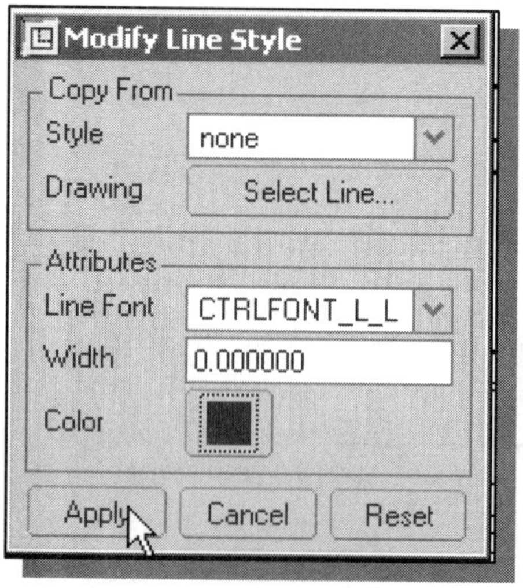

13. Set the Line Font to **CTRLFON_L_L** as shown.

14. Click on the **Apply** button to accept the settings.

15. Click on the **Close** button to exit the Modify Line Style option.

16. On your own, create the diameter dimension of the circle as shown. (Hint: double-click on the circle to switch from radius to diameter.)

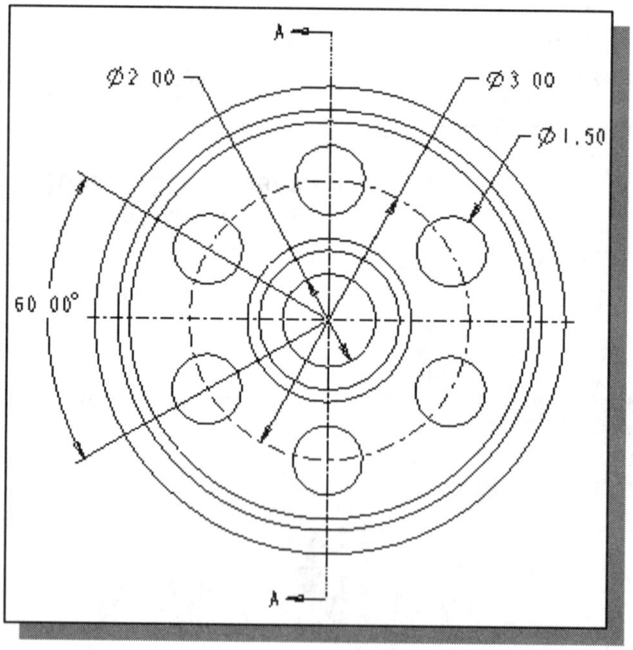

- The drawing is set to half scale and the size of the created circle reflects that.

Using the Drawing Options

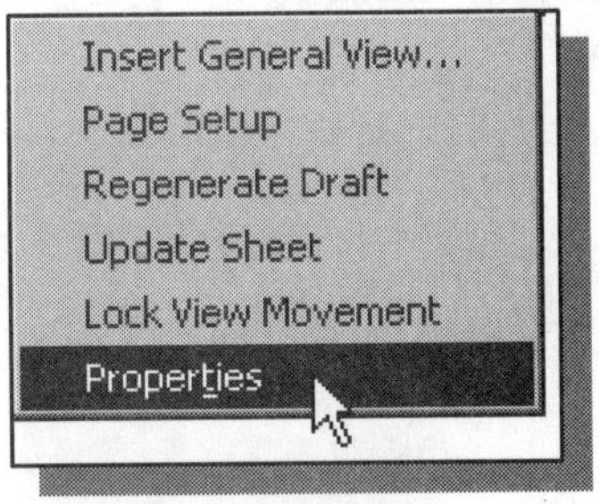

1. Deselect any pre-selected item.

2. Inside the display area, click once with the right-mouse-button to bring up the option menu

3. Select **Properties** in the option list as shown.

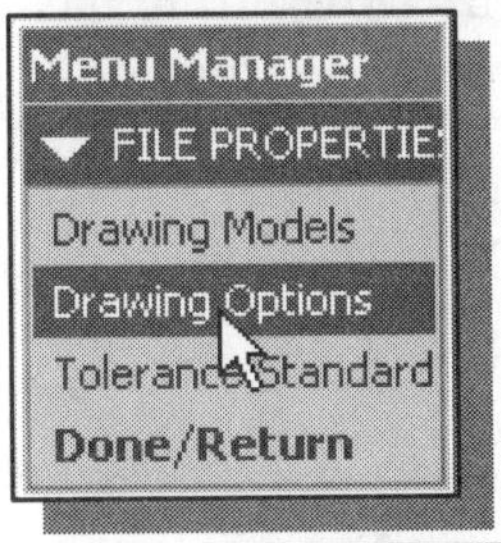

4. In the **Menu Manager**, select **Drawing Options**.

5. Scroll down the option list and set the **draft_scale** option to **0.5**, matching the drawing scale we set on page 7-16.

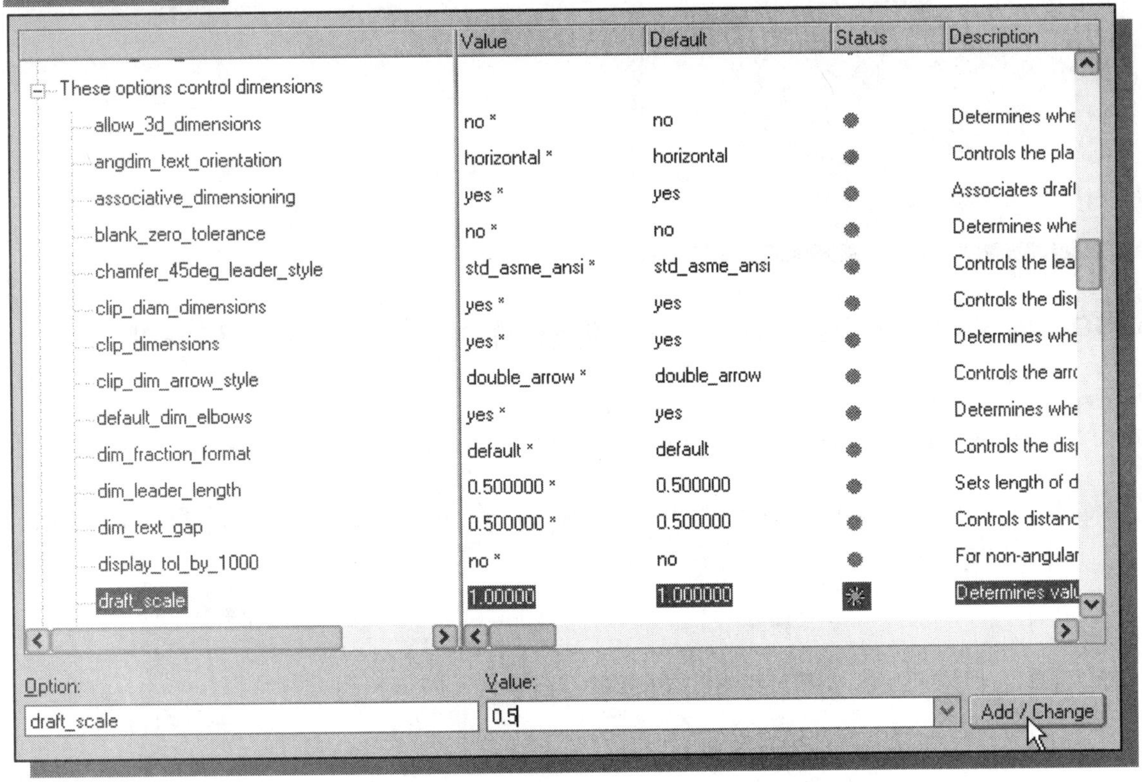

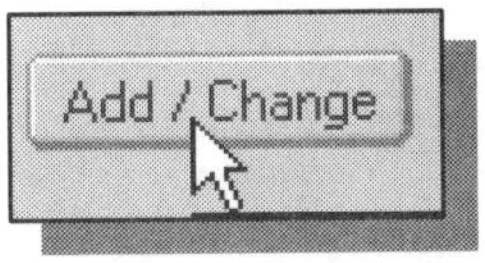

6. Click on the **Add/Change** button as shown in the figure above.

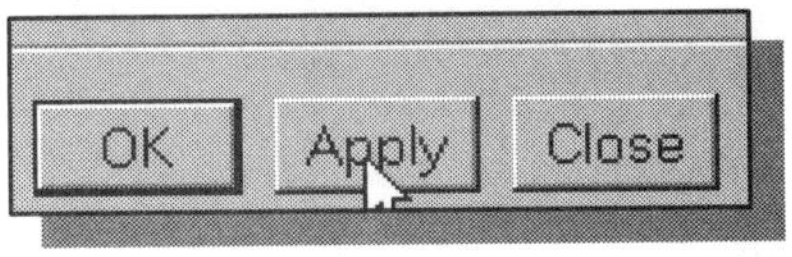

7. Click on the **Apply** button to accept the settings.

8. Click on the **Close** button to exit the Drawing Options command.

9. In the **Menu Manager**, click **Done/Return** to exit the **Properties** options.

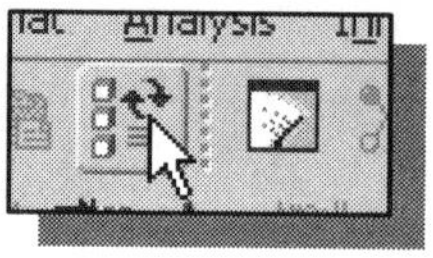

10. In the *Standard* toolbar, select **Regenerate.**

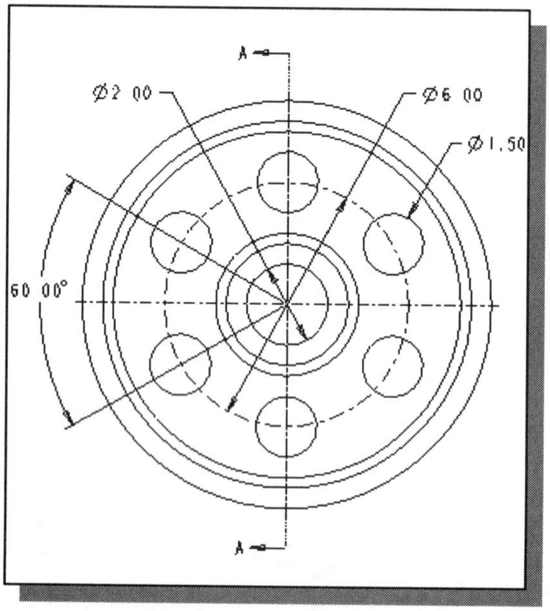

- The dimension for the centerline circle has been updated to reflect the drawing scale used.

Associative Functionality – Modifying Feature Dimensions

❖ *Pro/ENGINEER's associative functionality* allows us to change the design at any level, and the system reflects the change at all levels automatically.

1. Double-click with the left-mouse-button on the dimension text (**Ø 1.5**) of the hole pattern.

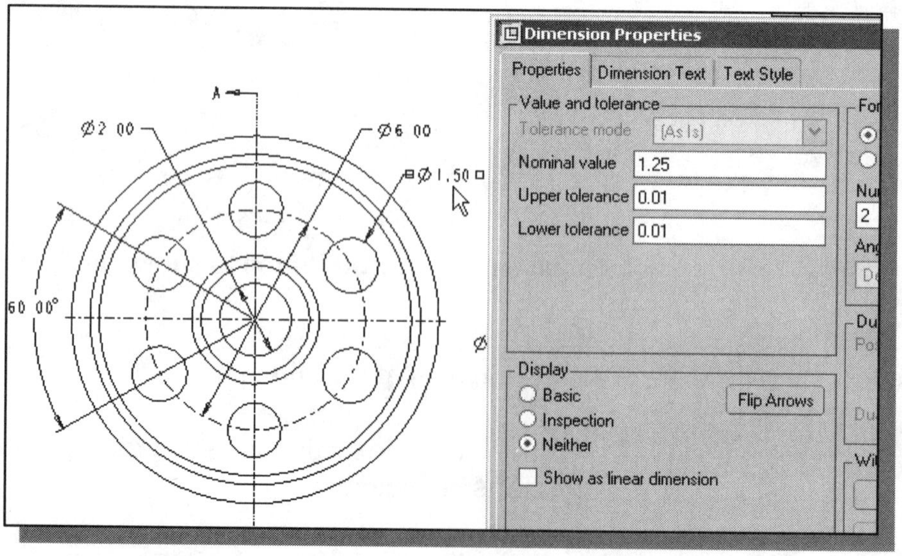

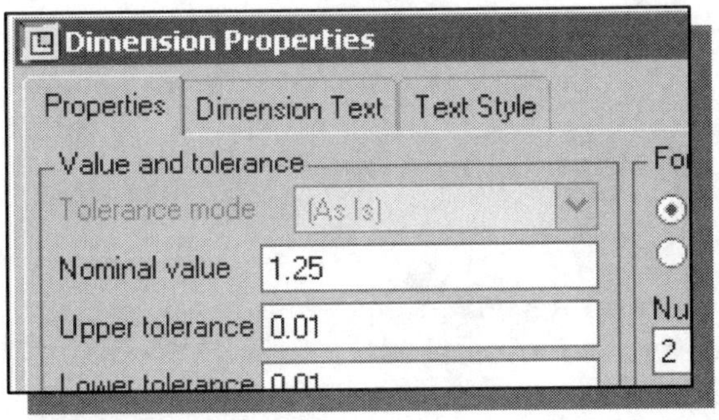

2. In the *Dimension Properties* window, enter **1.25** as the new **Nominal Value**.

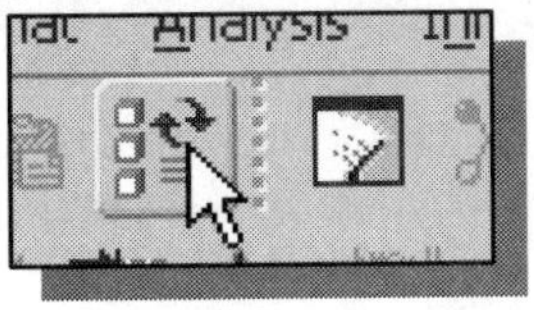

3. Pick **Regenerate** in the *Standard* toolbar. Note the hole size in the front view is updated.

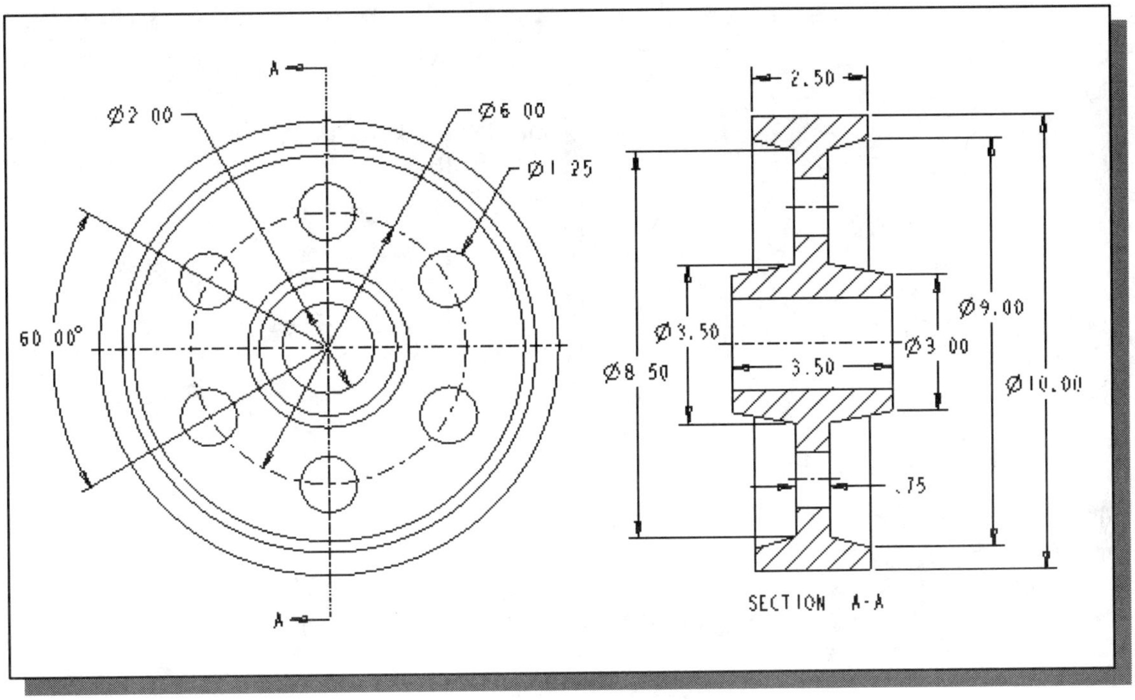

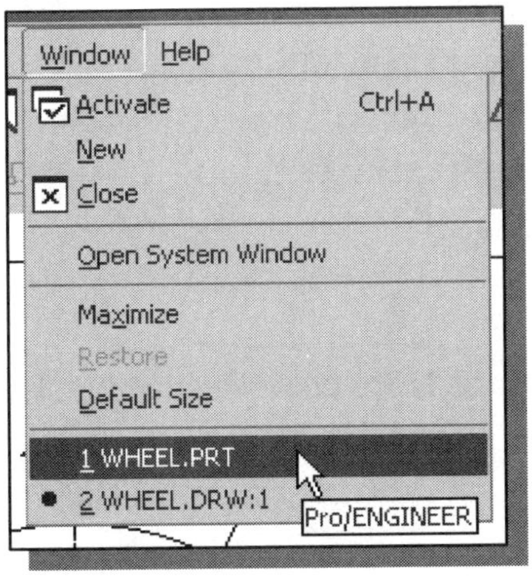

4. Select **Window** in the pull-down menu.

5. Pick **WHEEL.PRT.**

➢ On your own, confirm the model is also updated as the dimension was modified in the drawing mode.

Bi-Directional Associative Functionality

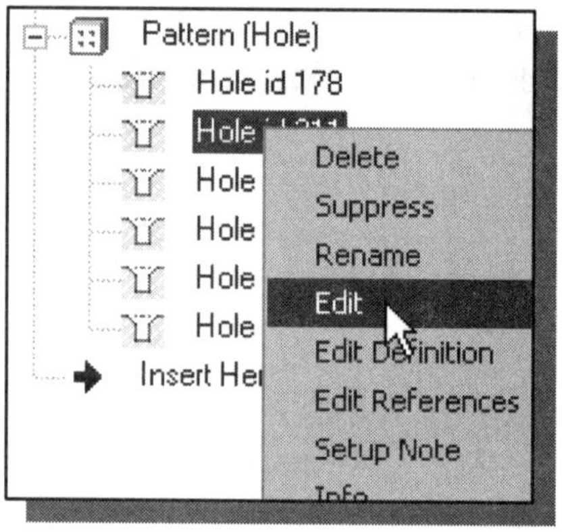

1. Expand the **Pattern** listing in the *Model Tree* window

2. Press down with the right-mouse-button on one of the hole features to bring up the option list.

3. Pick **Edit** in the option list.

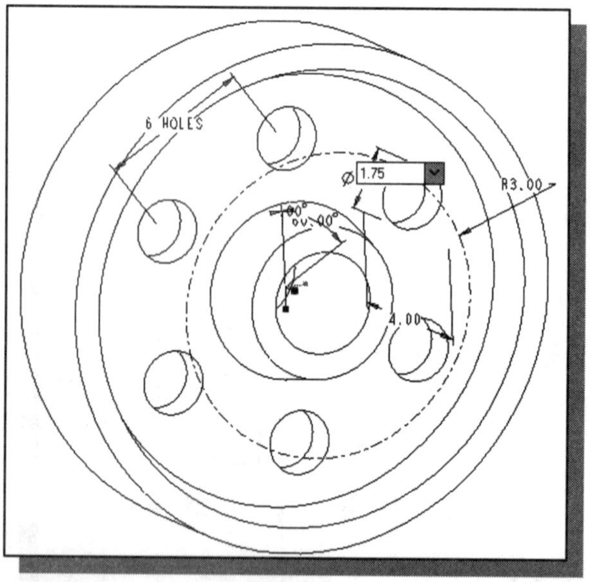

4. Modify the hole diameter to **1.75** as shown.

5. Pick **Regenerate** in the *Standard* toolbar.

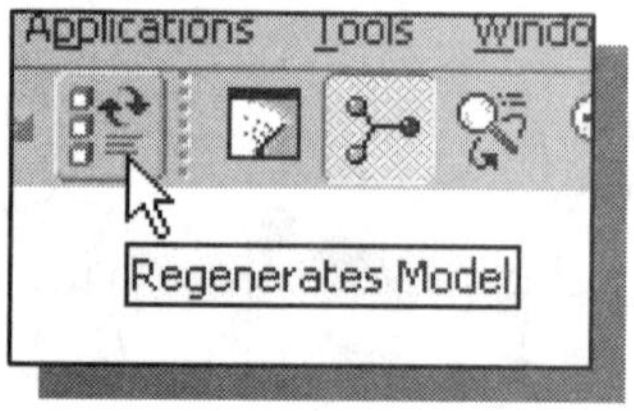

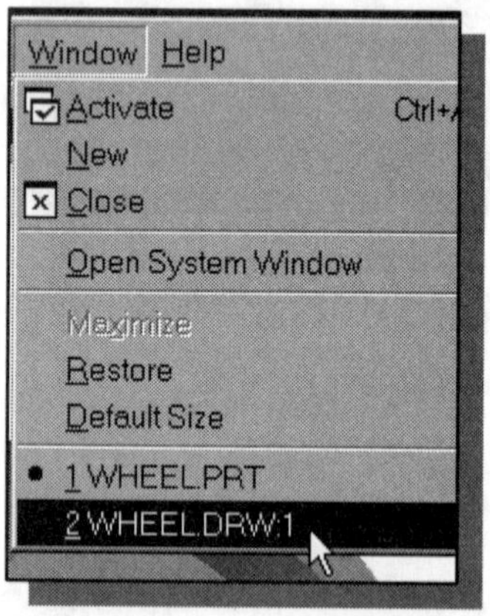

6. Select **Window** in the pull-down menus.

7. Pick **WHEEL.DRW:1.**

❖ Notice the drawing is updated automatically as the part was modified in the *Part Modeling* mode.

8. On your own, use the **Create Note** command to fill in the title block. Also, add the additional centerlines and an additional 3D view so that the completed drawing appears as shown on the next page.

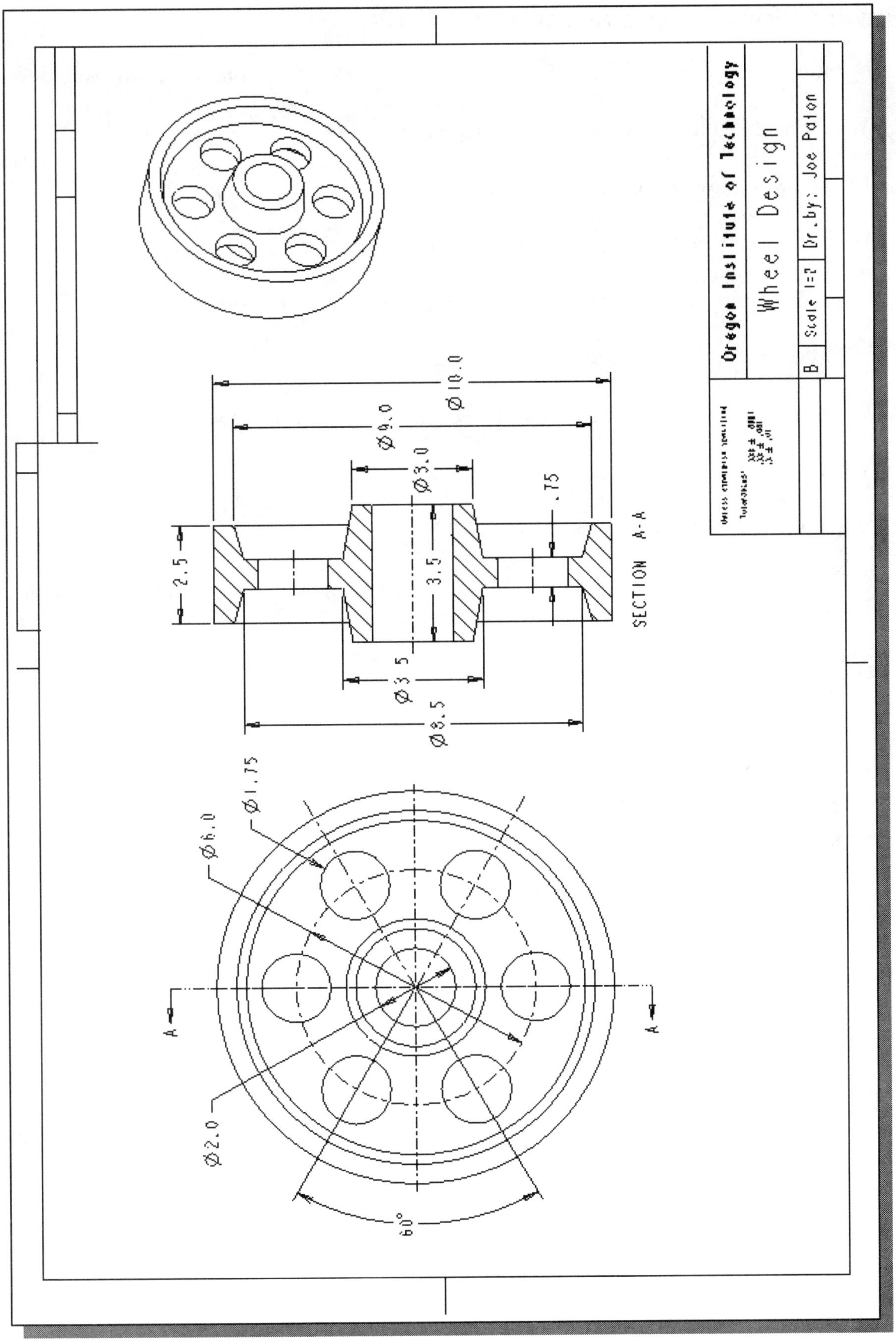
Oregon Institute of Technology
Wheel Design
B
Scale 1:2
Dr.by: Joe Paton
SECTION A-A
Ø10.0
Ø9.0
Ø3.0
.75
2.5
3.5
Ø3.5
Ø8.5
Ø1.75
Ø6.0
Ø2.0
60°
A
A

Questions:

1. What does *Pro/ENGINEER's bi-directional associative functionality* allow us to do?
2. What essential element is common to revolved features?
3. How do you move a view on the *drawing sheet*?
4. How do you move a dimension from one view to another?
5. What is the difference between *erasing* and *deleting* a dimension?
6. How do you reposition dimensions?
7. What are the required elements in order to generate a sectional view?
8. What is the first feature in a pattern called in *Pro/ENGINEER*?
9. What is a "radial hole" in *Pro/ENGINEER*?
10. Create sketches showing the steps you plan to use to create the two models shown on the next page:

Ex.1)

Ex.2)

Exercises: (All dimensions are in inches.)

1.

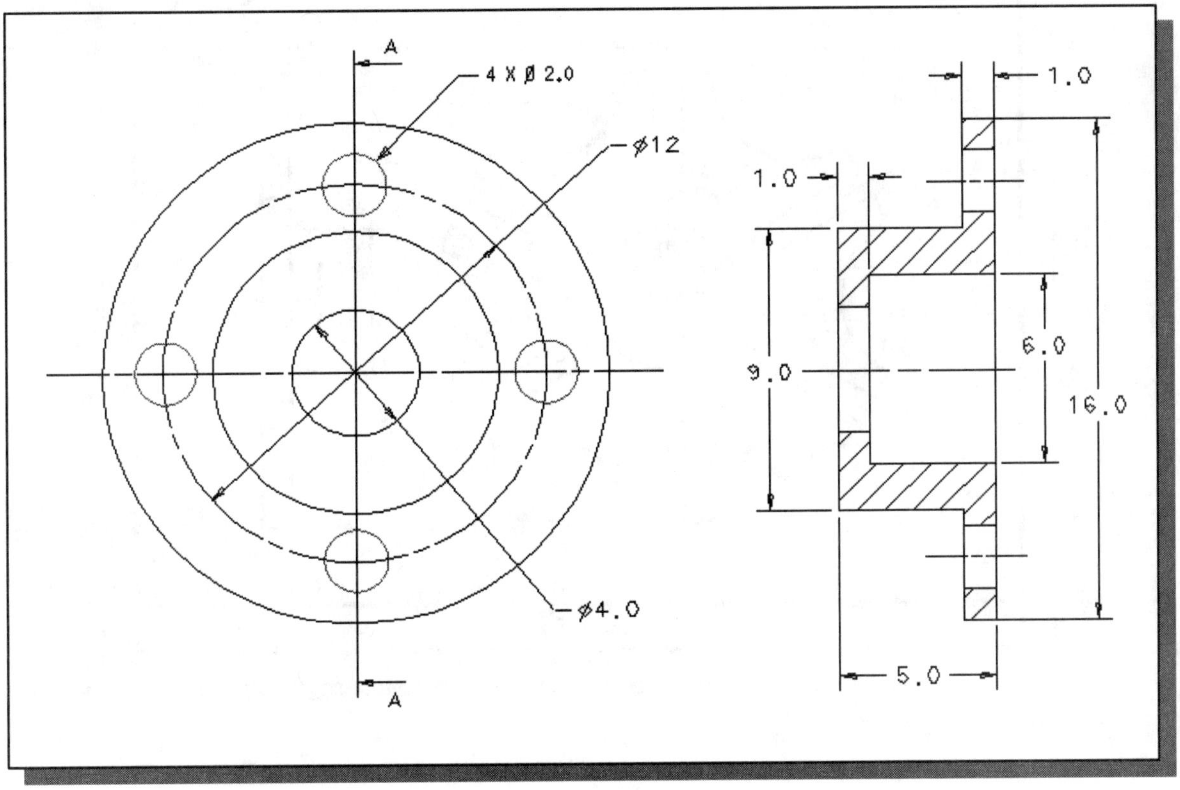

2. Plate thickness: 0.125 inch. Hint: Create a circular disk as the base feature.

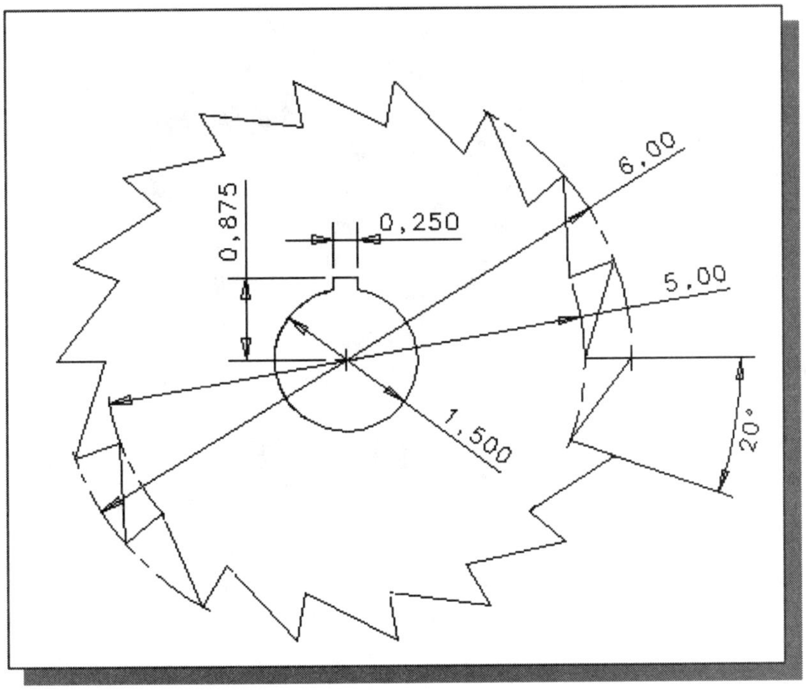

3.

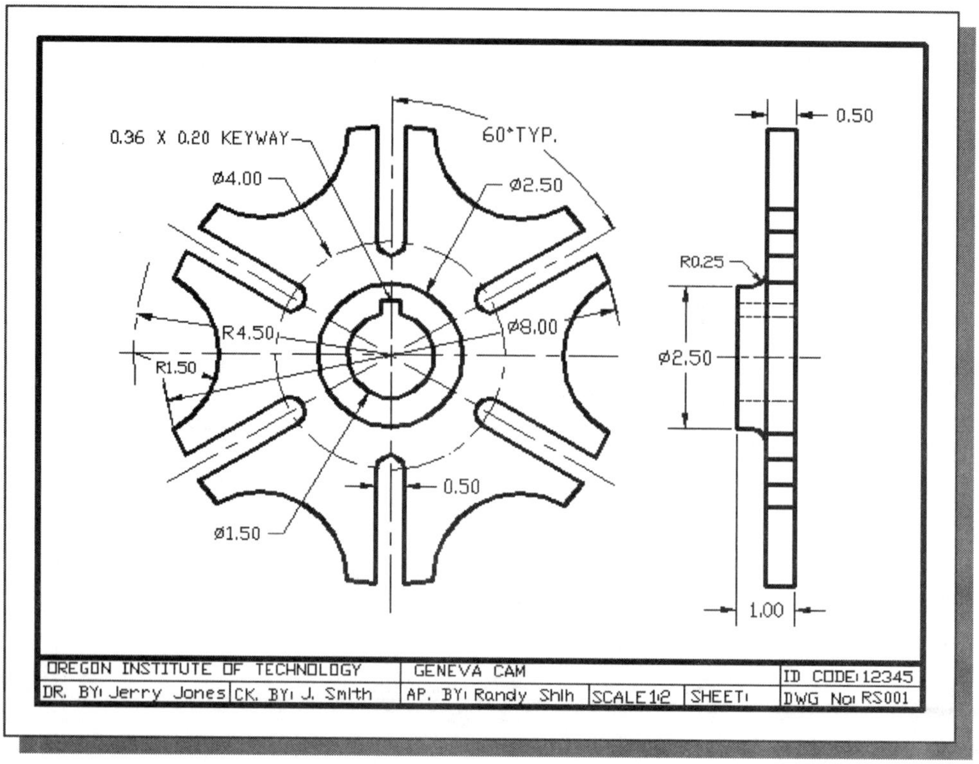
0.36 X 0.20 KEYWAY
60°TYP.
Ø4.00
Ø2.50
R4.50
Ø8.00
R1.50
0.50
Ø1.50
0.50
R0.25
Ø2.50
1.00
OREGON INSTITUTE OF TECHNOLOGY
GENEVA CAM
ID CODE: 12345
DR. BY: Jerry Jones
CK. BY: J. Smith
AP. BY: Randy Shih
SCALE 1:2
SHEET:
DWG No: RS001

4.

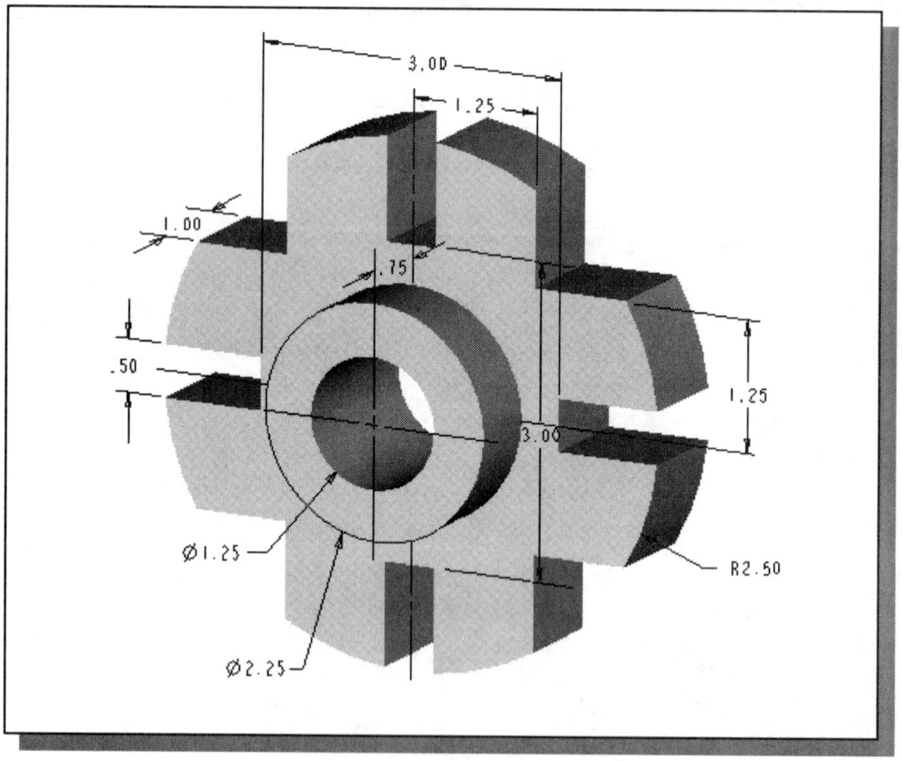
3.00
1.25
1.00
.75
.50
1.25
3.00
Ø1.25
R2.50
Ø2.25

Chapter 8
Three-Dimensional Construction Tools

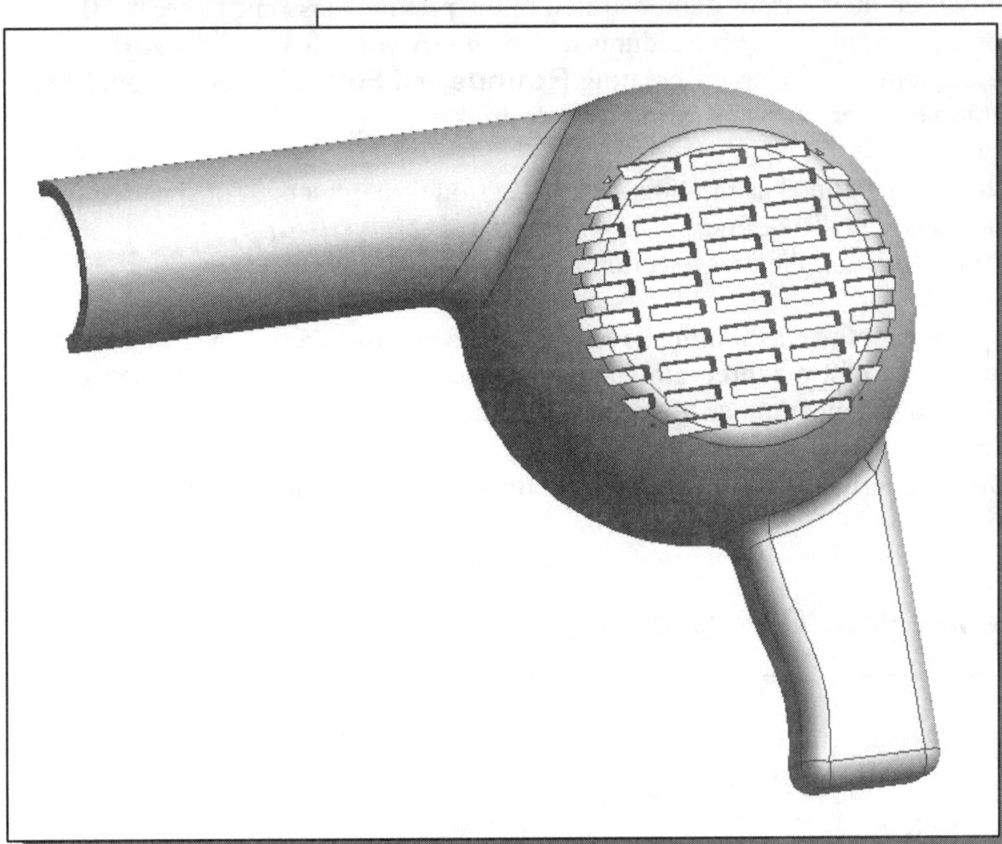

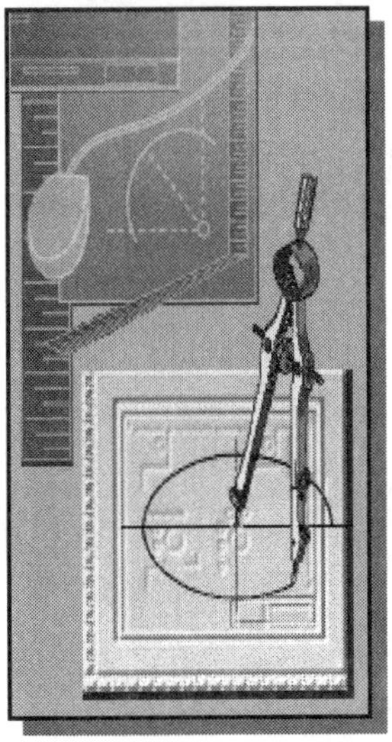

Learning Objectives

- Understand the Parent/Child relations between features.
- Create Swept Features.
- Setup Multiple Work planes.
- Create Lofted Features.
- Use the Shell Command.
- Create 3D Rounds and Fillets.

Introduction

Pro/ENGINEER provides an assortment of three-dimensional construction tools to make the creation of solid models easier and more efficient. **Extrude** and **Revolve** are the two most commonly used methods, as demonstrated in the previous lessons, to create 3D models. In this next example, the procedures to creating **Sweep** features, **Blend** features, **Shell** features, and also for creating **Rounds** and **Fillets** are examined. These features are common characteristics of molded designs.

The **Sweep** option is defined as moving a cross-section through a path in space to form a three-dimensional object. To define a *sweep feature* in *Pro/ENGINEER*, two sections are required: the trajectory and the cross-section.

The **Blend** option is defined as joining several planar sections together at the edges, with transitional surfaces, to form a three-dimensional object. In *Pro/ENGINEER*, each planar section must have the same number of segments/vertices.

The **Shell** option is defined as hollowing out the inside of a solid, leaving a shell of specified wall thickness.

A Thin-Walled Part: *Dryer Housing*

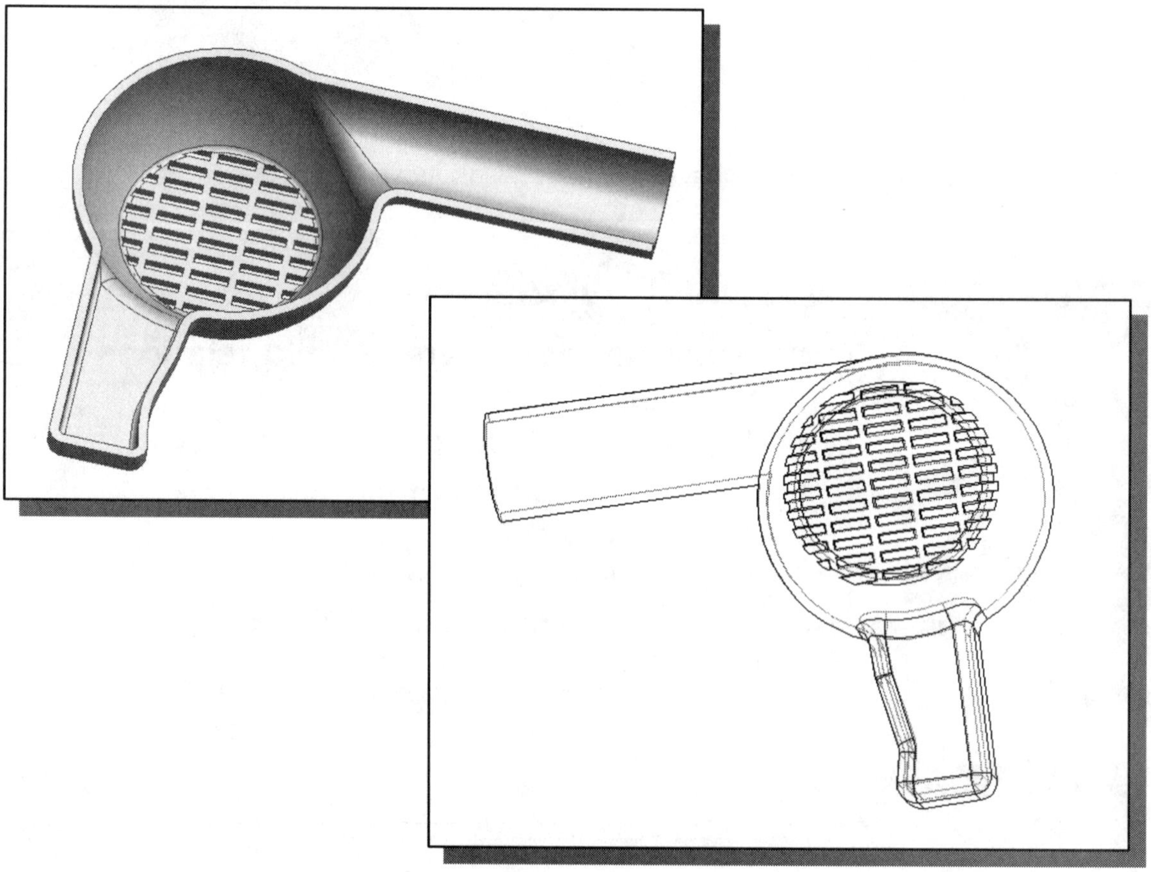

Modeling Strategy

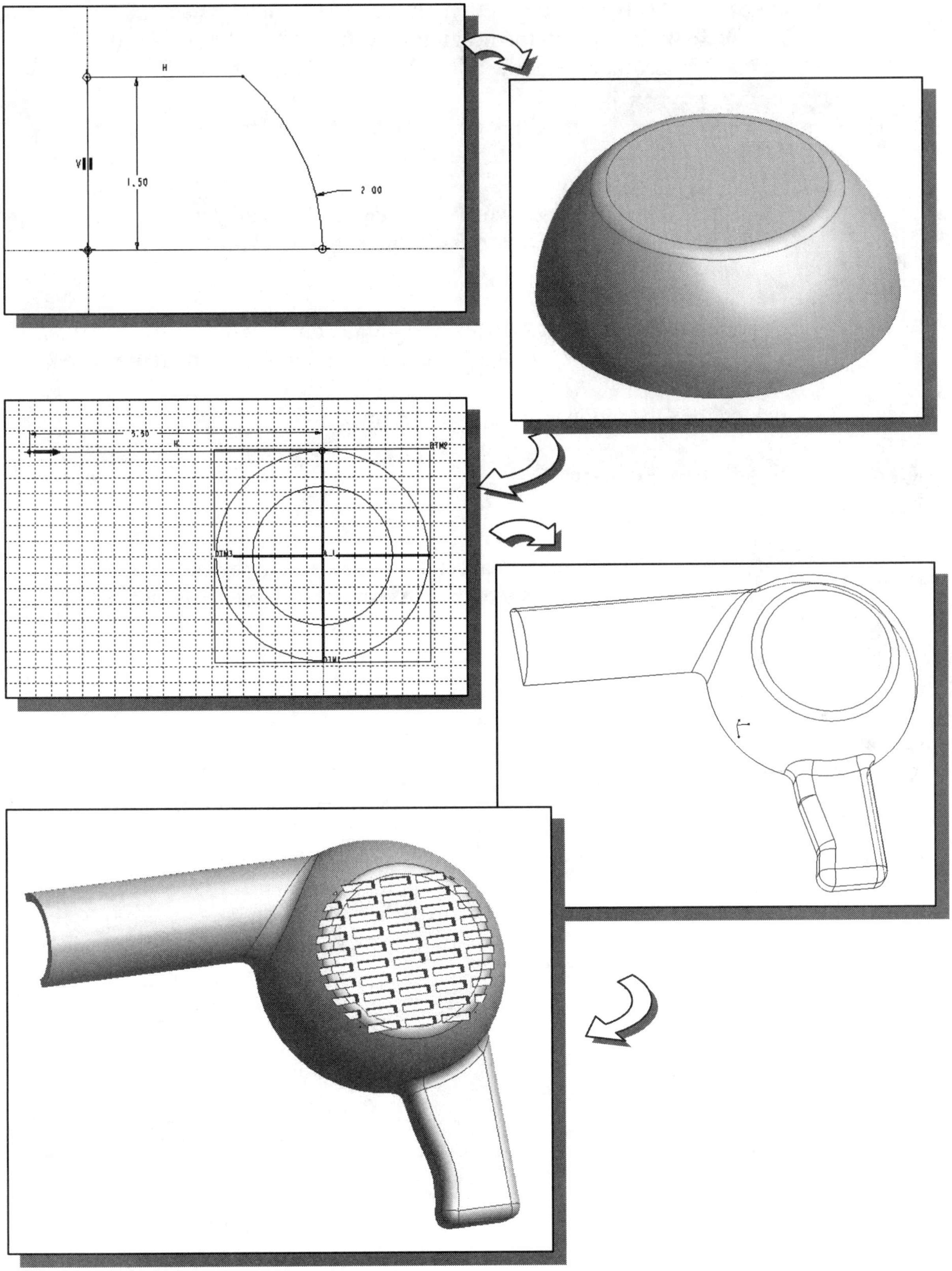

Starting *Pro/ENGINEER*

1. Select the **Pro/ENGINEER** option on the *Start* menu or select the **Pro/ENGINEER** icon on the desktop to start *Pro/ENGINEER*. The *Pro/ENGINEER* main window will appear on the screen.

2. Click on the **New** icon, located in the *Standard* toolbar as shown.

3. On your own, start a new solid part file using ***DryerHousing*** as the part Name.

4. Confirm the **Use default template** option is turned ***on*** so that the system units are set to the *Pro/ENGINEER* default settings (**Inch-lbm-Second**).

5. Click on the **OK** button to accept the settings.

Creating the Base Feature

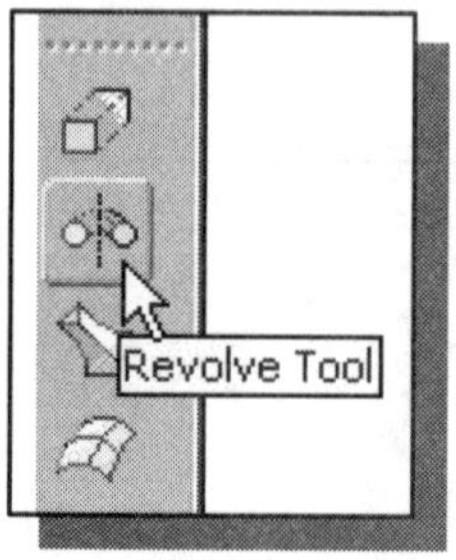

1. In the *Feature* toolbars (toolbars aligned to the right edge of the main window), select the **Revolve Tool** option as shown.

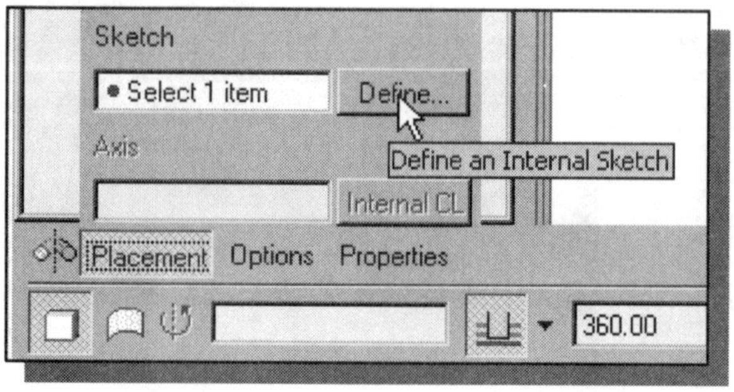

2. Click the **Placement** option and choose **Define** to create a new *Internal sketch.*

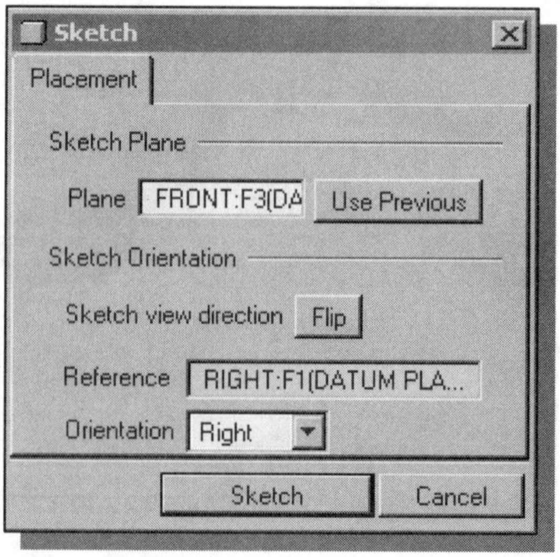

3. Set up the **Front** datum plane as the sketch plane with the **Right** datum plane facing the **right** edge of the computer screen, and pick **Sketch** to enter the *Sketcher* mode.

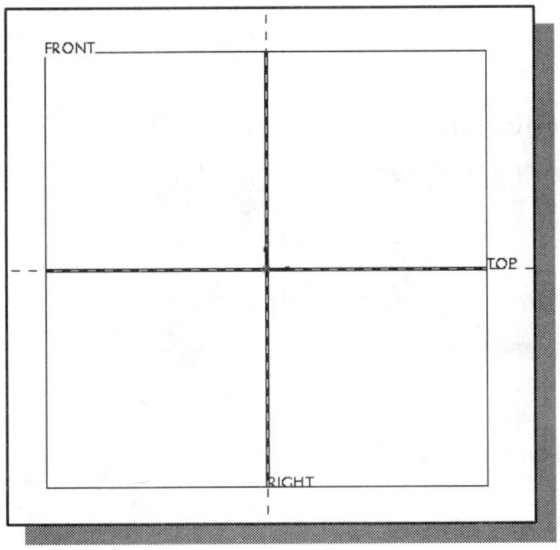

❖ Note the default selection of the two datum planes, **Right** and **Top**, as the references for the 2D sketch.

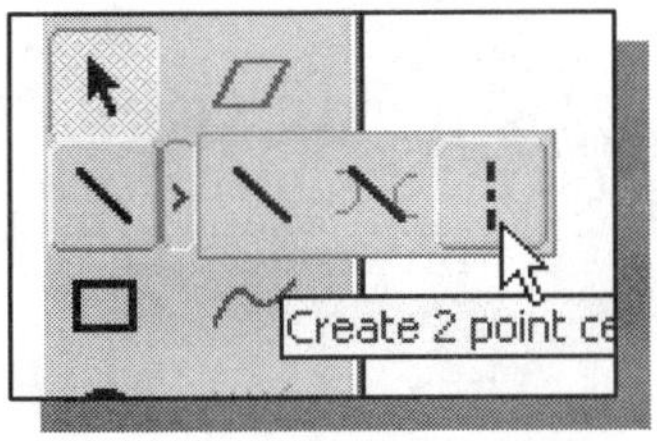

4. In the *Sketcher* toolbar, select **Create Centerline**.

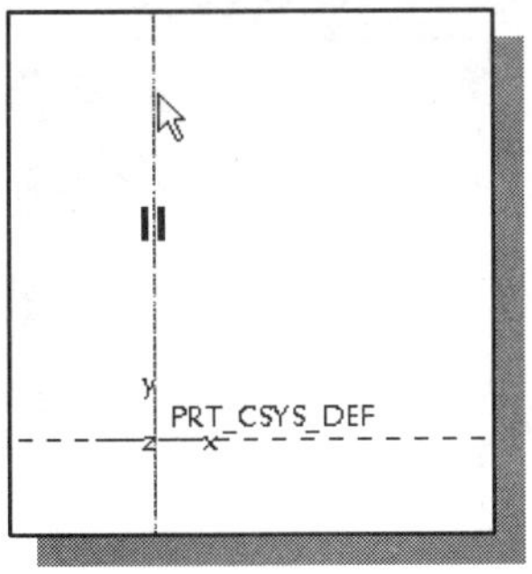

5. Create a vertical line by first selecting a point on the vertical axis (datum plane **Right**).

6. Select a second point aligned to the vertical axis (datum plane **Right**) to create a vertical centerline.

7. On your own, create and modify the 2D section (consisting of two horizontal lines, one vertical line, and an arc) as shown below. Note that the arc *radius* is **2.0**, and the *vertical distance* is **1.5**.

8. Click on the **Accept** icon to accept the completed section.

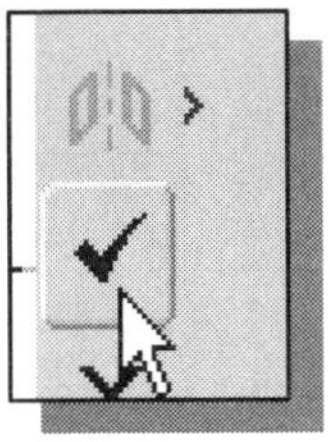

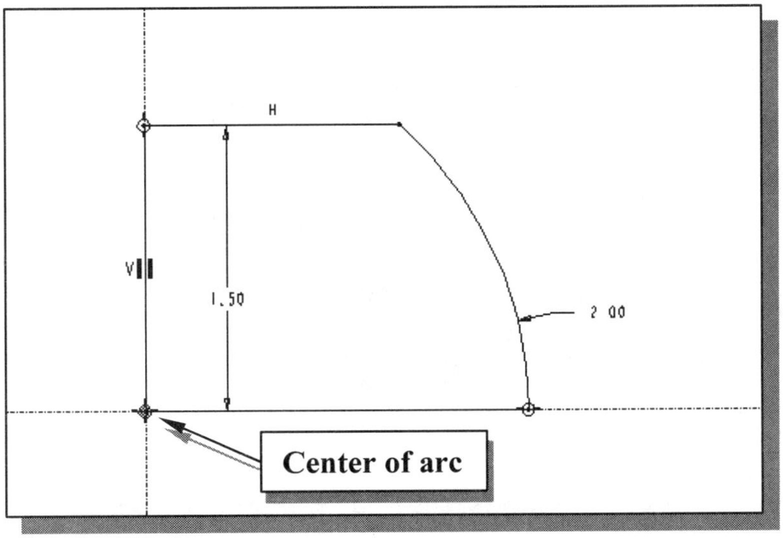

Create the Revolved Part

1. In the **Revolve** option menu, confirm the revolve angle is set to **360.**

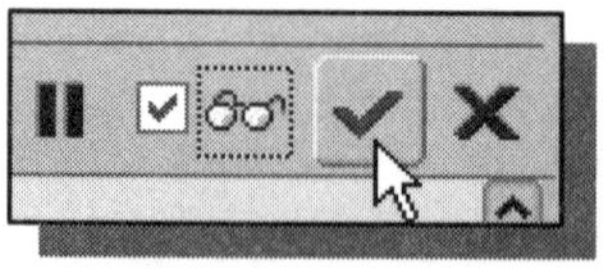

2. Pick **Accept** in the *Feature Option Dashboard* to create the solid feature.

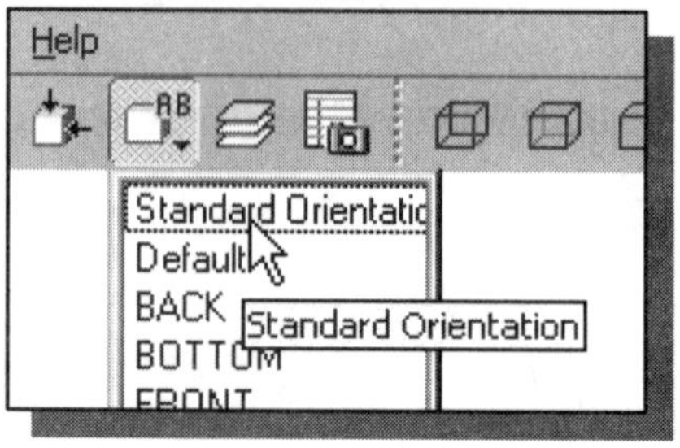

3. Click on the **Saved View List** icon and choose **Standard Orientation** to view the completed 3D solid feature in the direction of the preset default *trimetric* view.

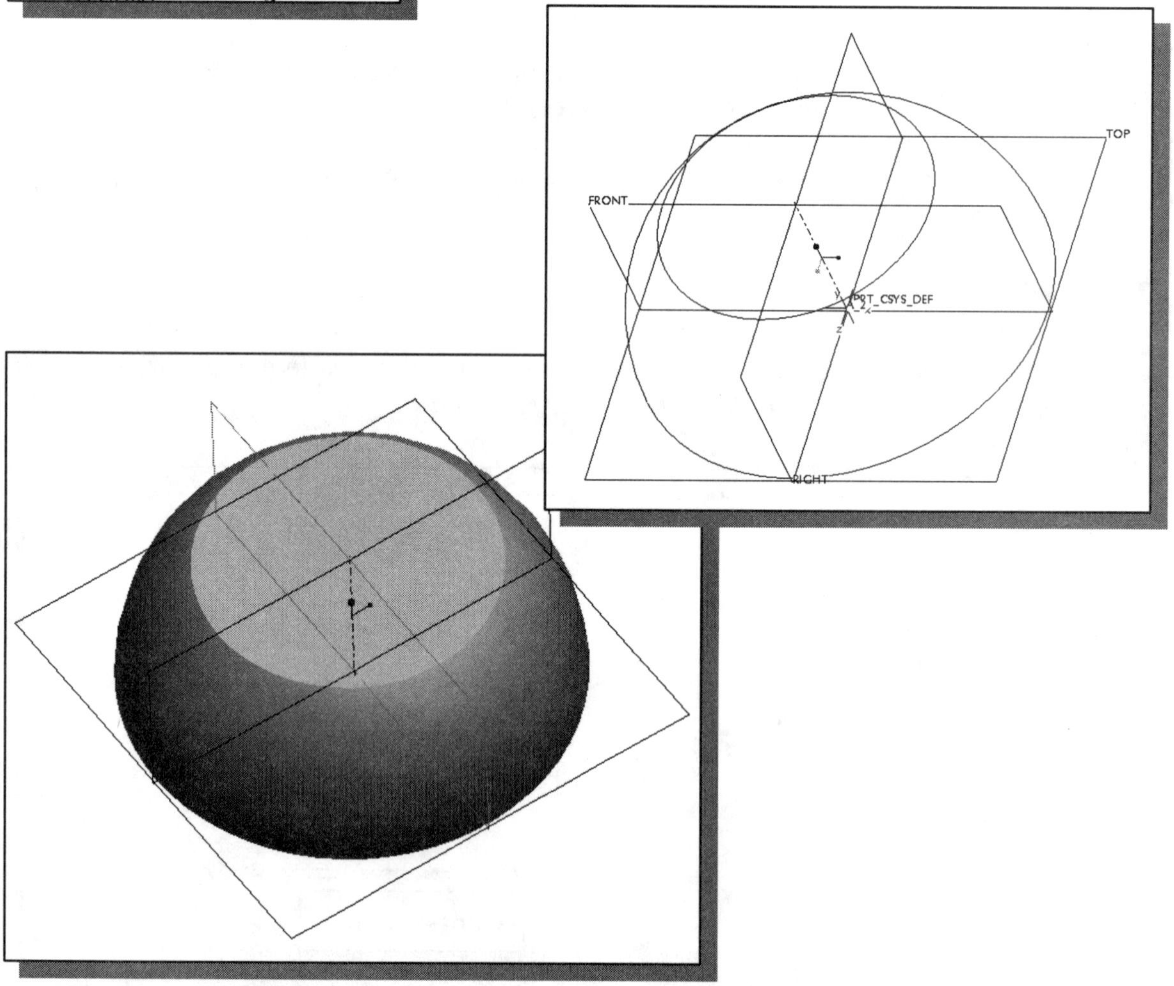

Creating the second Feature

1. In the *Feature* toolbars (toolbars aligned to the right edge of the main window), select the **Extrude Tool** option as shown.

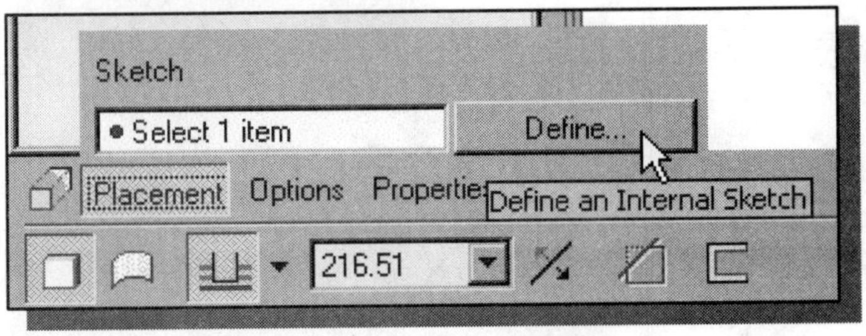

2. Click the **Placement** option and choose **Define** to begin creating a new *Internal sketch.*

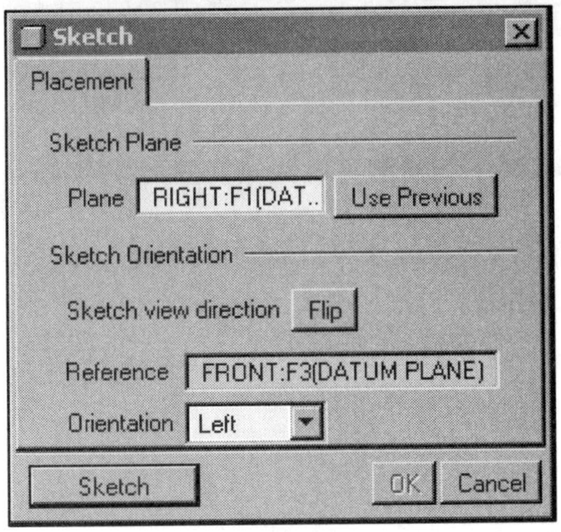

3. On your own, set up the **Right** datum plane as the sketch plane with the **Front** datum plane facing the **left** edge of the computer screen and pick **Sketch** to enter the *Sketcher* mode.

❖ Notice the use of the two datum planes, **Front** and **Top**, as the references for the 2D sketch.

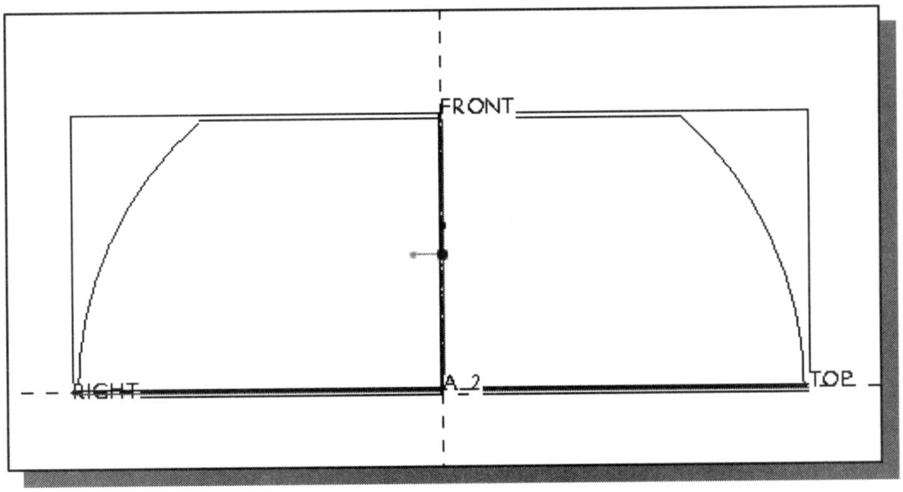

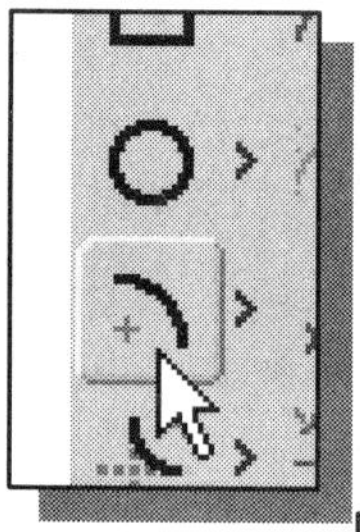

4. In the *Sketcher* toolbar, select **Center/Ends Arc**.

5. Place the center of the arc to the right of the vertical reference, and place both ends of the arc aligned to datum plane **TOP**.

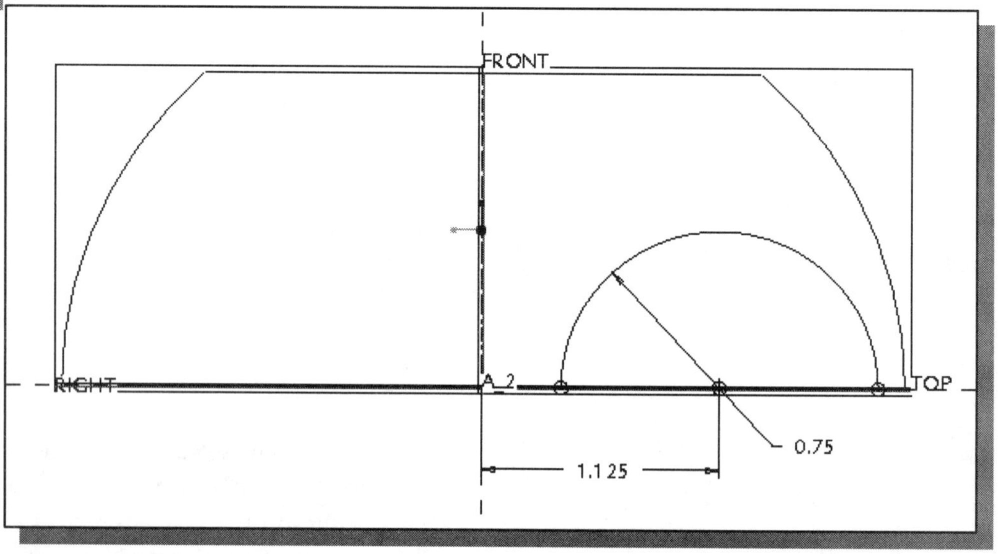

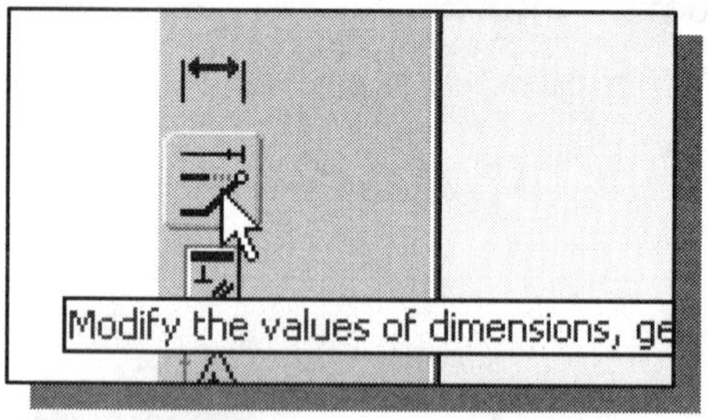

6. On your own, adjust the dimensions as shown in the above figure.

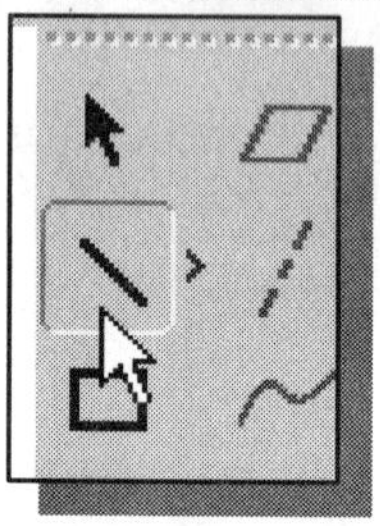

7. Pick **Line** in the *Sketcher* toolbar and create a line connecting the two ends of the arc.

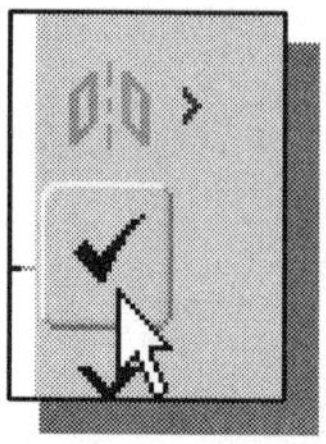

8. Click on the **Accept** icon to accept the completed *sweep section.*

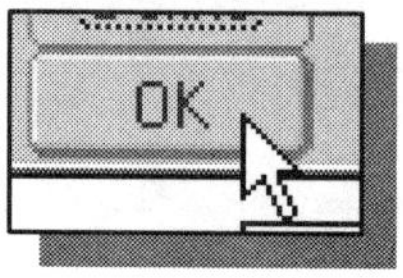

9. Click **OK** to close the sketch window.

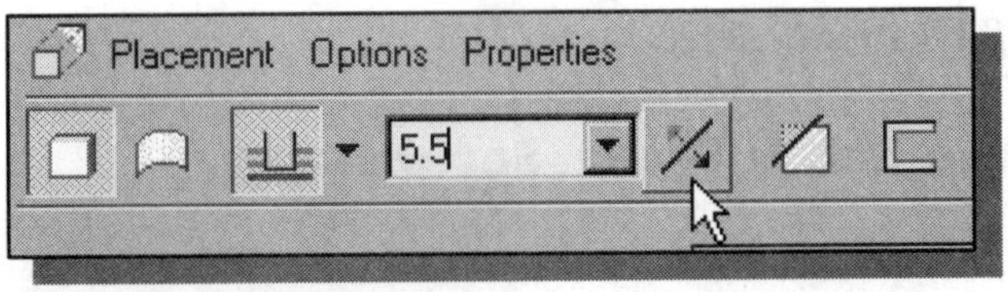

10. Set the extrude distance to 5.5 and change the extrusion direction by clicking once on the direction button.

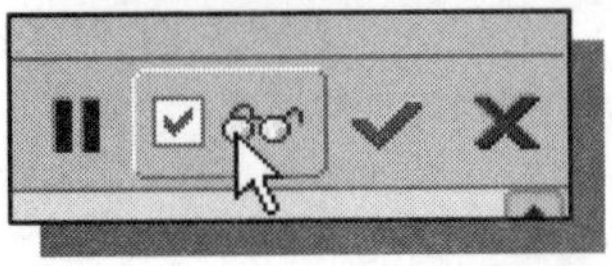

11. Pick **Preview** in the *Feature Option Dashboard* to examine the solid feature.

12. On your own, use the *Dynamic Viewing* functions to view the 3D model.

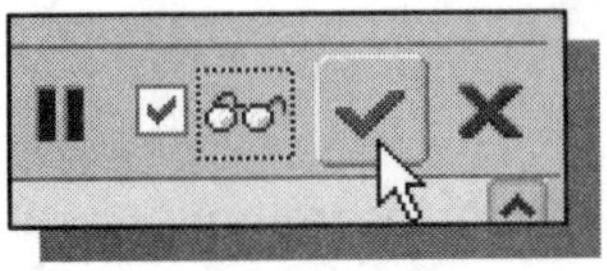

13. Pick **Accept** in the *Feature Option Dashboard* to create the solid feature.

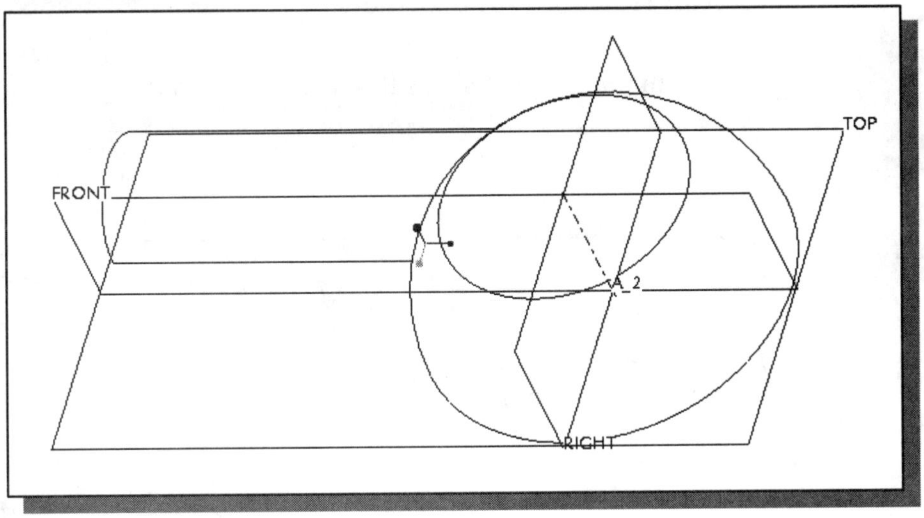

Create a Blend Feature

❖ A *blend* feature is a series of two-dimensional sections that are joined together at the edges with transitional surfaces to form a continuous solid.

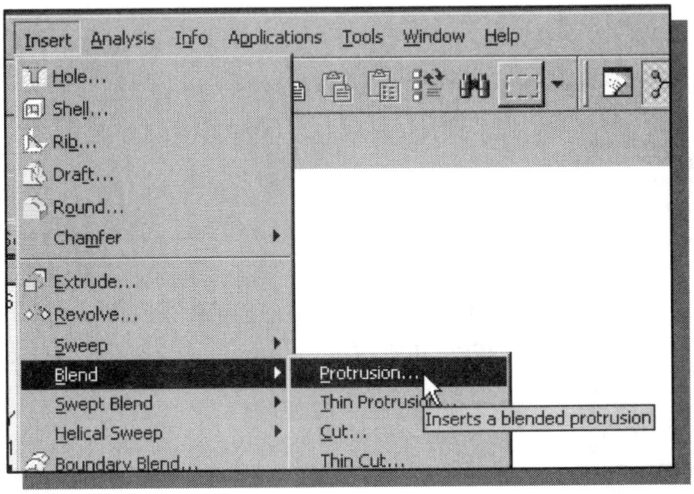

1. In the **Insert** pull-down menu, select **Blend → Protrusion**.

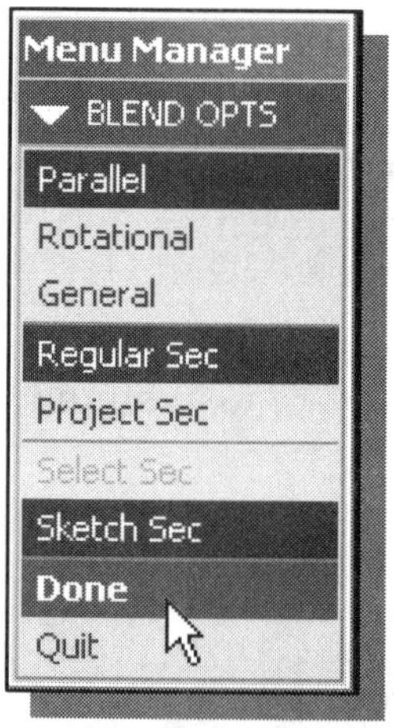

2. In the **BLEND OPTS** menu, select

   Parallel → Regular Sec → Sketch Sec → Done

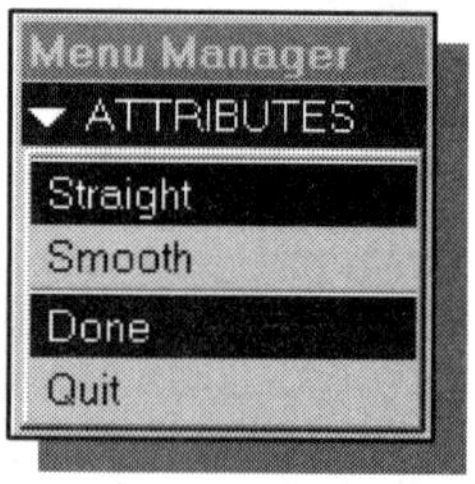

3. In the **ATTRIBUTES** submenu, select

   Straight → Done

4. Confirm that the options in the **SETUP SK PLN** submenu are set to **Setup New → Plane** as shown.

5. Pick datum plane **FRONT** as the *sketching plane.*

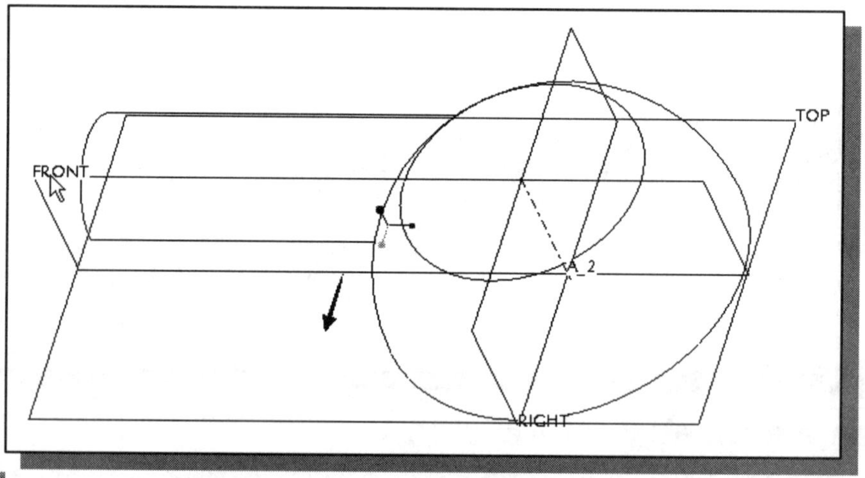

6. Confirm the direction arrow is pointing forward as shown in the figure above and click **Okay** in the **DIRECTION** submenu.

7. Pick **Right → Plane** in the **SKET VIEW** submenu.

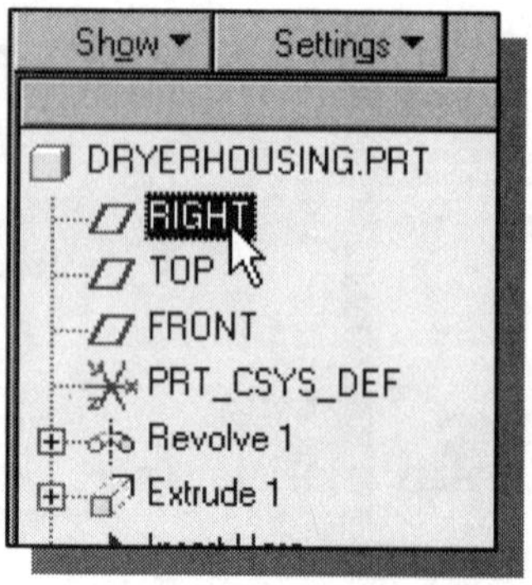

8. In the *Model Tree* area, select datum plane **Right** as the **right** side *reference plane.*

Creating 2D Blend Sections

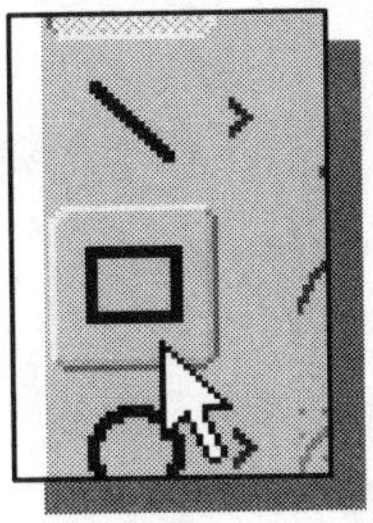

1. In the *Sketcher* menu, click the **Rectangle** icon with the left-mouse-button.

2. Place the first corner of the rectangle on the horizontal axis and create a rectangle as shown.

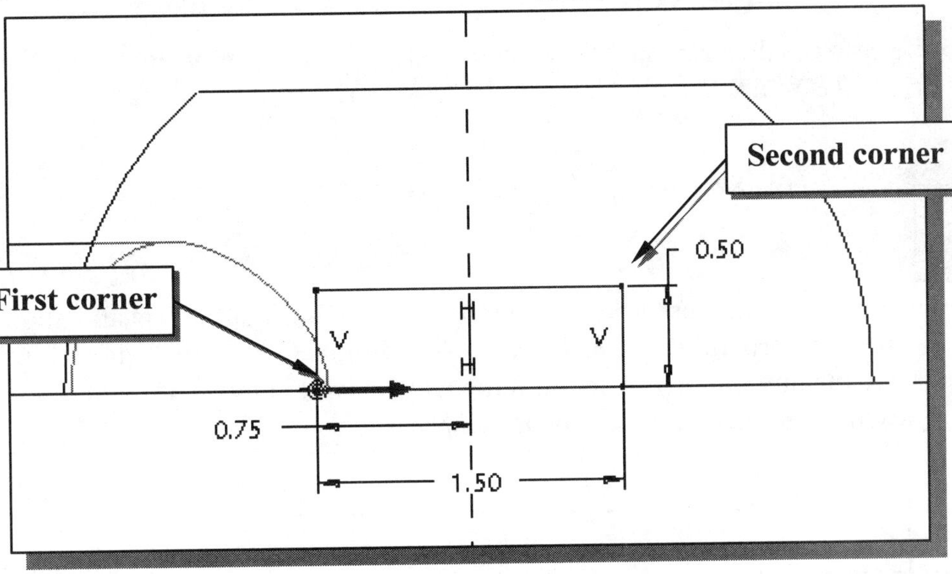

3. Use the **Relation** option and position the rectangle centered about the vertical axis.

4. Use **Modify** and **Regenerate** in the *Sketcher* menu to adjust the size of the rectangle to **1.50 x .50**.

❖ We have completed the first 2D section. We will next create the second 2D section. A *blend* feature is a series of two-dimensional sections that are joined together.

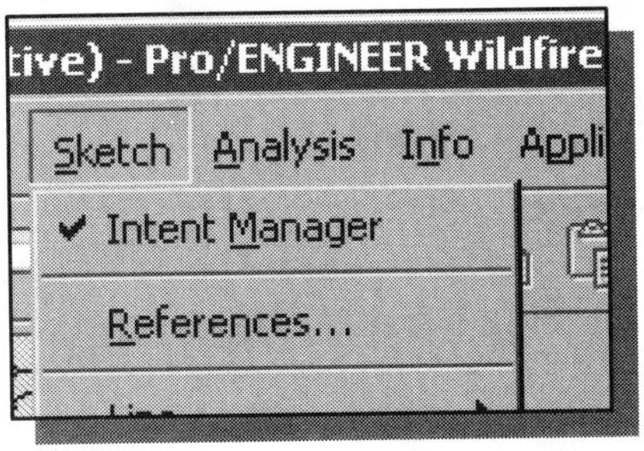

5. Select **Sketch** in the pull-down menu.

6. In the **Sketch** pull-down menu, select **Feature Tools → Toggle Section**

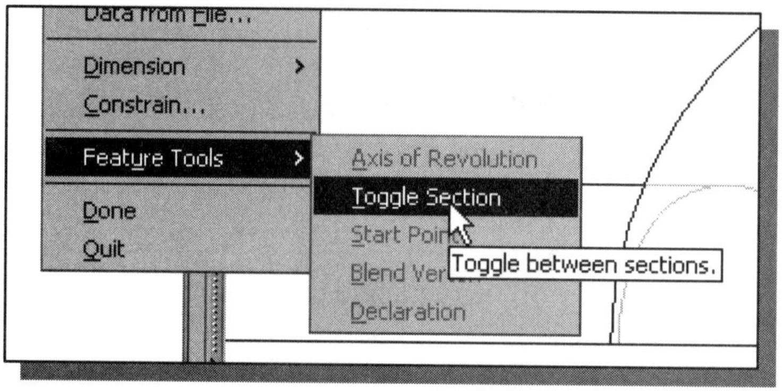

- The color of the previously sketched rectangle turned gray signifying that *Pro/ENGINEER Sketcher* is ready to create the second 2D section.

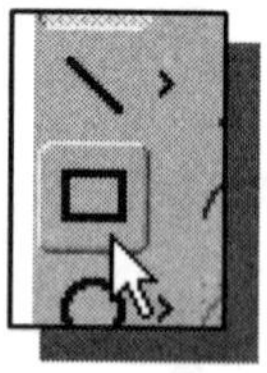

7. In the *Sketcher* toolbar, select **Rectangle**.

8. Place the first corner of the rectangle on the horizontal axis and align the second corner as shown. The arrowhead indicates the direction of the blend. By placing the rectangle the same way as we did the first section, the blend will be aligned at the corresponding corners.

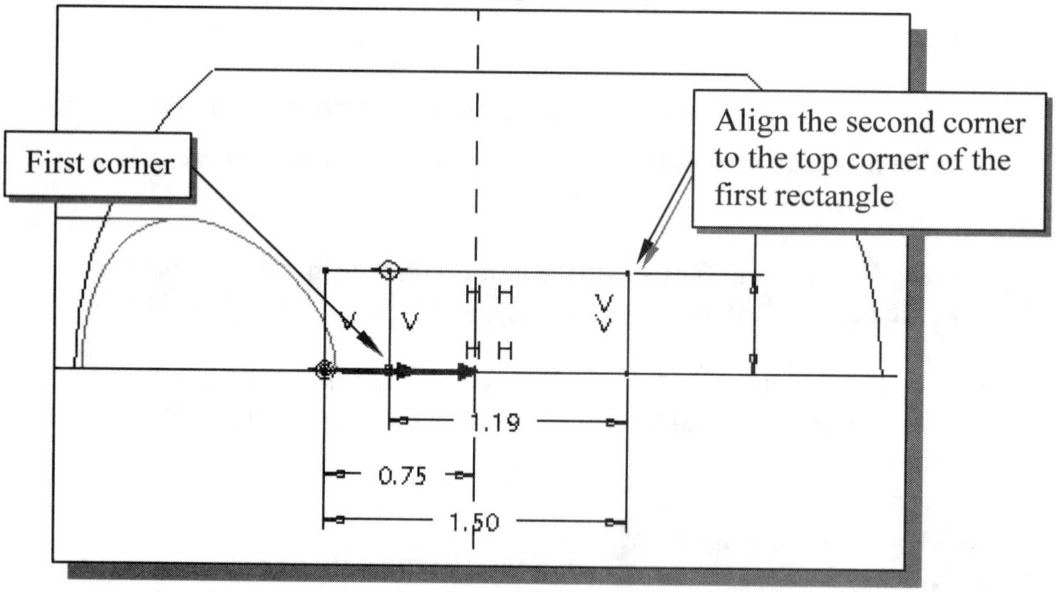

9. Modify the *width* of the second rectangle to **1.40**.

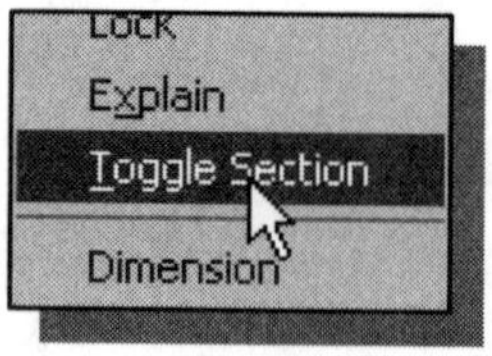

10. Inside the graphics area, press down the **right-mouse-button** to bring up the option menu.

11. In the popup list, select **Toggle Section**.

12. On your own, repeat the above steps and create the third rectangle (width:**1.2**) as shown.

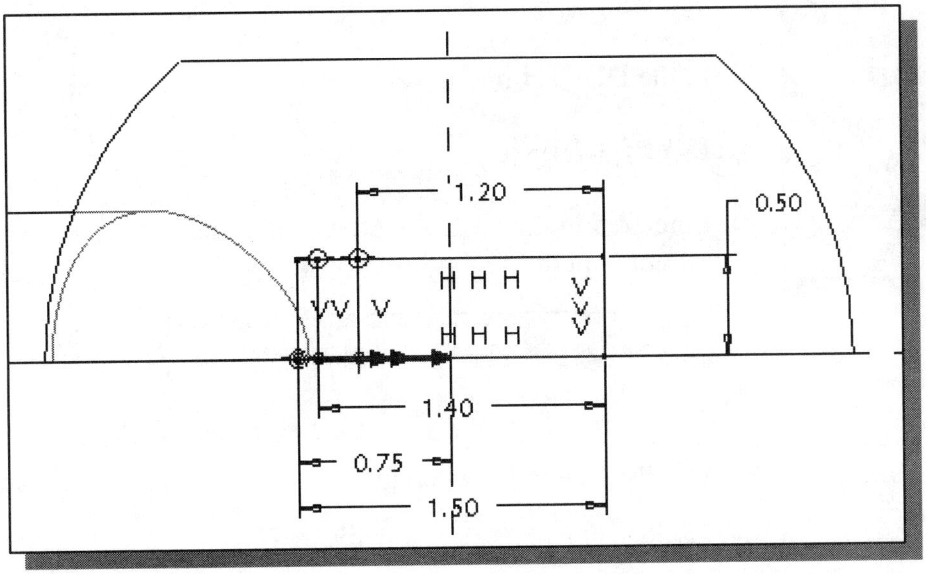

13. Reposition all of the existing dimensions by **drag-and-drop** with the left-mouse-button.

- Note that only ***one*** loop per subsection is allowed for the **Blend** operation.

14. On your own, repeat the above procedure and create another rectangle (width: **1.35**) as shown. Note that all four sections are aligned at the top-right corner. (Important: **Do not** click **TOGGLE SECTION** for the last section.)

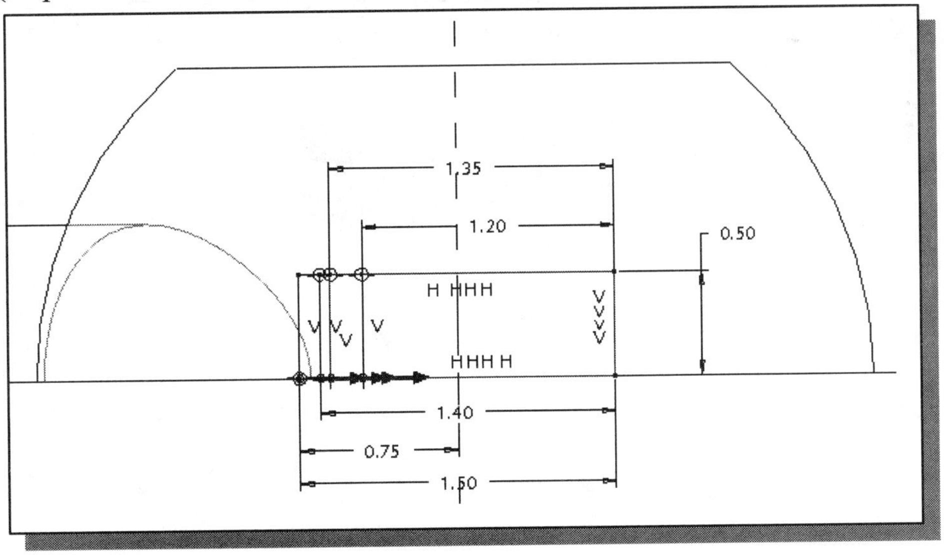

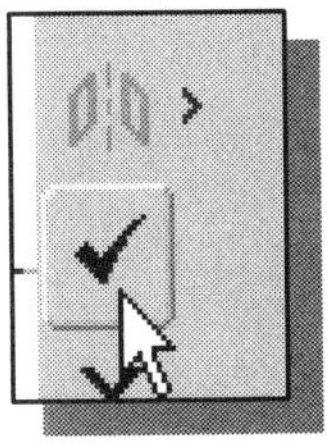

15. Click on the **Accept** icon to accept the completed *sweep section*.

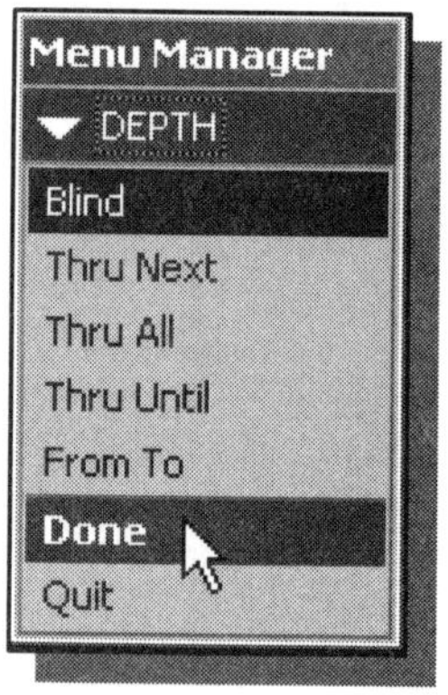

❖ Next, the distances in between the constructed 2D sections are defined.

16. In the **DEPTH** menu, select

    Blind → Done

17. Enter **2.5** in the message area for the distance in between the first section and the second section.

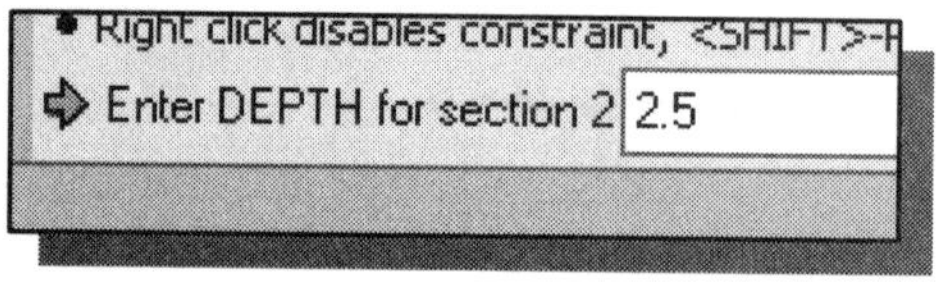

18. Enter **1.0** as the distance in between the second section and the third section.

19. Enter **0.75** in the message area for the distance in between the third section and the forth section.

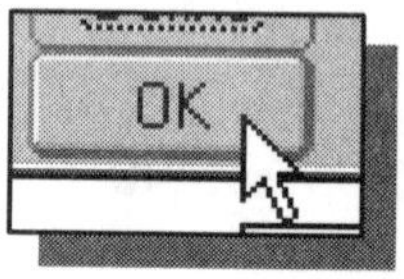

20. Notice that all elements are defined in the feature dialog window; click **OK** to create the solid feature.

21. On your own, use the *Dynamic Viewing* functions to view the completed 3D solid feature.

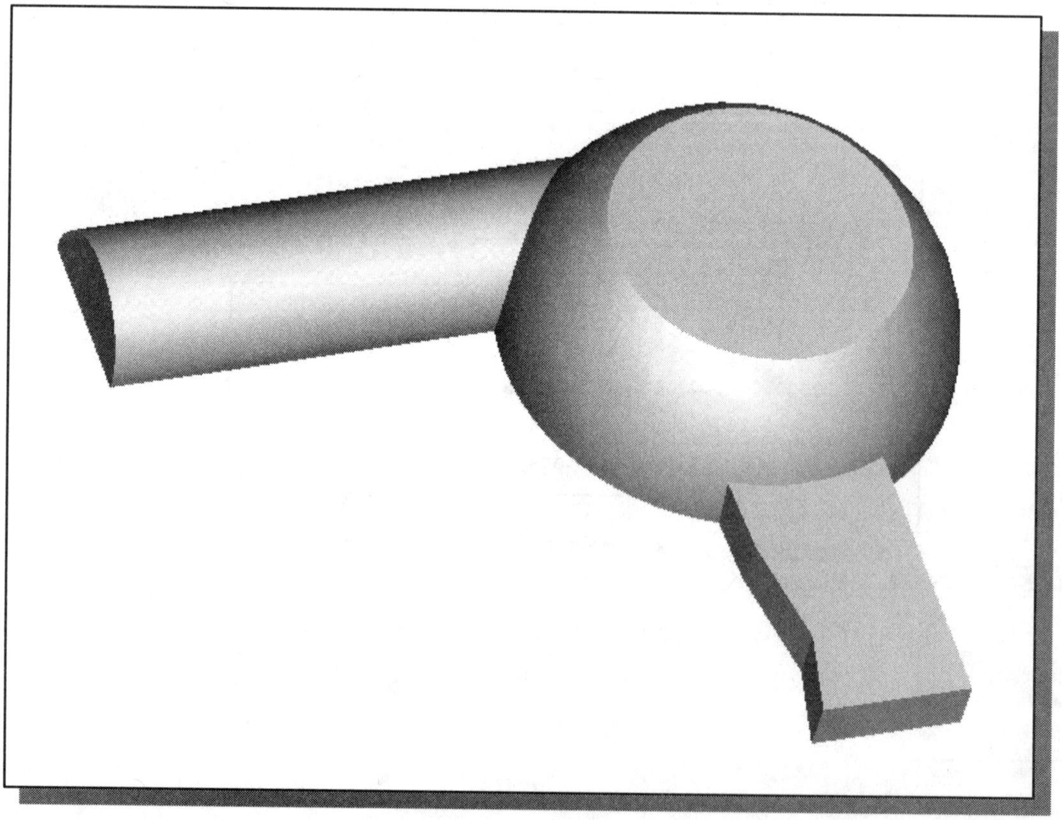

Create 3D Rounds and Fillets

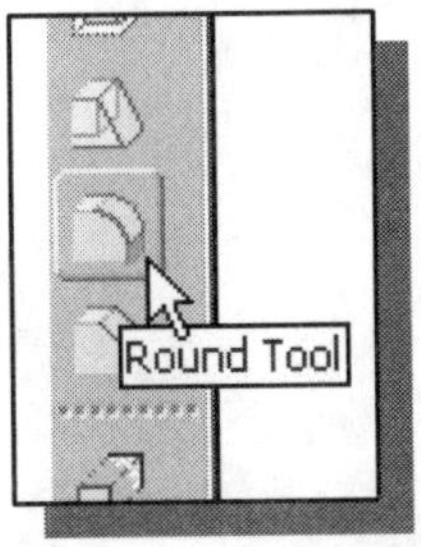

1. In the *Standard* toolbar, select **Round Tool**.

2. In the **Round** dashboard, set the radius to **0.25** as shown.

3. Pick the top circle, the intersection curve, and all the edges connected to the top surface of the handle; the highlighted edges shown in the figure. (Refer to the resulting feature shown on the next page.)

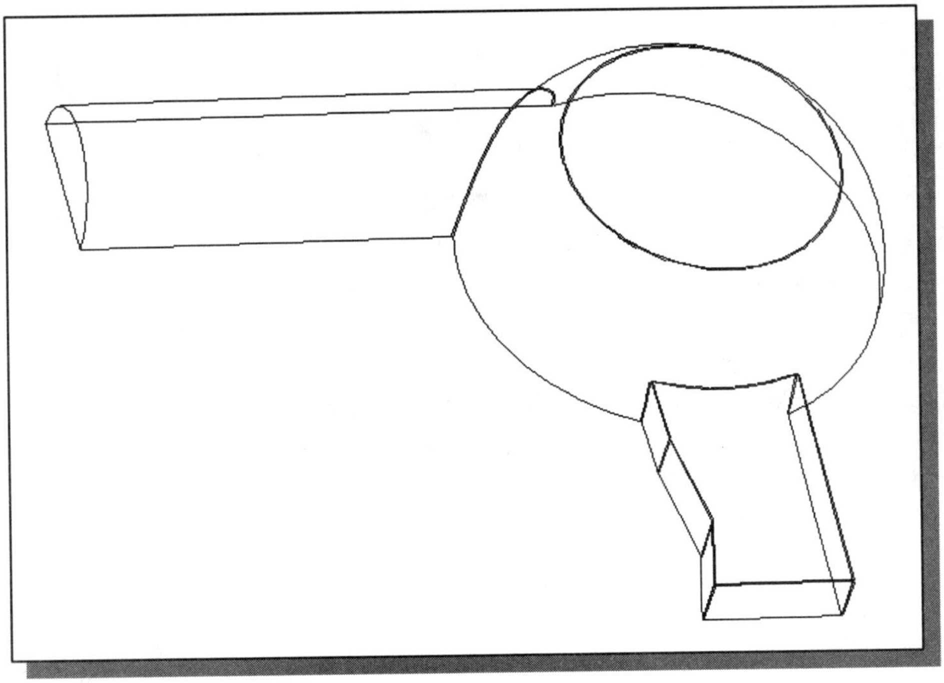

4. On your own, use the **Preview** option in the feature option menu to examine the feature.

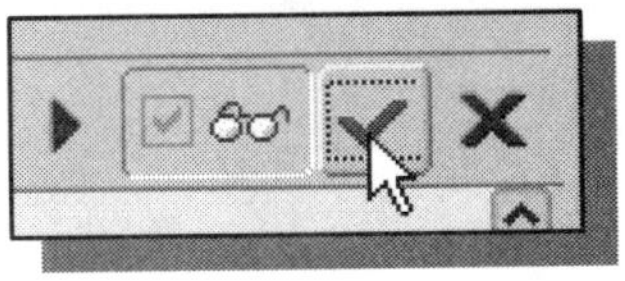

5. Pick **Accept** in the feature option menu to create blended feature.

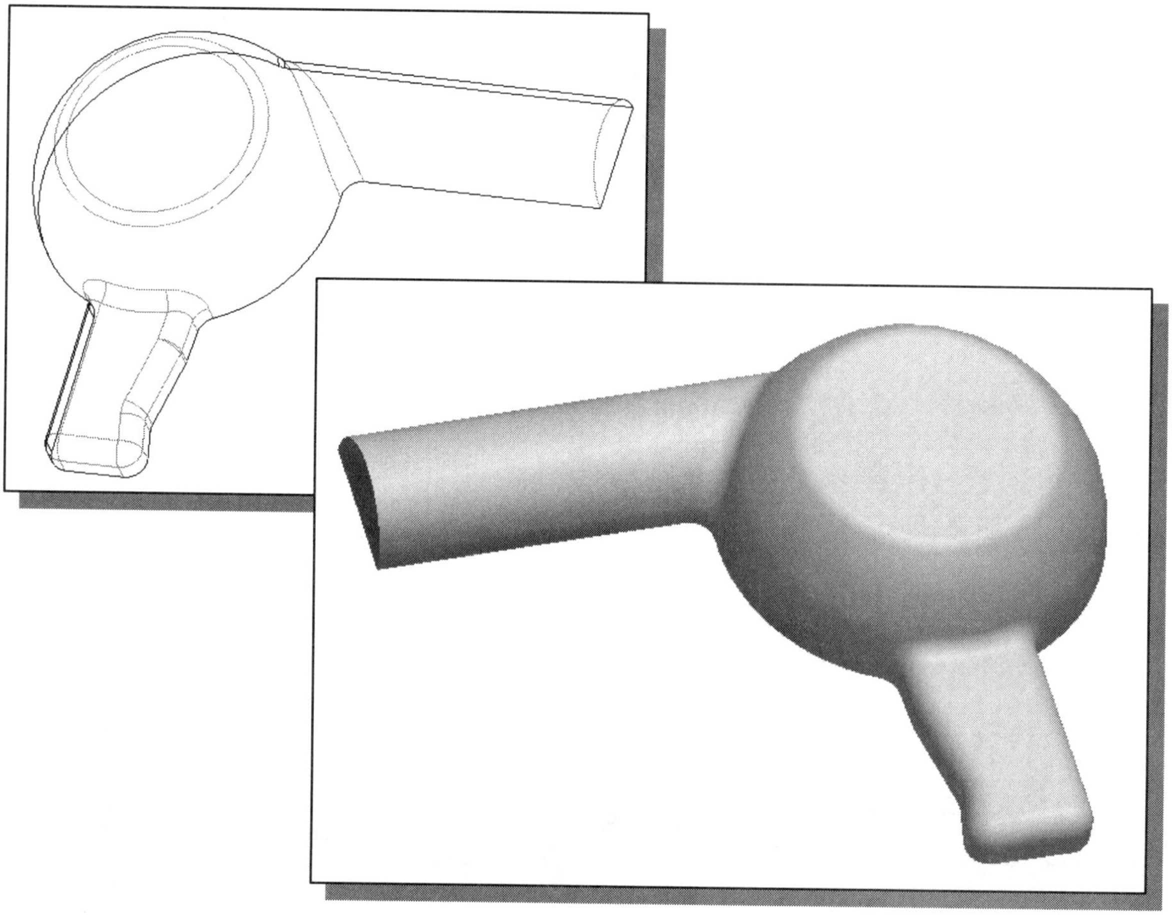

Creating a Shell Feature

❖ The **Shell** operation hollows out the inside of the solid, leaving a shell of a specified wall thickness. We can specify a surface or surfaces to be removed with the **Shell** operation. If no surfaces were selected to be removed, a "closed" shell is created, with the whole inside of the part hollowed out and no access to the hollow.

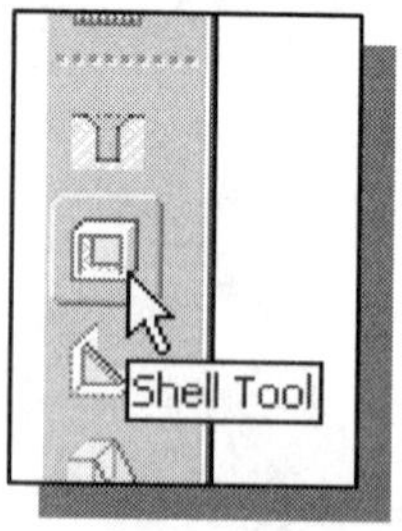

1. In the *Standard* toolbar, select **Shell Tool**.

2. In the feature option menu, enter **0.125** as the wall thickness.

3. The message "*Select surfaces to remove from the part*" is displayed in the message area. Use the *Dynamic Rotation* function to rotate the model and pick the ***two surfaces*** as shown. (Hold down the [**Ctrl**] key while selecting.)

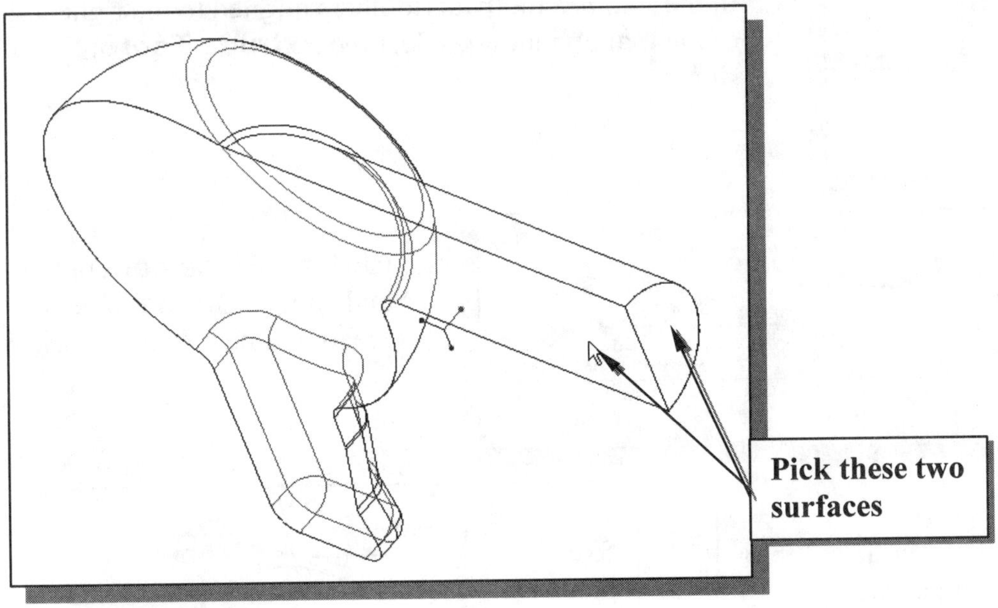

4. On your own, use the **Preview** option in the feature option menu to examine the feature.

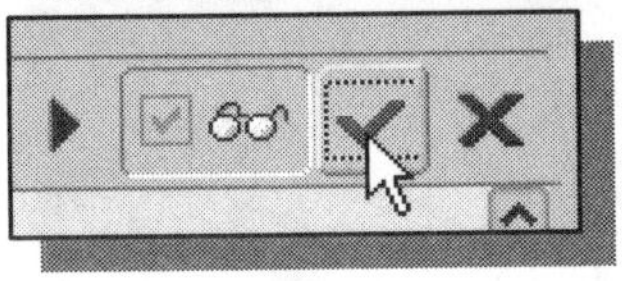

5. Pick **Accept** in the feature option menu to create the shell feature.

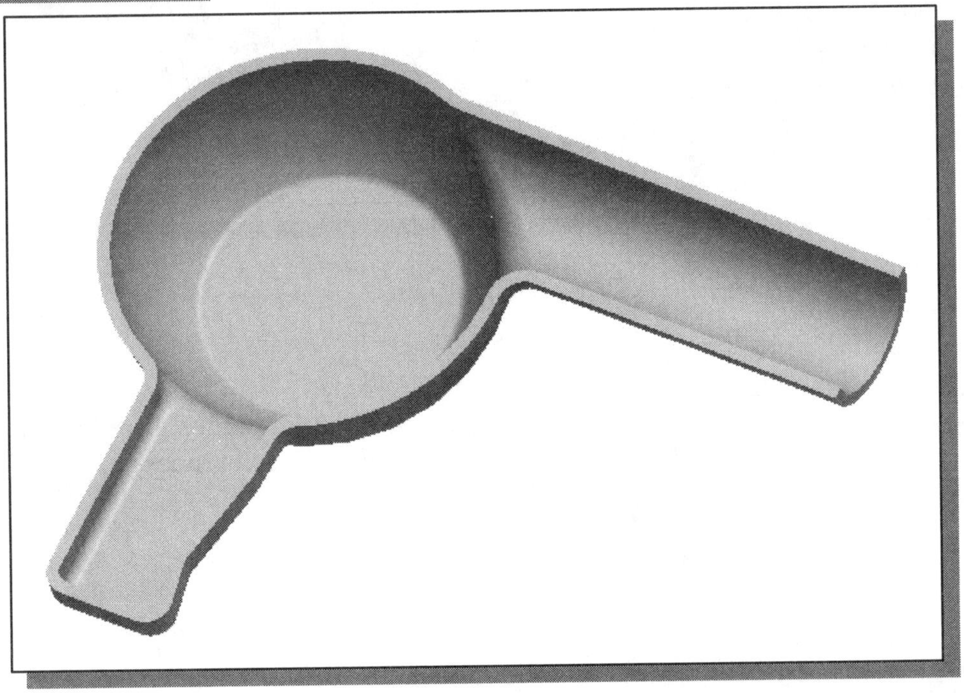

Create a Pattern Leader

1. In the *Feature* toolbars (toolbars aligned to the right edge of the main window), select the **Extrude Tool** option as shown.

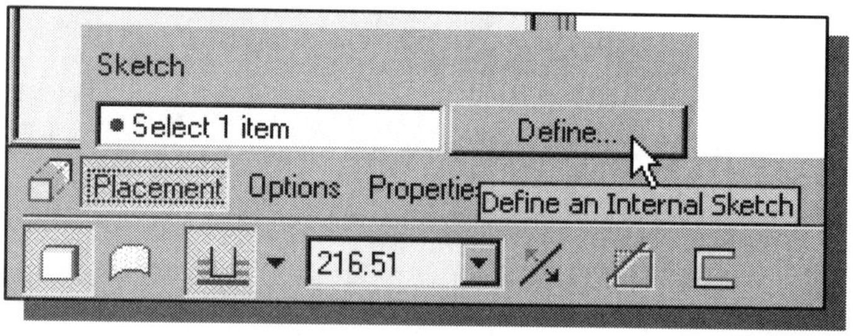

2. Click the **Placement** option and choose **Define** to begin creating a new *Internal sketch.*

3. Pick the top surface of the base feature as the *sketching plane,* with the arrow pointing towards the base of the solid model.

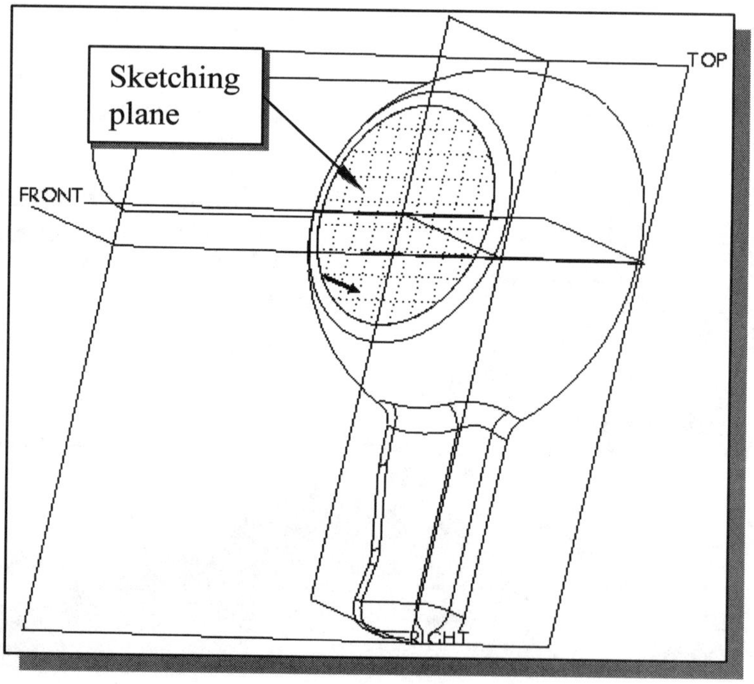

4. In the *Section Placement* window, set the datum plane **FRONT** as the **bottom** side *reference plane* and click **Sketch** to enter the *Sketcher* mode.

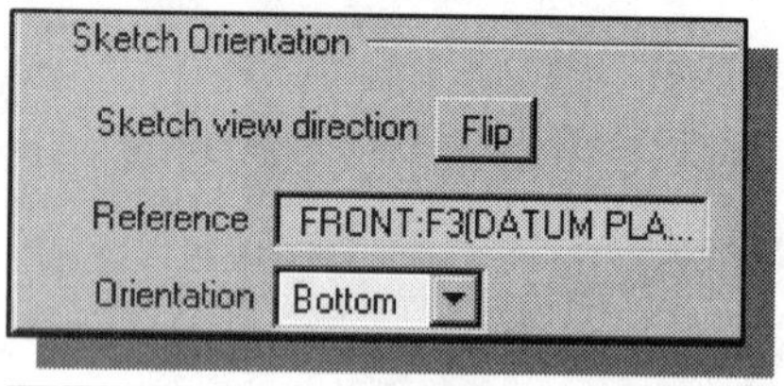

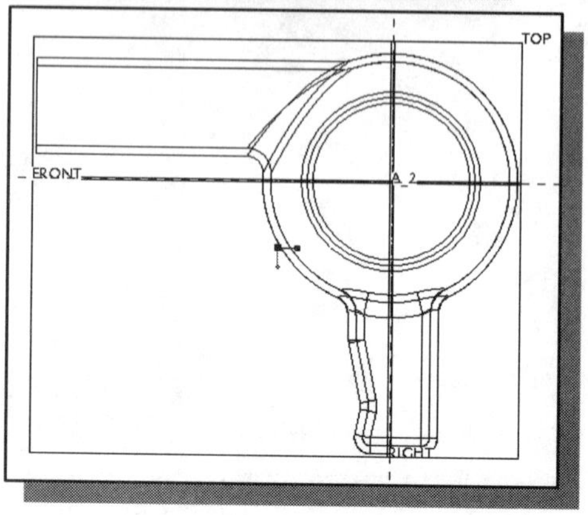

❖ Notice the two datum planes, Front and Right, are set as the default references for the 2D sketch.

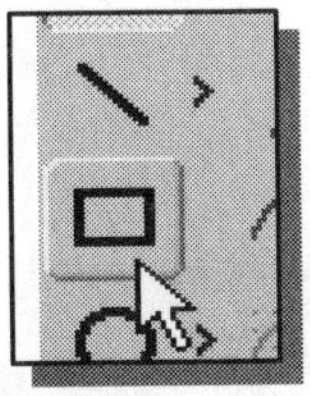

5. In the *Sketcher* toolbar, select **Rectangle**.

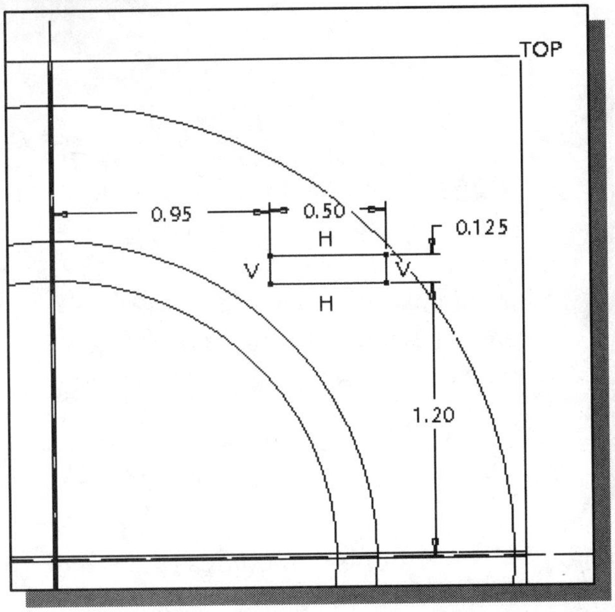

6. Create a rectangle toward the upper-right side of the solid model as shown.

7. On your own, modify the dimensions as shown in the figure. Rectangle size: **0.5 x 0.125**, location: **1.2** away from the horizontal axis and **0.95** away from the vertical axis.

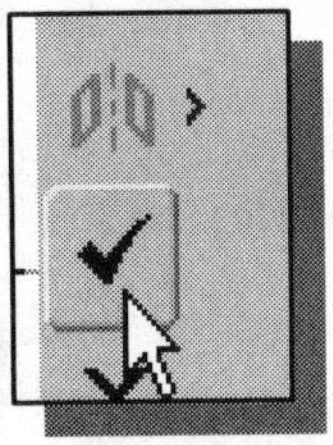

8. Click on the **Accept** icon to accept the completed *sweep section.*

9. Toggle ***on*** the **Remove Material** option and enter **0.13** as the depth of the cut.

10. Click **Flip direction** once to flip the cut direction to cut into the solid model.

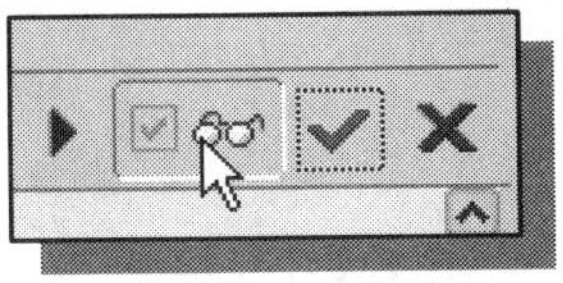

11. On your own, use the **Preview** option in the feature option menu to examine the feature.

- Notice in the message area, the message "*WARNING: EXTRUDE_2 is entirely outside the model; model unchanged.*" is displayed.

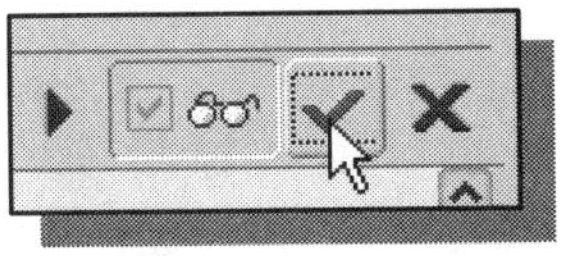

12. Pick **Accept** in the feature option menu to create the shell feature.

- Even though the feature is entirely outside the model, it is created successfully. We will next duplicate this feature to form the necessary pattern on the model.

Create a Rectangular Pattern

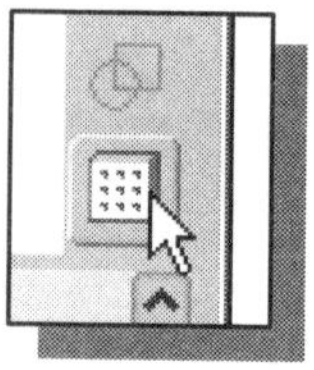

1. In the *Feature* toolbar, select **Pattern**. (Note that the **Cut** feature is currently highlighted in the *Model Tree* area, which means it is pre-selected.)

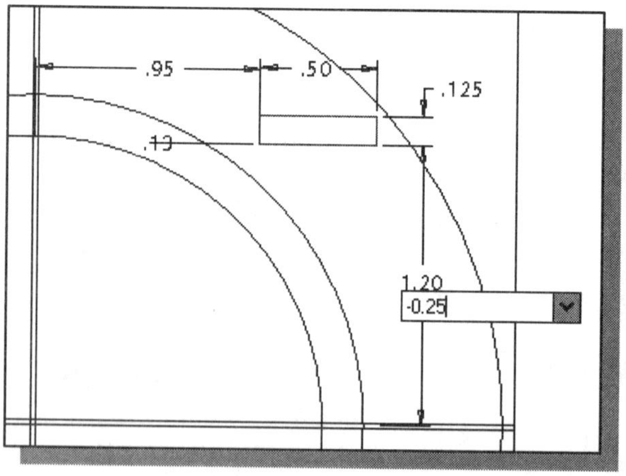

2. Pick the vertical location dimension (**1.20**).

3. In the message area, enter **-0.25** to place the copies of the feature below the pattern leader.

4. In the feature option menu area, enter **11** as the number of copies (including the original) needed in the vertical direction.

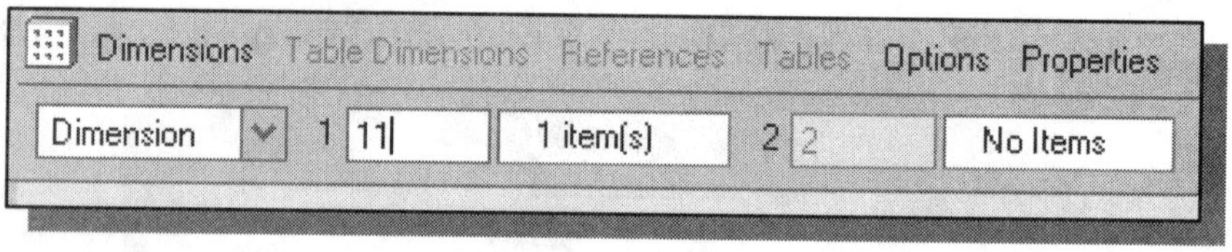

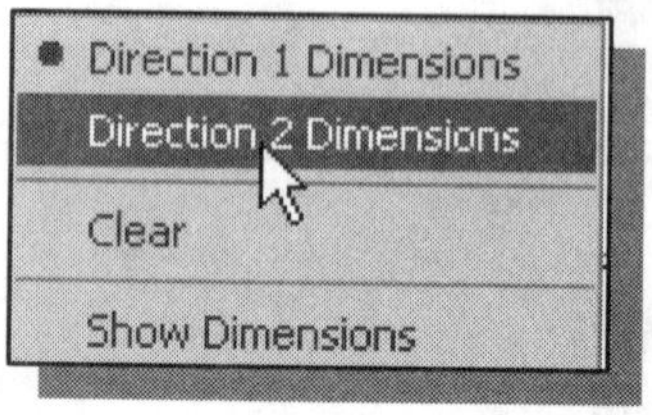

5. Inside the display area, click once with the right-mouse-button to bring up the option menu.

6. Select **Direction 2 Dimensions** in the option list.

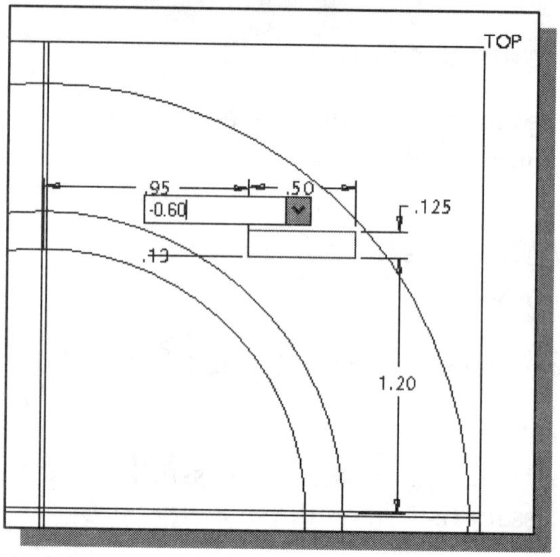

7. Pick the horizontal location dimension (**0.95**).

8. In the message area, enter **-0.60** to place the copies of the feature to the left of the pattern leader.

9. In the feature option menu area, enter **5** as the number of copies (including the original) needed in the second direction.

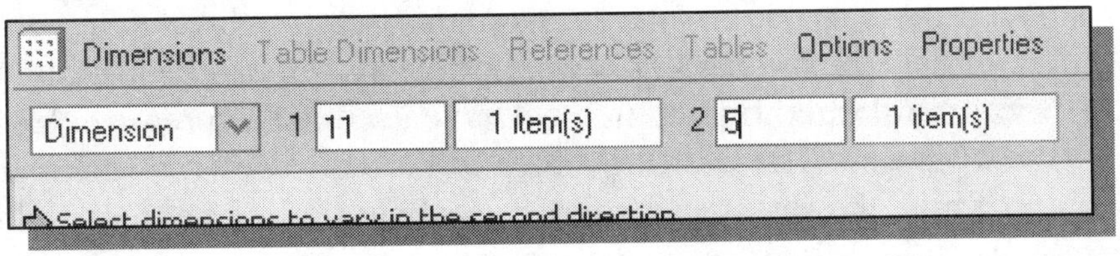

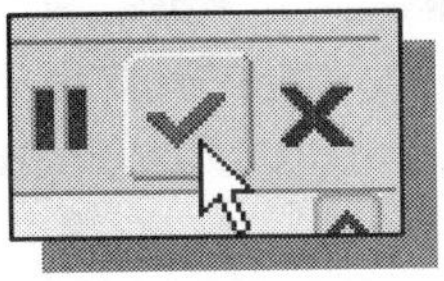

10. Click **Accept** to create the pattern.

- *Pro/ENGINEER* will now duplicate the *cut* feature and generate the rectangular cut pattern.

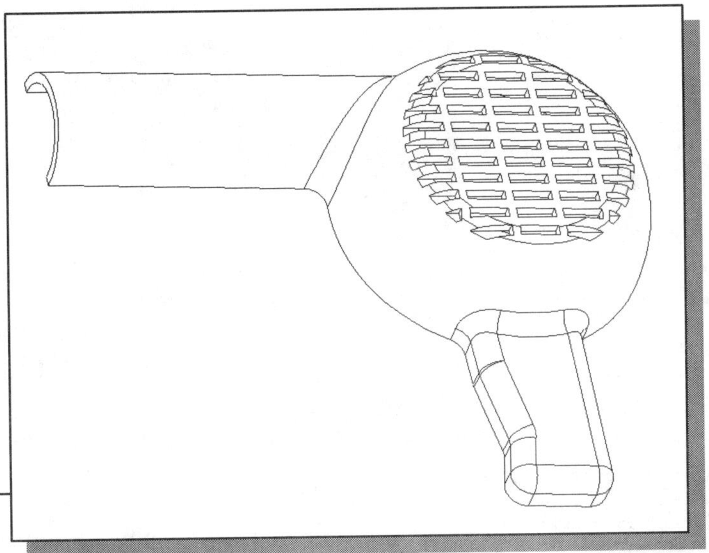

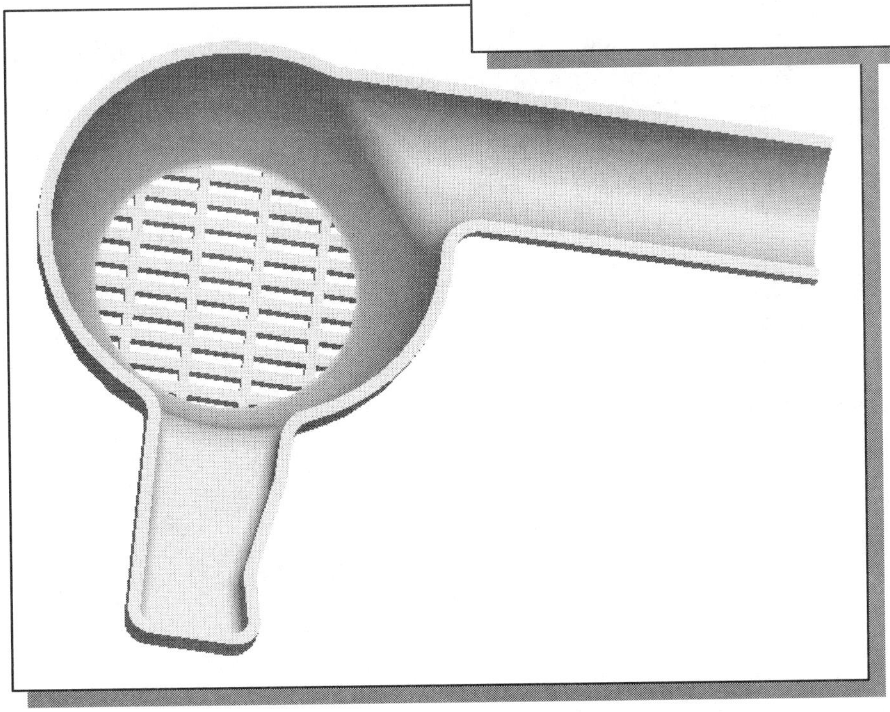

Creating a Swept Feature

❖ A ***sweep*** operation is defined as moving a planar section through a path in space to form a three-dimensional object. The path of the sweep can be a straight line or a curve. The **Extrude** operation, which we have used in the previous lessons, is a specific type of sweep. The Extrude operation is also known as the ***linear sweep*** operation, in which the sweep control path is always a line perpendicular to the two-dimensional section. Linear sweeps of unchanging shape result in what are generally called ***prismatic solids***; which means solids with a constant cross-section from end to end.

1. In the **Insert** pull-down menu, select **Sweep → Cut** as shown.

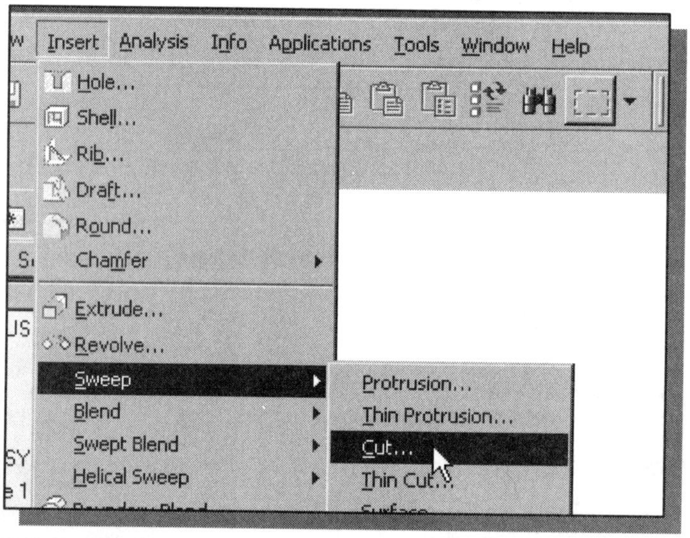

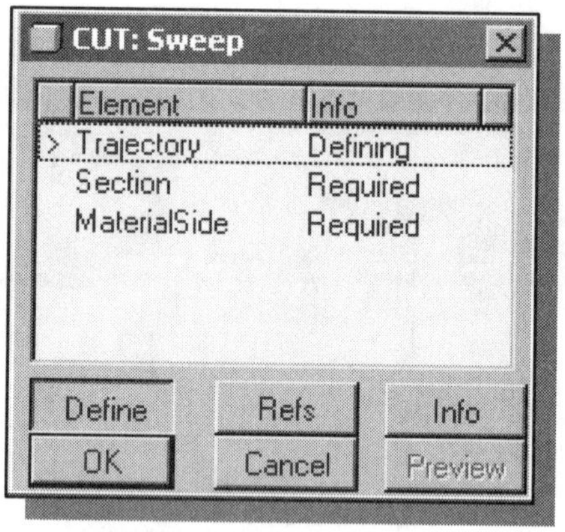

❖ The feature dialog window appears in the upper right corner of the screen. Three elements are listed in the dialog window: Trajectory, Section and MaterialSide. Note that all three elements are required to create the feature. The little marker next to the element indicates the definition that is active.

➢ **Define the Sweep Trajectory**

The *sweep trajectory* is the *sweep path* that the *sweep section* is moved along.

2. Pick **Select Traj** in the **SWEEP TRAJ** menu.

3. Set the **Chain** option to **Tangent Chain** as shown.

4. Select the **top edge** of the model as shown in the figure below. Note that an arrowhead is displayed indicating the sweep direction.

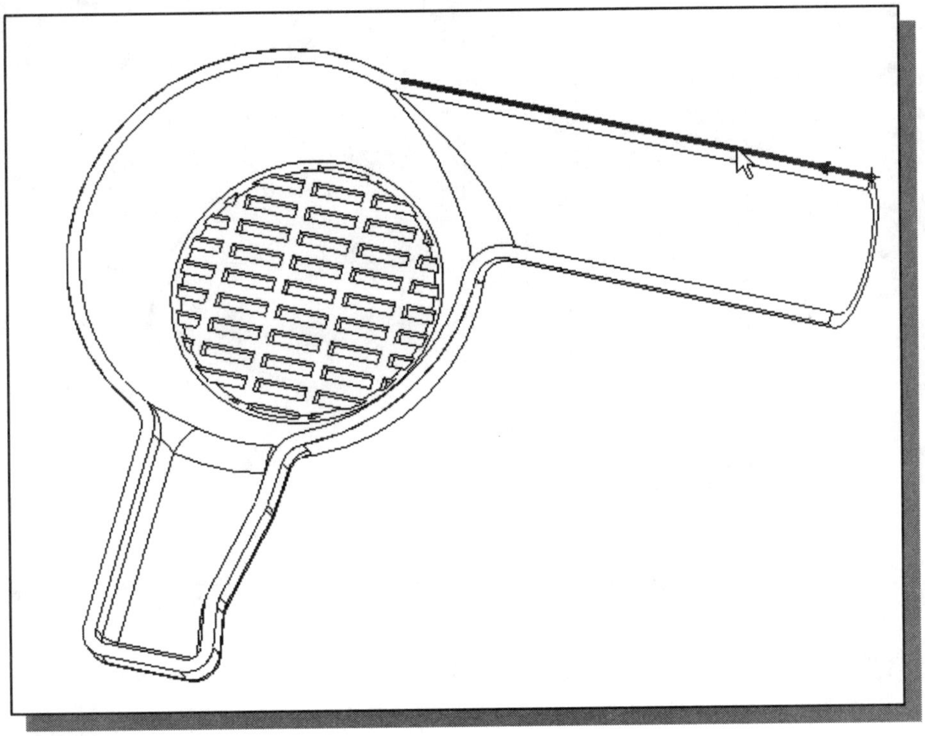

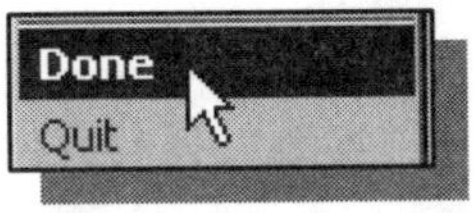

5. Click **Done** to accept the selected *sweep trajectory*.

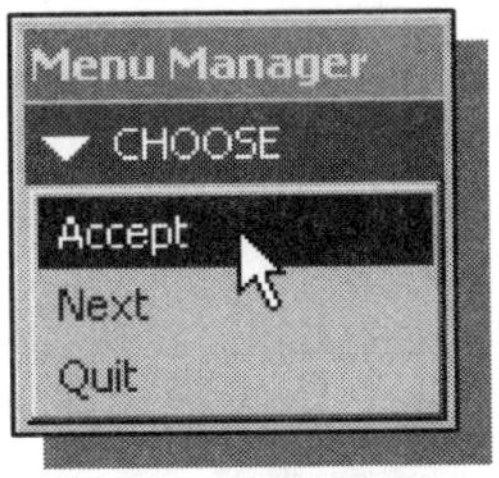

6. Click **Accept** to use the default surface as the normal surface of the cross section.

7. Click **Okay** to use the default direction to orient the cross section.

➢ Define the Sweep Section

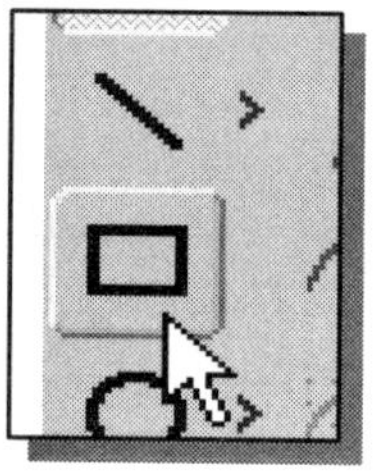

1. In the *Sketcher* toolbar, select **Rectangle**.

2. Create a rectangle, **0.125** × **0.0625**, aligned to the top corner of the model as shown in the figure below. (Hint: First rotate dynamically to confirm the orientation of the model.)

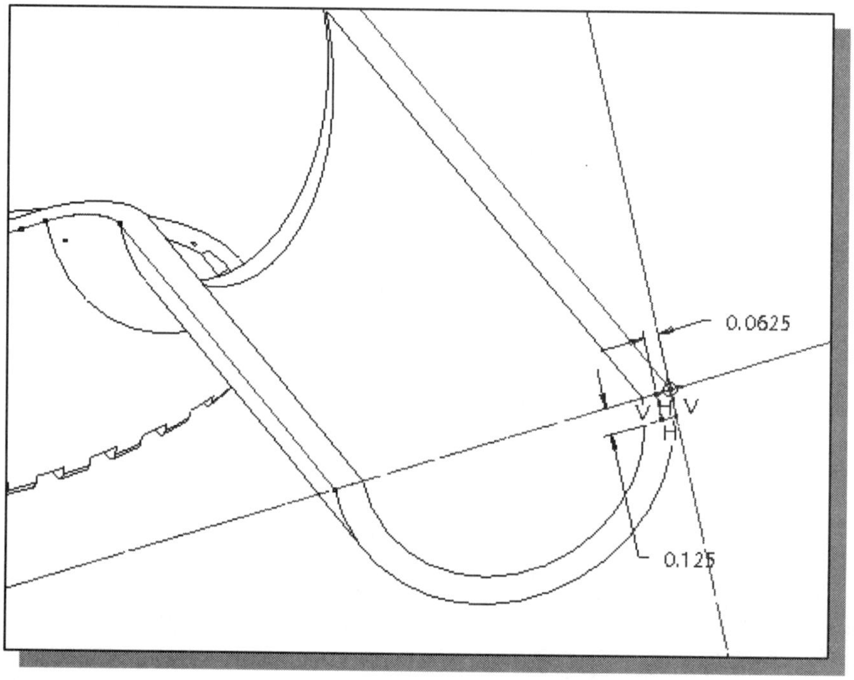

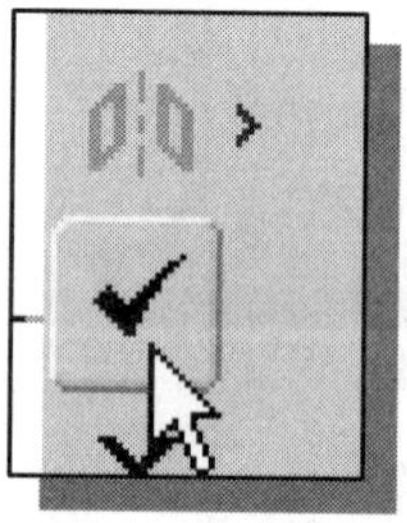

3. Click on the **Accept** icon to accept the completed *sweep section.*

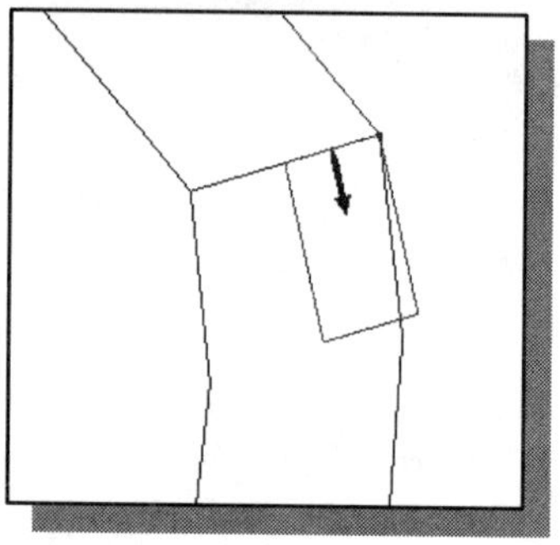

4. Confirm the arrow points toward the center of the small rectangle. Click **Okay** to accept the MaterialSide as shown.

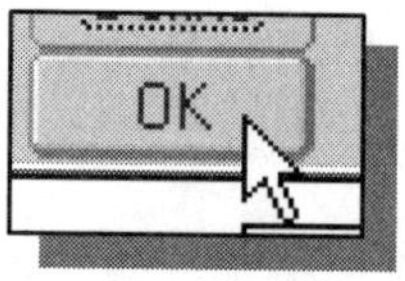

5. Notice that all elements are defined in the feature dialog window, click **OK** to create the solid feature.

6. On your own, use the *Dynamic Viewing* functions to view the constructed 3D model.

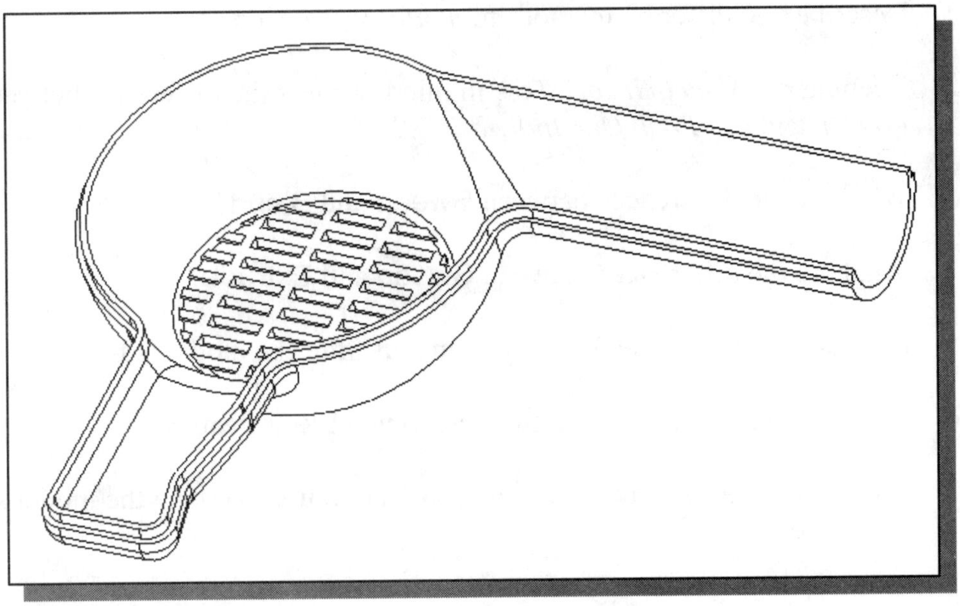

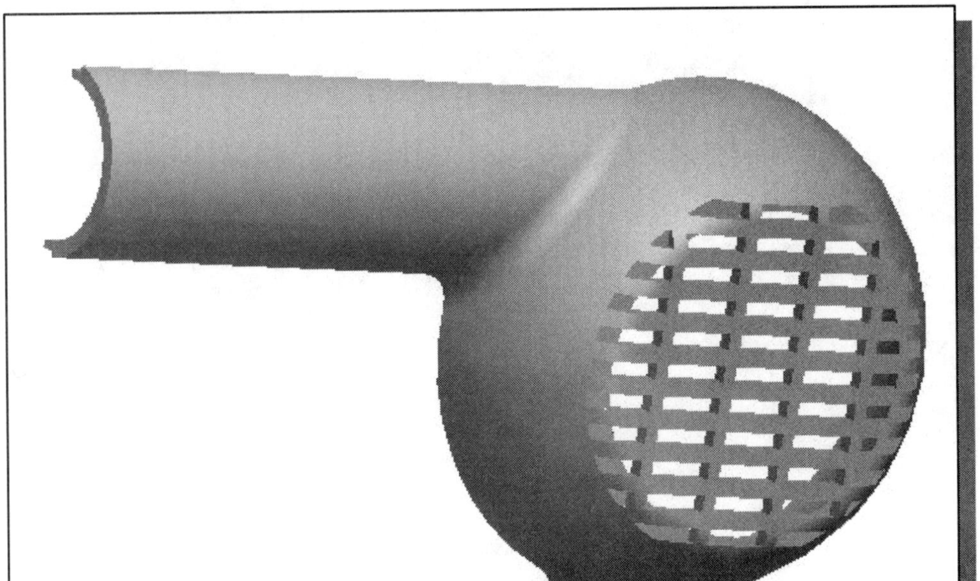

Questions:

1. Describe two different methods to *modify* dimensions.

2. Keeping the *Model History Tree* in mind, what is the difference between *Cut with a Pattern* and *Cut Each One Individually*?

3. What are the differences between **Sweep** and **Blend**?

4. What are the differences between **Sweep** and **Extrude**?

5. How do we modify the pattern parameters after the model is built?

6. Describe the elements required in creating a **Blend** feature.

7. Create sketches showing the steps you plan to use to create the model shown on the next page:

Exercises:

1. Dimensions are in inches (Wall thickness: 0.125)

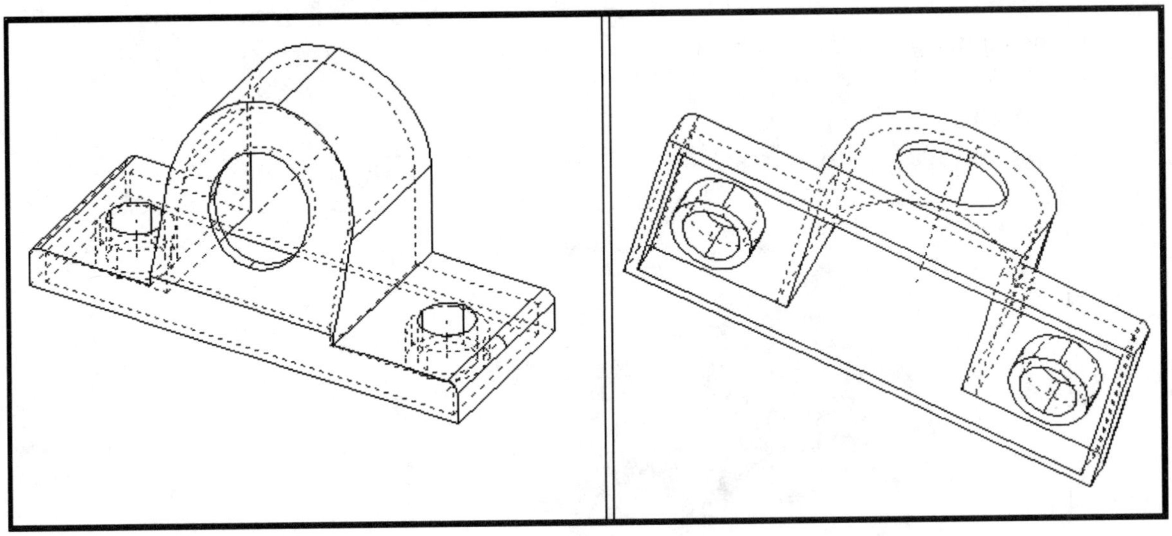

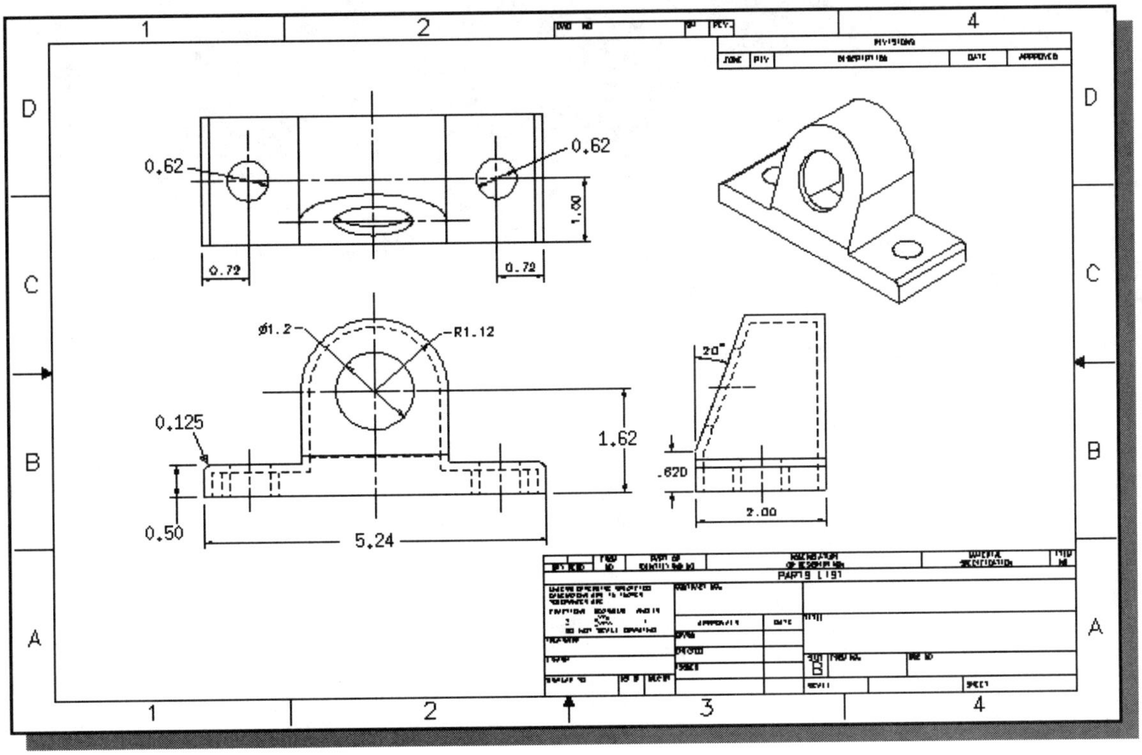

2. Using the same dimensions given in the tutorial, construct the other half of the dryer housing. Plan ahead, and consider how you would create the matching half of the design? (Save both parts so that you can create an assembly model once you have completed chapter 10.)

3. Dimensions are in inches. Hint: Use the rib tool.

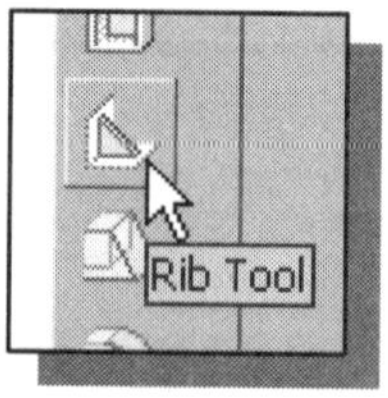

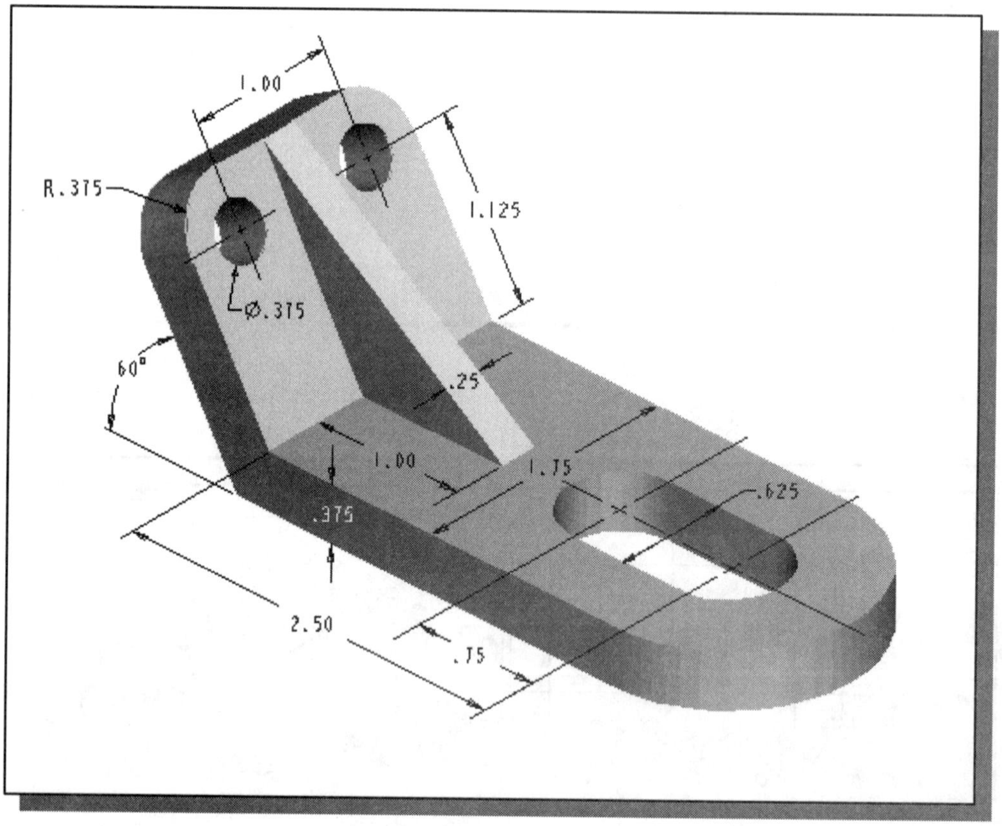

Chapter 9
Advanced Modeling Techniques

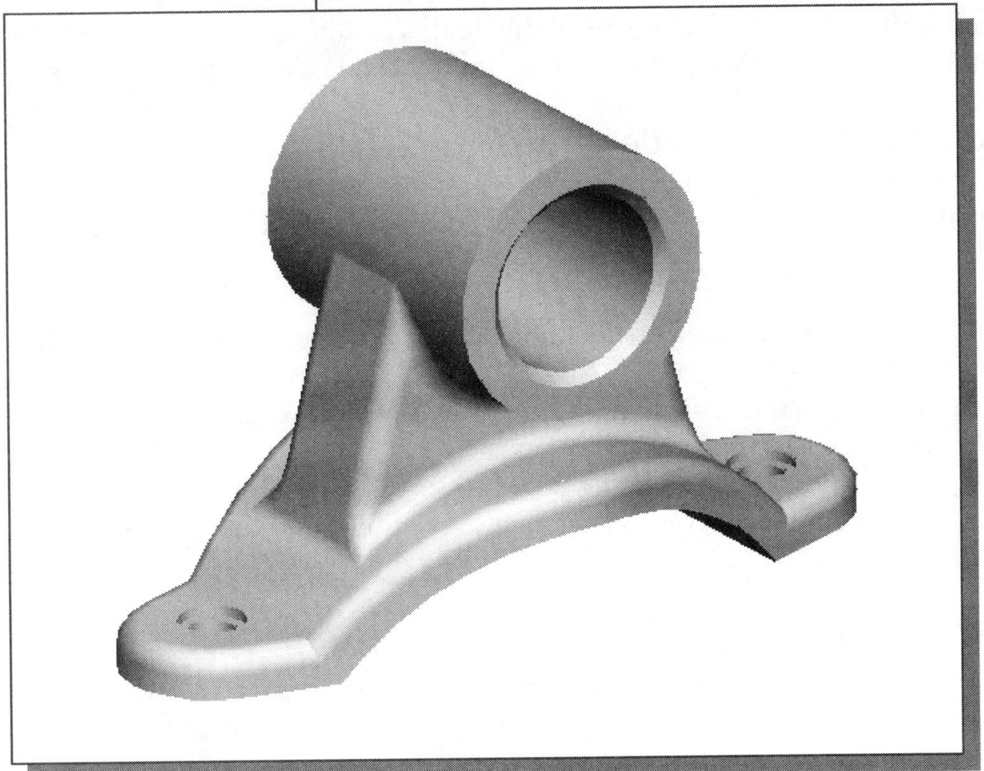

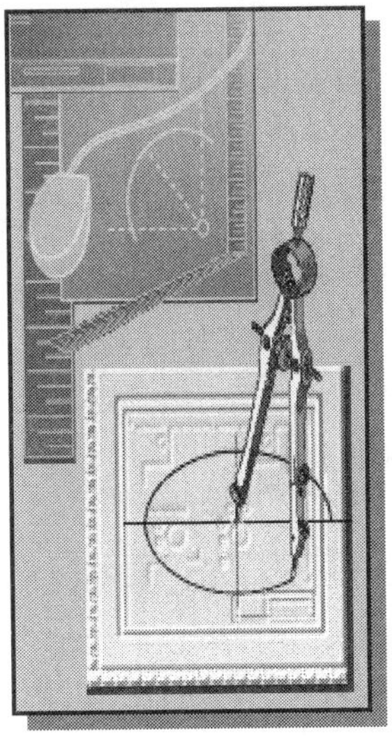

Learning Objectives

- Perform the Boolean - Intersect Operation.
- Apply Draft Angles to surfaces.
- Create complex 3D Rounds and Fillets.
- Using the Feature Copy command.
- Adjusting the Model Color.
- Use the CHAMFER command.
- Review the different aspects of Part Modeling.

Introduction

This chapter is intended to provide a summary of the basic construction techniques presented in the previous chapters. We will also demonstrate the more advanced construction techniques such as using the *Boolean-Intersect operation*, the procedure to create *drafted surfaces*, *sketched holes, chamfers,* and *complex 3D rounds*; these are common characteristics to casting and molded parts. In this lesson, we will also explore some of the variations of using the basic techniques introduced in the previous chapters.

Summary of Modeling Considerations

- **Design Intent** – determine the functionality of the design; select features that are central to the design.
- **Order of Features** – consider the parent/child relationships necessary for all features.
- **Dimensional and Geometric Constraints** – the way in which the constraints are applied determines how the components are updated.
- **Relations** – consider the orientation and parametric relationships required between features and in an assembly.

The *Bracket* Design

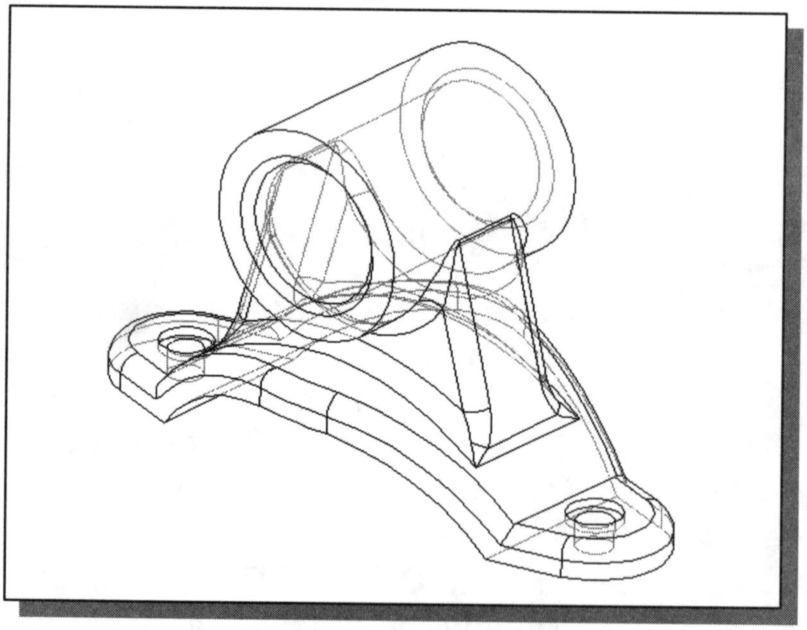

❖ Based on your knowledge of *Pro/ENGINEER*, how many features would you use to create the design? Which feature would you choose as the **base feature** of the model? Identify the more difficult features in the design and consider the possibilities in creating the design. You are encouraged to create a similar design on your own prior to following through the tutorial.

Modeling Strategy

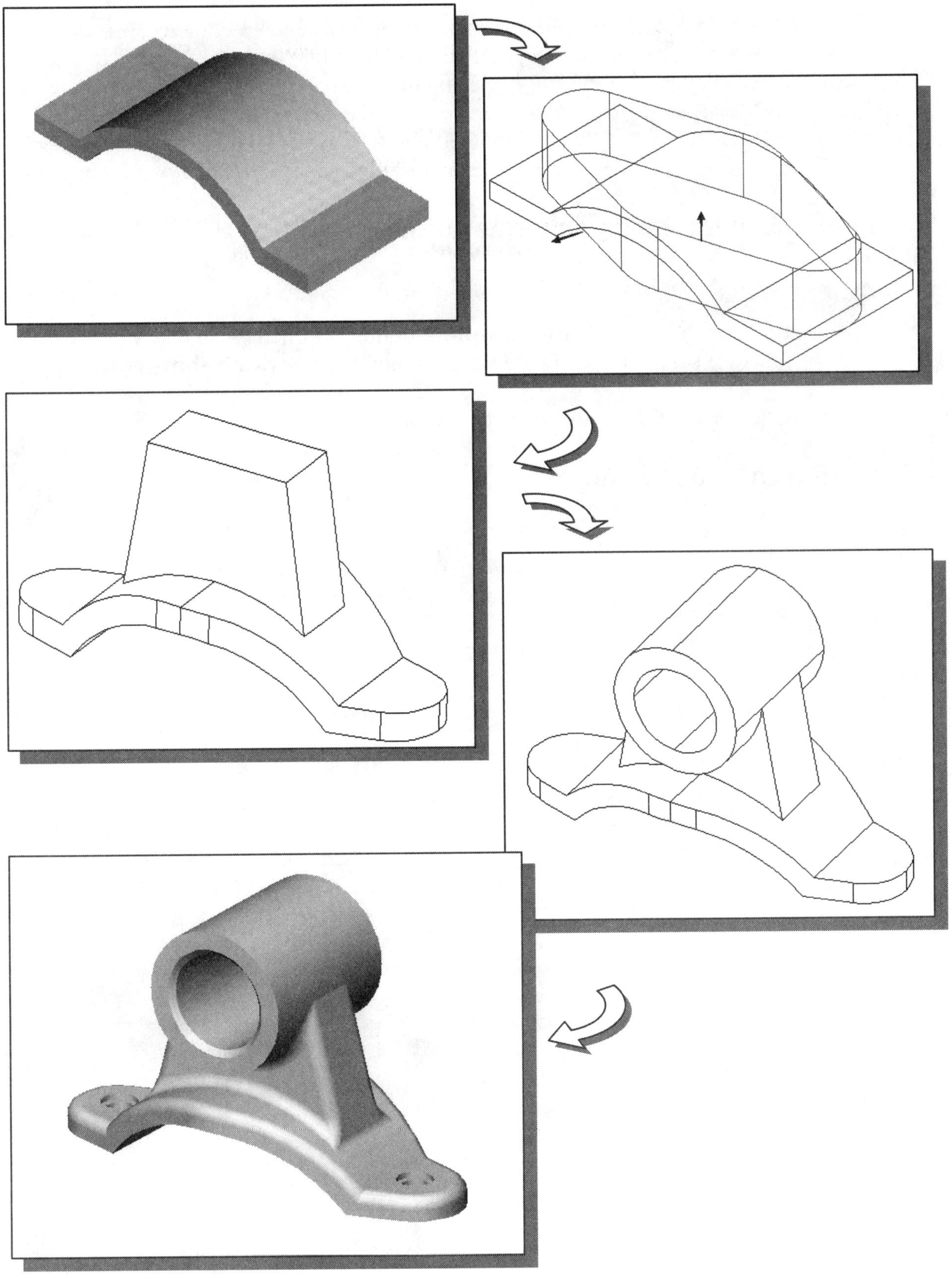

Starting *Pro/ENGINEER*

1. Select the **Pro/ENGINEER** option on the *Start* menu or select the **Pro/ENGINEER** icon on the desktop to start *Pro/ENGINEER*. The *Pro/ENGINEER* main window will appear on the screen.

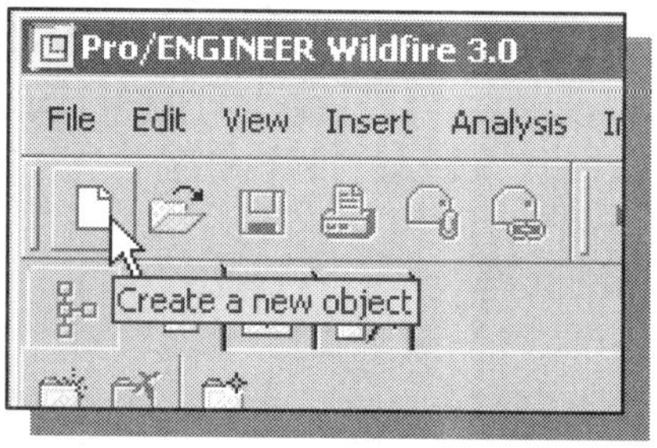

2. Click on the **New** icon, located in the *Standard* toolbar as shown.

3. On your own, start a new solid part file using ***Bracket*** as the part Name.

4. Confirm the **Use default template** option is turned ***on*** so that the system units are set to the *Pro/ENGINEER* default settings (**Inch-lbm-Second**).

5. Click on the **OK** button to accept the settings.

Creating the Base Feature

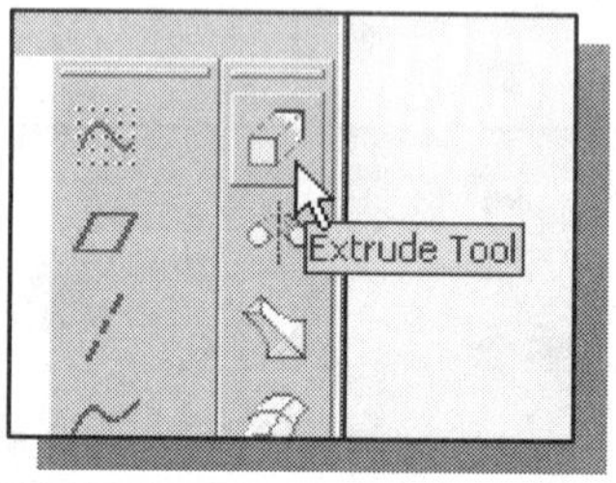

1. In the *Feature* toolbars (toolbars aligned to the right edge of the main window), select the **Extrude Tool** option as shown.

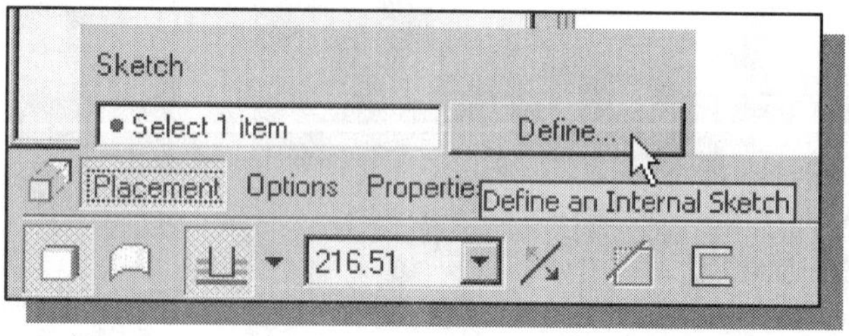

2. Click the **Placement** option and choose **Define** to begin creating a new *Internal sketch.*

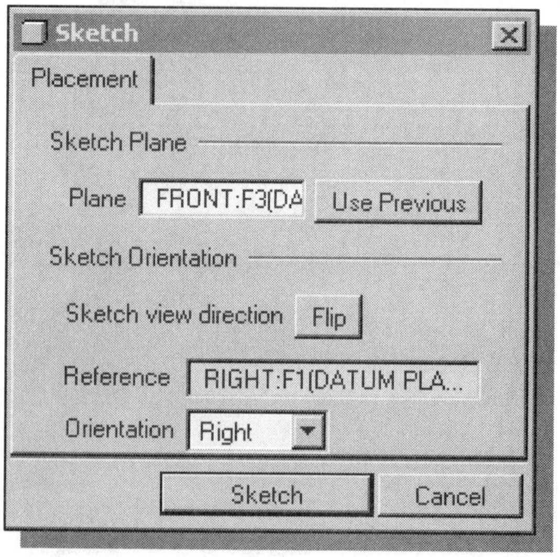

3. On your own, set up the **Front** datum plane as the sketch plane with the **Right** datum plane facing the **right** edge of the computer screen, and pick **Sketch** to enter the *Sketcher* mode.

❖ Notice the use of the two datum planes, **Right** and **Top**, as the references for the 2D sketch.

4. Create the 2D section as shown in the below figure. Note the sketch is symmetrical with respect to the vertical axis and the coincident center point of the two arcs is aligned to the vertical axis.

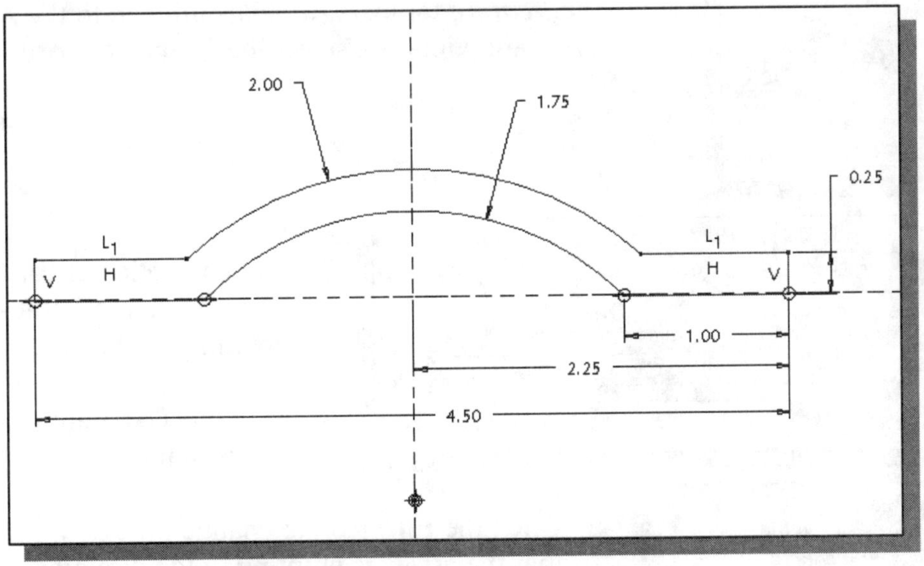

5. Complete the protrusion feature by extruding in ***both directions*** of the sketching plane and a depth of extrusion of **2.0**.

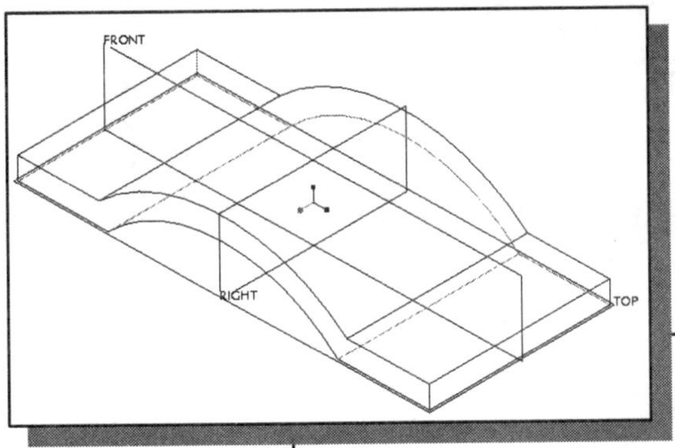

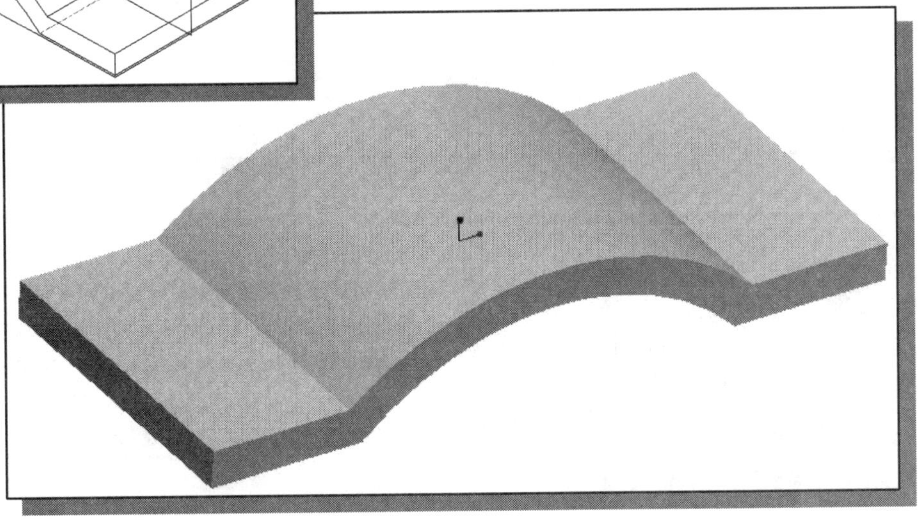

Using the CUT Option to Create a *Boolean INTERSECT*

1. In the *Feature* toolbars (toolbars aligned to the right edge of the main window), select the **Extrude Tool** option as shown.

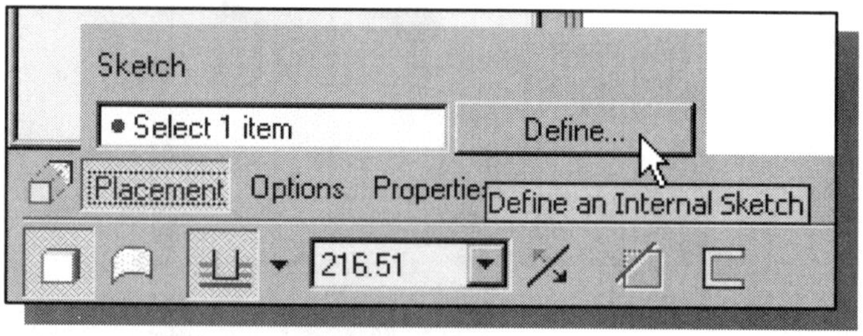

2. Click the **Placement** option and choose **Define** to begin creating a new *Internal sketch*.

3. Pick the **Top** datum plane as the sketch plane.

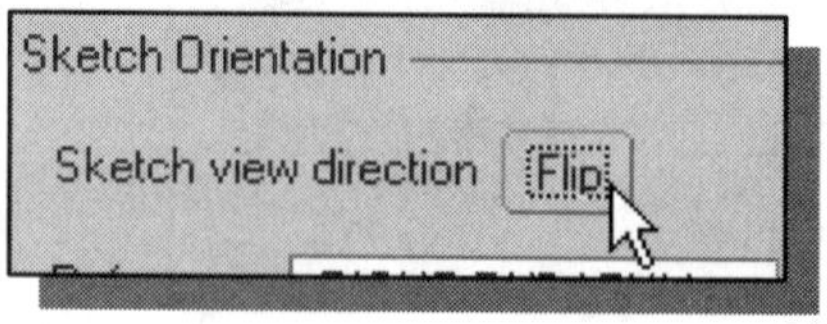

4. Click the **Flip** icon to change the arrow direction so that the arrow is pointing in the direction shown in the figure below.

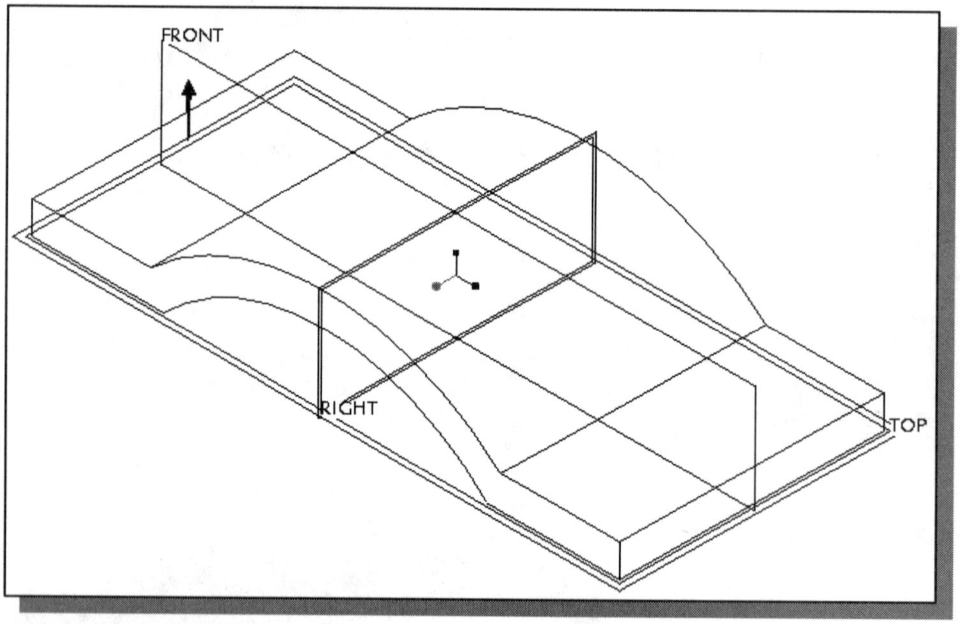

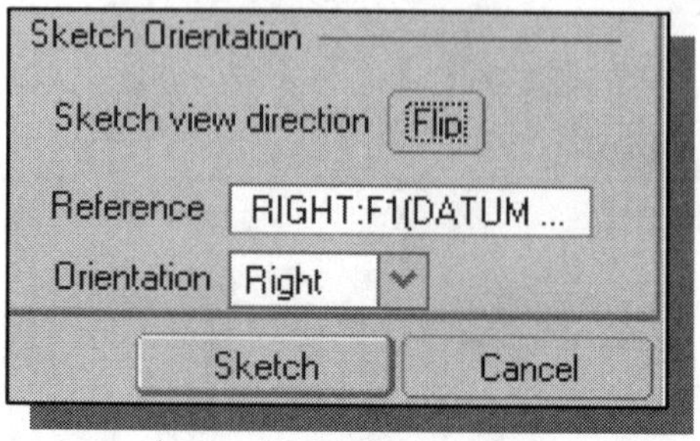

5. Confirm the **Right** datum plane is facing the **right** edge of the computer screen and pick **Sketch** to enter the *Sketcher* mode.

6. On your own, add the four outer edges of the base feature as additional references for the 2D sketch to be created.

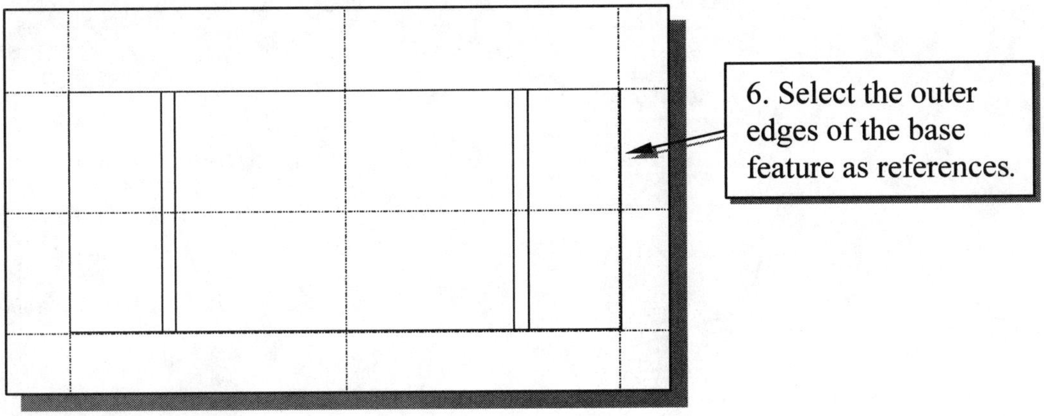

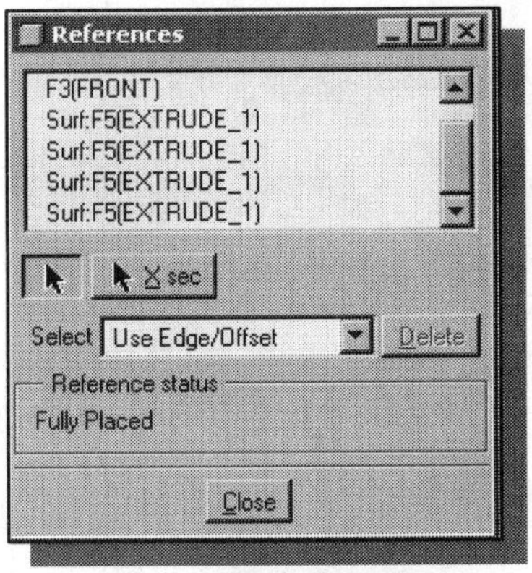

7. In the *References* window, confirm datum planes **Right**, **Front** and the four edges we selected are listed as references for the 2D sketch. Click the **Close** button to accept the settings

8. In the *Sketcher* toolbar, select **Circle** as shown. The default option is to create a circle by specifying the center point and a point through which the circle will pass.

9. Create the three circles as shown. Note all three centers are on the horizontal axis and tangent to the outer edges of the base feature. No dimension is needed for the middle circle since the center is aligned to the origin.

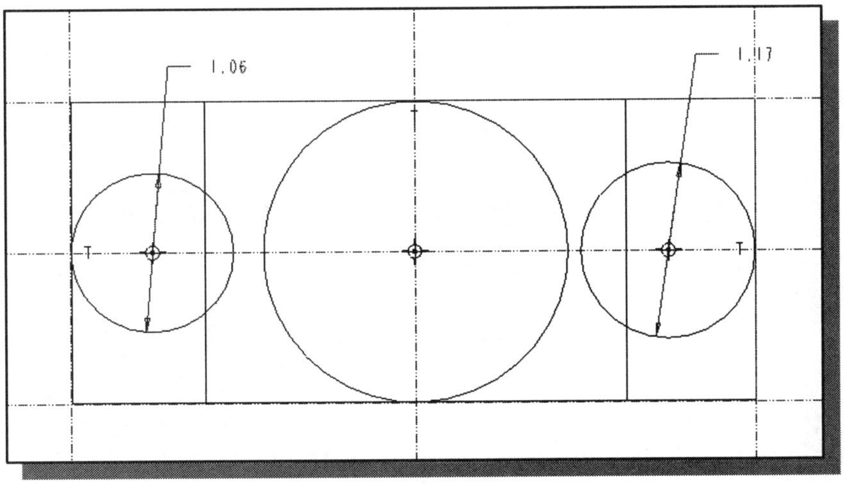

10. Add the four tangent lines and set the two smaller circles to be the same diameter as shown. (Hint: Use the *tangent* constraint.)

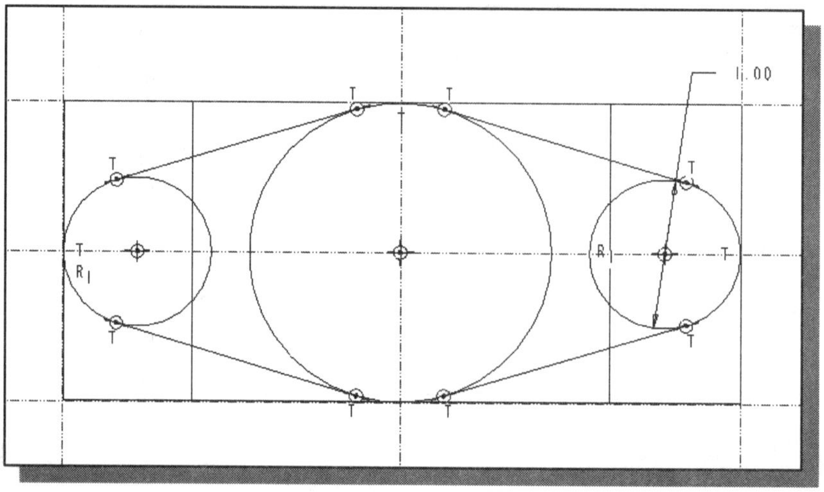

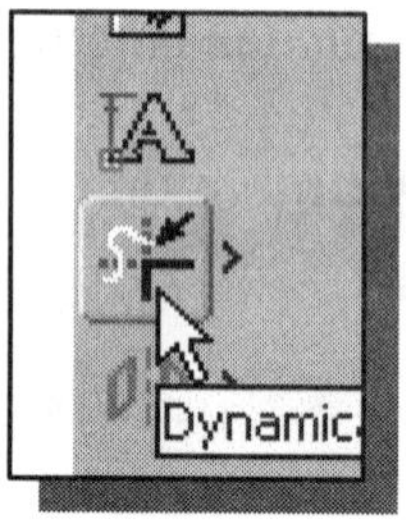

11. Select **Dynamic Trim** in the *Sketcher* toolbar. The Dynamic Trim function allows us to quickly remove unwanted portions of entities by simply clicking on them.

12. On your own, modify the sketched geometry (radius **0.5**) as shown in the figure below.

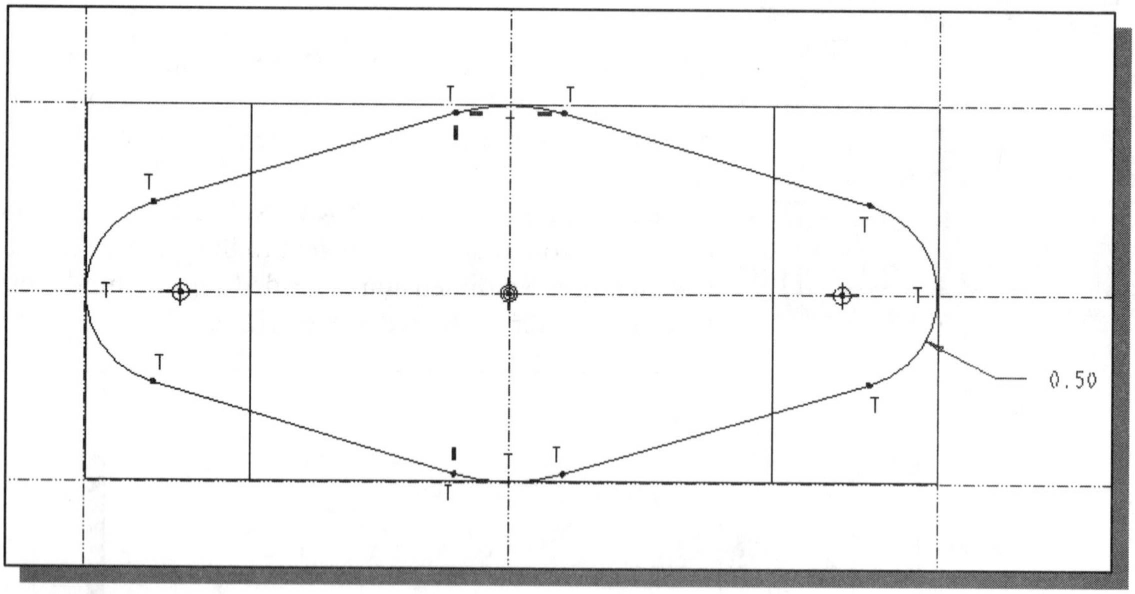

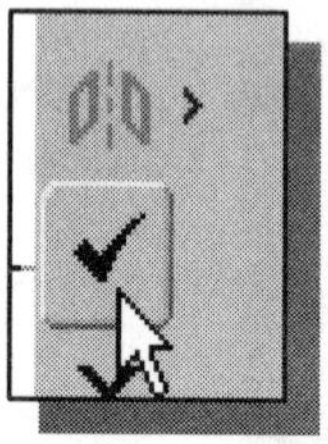

13. Click on the **Accept** icon to accept the completed section.

14. In the feature option menu, switch ***on*** the **Remove Material** option and set the *extrude* option to **Thru All** as shown.

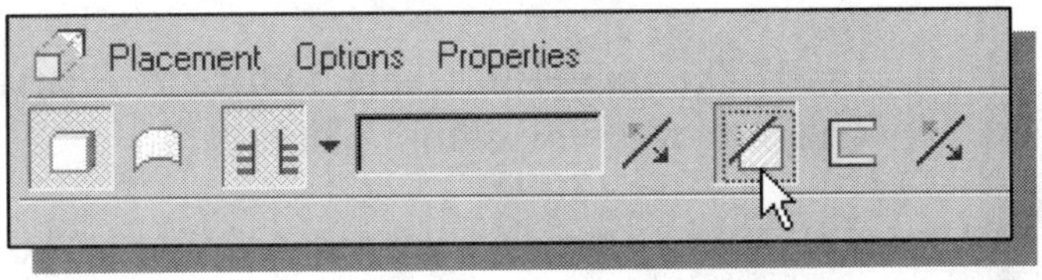

15. On your own, use the two **Flip-Direction** options to set the extrude *direction* and the **Remove Material** direction as shown in the below figure.

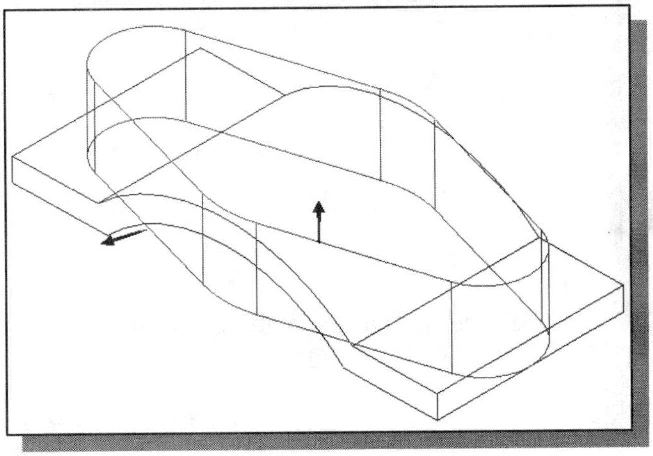

16. On your own, use the **Preview** option in the feature option menu to examine the feature.

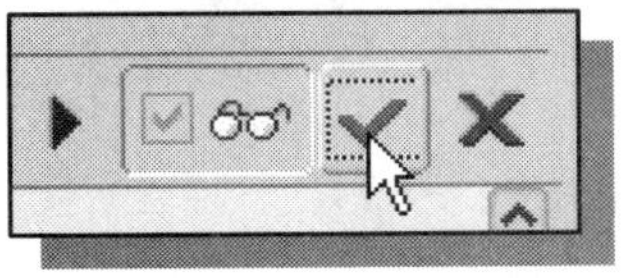

17. Pick **Accept** in the feature option menu to create blended feature.

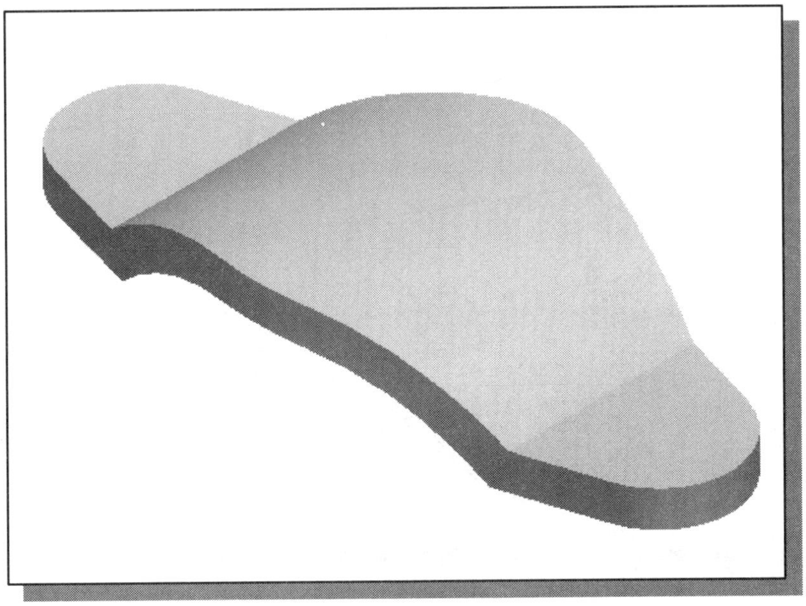

Creating a Reference Plane

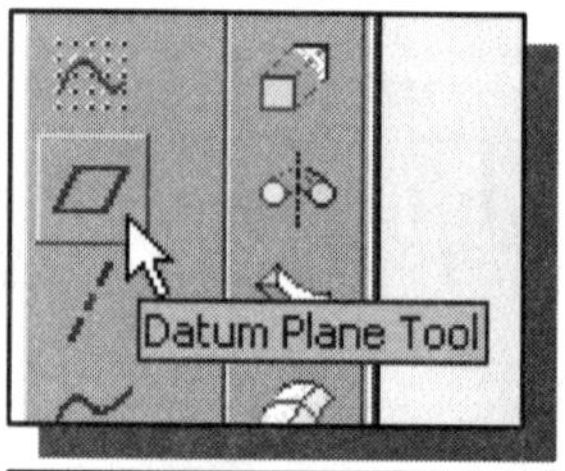

1. In the *Datum* toolbar, select **Datum Plane Tool** to start the creation of a new datum plane.

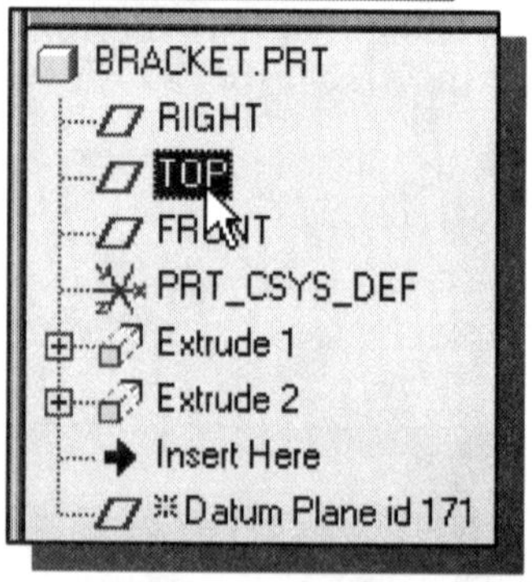

2. Pick datum plane **TOP** in the *Model Tree* window.

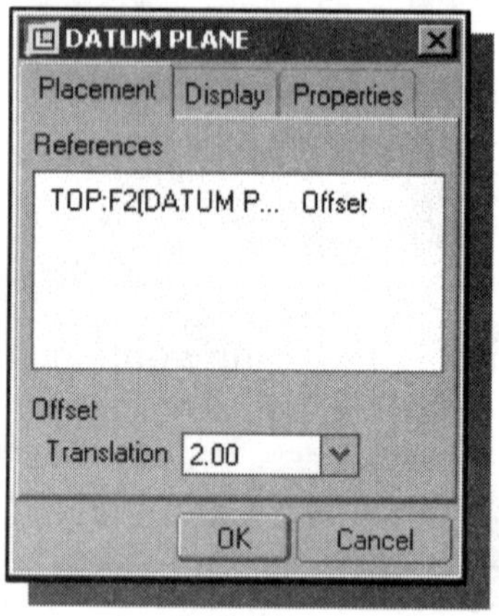

3. Place the datum plane at an offset distance of **2.0** from datum plane **TOP**.

4. Pick **OK** in the section dialog box.

❖ Datum plane DTM1 is exactly 2.0 inches above datum plane TOP.

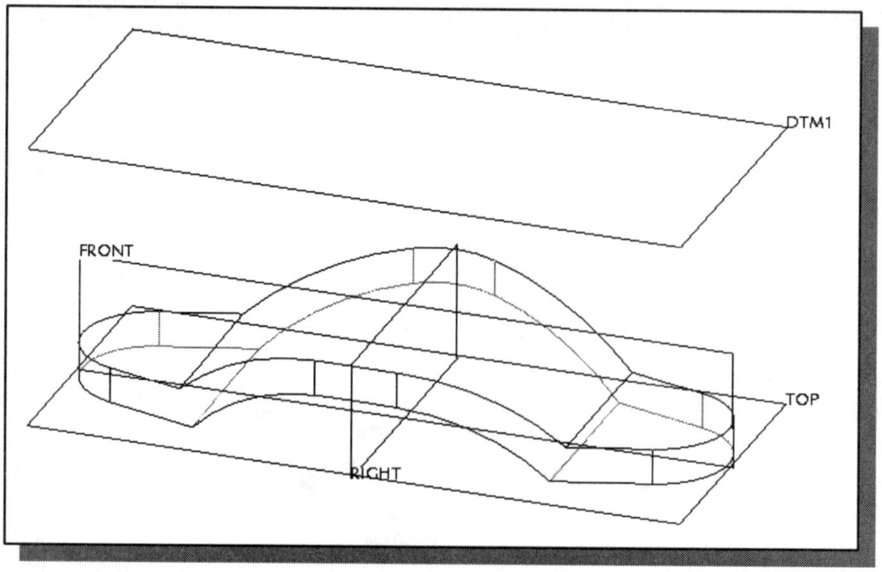

❖ In *Pro/ENGINEER*, **datum planes** can be constructed with the following options:

- **Through**: Create a datum plane through the selected references.
- **Normal**: Create a datum plane perpendicular to the selected reference.
- **Parallel**: Create a datum plane parallel to the selected reference.
- **Offset**: Create a datum plane by specifying a distance or a specific location away from the selected reference.
- **Angle**: Create a datum plane by specifying an angle relative to the selected reference.
- **Tangent**: Create a datum plane tangent to the selected reference.

Creating Another Extrusion Feature

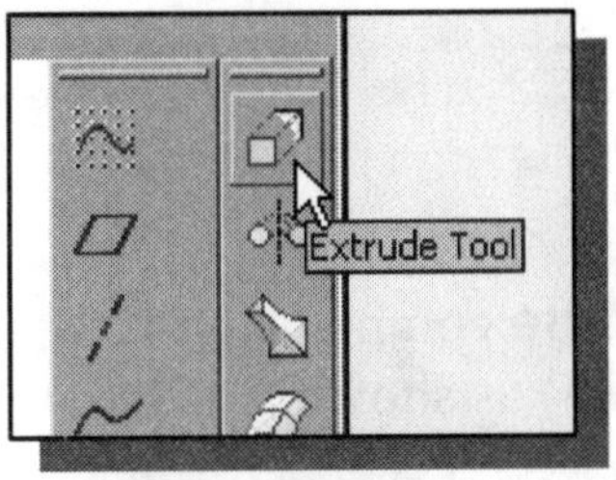

1. In the *Feature* toolbars (toolbars aligned to the right edge of the main window), select the **Extrude Tool** option as shown.

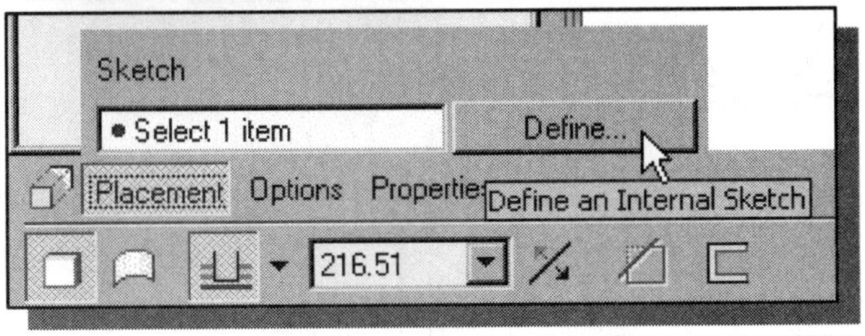

2. Click the **Placement** option and choose **Define**.

3. Pick the **DTM1** datum plane as the sketch plane.

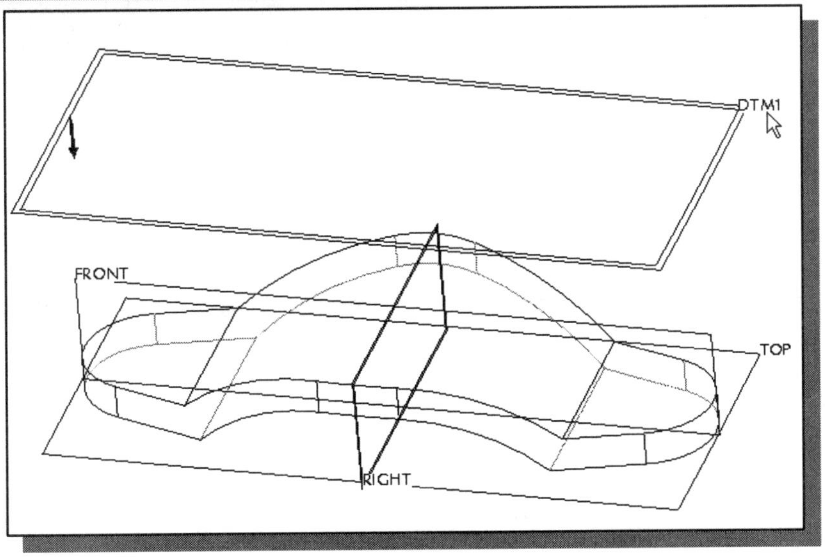

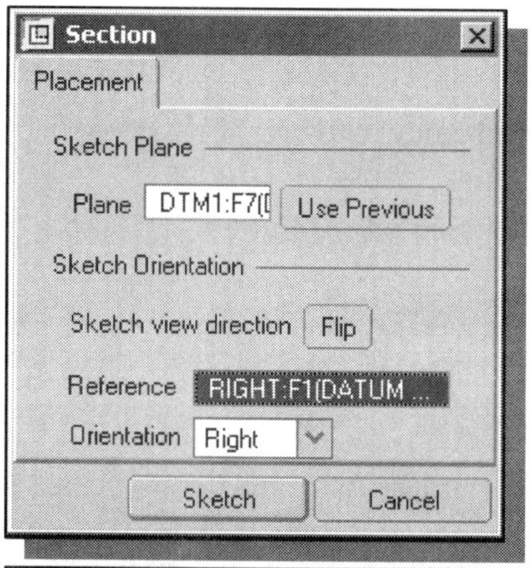

4. Confirm the view direction points downward and datum plane **Right** is set as the *referencing plane,* orientation **right,** as shown.

5. Click **Sketch** to enter the *Sketcher* mode.

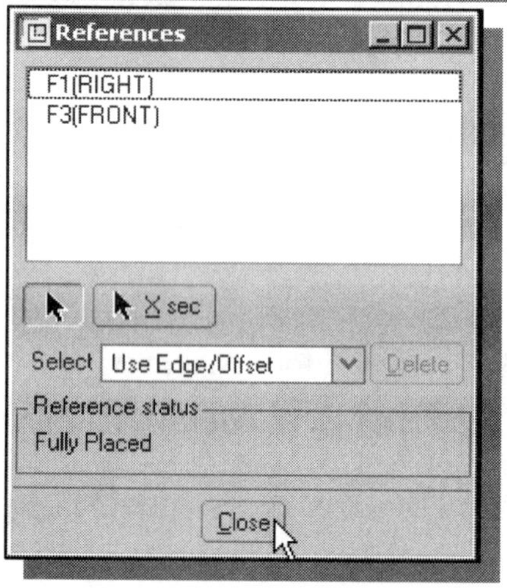

6. On your own, open the *References* window and confirm datum planes **Right** and **Front** are listed as references for the 2D sketch. Click the **Close** button to accept the settings

7. On your own, use the **Rectangle** command and create the rectangle (**1.5 x 0.75**) as shown.

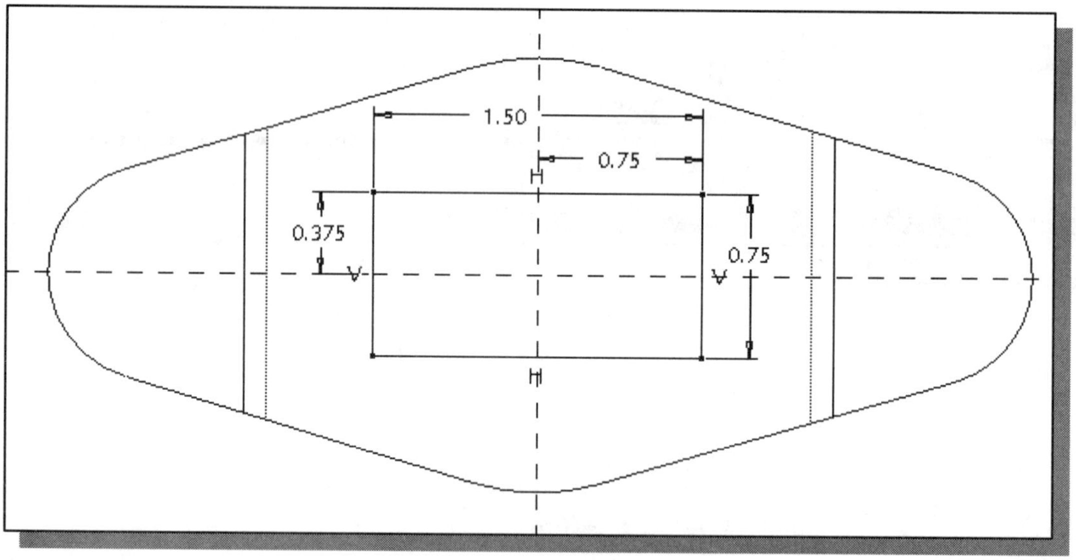

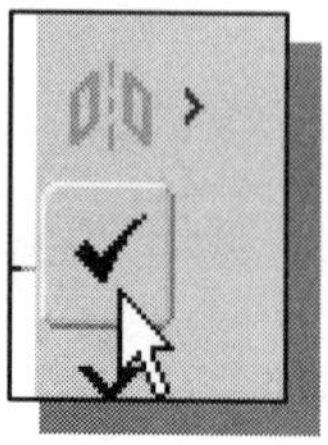

8. Click on the **Accept** icon to accept the completed section.

9. Click the **Flip** icon once to change the extrude direction so that the arrow points toward the solid model as shown in the figure below.

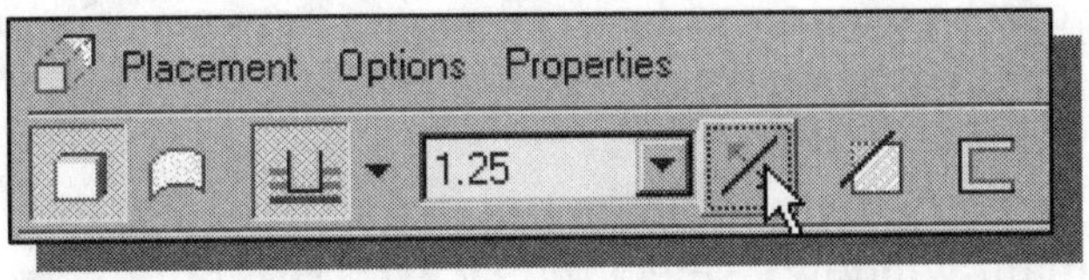

10. Set the *depth* option to **Extrude to Selected** as shown.

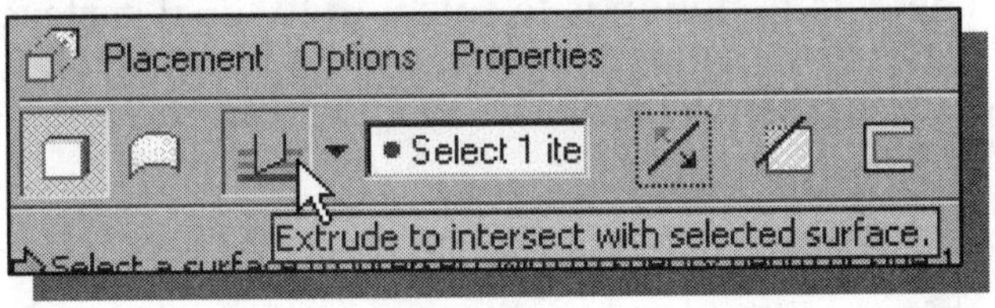

11. Pick the top surface of the solid model as shown.

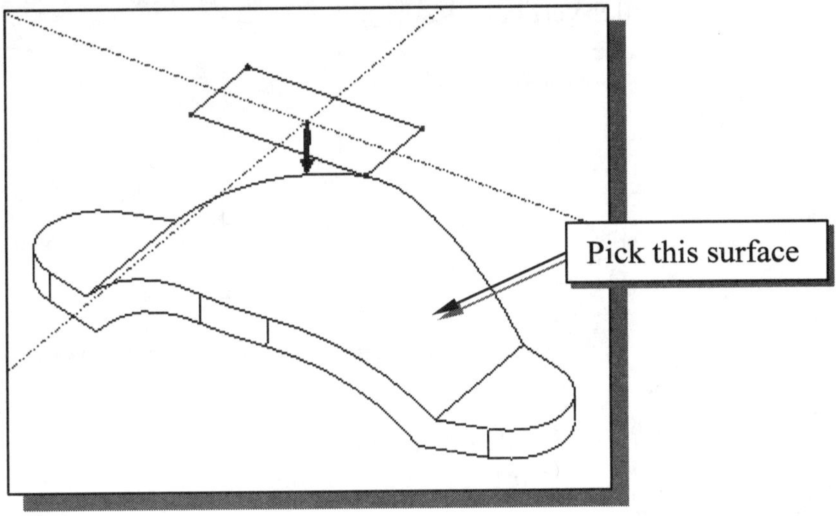

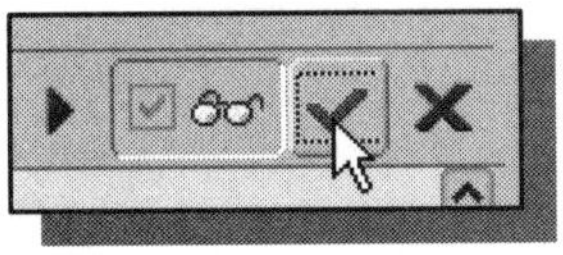

12. Pick **Accept** in the feature option menu to create the feature.

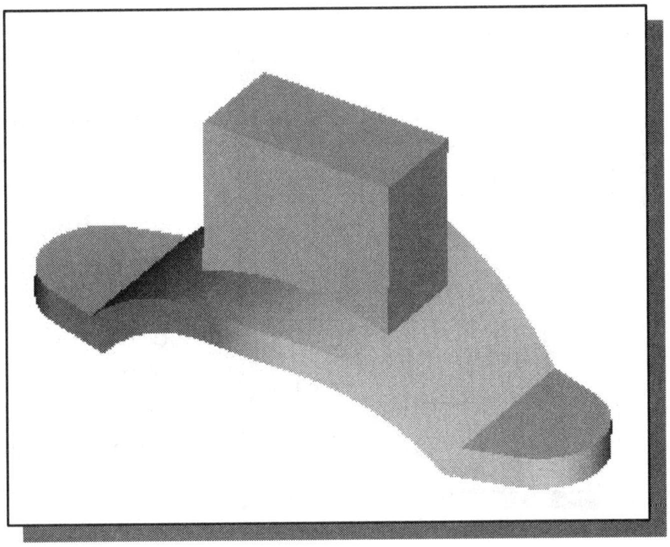

Using the Draft Tool

❖ The Draft feature can be used to add a draft angle, between -30° and +30°, to selected surfaces.

1. In the *Feature* toolbars (toolbars aligned to the right edge of the main window), select the **Draft Tool** option as shown.

2. The default option is to select individual surfaces. Hold down the [**Ctrl**] key and pick the two smaller vertical surfaces.

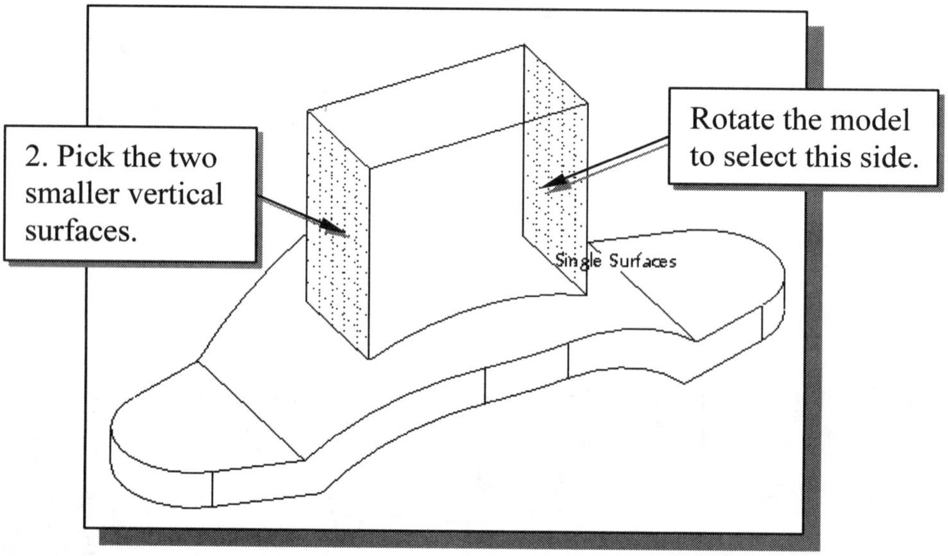

3. Inside the graphics area, press down the **right-mouse-button** to bring up the option menu.

4. In the popup list, select **Draft Hinges**.

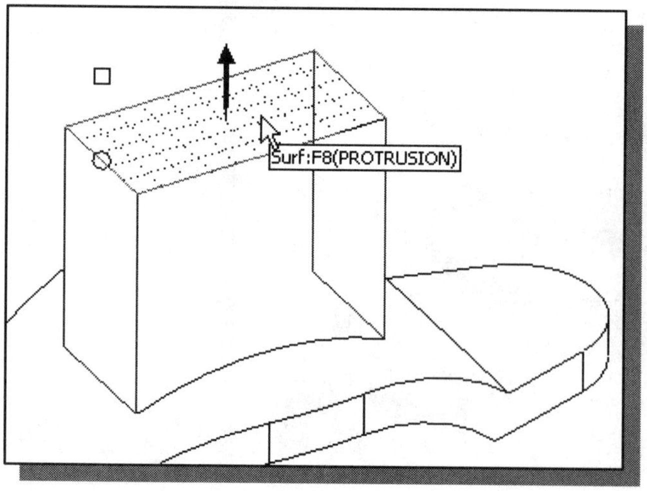

5. Pick the top surface of the last feature as the *draft hinge*.

❖ Draft hinges are curves on the draft surfaces that the surfaces are pivoted about. Draft hinges can be defined by selecting a plane, in which case the draft surfaces are pivoted about their intersection with this plane.

6. In the graphics area, use the *adjust handle* and set the draft angle to **10°** as shown.

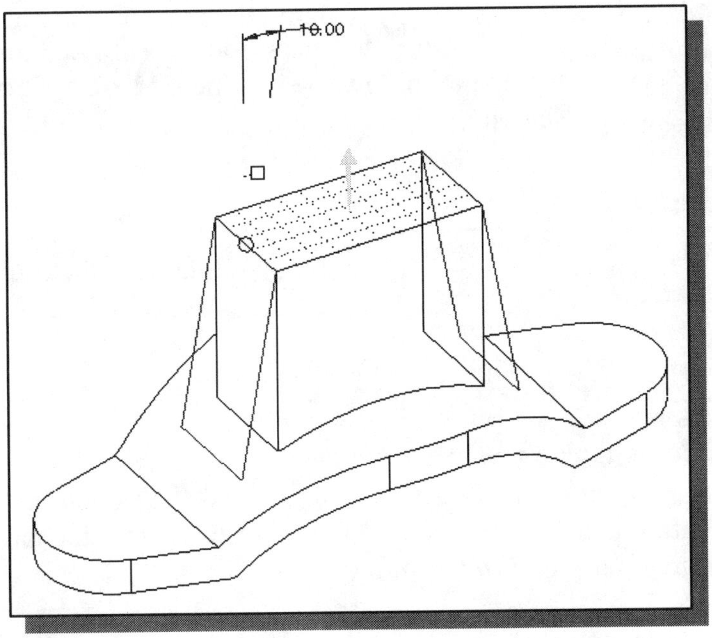

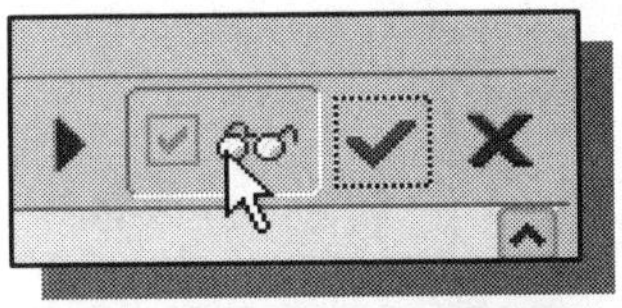

7. On your own, use the **Preview** option in the feature option menu to examine the feature.

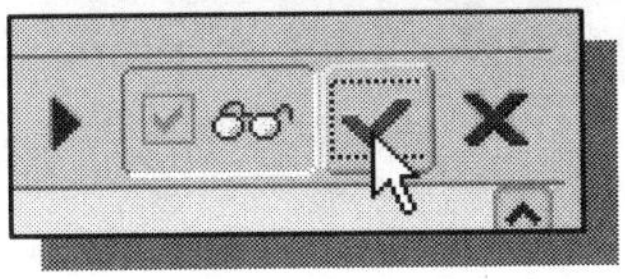

8. Pick **Accept** in the feature option menu to create drafted feature.

Create the Top Cylindrical Feature

1. In the *Feature* toolbars (toolbars aligned to the right edge of the main window), select the **Extrude Tool** option as shown.

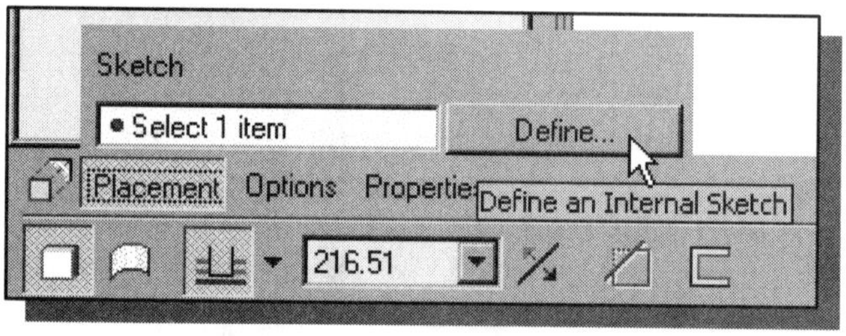

2. Click the **Placement** option and choose **Define**.

3. On your own, set up the **Front** datum plane as the sketch plane with the **Right** datum plane facing the **right** edge of the computer screen, and pick **Sketch** to enter the *Sketcher* mode.

4. On your own, open the references window..

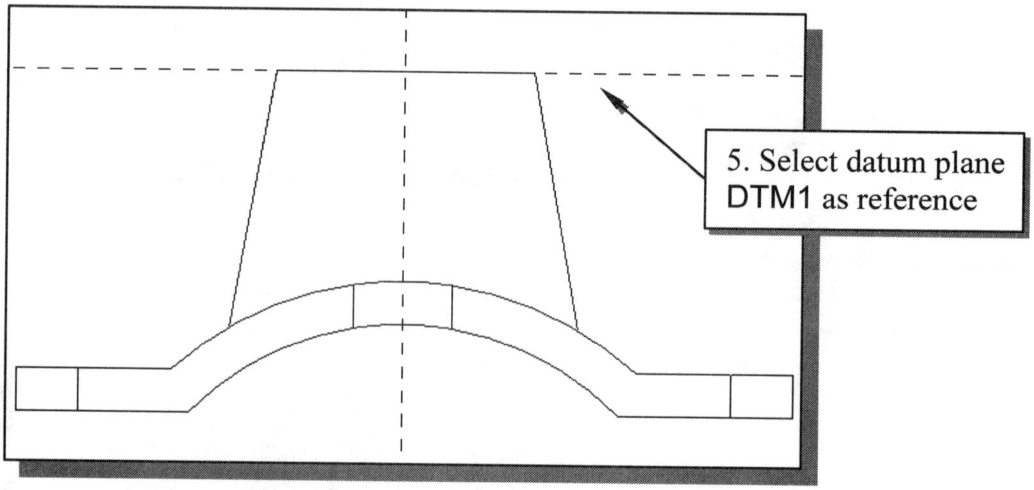

5. In the *References* window, remove the datum plane **Top** from the reference list; confirm datum planes **Right** and **DTM1** are listed as references for the 2D sketch. Click the **Close** button to accept the settings

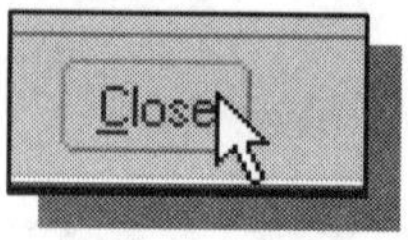

6. In the *Sketcher* toolbar, select **Circle** as shown. The default option is to create a circle by specifying the center point and a point through which the circle will pass.

7. Create a **Circle** (diameter **1.50**) that is aligned to the two reference planes as shown.

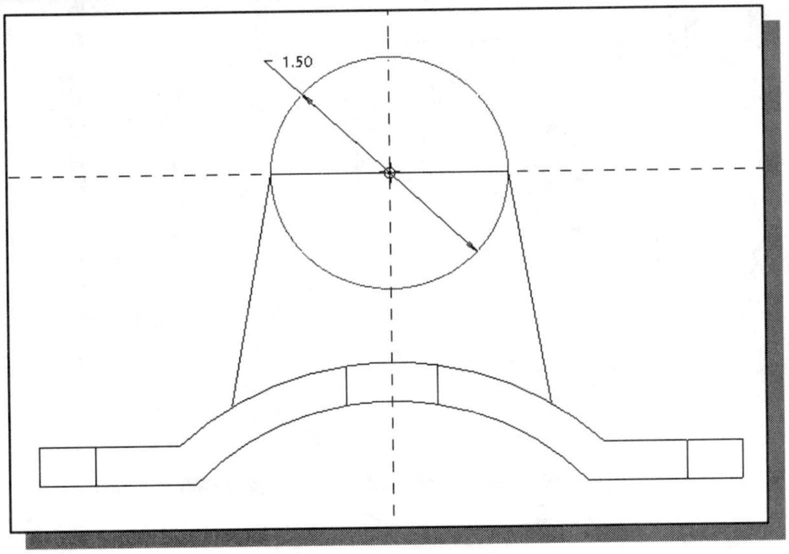

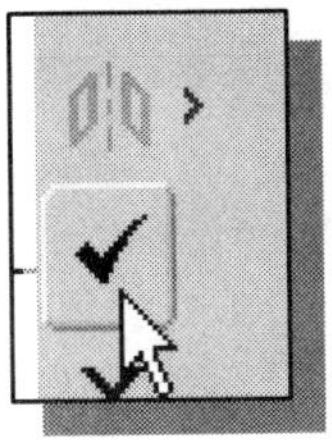

8. Click on the **Accept** icon to accept the completed section.

9. Complete the cylinder (length **1.75**) using the **Both Sides** extrusion option. The complete model should appear as shown.

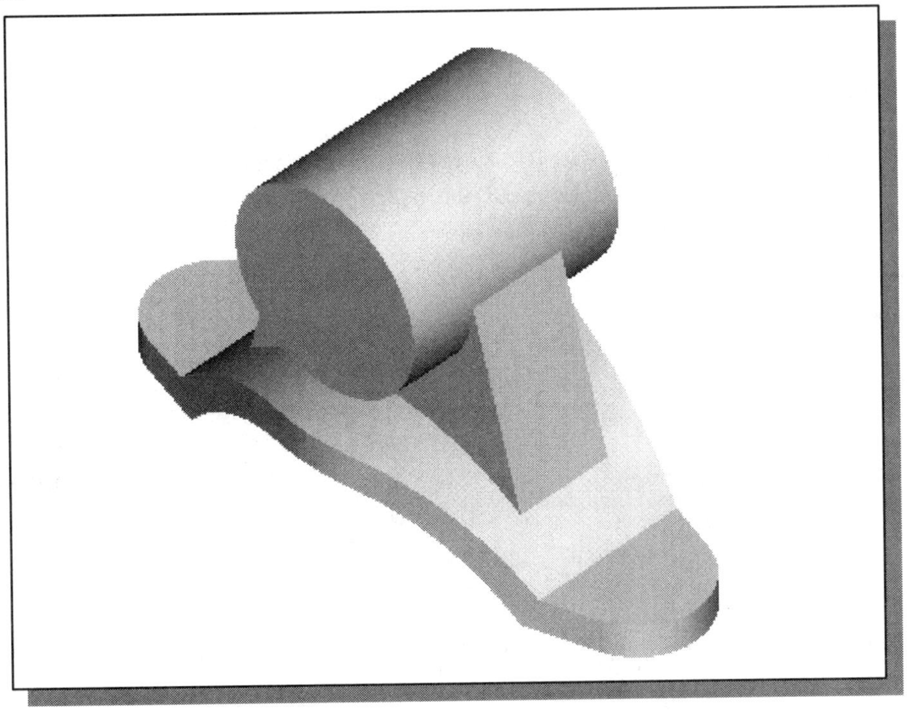

Create a Hole through the Cylinder

➤ On your own, create a **1.0** diameter hole through the cylinder we just created. The completed solid model should appear as shown in the figure below.

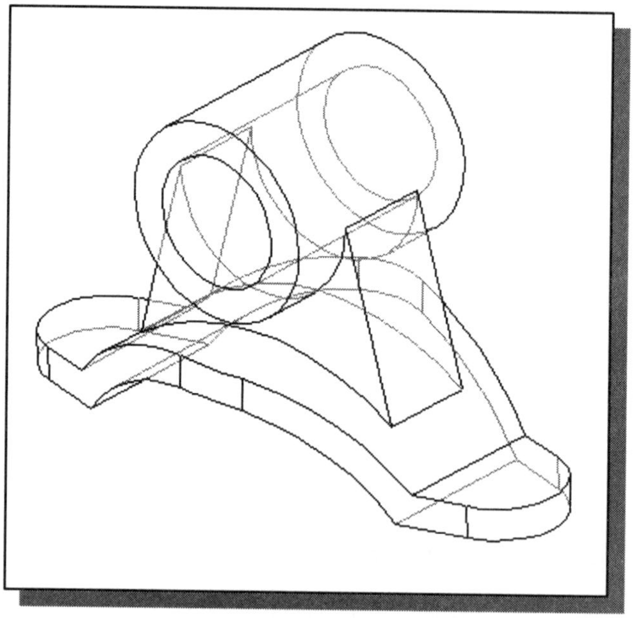

Create a Sketched Hole on the Base Feature

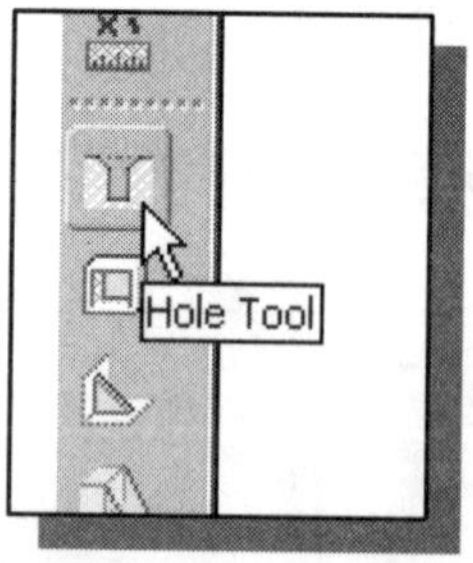

1. In the *Feature* toolbars (toolbars aligned to the right edge of the main window), select the **Hole Tool** option as shown.

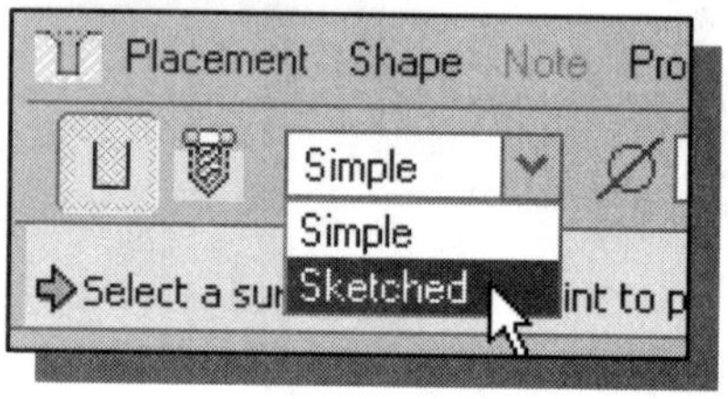

2. In the *Feature Option Dashboard*, select the **Sketched** option as shown.

3. In the *Feature Option Dashboard*, click **Activate Sketcher**.

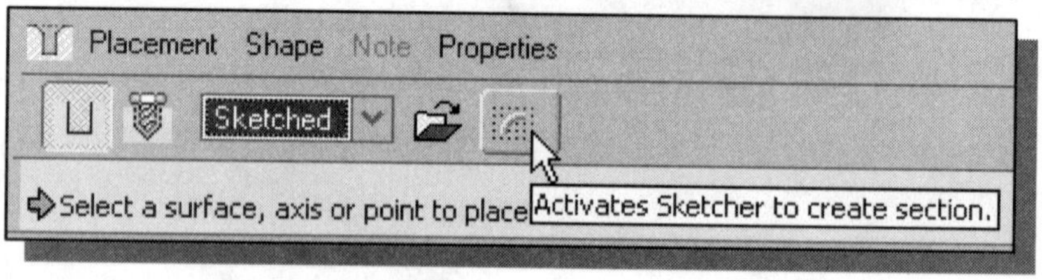

❖ For a sketched hole feature we can use existing section files or create the sketch on the fly.

❖ The *sketched hole* option allows us to create a 2D section, which is then used to revolve about an axis and form the hole (similar to a revolved cut feature).

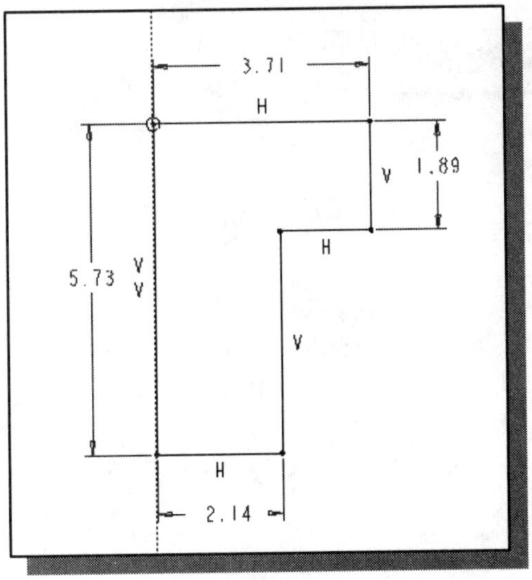

4. Create a rough sketch as shown (six regular lines and a vertical **centerline**).

5. Create and modify the dimensions as shown in the figure below.

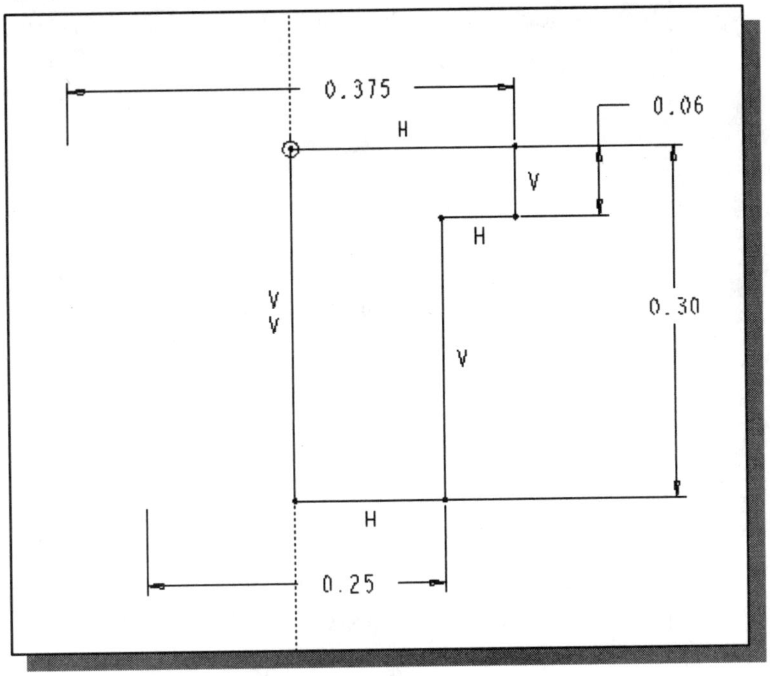

➢ **To dimension a diameter dimension for a revolved section: (1) pick the entity; (2) pick the centerline to use as the axis of revolution; (3) pick the entity again; then (4) place the dimension.**

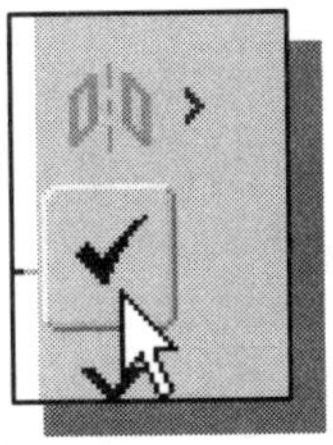

6. Click on the **Accept** icon to accept the completed section.

7. Pick the top face of the base as shown as the *primary placement reference.*

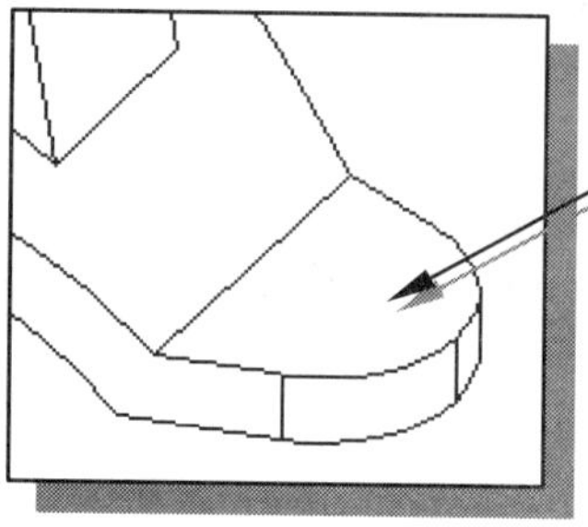

8. Inside the graphics area, press down the **right-mouse-button** to bring up the option menu.

9. In the popup list, select **Secondary References Collector**.

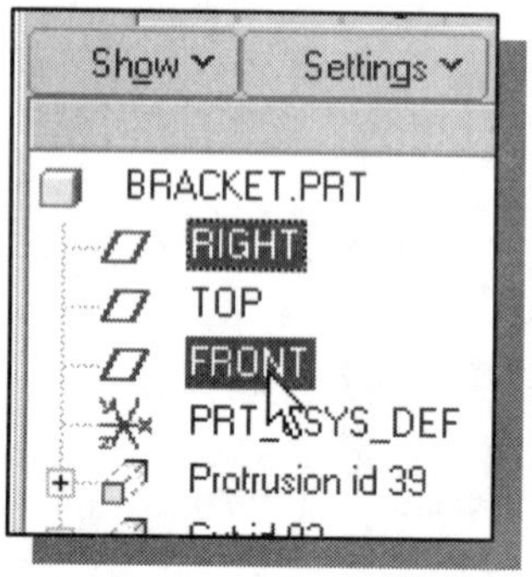

10. Inside the *Model Tree* area, press down the [**Ctrl**] key and select datum planes **Right** and **Front** as the two *secondary references.*

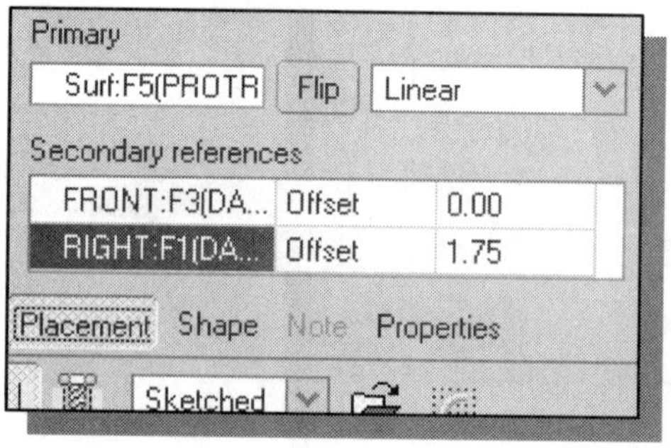

11. On your own, setup the **Offset** *distances* to **0.00** and **1.75** as shown.

12. Pick **Accept** in the **Hole** feature menu to create the placed hole feature.

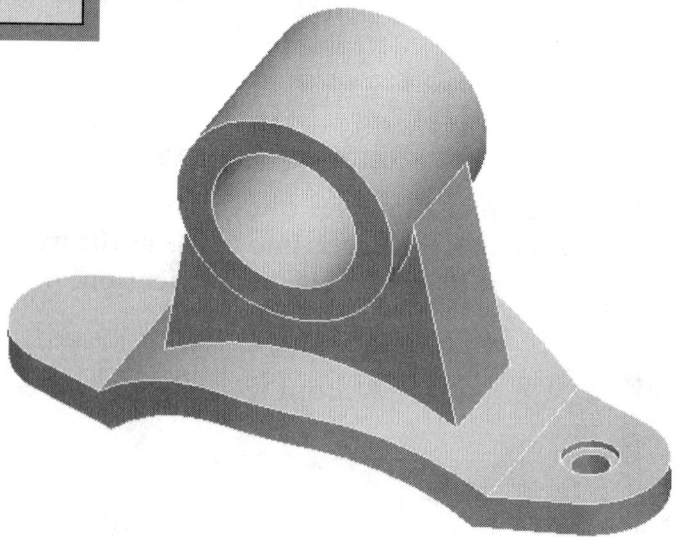

Using the Copy Function to Create a Mirror Feature

1. Pick **Feature Operations** in the **Edit** pull-down menu.

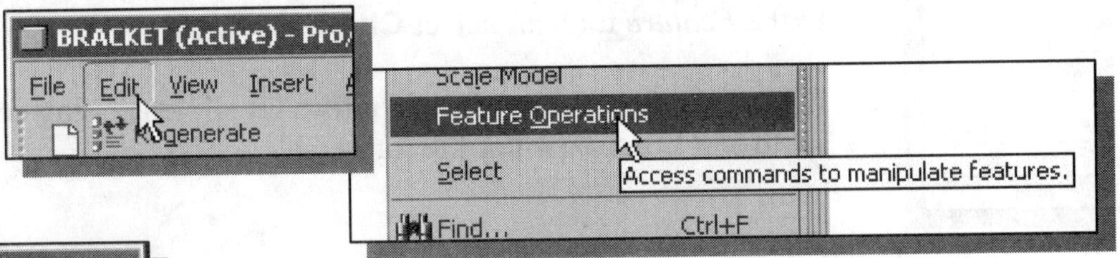

2. In the **FEAT** menu, select **COPY**.

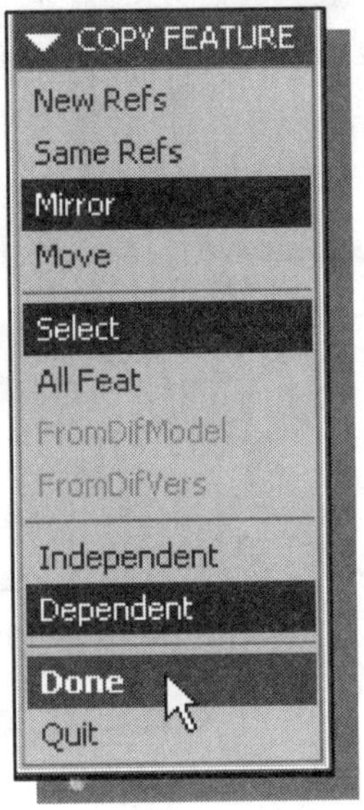

3. In the **COPY FEATURE** menu, select

   Mirror → Select → Dependent → Done

4. Pick the last feature we just created.

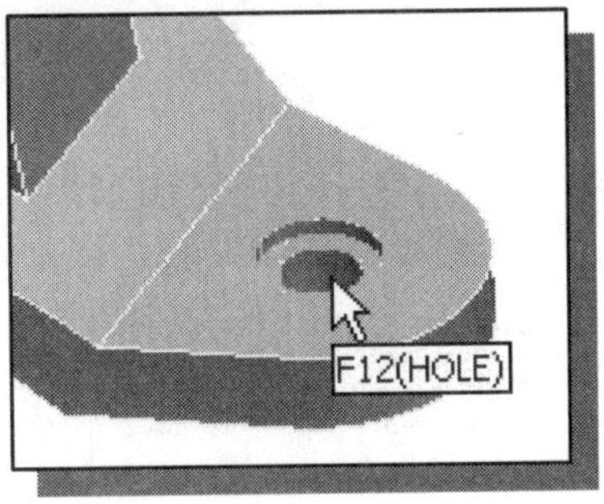

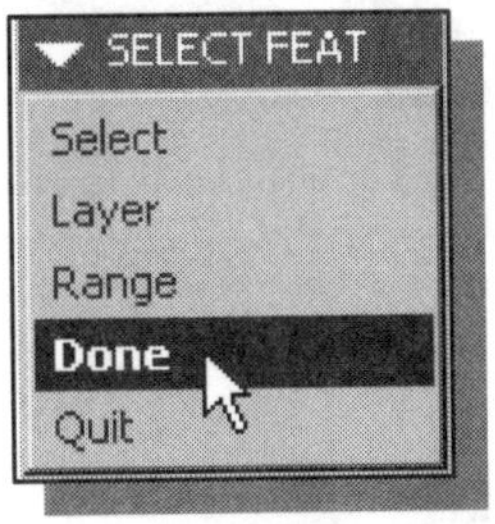

5. In the **Select Feat** menu, pick **Done** to continue.

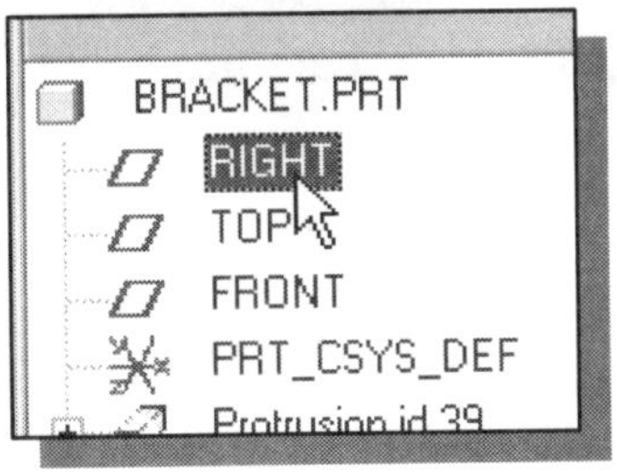

6. Pick datum plane **Right** as the *reference plane* to create the mirrored feature.

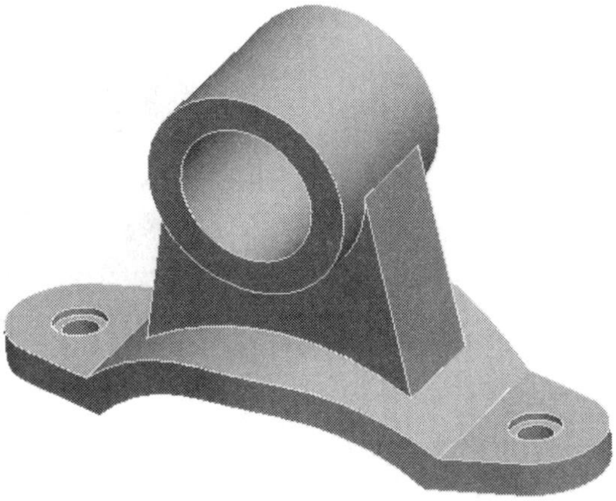

7. Click **Done** in the **Feature** menu to end the **Feature Operations** options.

Creating Chamfers

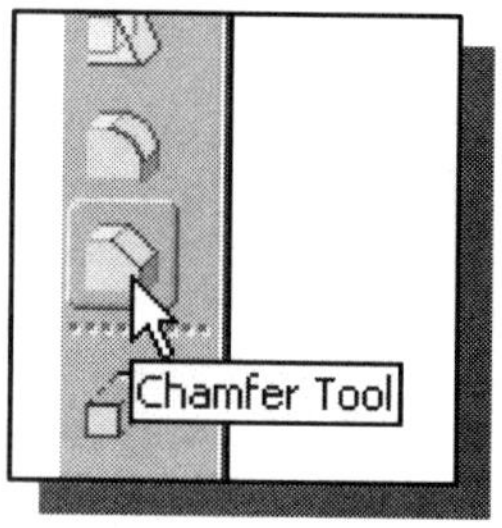

1. In the *Feature* toolbar, select **Chamfer Tool**.
2. In the chamfer feature menu, confirm the dimensioning scheme is set to D x D as shown in the figure below.
3. In the chamfer feature menu, enter **1/16** as the chamfer *dimension*.

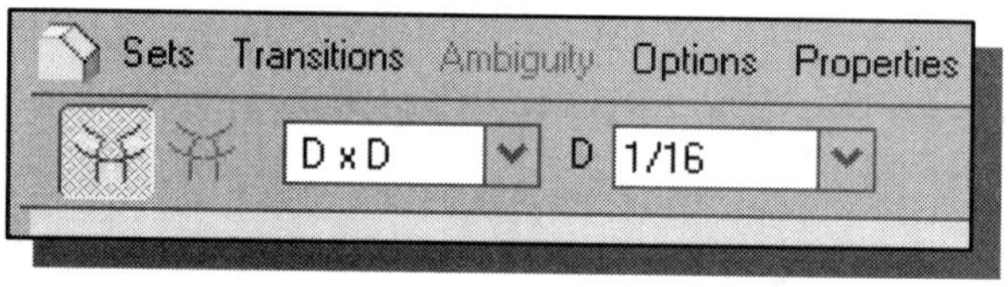

4. Pick the front and back inside circles as shown.

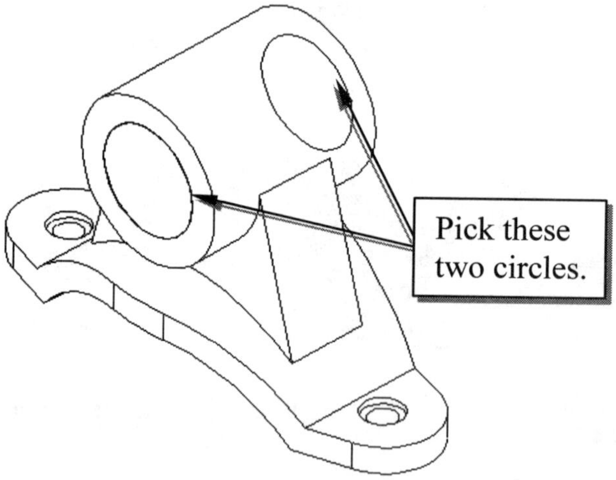

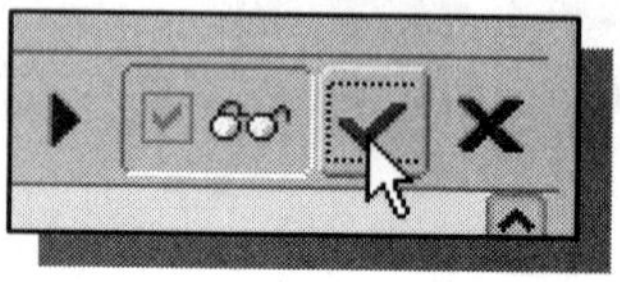

5. Pick **Accept** in the feature option menu to create the chamfer feature.

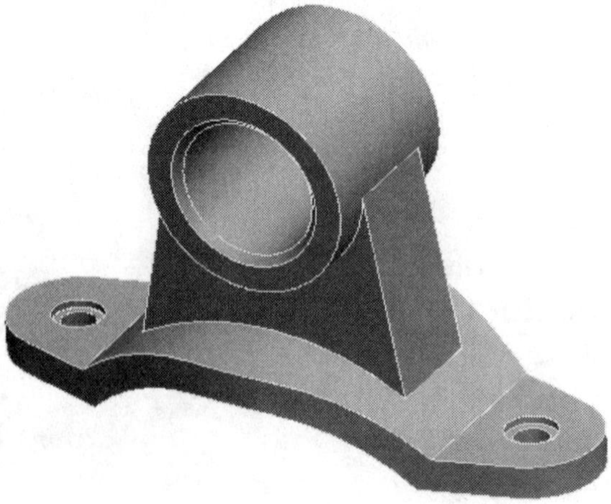

Adding *Simple* Rounds and Fillets – Edges

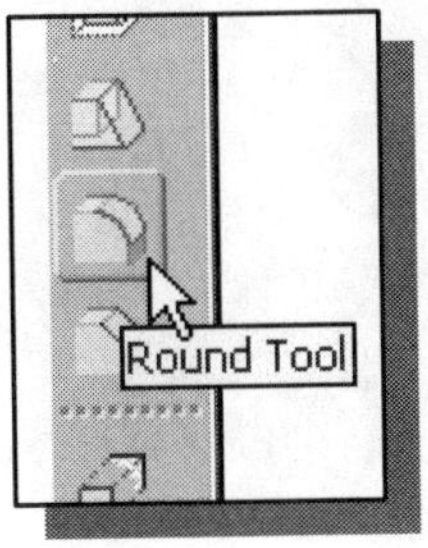

1. In the *Standard* toolbar, select **Round Tool**.

2. In the round option menu, set the *radius* to **0.125** as shown.

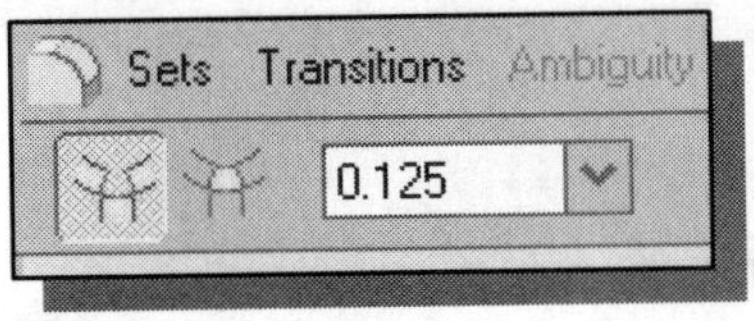

3. Pick the top edges of the base feature as shown below.

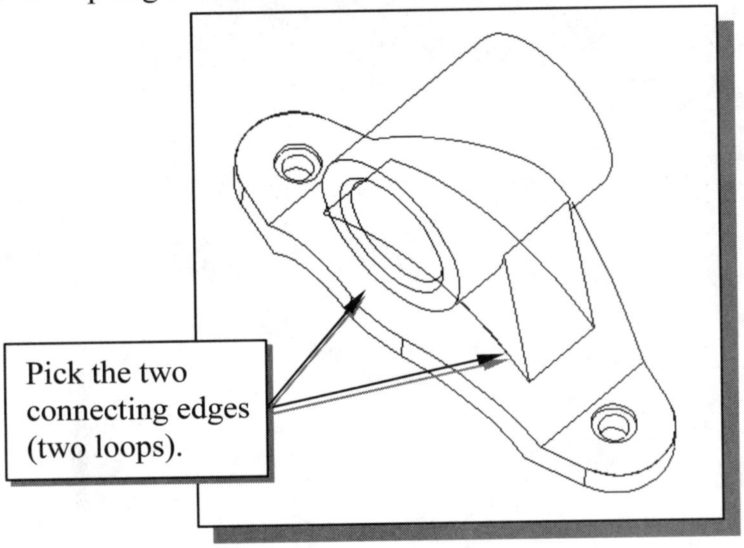

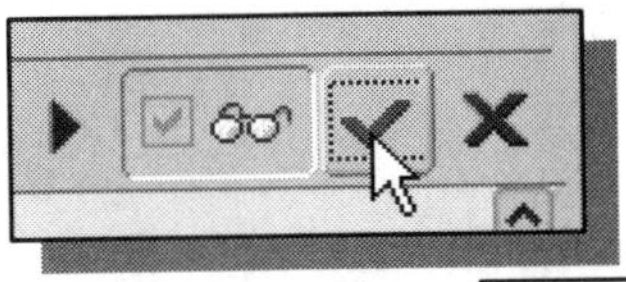

4. Pick **Accept** in the feature option menu to create the round feature.

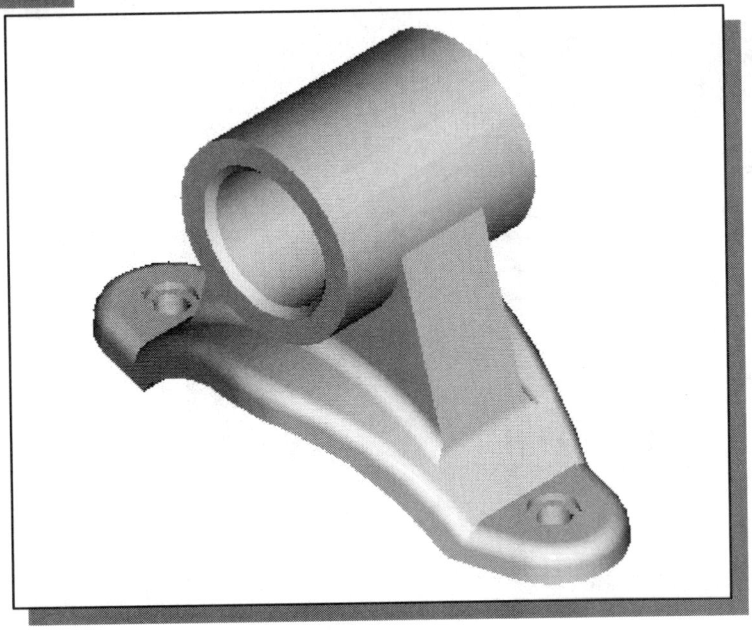

Adding *Advanced* Rounds and Fillets – surfaces

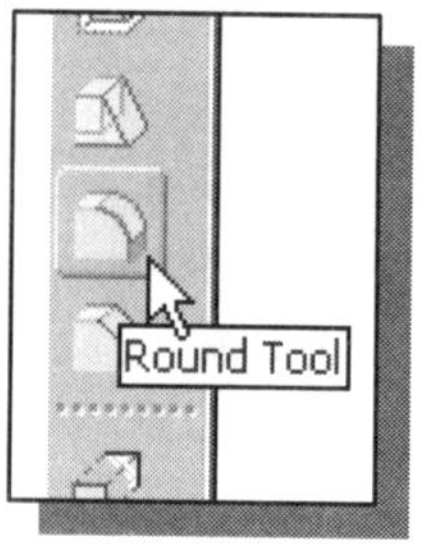

1. In the *Standard* toolbar, select **Round Tool**.

2. In the round option menu, set the *radius* to **0.125** as shown.

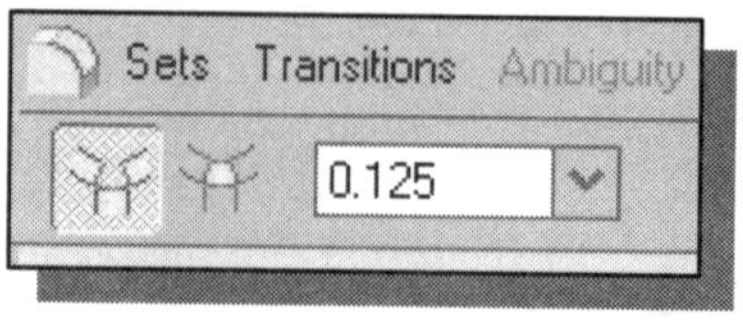

❖ We will create multiple sets of rounds in between the four vertical surfaces.

3. Pick the two surfaces as shown. (Hint: use the [**Ctrl**] key.)

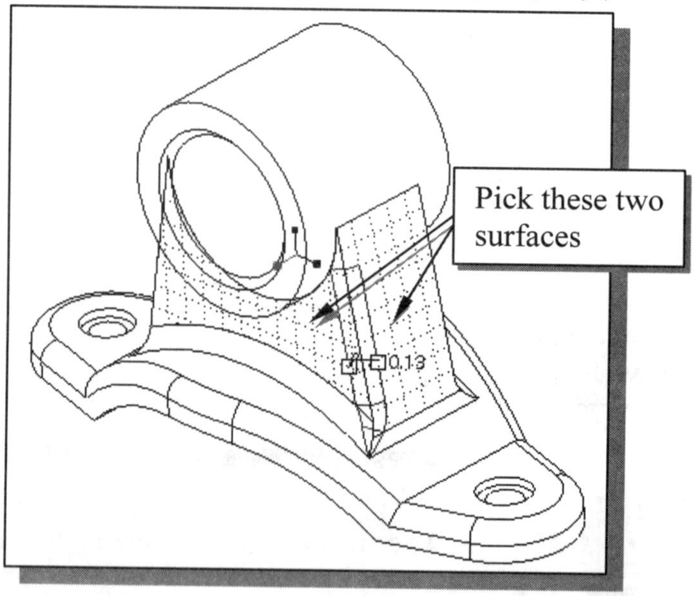

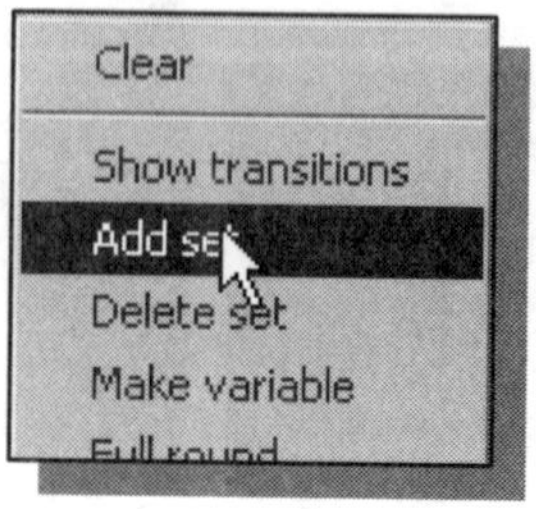

4. Inside the graphics area, press down the **right-mouse-button** to bring up the option menu.

5. In the popup list, select **Add Set**.

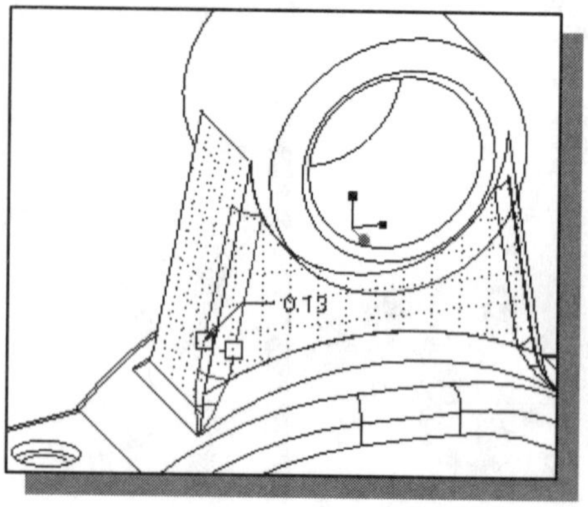

6. Pick the two adjacent surfaces to form the round as shown.

7. Repeat the above steps and create the four rounded corners as shown in the figure below.

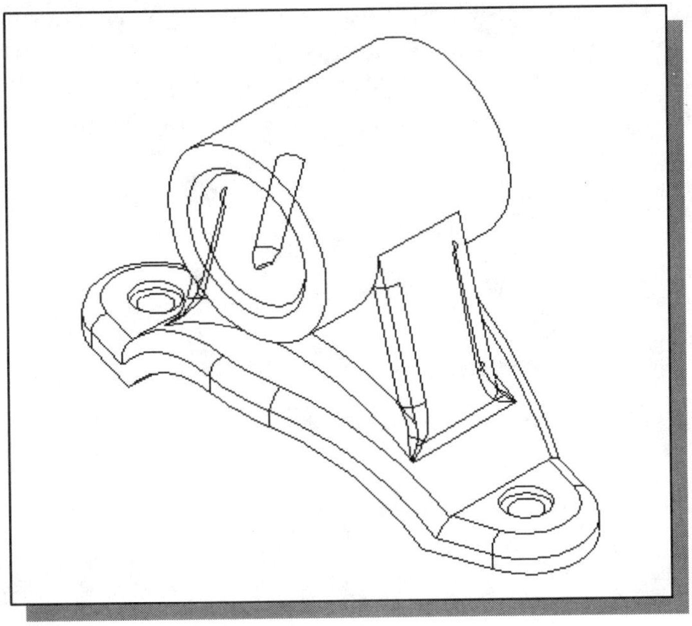

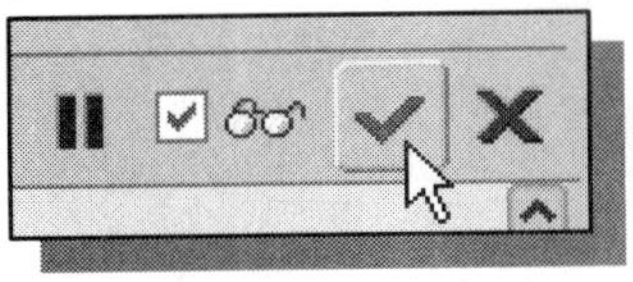

8. Pick **Accept** in the feature option menu to create the feature.

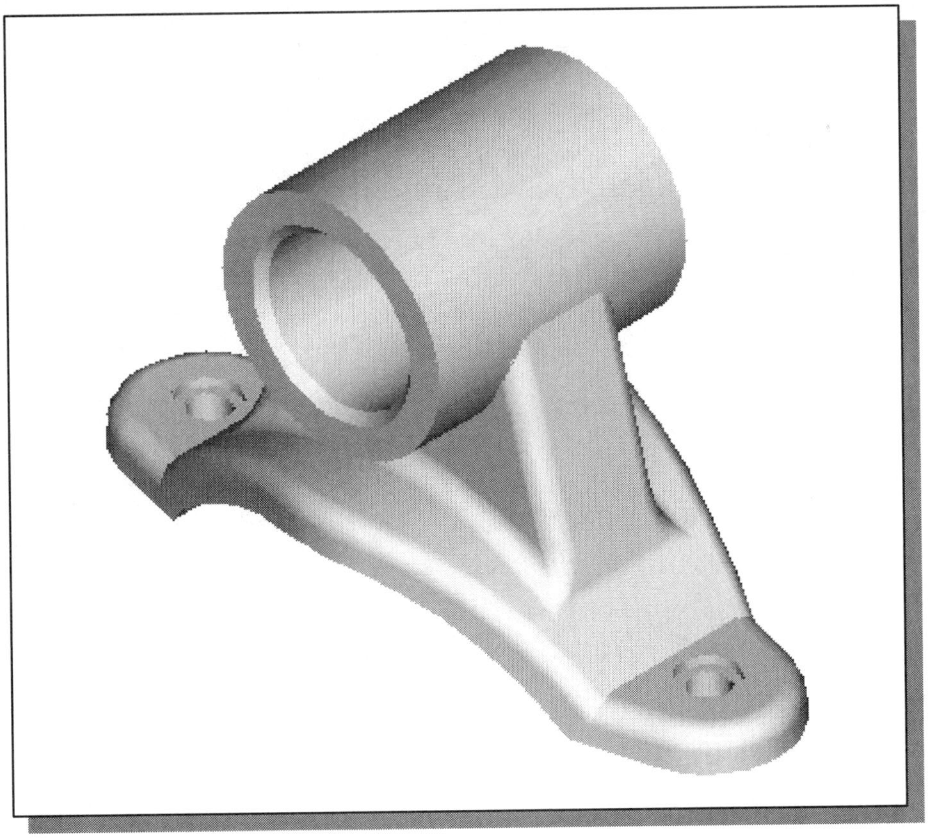

Adding the Last Feature

❖ The last feature will be the fillets around the cylinder. Note that 3D rounds and fillets are complex geometry and it may be necessary to create them individually.

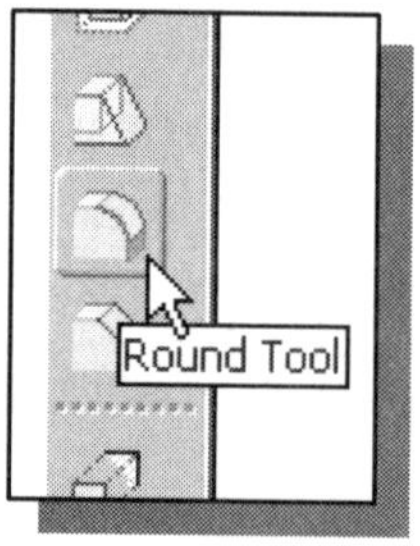

1. In the *Standard* toolbar, select **Round Tool**.

2. In the round option menu, set the *radius* to **0.125** as shown.

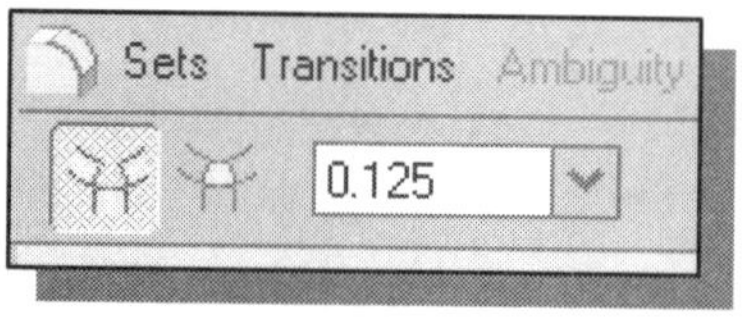

3. Pick the two surfaces (one cylindrical and one vertical) as shown.

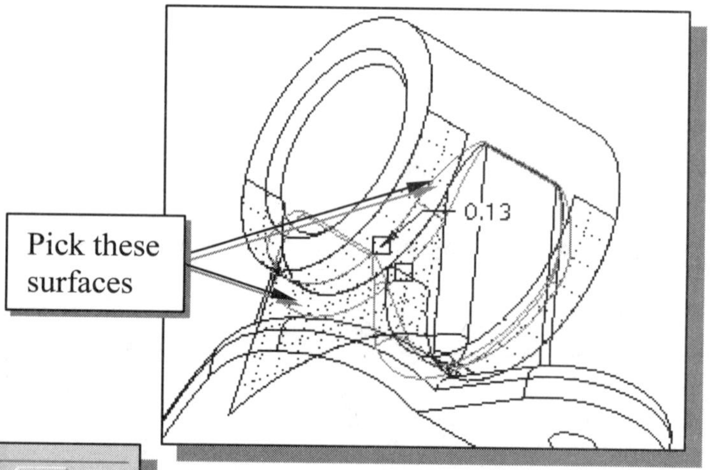

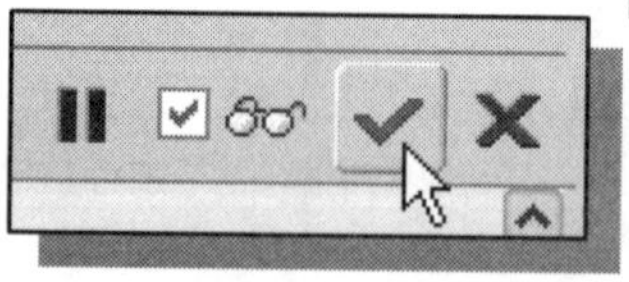

4. Pick **Accept** in the feature option menu to create the feature.

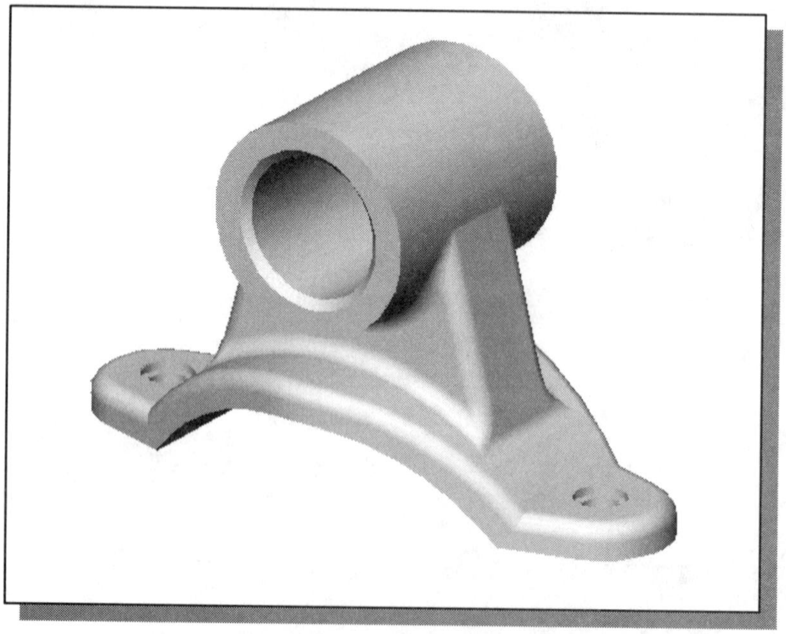

Changing Model Color

❖ In *Pro/ENGINEER*, changing model color is a two-step procedure: (1) define the color we want, and (2) apply the color to the model.

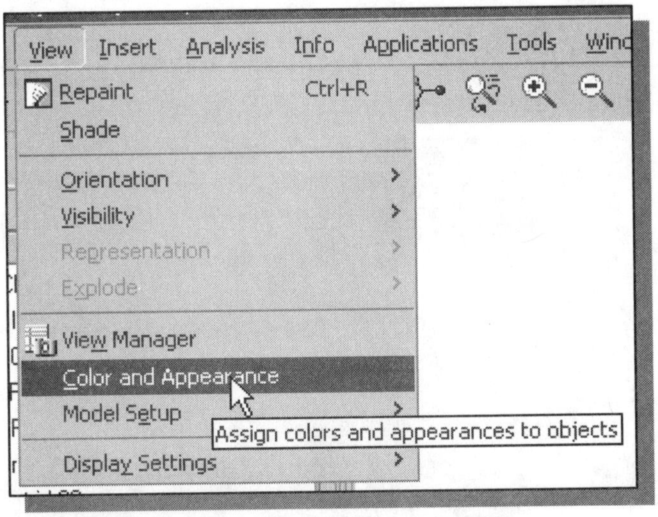

1. Select **Color and Appearance** in the **View** pull-down menu.

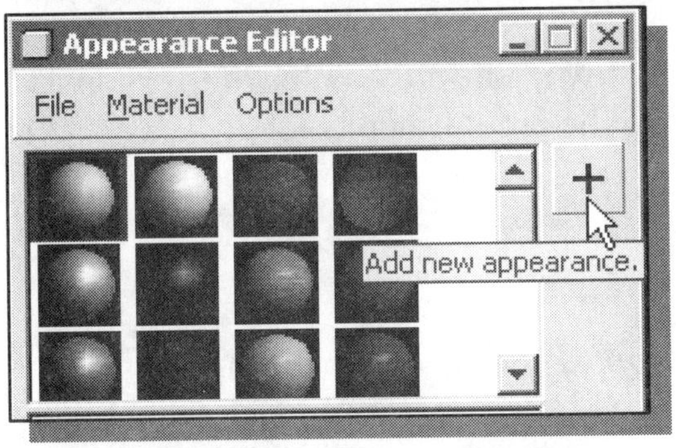

2. In the *Appearance Editor* window, several pre-defined colors are available. Pick the **Add** icon to define a new appearance setting.

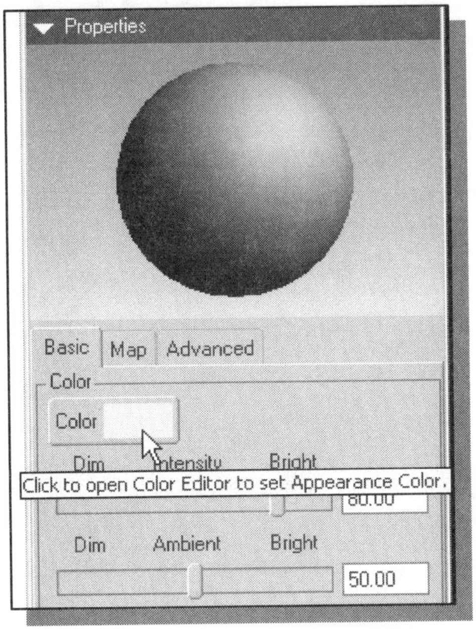

3. In the *Appearance Properties* window, click the **Edit Color** button to open the *Color Editor*.

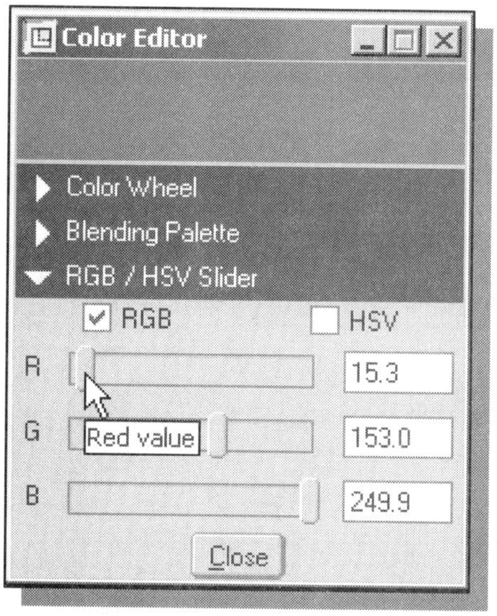

4. In the *Color Editor* window, control the amount of **red**, **green** and **blue** in the color being defined by dragging the RGB sliders.

5. Pick **Close**, in the *Color Editor* window, to accept the settings.

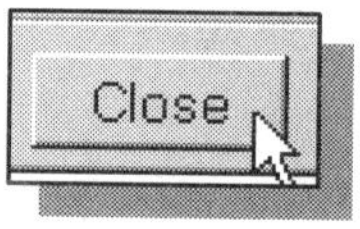

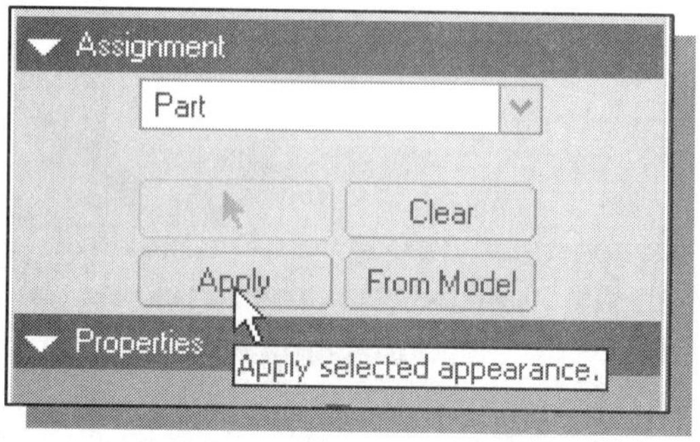

6. In the *Assignment* window, pick the **APPLY** button to apply the new color to the part.

7. On your own, adjust the color of the model by using the pre-defined colors.

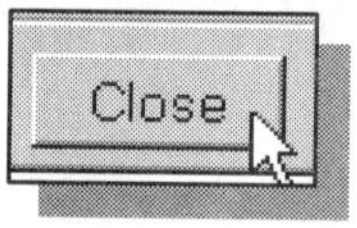

8. Pick the **Close** icon to exit the *Appearance Editor* window.

Saving the Part

❖ From the icon panel, select the **File** pull-down menu and pick the **Save** option. Notice that you can also use the **Ctrl-S** combination (press down the [**Ctrl**] key and hit the [**S**] key once) to save the part. In the message window, the name of the file is displayed. Press the **ENTER** key to accept the file name, ***Bracket***.

- On your own, create and complete a multi-view part drawing of the design.

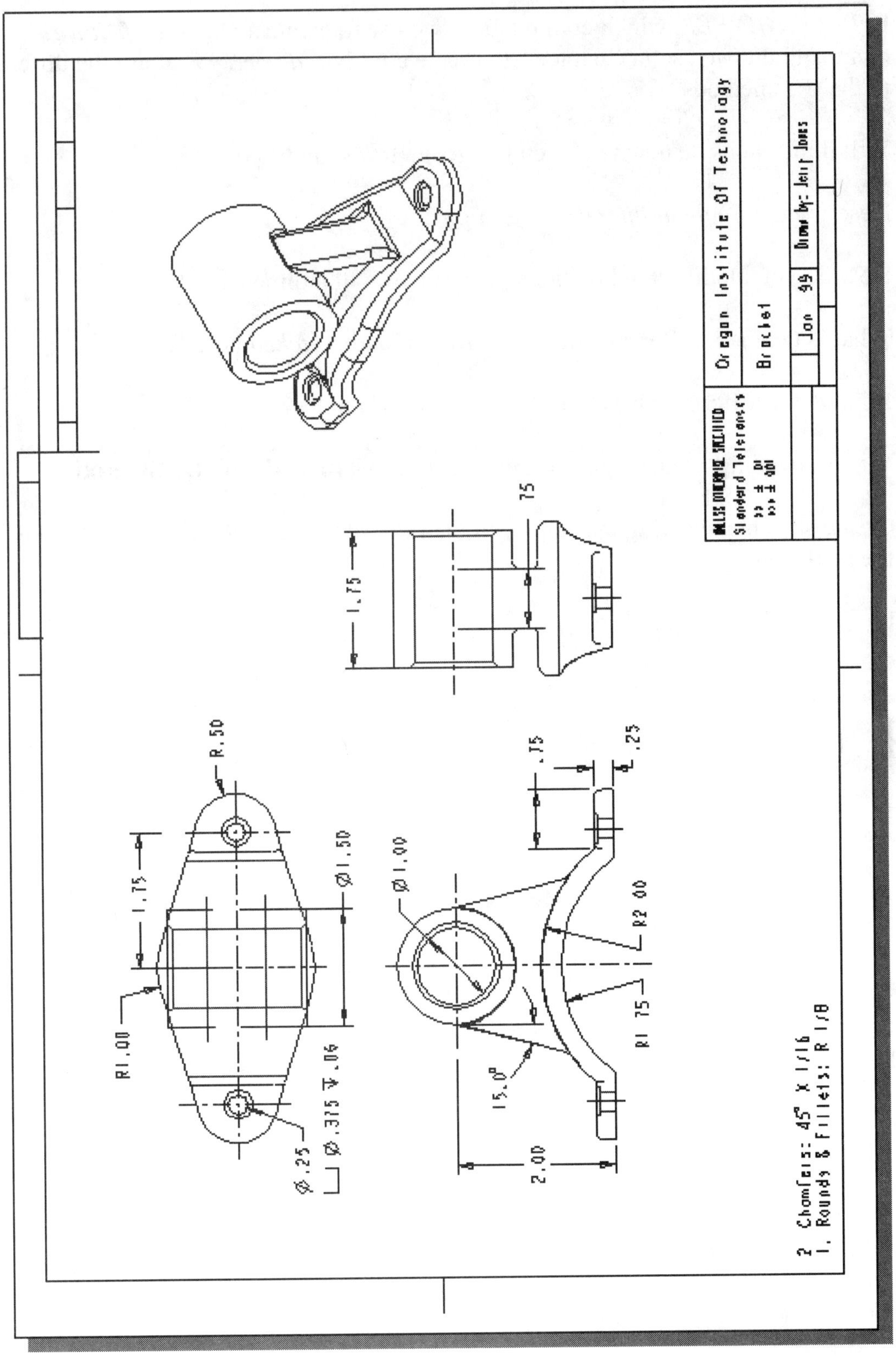

Questions:

1. In *Pro/ENGINEER*, which command do we use to create a *mirrored feature*? Which command did we use in Chapter 7 to create a *mirrored part*? What are the differences in the two methods?

2. Will dimensions be updated when the geometry is modified?

3. Describe the *Boolean Intersect* operation.

4. How do you modify the directions of the arrows of a dimension?

5. What is the main difference between *straight* and *sketched* holes?

6. What is the difference between *coaxial* and *radii* holes?

7. When extruding, what is the difference between **Thru All** and **Until Next**?

8. Create sketches showing the steps you plan to use to create the two models shown on the next page:

Ex.1)

Ex.2)

Exercises:

1. Dimensions are in inches.

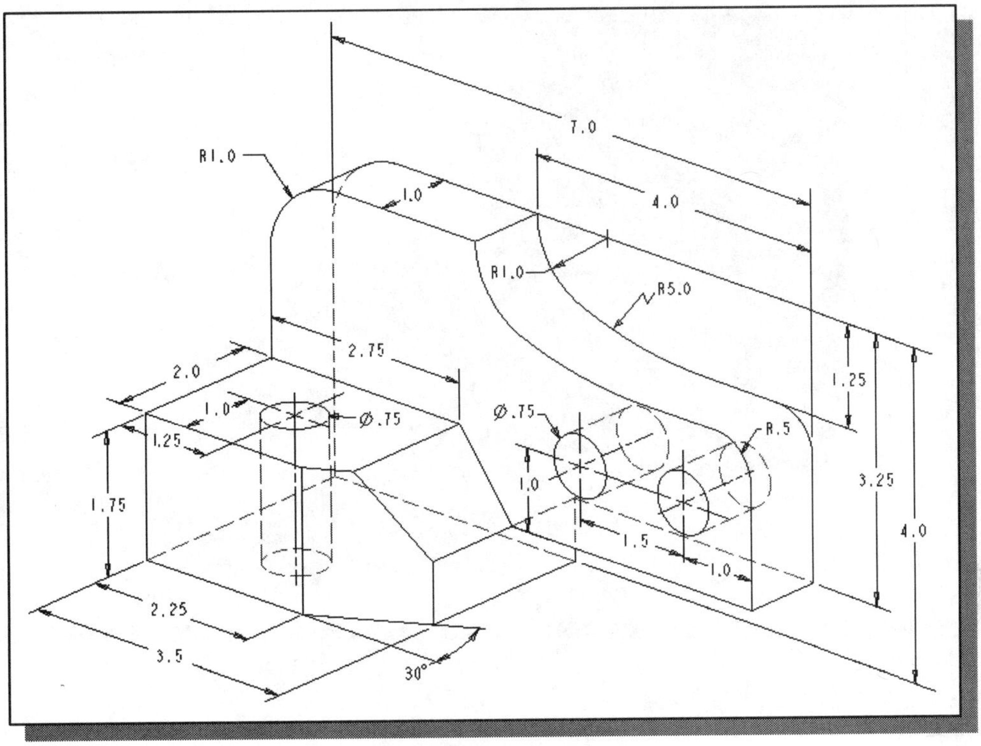

3. Dimensions are in inches.

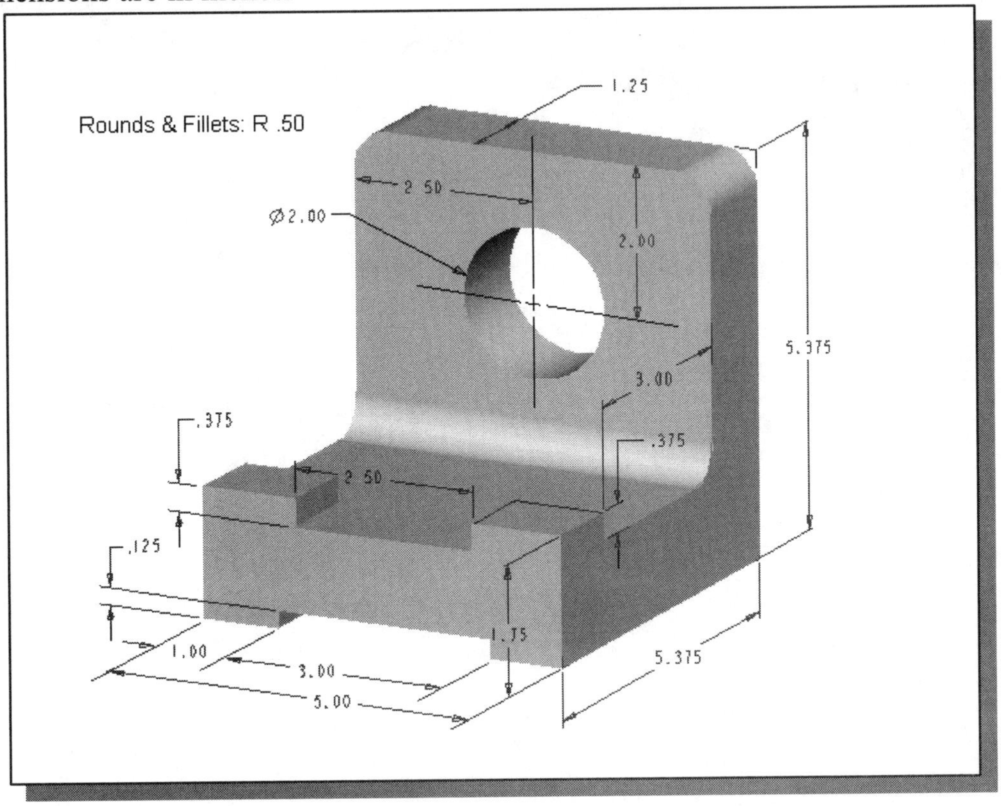

4. Dimensions are in inches.

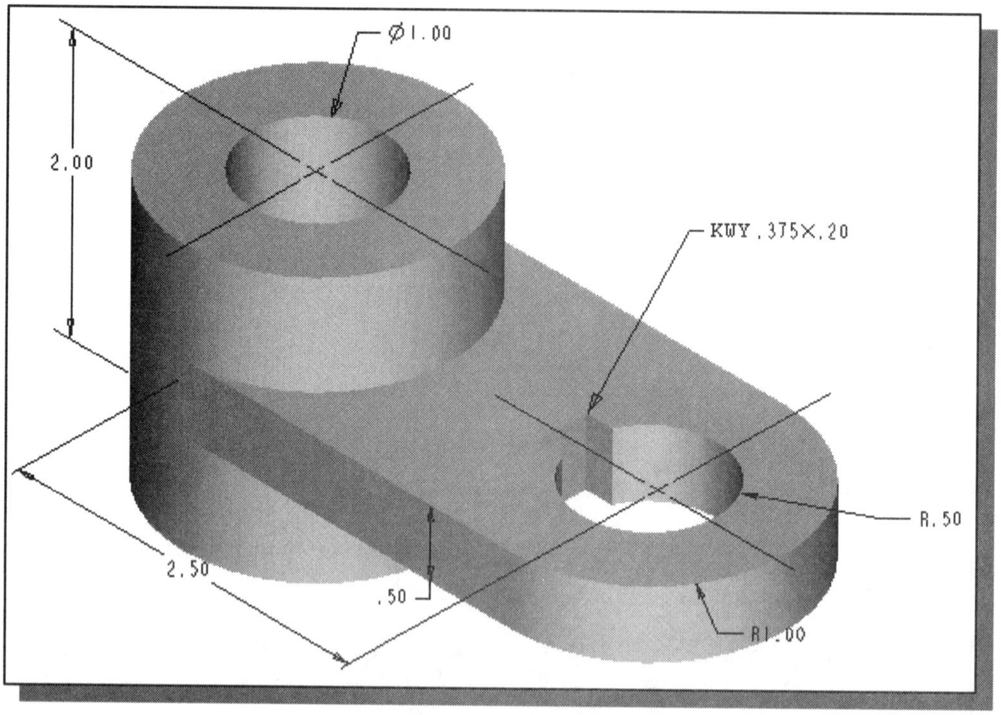

Chapter 10
Assembly Modeling - Putting It All Together

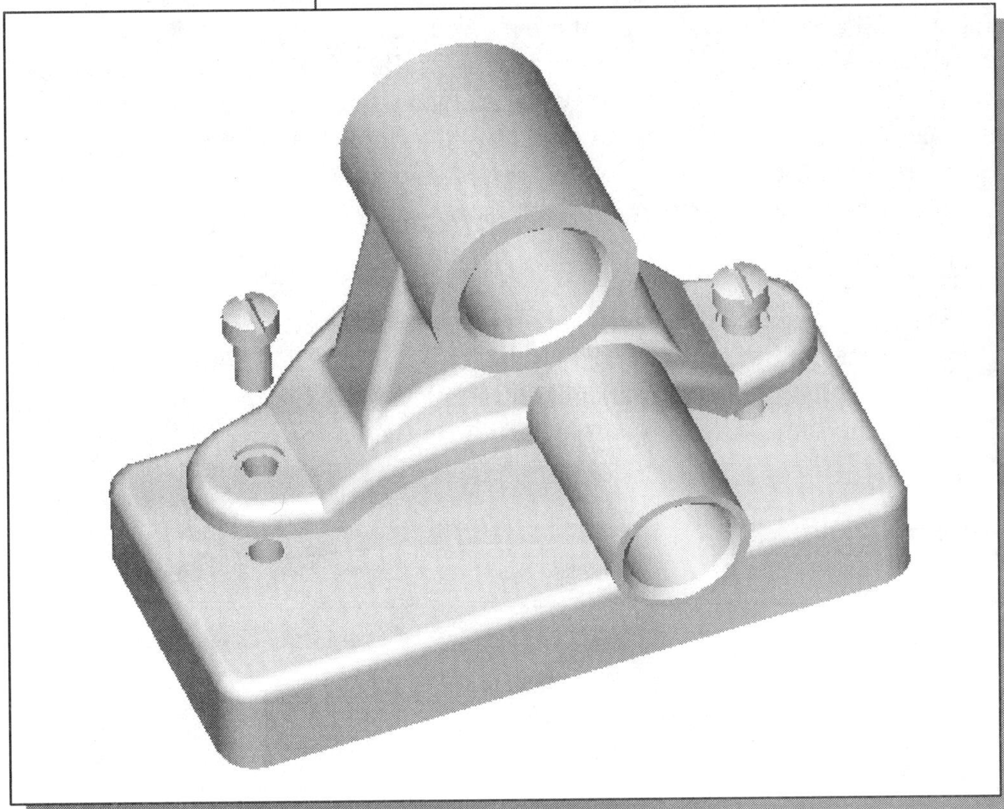

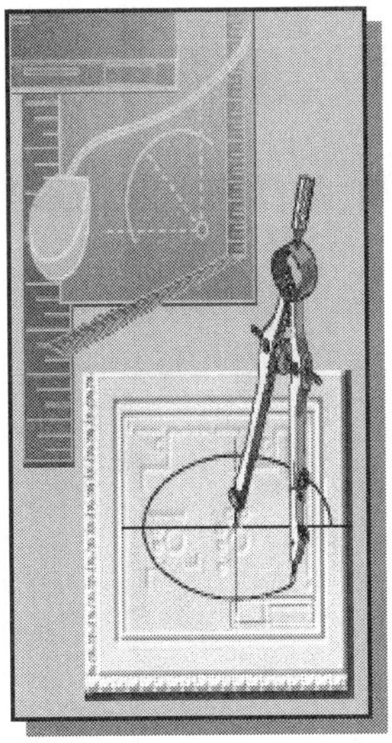

Learning Objectives

- Understand the Assembly Modeling Methodology
- Acquire Parts in the Assembly Mode
- Understand and Utilize Assembly Constraints.
- Create Subassemblies.
- Understand the Bi-directional Full Associative Functionality
- Create Exploded Assemblies.

Introduction

In the previous lessons, we have gone over the fundamentals of creating basic parts and drawings. In this lesson, we will demonstrate how to create and modify assembly models. *Pro/ENGINEER* provides full associative functionality in all *Pro/ENGINEER* modules, including assemblies. When we change a part model, *Pro/ENGINEER* will automatically reflect the changes in all assemblies that use the part. We can also modify a part in an assembly. The bi-directional full associative functionality allows us to make very flexible model modifications in *Pro/ENGINEER*. Many parallels exist between assembly design and part design. The main task in creating an assembly is establishing the assembly relationships between parts. In this lesson, we will construct an assembly model using the ***Bracket*** part, which was created in the previous lesson.

To assemble parts into an assembly, we will need to consider the assembly relationships between parts. It is a good practice to assemble parts based on the way they would be assembled in the actual design. We should also consider breaking down the assembly into smaller subassemblies, which helps the management of parts. In *Pro/ENGINEER*, a subassembly is treated the same way as a single part during assembling.

The *BRACKET* Assembly:

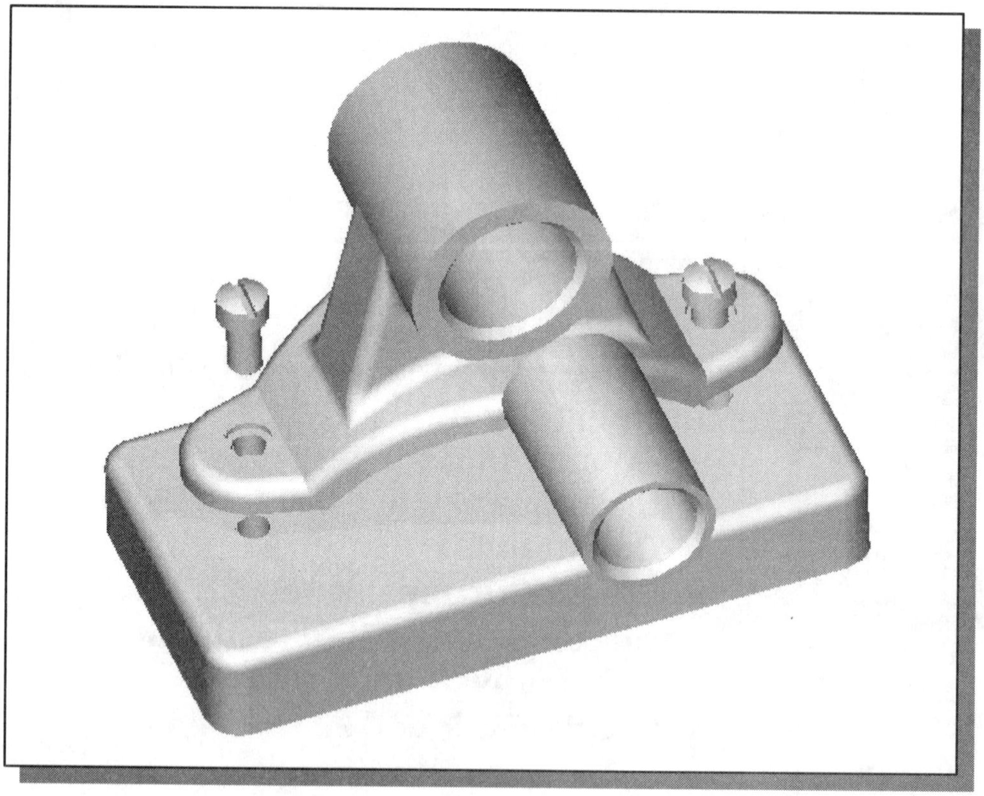

Assembly Modeling Methodology

The *Pro/ENGINEER Assembly Modeler* provides tools and functions that allow us to create 3D parametric assembly models. An assembly model is a 3D model with any combination of multiple part models. *Parametric assembly constraints* can be used to control relationships between parts in an assembly model.

Pro/ENGINEER can work with any of the assembly modeling methodologies:

The Bottom Up approach

The first step in the *bottom up* assembly modeling approach is to create the individual parts. The parts are then pulled together into an assembly. This approach is typically used for smaller projects with very few team members.

The Top Down approach

The first step in the *top down* assembly modeling approach is to create the assembly model of the project. Initially, individual parts are represented by names or symbolically. The details of the individual parts are added as the project gets further along. This approach is typically used for larger projects or during the conceptual design stage. Members of the project team can then concentrate on the particular section of the project to which he/she is assigned.

The Middle Out approach

The *middle out* assembly modeling approach is a mixture of the bottom-up and top-down methods. This type of assembly model is usually constructed with most of the parts already created and additional parts are designed and created using the assembly for construction information. Some requirements are known and some standard components are used, but new designs must also be produced to meet specific objectives. This combined strategy is a very flexible approach to creating assembly models.

The different assembly modeling approaches described above can be used as guidelines to manage design projects. Keep in mind that we can start modeling our assembly using one approach and then switch to a different approach without any problems.

In this chapter, the *bottom up* assembly modeling approach is illustrated. All of the parts (components) required to form the assembly are created first. *Pro/ENGINEER's Assembly Modeling* tools allow us to create complex assemblies by using components that are created in separate part files or in the same part file. A component can be a subassembly or a single part, where features and parts can be modified at any time. The sketches and sections used to build part features can be fully or partially constrained. Partially constrained features may be adaptive, which means the size or shape of the associated parts are adjusted in an assembly when the parts are constrained to other parts. The basic concept and procedure of using the adaptive assembly approach is demonstrated in the tutorial.

Additional Parts

- Besides the ***Bracket***, we will need three additional parts: (1)***Base Plate***, (2)***Bushing*** and (3) ***Cap Screw***. Create the three parts as shown. Notice the positions of the parts in relation to the *datum planes*.

(1) ***Base Plate***

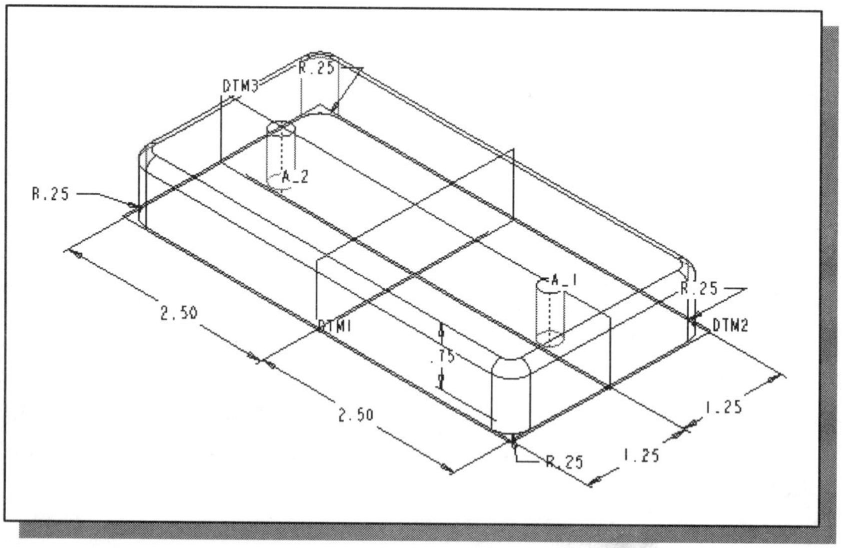

Hole sizes: Ø 0.25 and 0.6 deep.
Center to center distance of the two holes: 3.5.
Rounds: R 0.125

(2) ***Bushing***

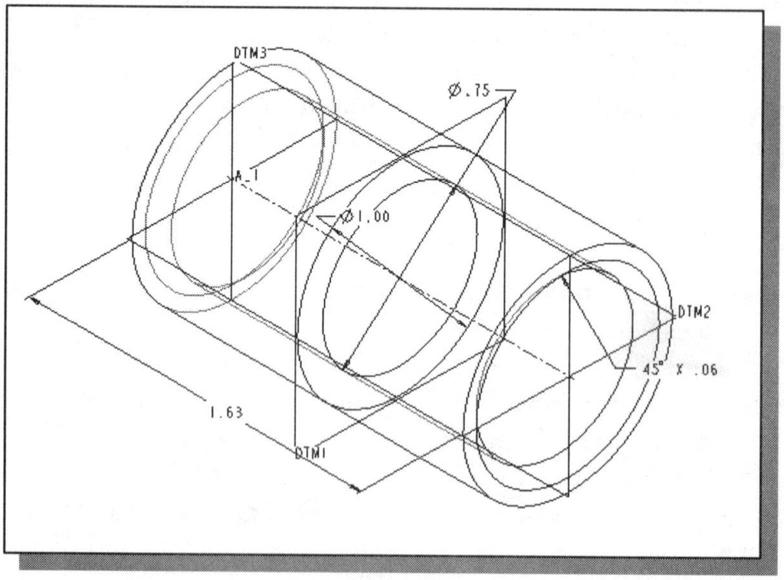

ID: Ø 0.75, OD: Ø 1.00, Length: 1.625
Chamfers: 45° × 0.0625

(3) ***Cap Screw***

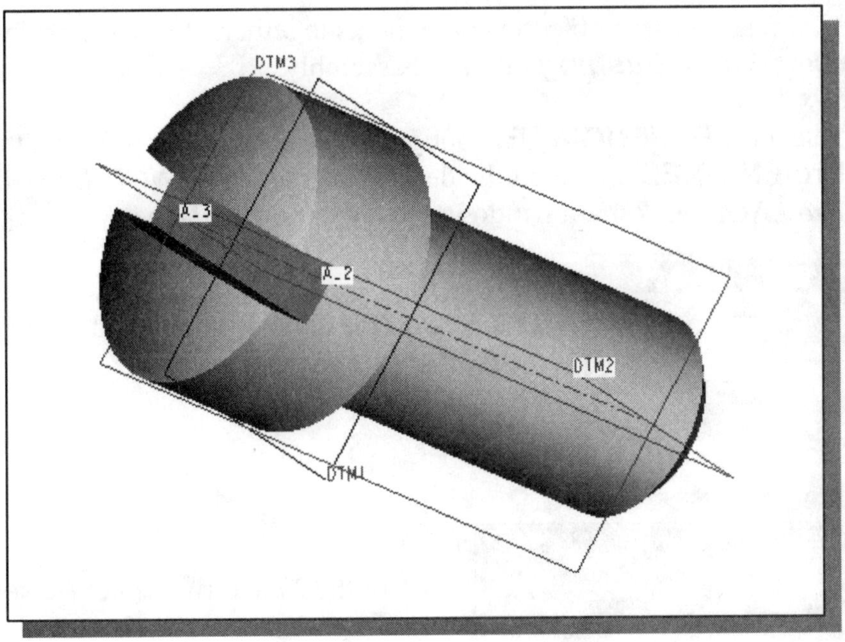

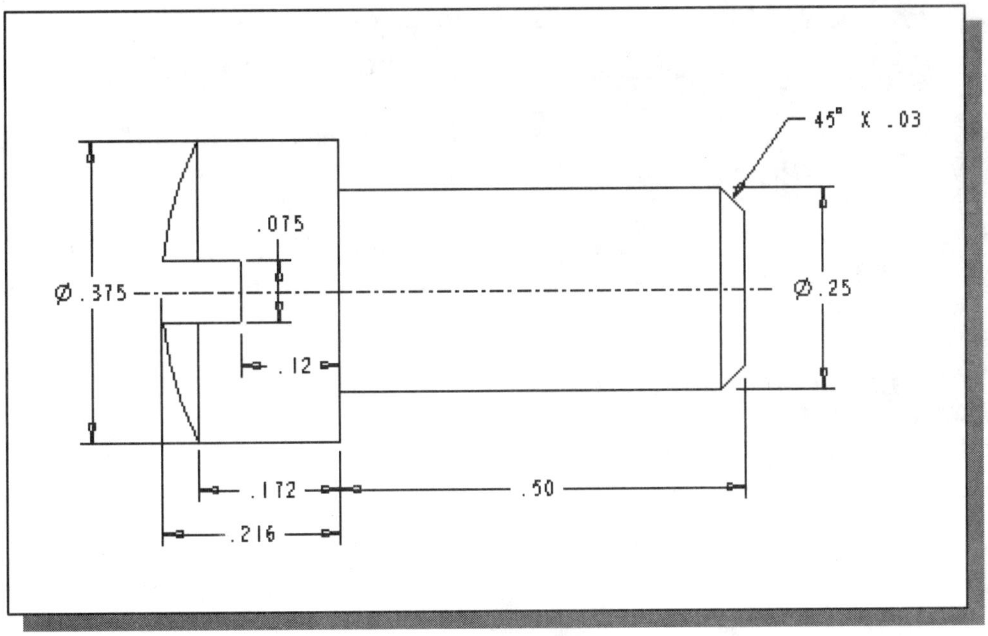

- We will omit the threads in this model. Threads are complex three-dimensional curves. You are encouraged to experiment with the thread option: **Insert → Helical Sweep**. Create the three parts and save the parts as separate part-files (***BasePlate***, ***Bushing***, and ***CapScrew***).

Create a Subassembly

❖ We are now ready to assemble the components together. We will start by assembling the ***Bracket*** and the ***Bushing*** into a subassembly.

1. Select the **Pro/ENGINEER** option on the *Start* menu or select the **Pro/ENGINEER** icon on the desktop to start *Pro/ENGINEER*. The *Pro/ENGINEER* main window will appear on the screen.

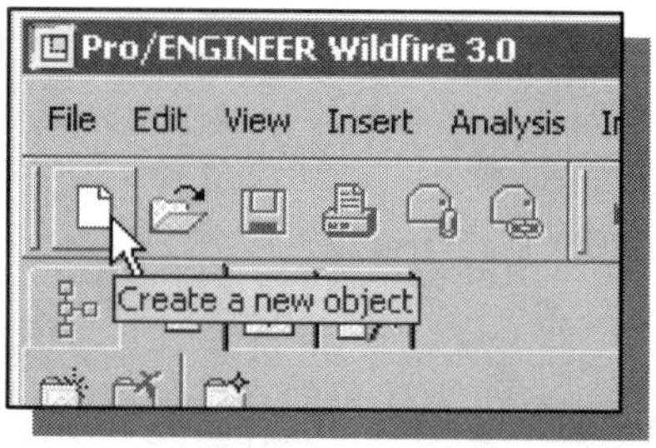

2. Click on the **New** icon, located in the *Standard* toolbar as shown.

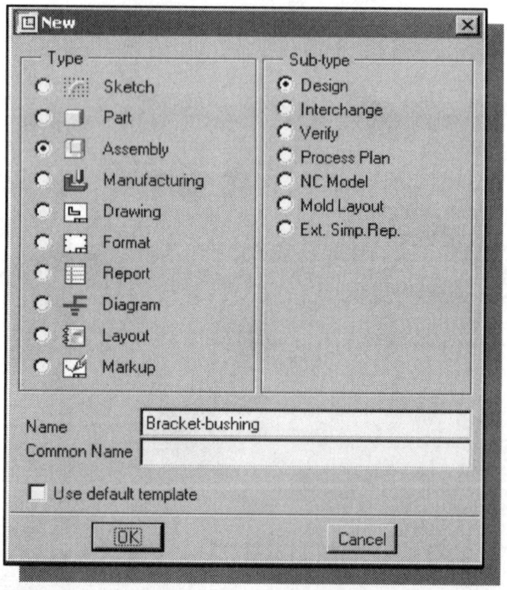

3. In the *New* form, select **Assembly** in the **Type** list.

4. Enter ***Bracket-bushing*** as the **Assembly** model file **Name**.

5. Turn *off* the **Use default template** option.

6. Click the **OK** button to continue. The *New File Options* window appears on the screen.

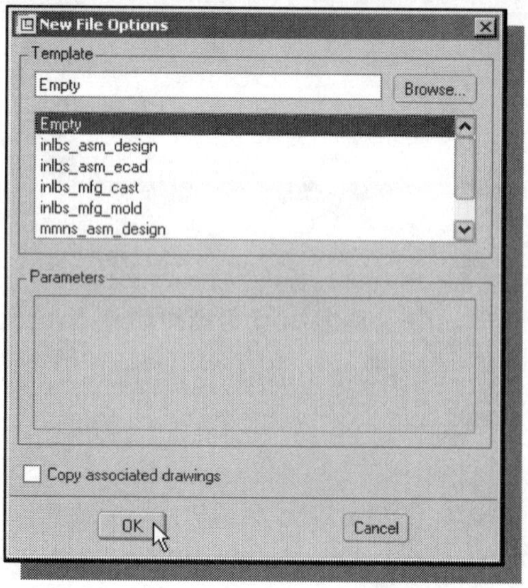

7. In the *New File Options* dialog box, select **Empty** in the option list to not use any template file.

8. Click the **OK** button to accept the settings and enter the *Pro/ENGINEER Assembly Modeling* mode.

Retrieving the Bracket Component

1. Pick **Add Component** in the *Assembly* toolbar to begin placing different parts into the assembly model.

❖ The *Open* window showing a list of part model files appears on the screen.

2. Select the ***Bracket*** (part file: *bracket.prt*) in the list window and pick **Open** to retrieve the model.

❖ The ***Bracket*** model is displayed in the graphics area.

Retrieving the *Bushing* Component

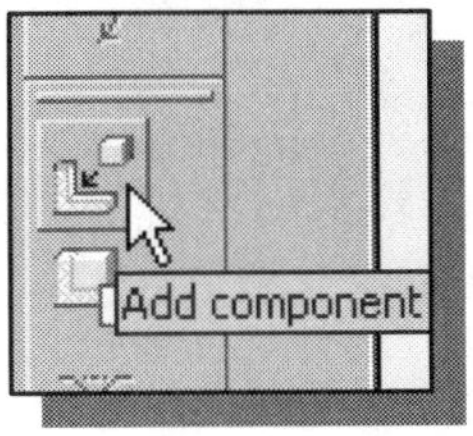

1. Pick **Add Component** in the *Assembly* toolbar to begin placing different parts into the assembly model.

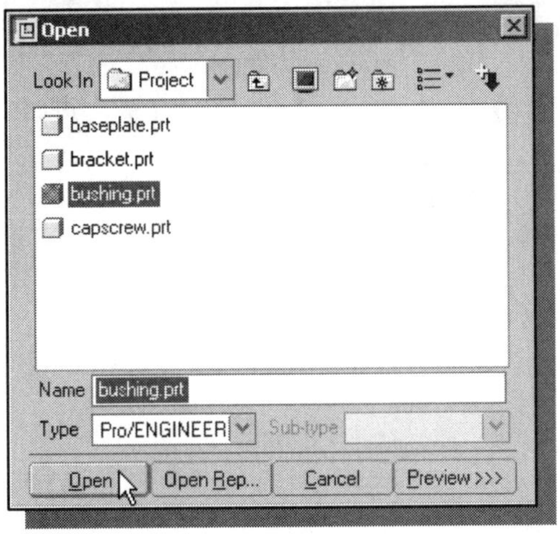

2. Select the ***Bushing*** (part file: *bushing.prt*) in the list window and pick **Open** to retrieve the model.

❖ In the display area, the ***Bushing*** is placed next to the ***Bracket*** and the *Component Placement* window appears.

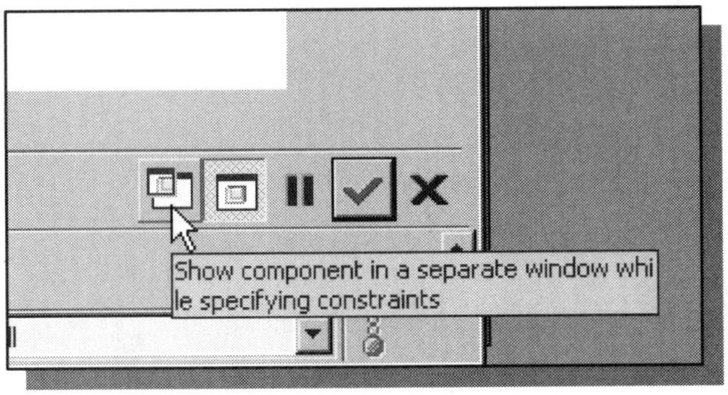

❖ The default setting of *Pro/ENGINEER* will display the new component in the *Assembly* window. Note that the new component can also be displayed in a separate window by selecting the first icon, **Separate Window,** the *Component Placement* window.

Placement Constraints

To assemble components into an assembly, we need to establish the assembly relationships between components. It is a good practice to assemble components the way they would be assembled in the actual design. ***Placement constraints*** create a parent/child relationship that allows us to capture the design intent of the assembly. Because the component that we are placing actually becomes a child to the already assembled components, we must use caution when choosing constraint types and references to make sure they reflect the intent.

The following is a list of the most commonly used placement constraints:

- **Mate** – Selected surfaces point in opposite directions and become coplanar.

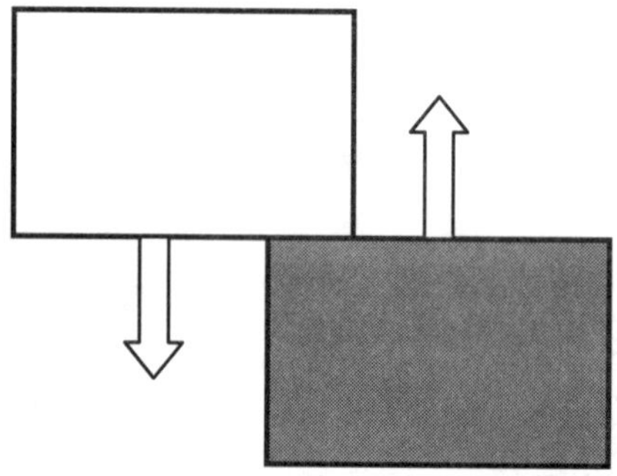

- **Mate Offset** – Selected surfaces point in opposite directions and are offset by a specified distance. We can modify the offset dimension at any time.

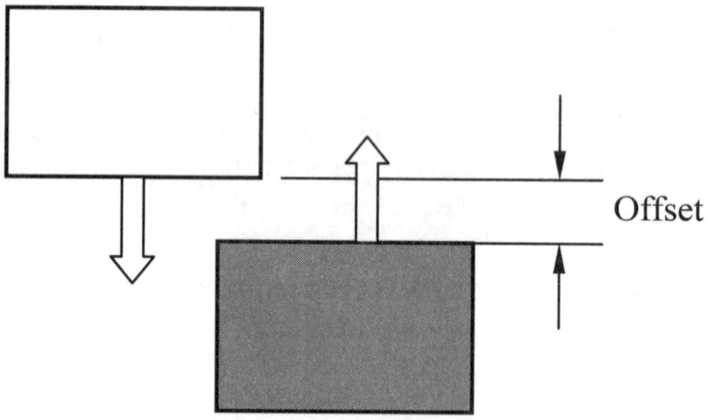

- **Align** – Selected surfaces point in the same direction and are made coplanar. We can use the **Align** constraint for axes to make them co-axial.

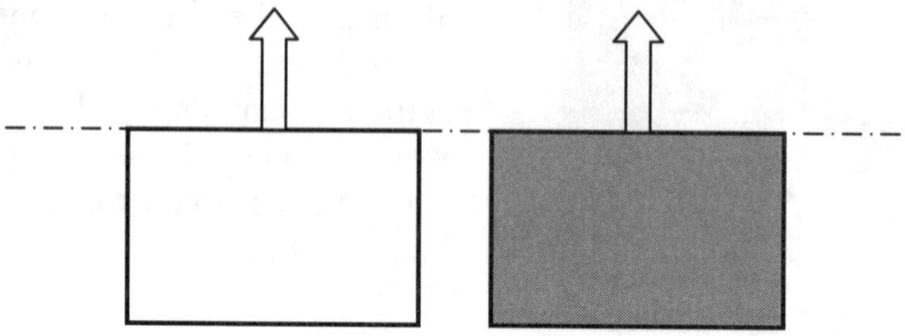

- **Align Offset** – Selected surfaces point in the same direction and are offset by a specified distance. We can modify the offset dimension at any time.

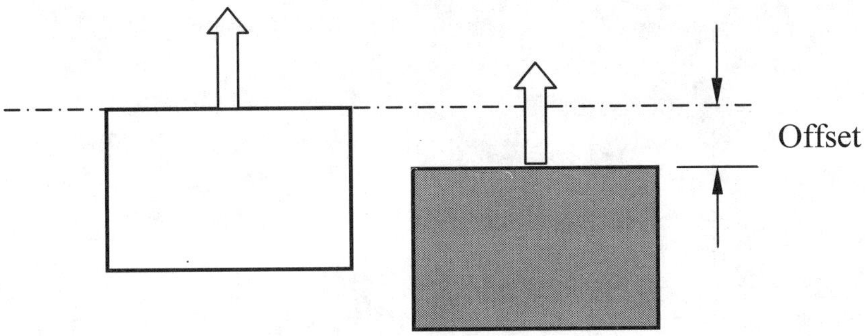

- **Insert** – Selected cylindrical surfaces become co-axial. The surfaces do not need to be full 360-degree cylinders.

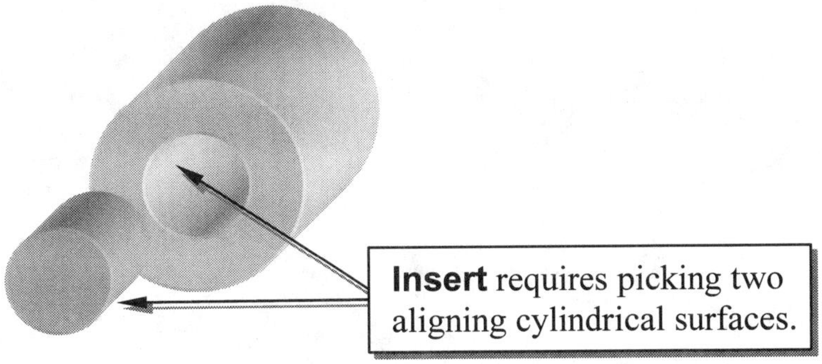

- **Orient** - Selected surfaces point in the same direction and are parallel. This constraint is similar to the **Align** constraint, except *Pro/ENGINEER* does not associate a specific offset value to this constraint.

❖ Note that besides surfaces, Pro/E also allows the use of points and edges.

Placing the *Bushing*

1. In the toolbar, switch off the display of the datum planes, datum axis and coordinate systems.

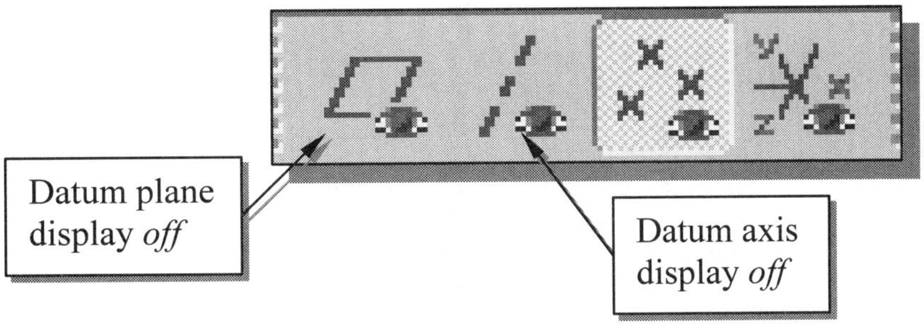

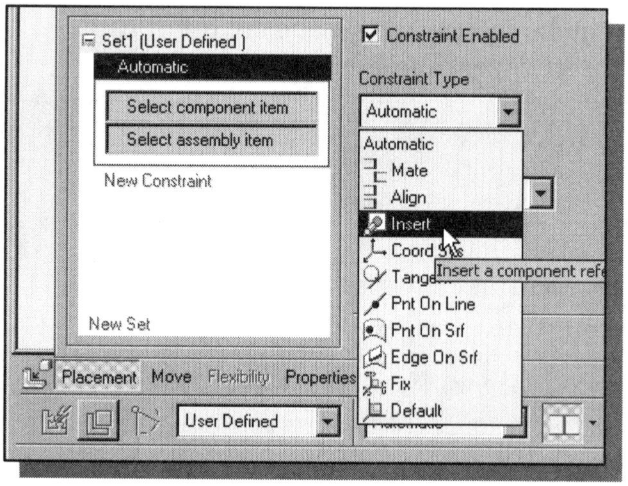

2. In the *Constraint Options Dashboard,* click the **Placement** option as shown.

3. Pick **Insert** from the **Constraints Type** list as shown.

4. Pick the **outer cylindrical surface** of the ***Bushing***.

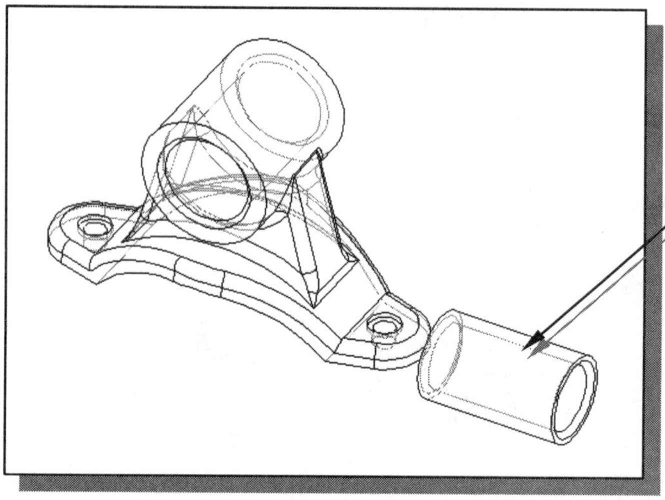

❖ Notice in the *Component Placement* window, the selected surface is identified. We can redefine the selection by clicking on the **Arrow** icon in front of the selected reference.

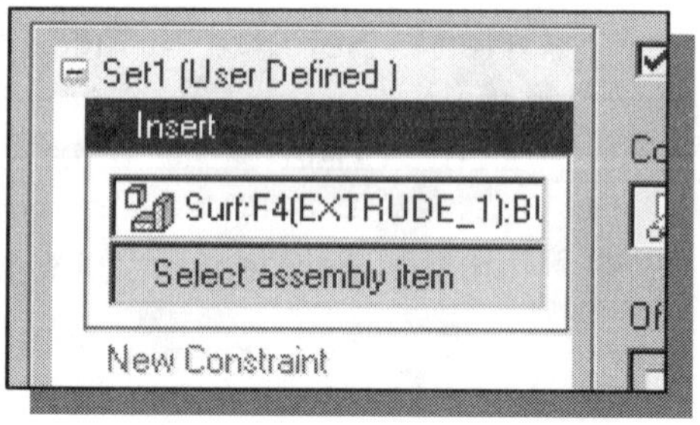

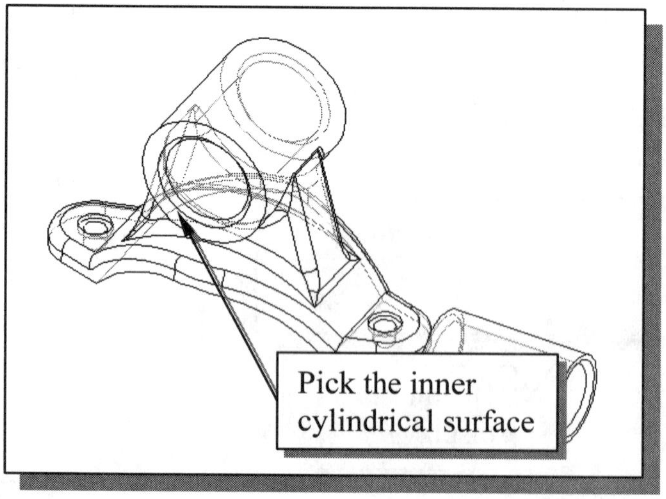

5. Pick the **inner cylindrical surface** of the ***Bracket***.

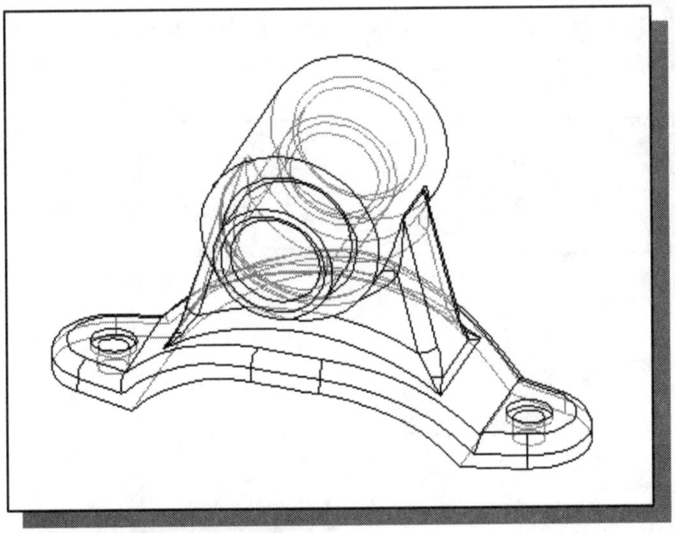

❖ In the display area, the ***Bushing*** is moved to a new location. Use the *Dynamic Viewing* functions to confirm that the centers are aligned.

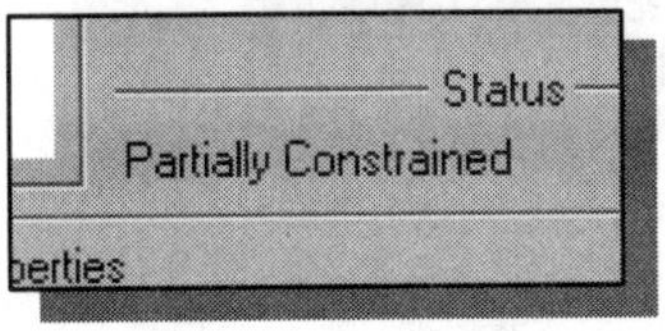

❖ At the bottom of the *Component Placement* window, *Pro/E* indicates the status of ***Bushing*** placement as ***partially constrained***. This is because the ***Bushing*** can still slide along the aligned axis.

❖ In general, we need to specify two or three constraints to fully assemble a component. We will add another placement constraint, the **Mate** constraint.

Create the Second Placement Constraint

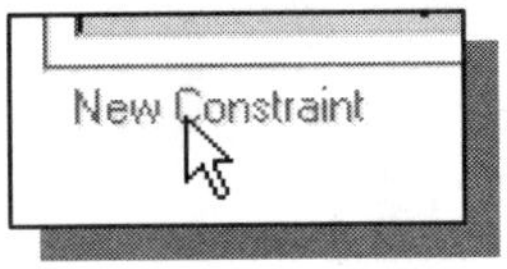

1. Click **New Constraint** to add an additional constraint.

2. Pick **Mate** from the **Constraints Type** list in the *Component Placement* window.

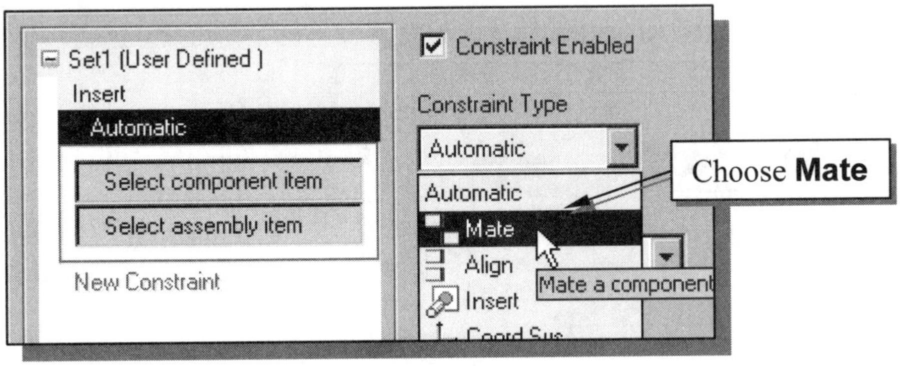

3. In the toolbar, switch ***on*** the display of the datum planes.

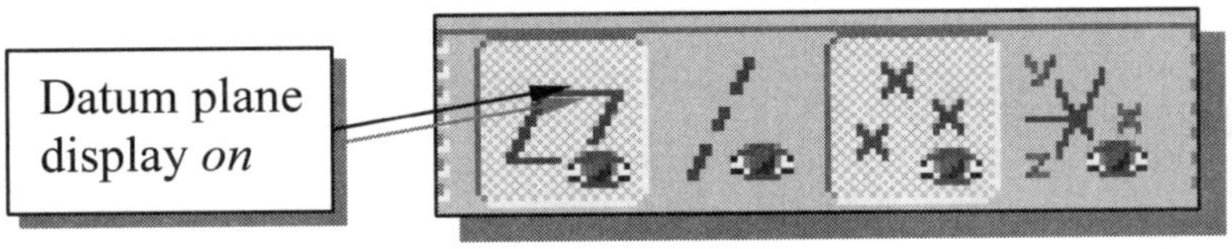

4. Pick the *reference plane* that is perpendicular to the ***Bushing* axis,** as shown in the figure below. (If necessary, set the option list to help picking the proper surface.)

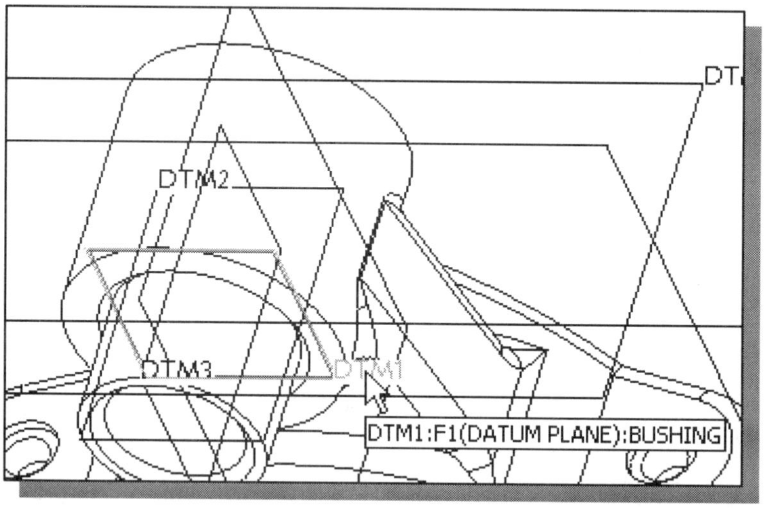

5. Pick **Front**, the *reference plane* associated with the ***Bracket*** part.

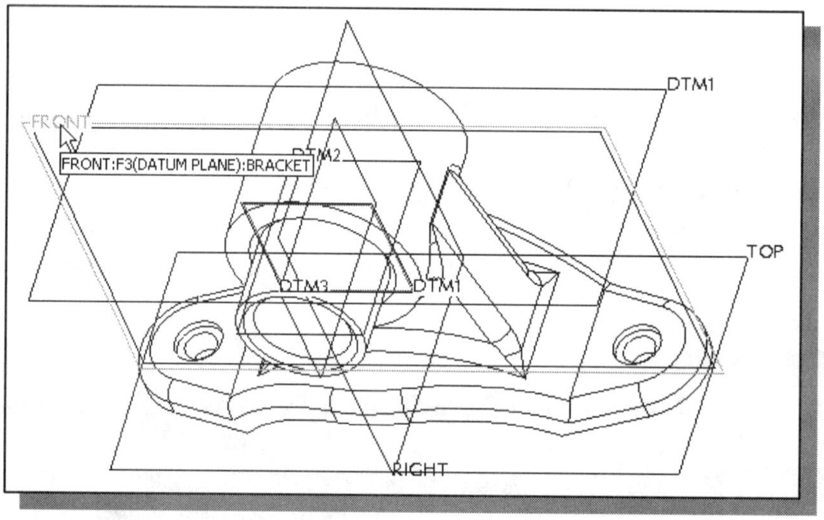

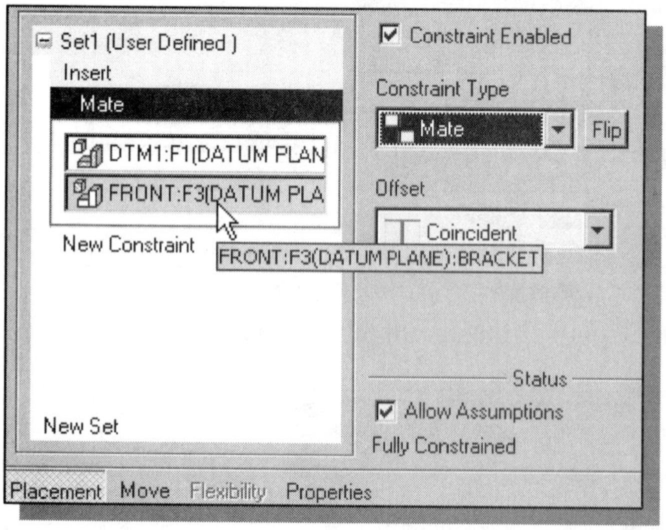

❖ In the *Component Placement* window, the components of the placement constraints are displayed. We can redefine the selection by clicking on the **info. box**.

Near the bottom of the display, *Pro/ENGINEER* indicates the status of placement of the ***Busing*** as ***fully constrained***.

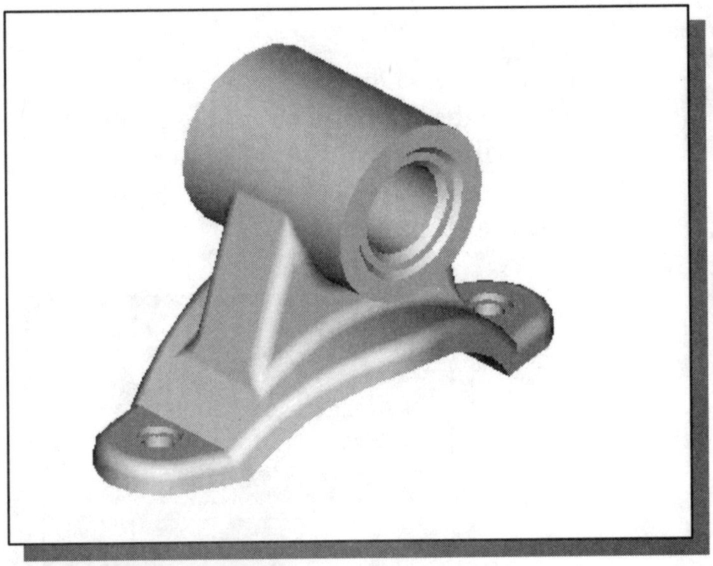

6. Pick **Accept** to complete the placement of the bushing component.

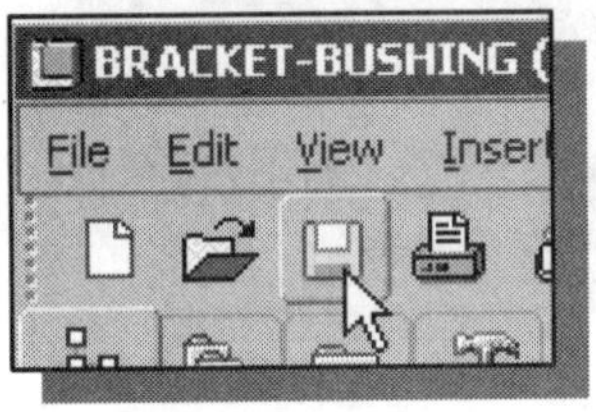

7. Pick **Save** in the toolbar.

8. Hit the **ENTER** key to accept the default filename and save the subassembly on disk.

Create the Bracket Assembly Model

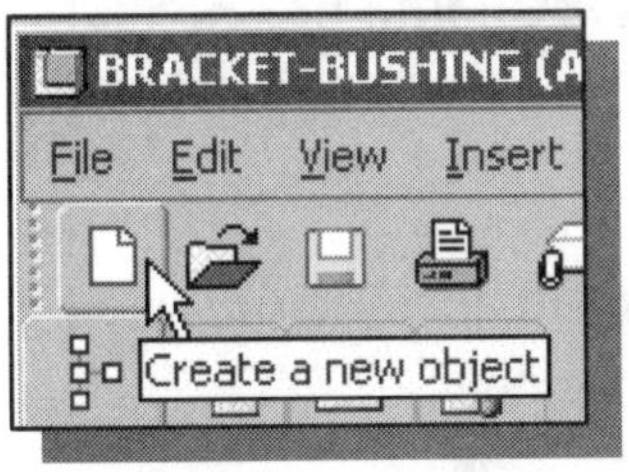

1. Pick the **New Object** icon in the toolbar. (We can also use the key combination **Ctrl-N** to start a new object.)

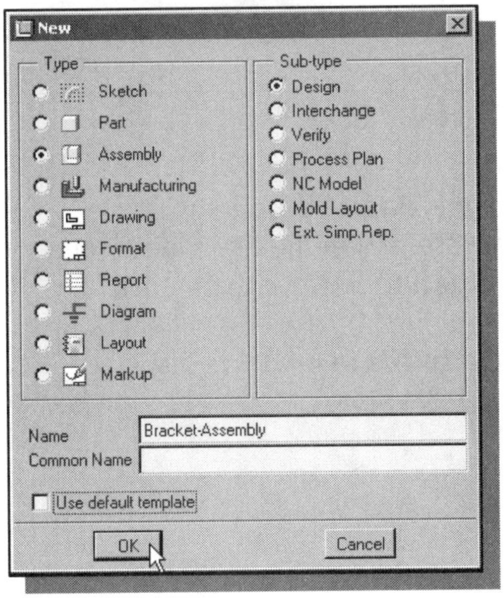

2. In the *New* form, select **Assembly** in the **Type** list.

3. Enter ***Bracket-Assembly*** as the **Assembly** model file Name.

4. Turn ***off*** the **Use default template** option.

5. Click the **OK** button to continue.

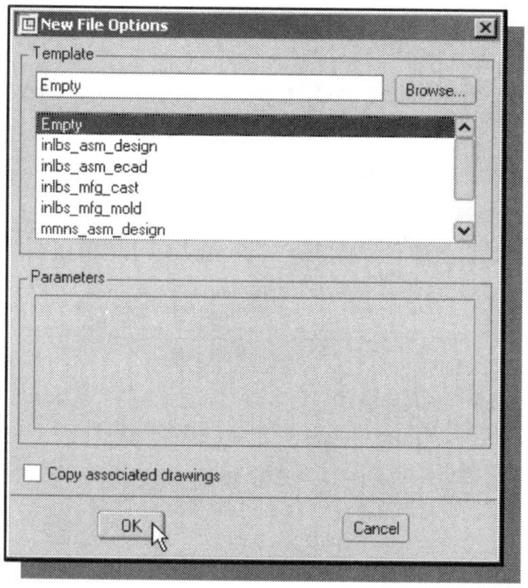

6. In the *New File Options* dialog box, select **Empty** in the option list to not use any template file.

7. Pick **OK** to continue.

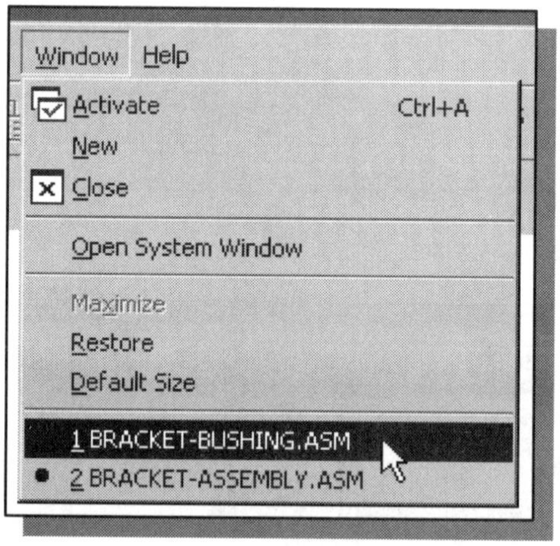

- Note that we can switch to the ***Bracket-Bushing*** *subassembly* through the **Window** pull-down menu.

Base Component

❖ In creating an assembly model in *Pro/ENGINEER*, we will need to decide which part to use as the first component. In most cases, this *base component* should be one that is **not likely to be removed** from the assembly. For our project, we will use the ***Base Plate*** as the base component.

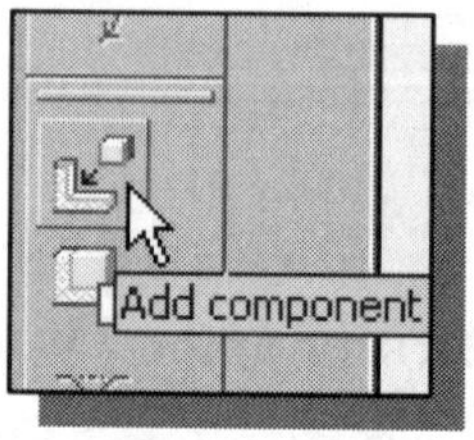

1. Pick **Add Component** in the *Assembly* toolbar to begin placing different parts into the assembly model.

❖ The *Open* window showing a list of part model files appears on the screen.

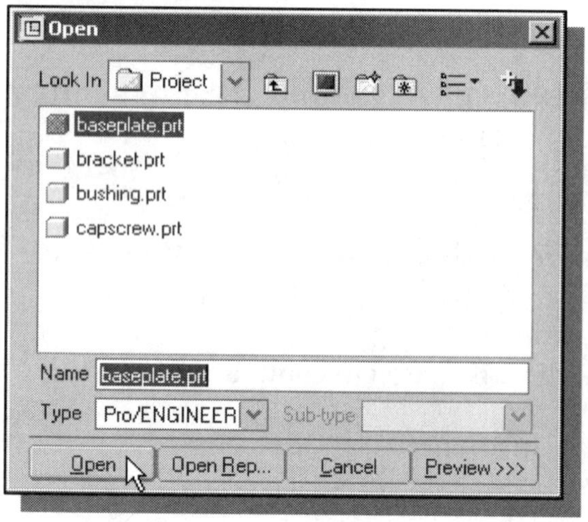

2. Select the ***Base Plate*** (part file: *baseplate.prt*) in the list window and pick **Open** to retrieve the model.

❖ The ***Base Plate*** model is displayed in the graphics area.

Retrieving the *Bracket-Bushing* Subassembly

1. Pick **Add Component** in the *Assembly* toolbar to begin placing different parts into the assembly model.

❖ The *Open* window showing a list of part model files appears on the screen.

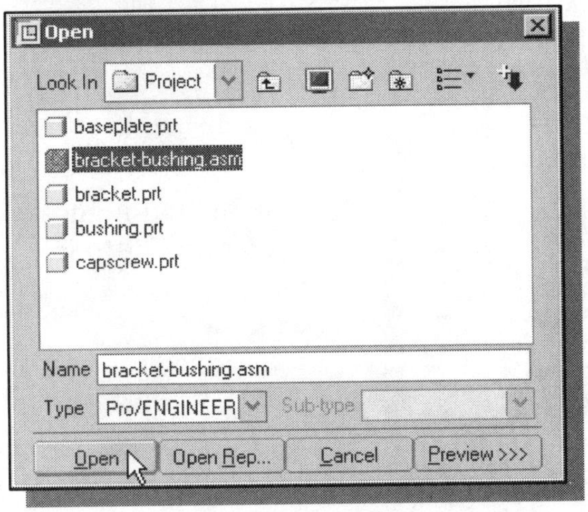

2. Select the ***Bracket-Bushing*** (assembly file: *bracket-bushing.asm*) in the list window and pick **OPEN** to retrieve the model.

Placing the *Bracket-Bushing* Subassembly

1. In the toolbar, switch ***off*** the display of the datum planes and switch ***on*** the display of the datum axis.

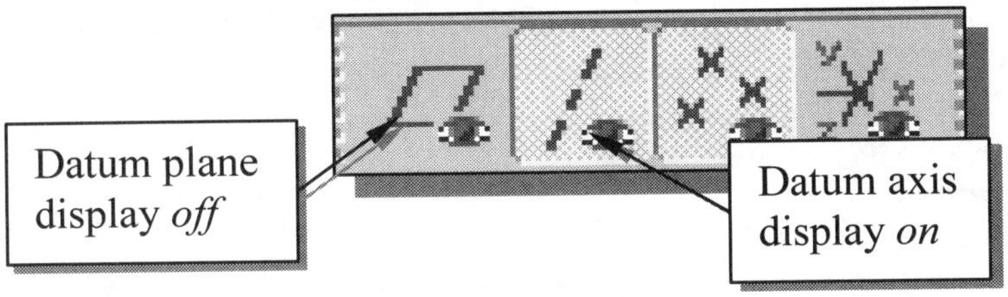

2. Pick the **bottom surface** of the ***Bracket*** as shown.

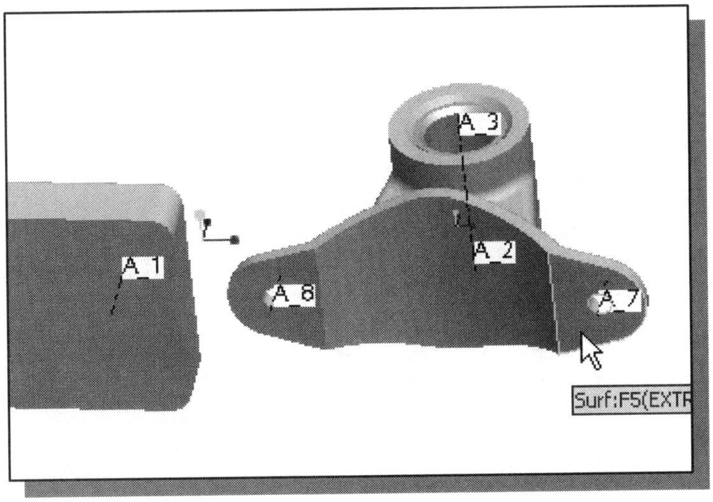

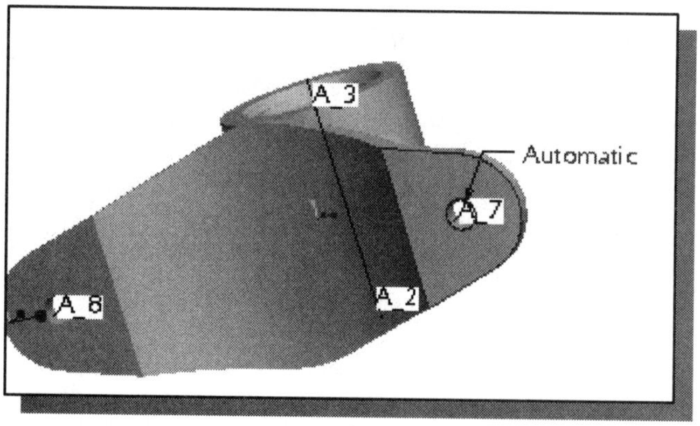

❖ Notice the selected surface is identified by a leader and the text "Automatic", indicates the default option is used.

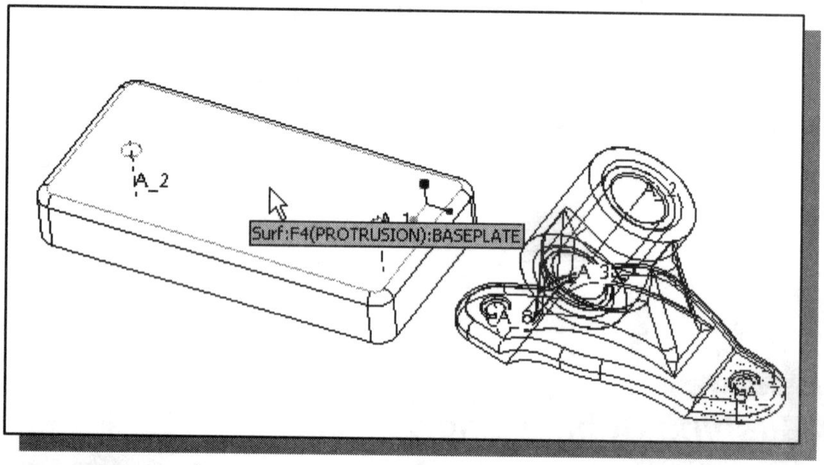

3. Pick the **top plane** of the ***Base Plate*** in the *Assembly* window.

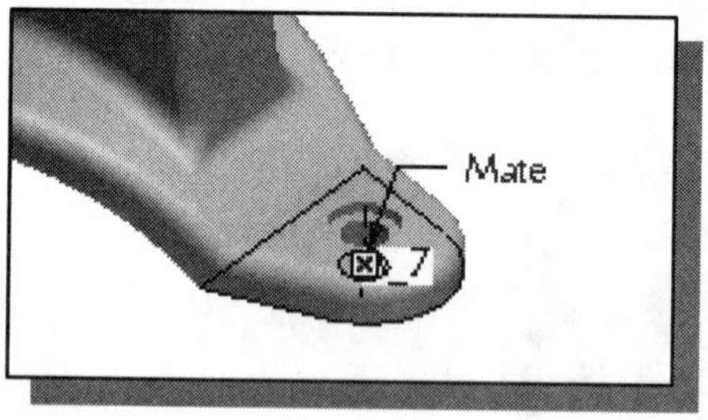

❖ Note that *Pro/ENGINEER* automatically determines the selected entities can be constrained with the **Mate** option.

4. Pick the corresponding axes, A_7 and A_1, to align the *bolt holes*.

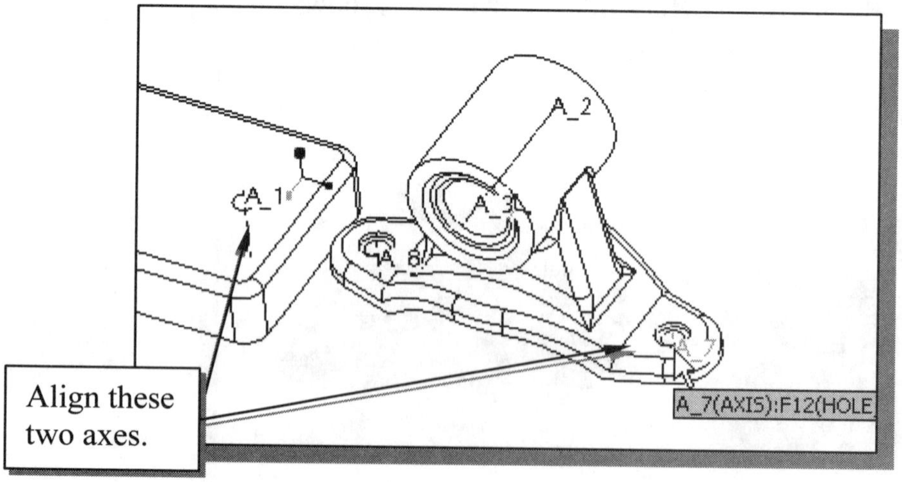

❖ At the bottom of the *Component Placement* window, *Pro/ENGINEER* indicates the status of the placement of the ***Bracket-Bushing*** as ***fully constrained***.

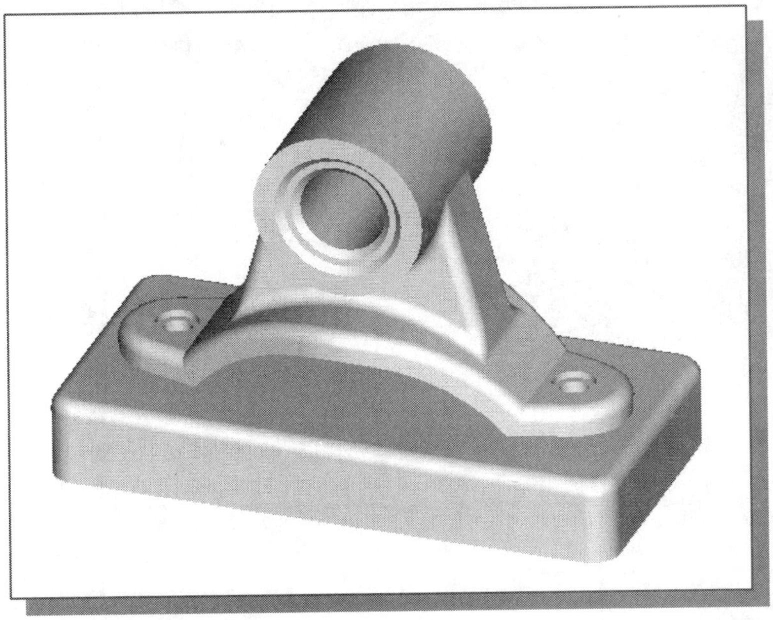

5. Pick **Accept** to complete the placement of the subassembly.

Placing the Cap Screws

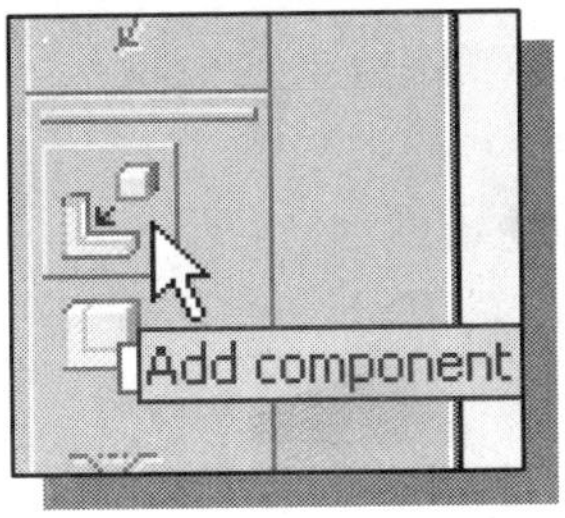

1. Pick **Add Component** in the *Assembly* toolbar to begin placing different parts into the assembly model.

2. Select the ***Cap-Screw*** (part file: *capscrew.prt*) in the list window and pick **Open** to retrieve the model.

3. Pick the two ***planar surfaces*** as shown.

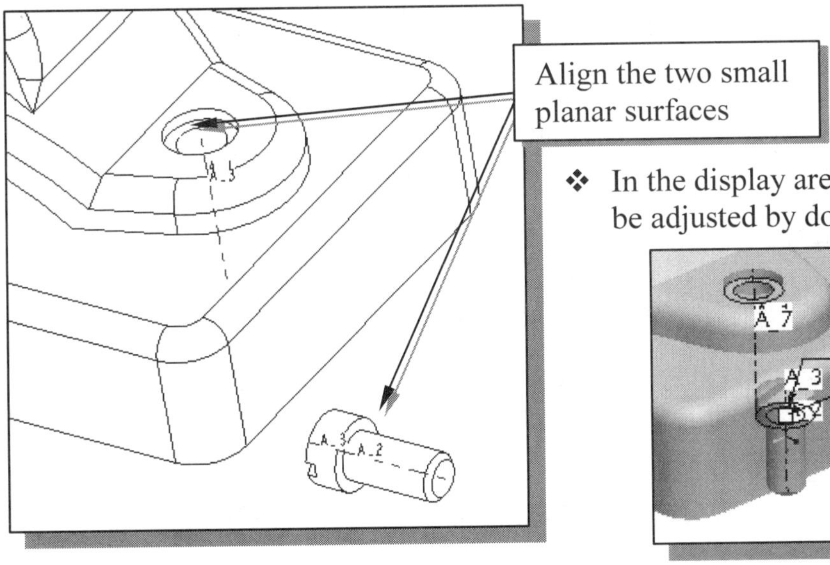

❖ In the display area, the offset distance can be adjusted by double-clicking.

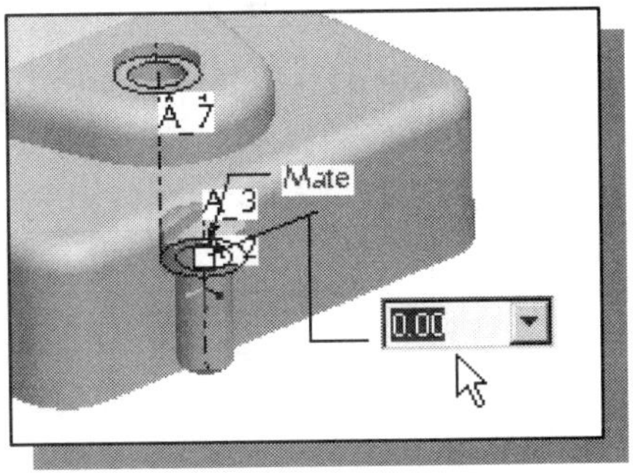

4. Pick the two cylindrical surfaces as shown.

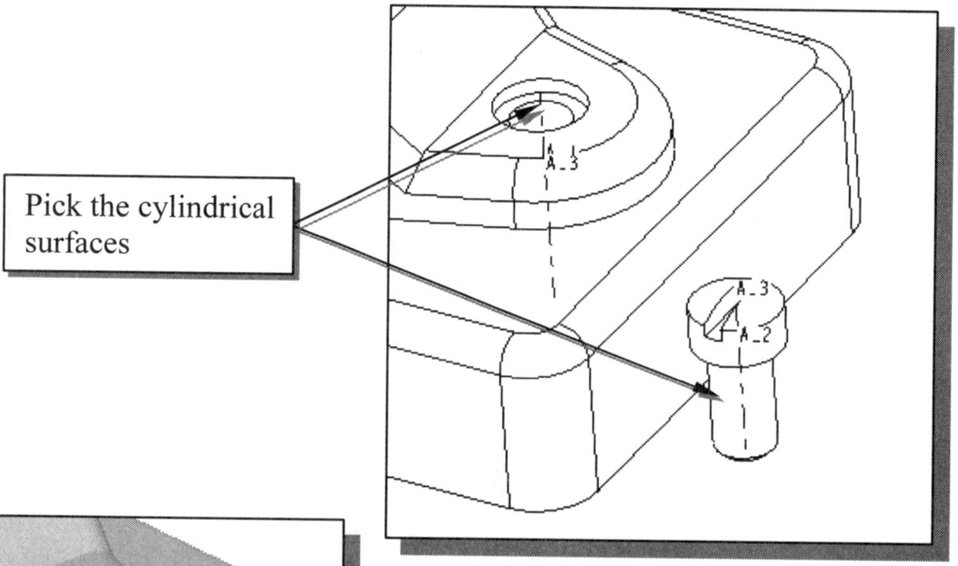

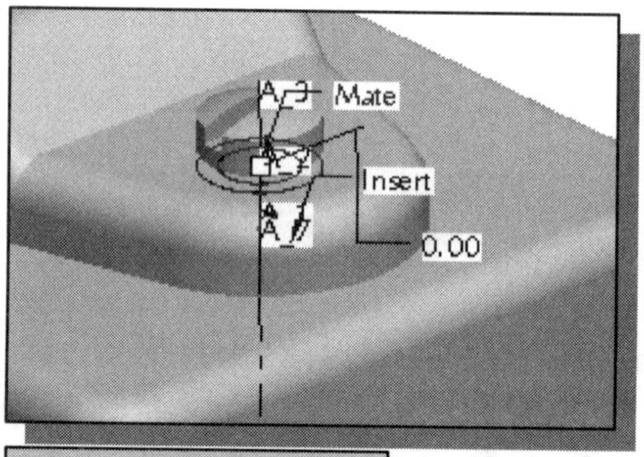

- At the bottom of the *Constraint control panel*, *Pro/ENGINEER* indicates the status of placement of the **Cap-screw** as ***fully constrained***.

5. Pick **Accept** to complete the placement of the subassembly

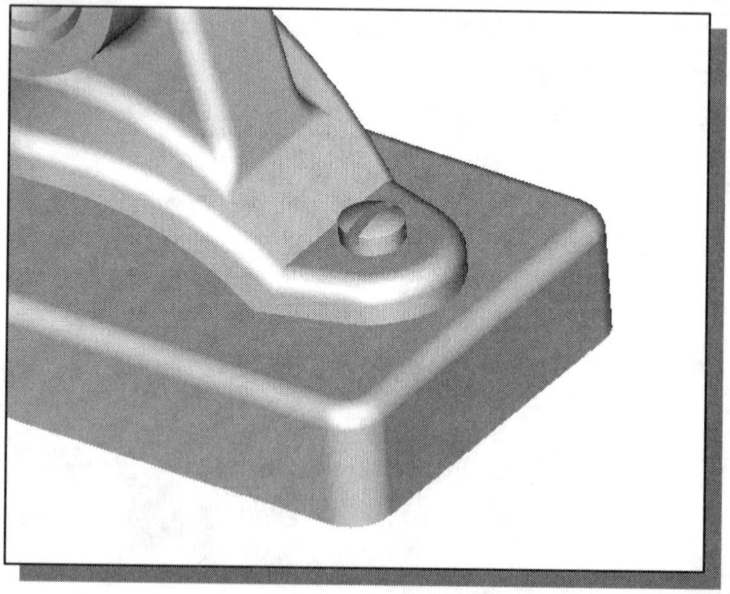

6. Repeat the steps and place another ***Cap-Screw*** into the assembly.

❖ Note that in an assembly model, multiple copies of the same part can be placed individually using different constraints. As the assembled part is modified, all copies of the same part are updated automatically. This approach of assembly modeling is extremely flexible and effective.

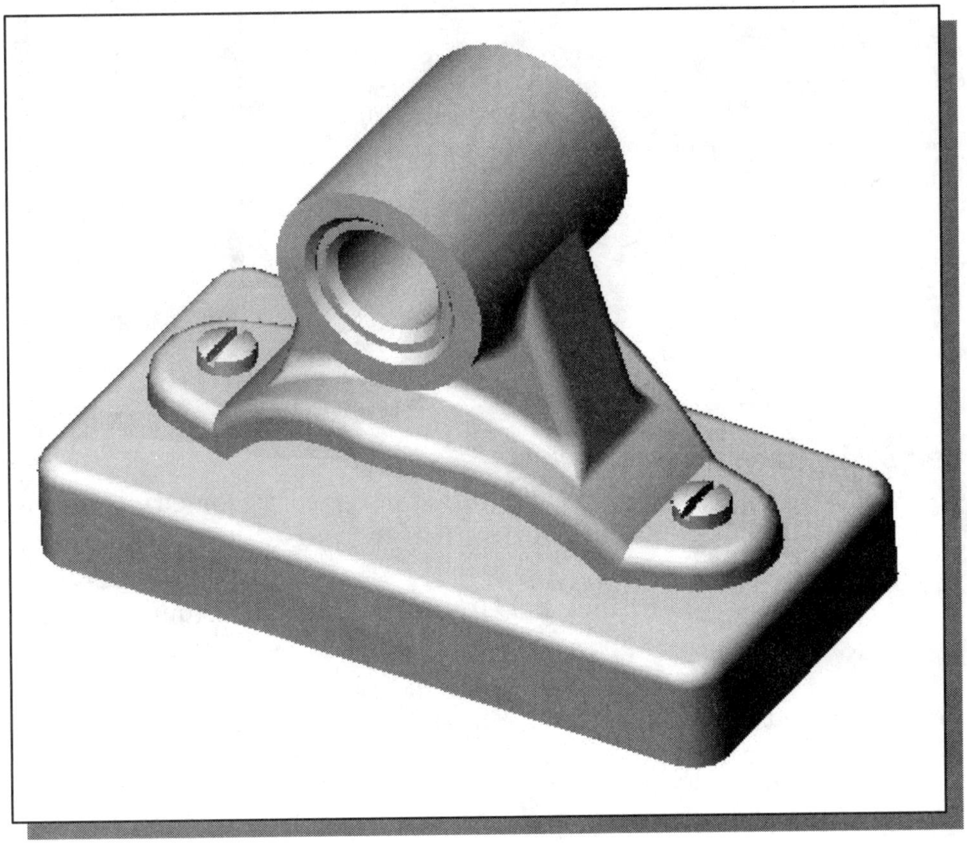

Exploding the Assembly

❖ Exploded assemblies are often used in design presentations, catalogs, sales literature, and in the shop to show all of the parts of an assembly and how they fit together.

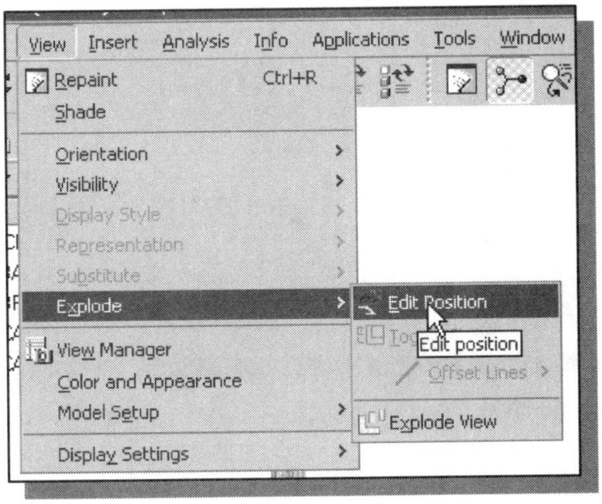

1. Select **View** in the pull-down menus and select:

   Explode → Edit Position

2. The message "*Select an axis or straight edge as the motion reference.*" is displayed in the message area. Pick the **center axis** of the ***Bushing***.

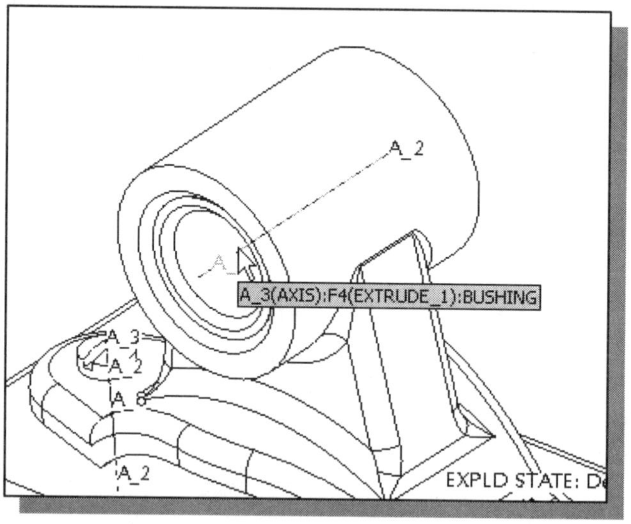

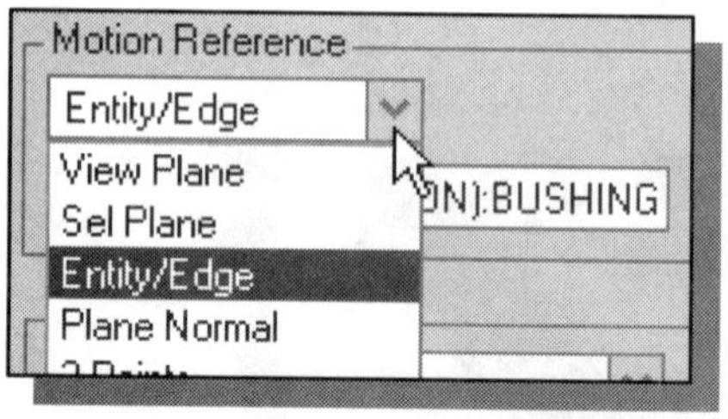

❖ Note that different types of geometry can be used as the **Motion Reference**.

3. Pick the ***Bushing*** part by selecting any edge of the part.

4. Move the mouse to reposition the ***Bushing*** part. Left-click once to accept the new location for the part.

5. Pick one of the ***Cap-Screws*** and notice it can also be moved in the same direction as the ***Bushing***.

6. Pick **OK** in the *Explode Position* window to end the command.

7. Select **View** in the pull-down menus and pick **EXPLODE → Unexplode** to restore the components to their assembled status.

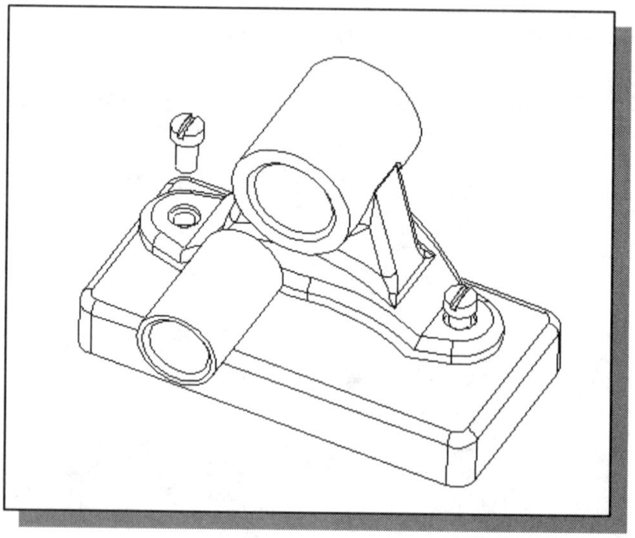

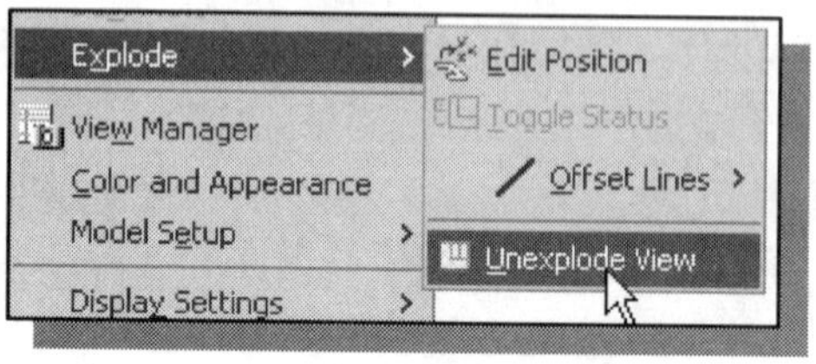

Bi-directional Associative Functionality

❖ The *bi-directional associative functionality* of *Pro/ENGINEER* allows us to change the design at any level, and the system reflects the change at all levels automatically.

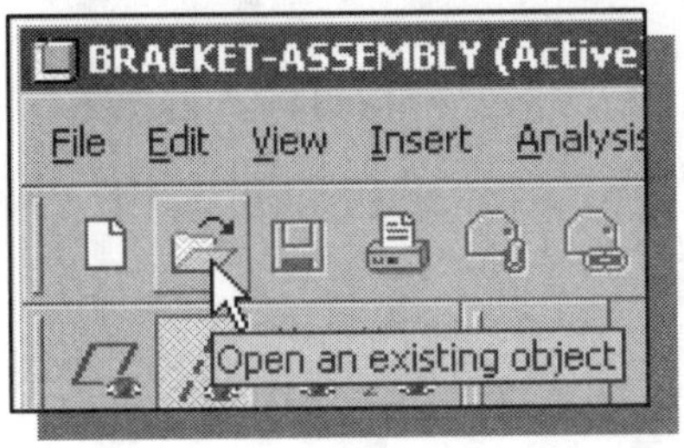

1. Pick the **Open** icon in the toolbar. (We can also use the key combination **Ctrl-O**.)

2. In the *Open* form, select the ***Bushing*** part file *(bushing.prt)* in the list window. Pick **Open** to retrieve the model.

3. On your own, modify the inside diameter to **0.5.** Confirm the part is updated before continuing to the next step.

❖ The associative functionality of *Pro/ENGINEER* assures the model is updated in all levels, including the assembly models.

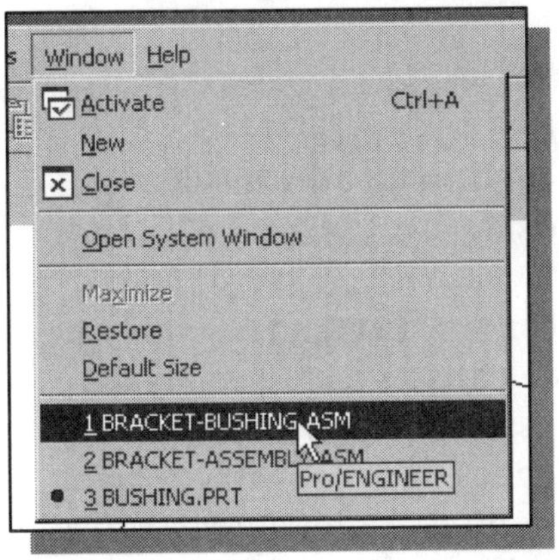

4. Select **Window** in the pull-down menus and pick **Bracket-bushing.ASM**. Notice the size of the ***Bushing*** is updated in the subassembly.

5. Select **Window** in the pull-down menus and pick **Bracket-Assembly.ASM**. Notice the size of the ***Bushing*** is updated in *Assembly* mode as well.

➢ We will next modify the dimension value in the *Bracket Assembly* window.

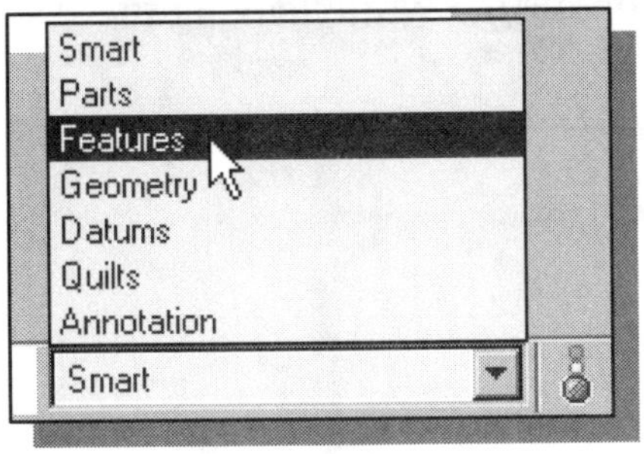

6. Pick **Features** in the *Selection filter* list as shown.

7. Double-click on the inside cylinder of the ***Bushing*** part to enter the ***Edit feature mode***.

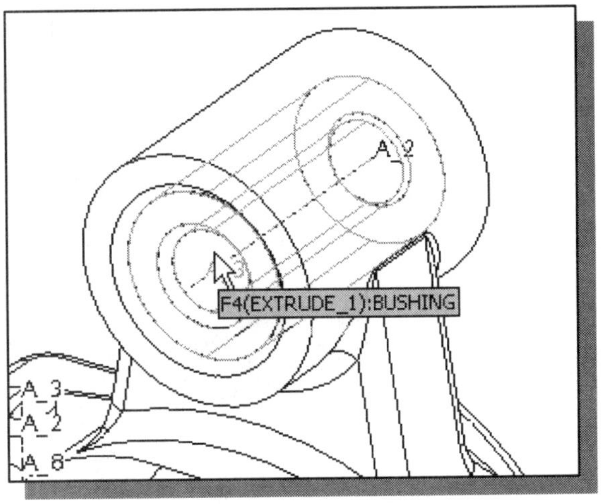

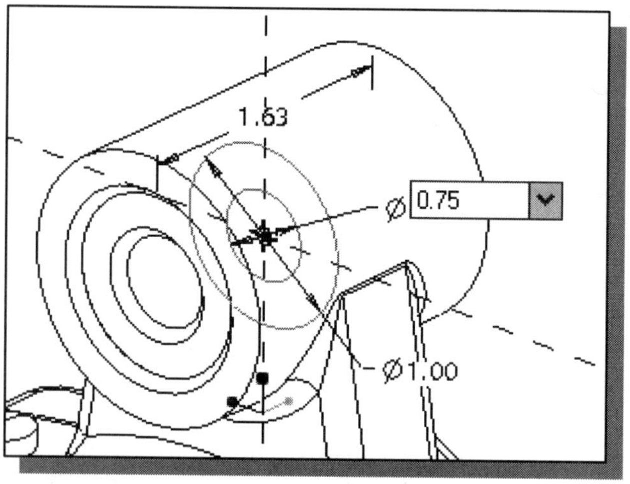

8. Modify the inside diameter by double clicking on the dimension text (**Ø 0.5**).

9. In the *value box*, enter **0.75** as the new diameter.

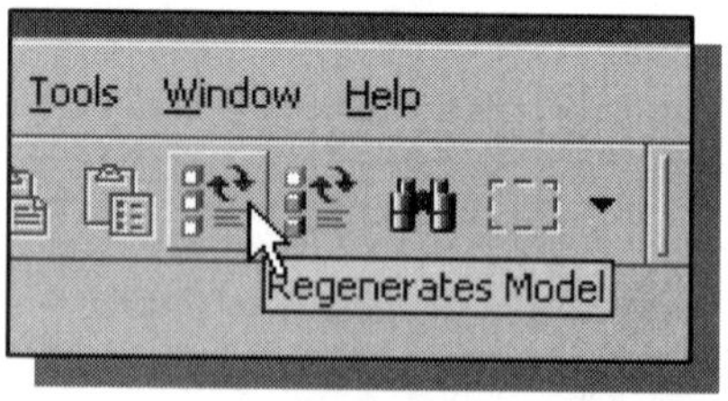

10. In the Standard toolbar area, select **Regenerate**.

11. On your own, switch to the other windows and confirm that the inside diameter of the ***Bushing*** part is updated in all models.

❖ With the full bi-directional associative functionality, when we **Save** an assembly in *Pro/ENGINEER*, any changes that we made to any of the parts in the assembly are also saved. When we save a part with any changes, the assembly is automatically updated. Consequently, we can experiment with any changes in a part without being concerned about saving the changes in the assembly. The *bi-directional associative functionality* allows us greater freedom to concentrate on design while leaving the tedious tasks to *Pro/ENGINEER*.

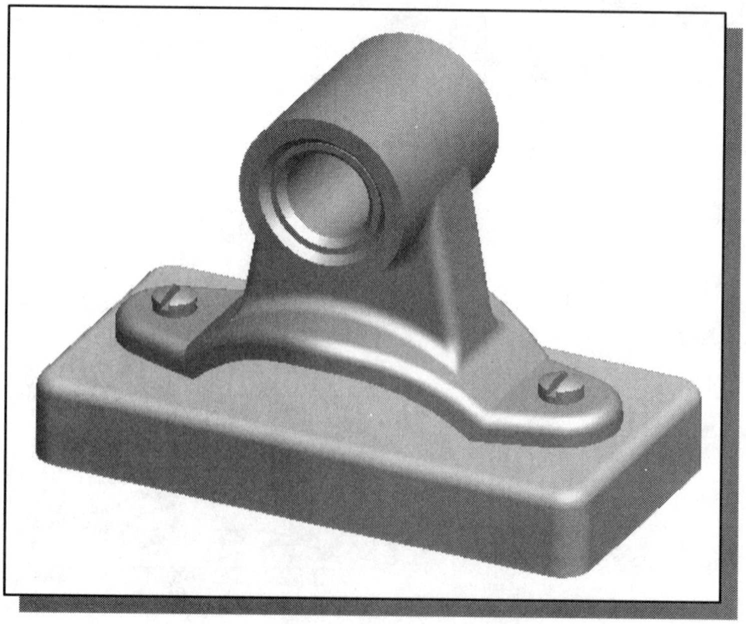

Conclusion

Design includes all activities involved, from the original concept to the finished product. Design is the process by which products are created and modified. For many years designers sought ways to describe and analyze three-dimensional designs without building physical models. With the advancements in computer technology, the creation of parametric models on computers offers a wide range of benefits. Parametric models are easier to interpret and can be easily altered. Parametric models can be analyzed using finite element analysis software, and simulation of real-life loads can be applied to the models and the results graphically displayed. The finalized solid models can also be used directly by manufacturing equipment to manufacture the product.

Throughout this text, various modeling techniques have been presented. Mastering these techniques will enable you to create intelligent and flexible solid models. The goal is to make use of the tools provided by *Pro/ENGINEER* and to successfully capture the ***design intent*** of the product. In many instances, only a single approach to the modeling tasks was presented; you are encouraged to repeat all of the tutorials and develop different ways of thinking in accomplishing the same tasks. We have only scratched the surface of *Pro/ENGINEER's* functionality. The more time you spend using the system, the easier it will be to perform parametric modeling with *Pro/ENGINEER*.

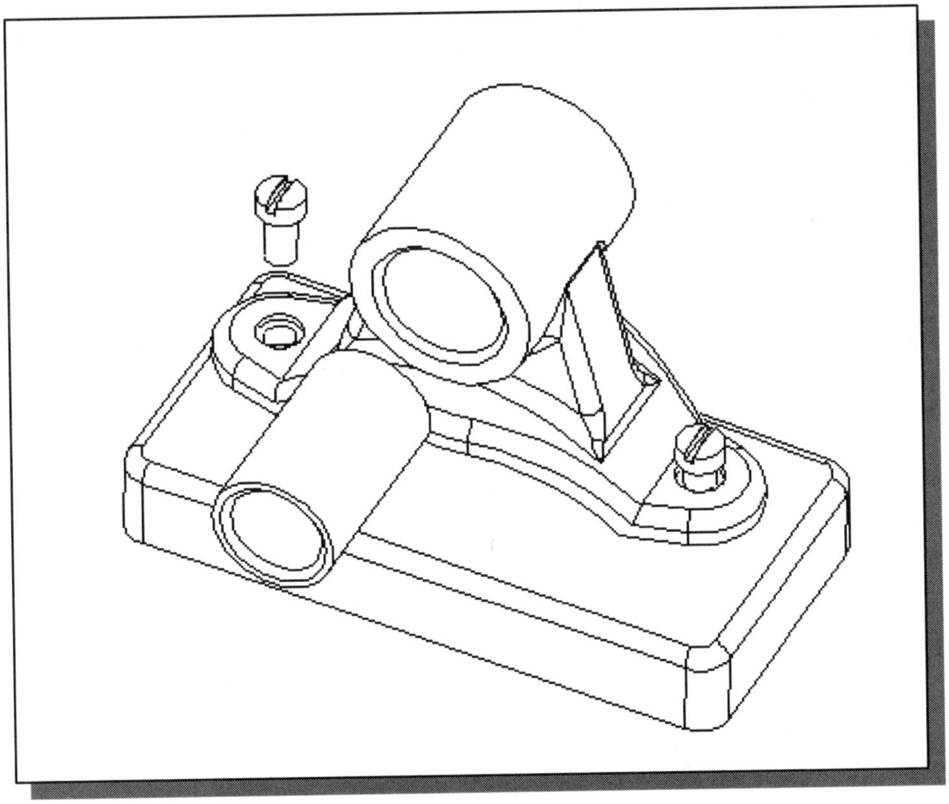

Questions:

1. What is the purpose of using *placement constraints*?
2. List four of the commonly used placement constraints.
3. Describe the difference between the **MATE** placement constraint and the **MATE OFFSET** placement constraint.
4. In an assembly, can we place more than one copy of a part?
5. How should we determine the assembling order of different parts in an assembly model?
6. How do we create an exploded assembly in *Pro/ENGINEER*?
7. In *Pro/ENGINEER*, how do we adjust the locations of parts in an exploded assembly?
8. Create sketches showing the steps you plan to use to create the four parts required for the assembly shown on the next page:

Ex.1)

Ex.2)

Ex.3)

Ex.4)

Exercise:

1. **Wheel Assembly** (Create a set of detail and assembly drawings. All Dimensions are in mm.)

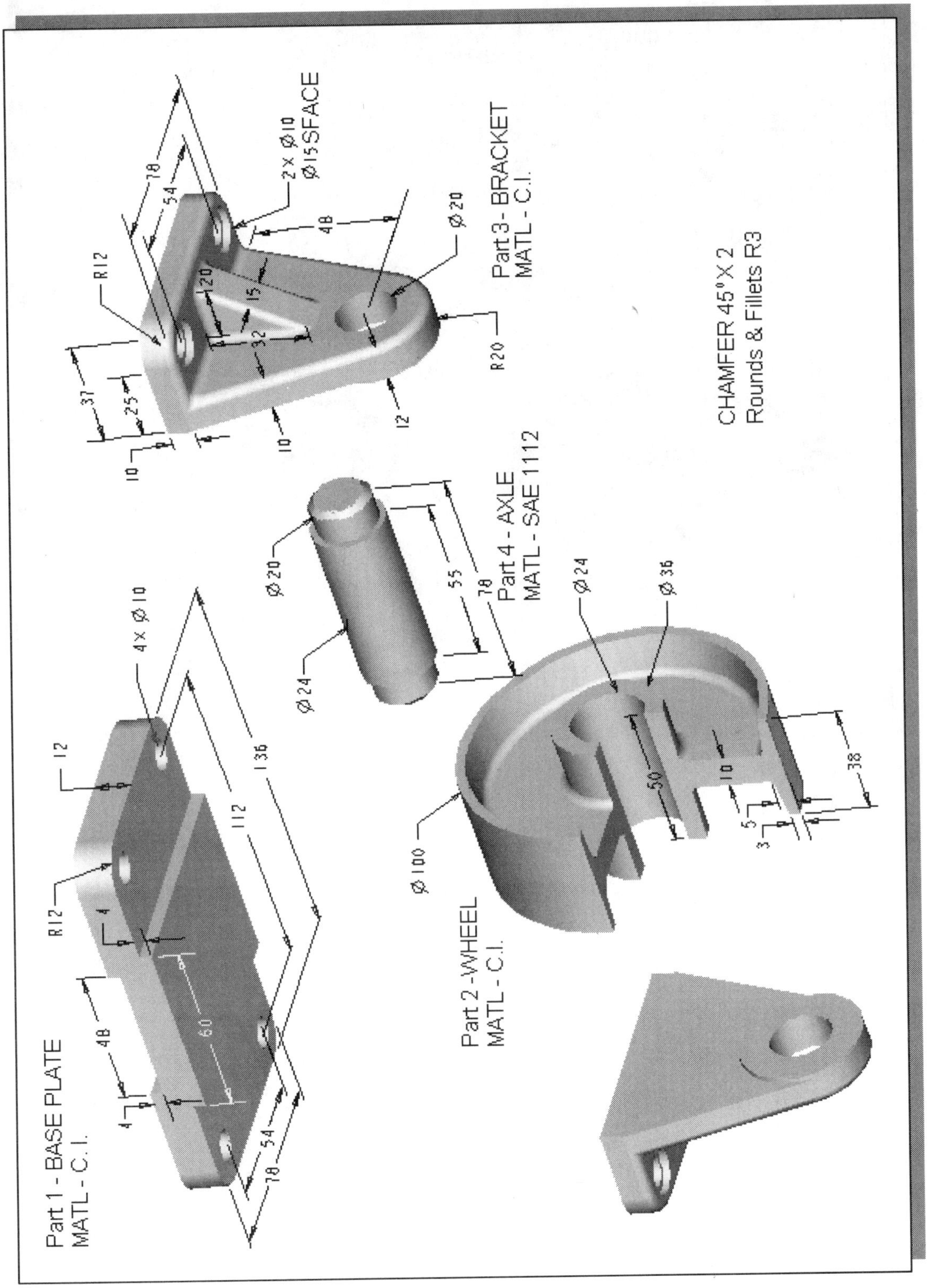

2. **Leveling Assembly** (Create a set of detail and assembly drawings. All Dimensions are in mm.)

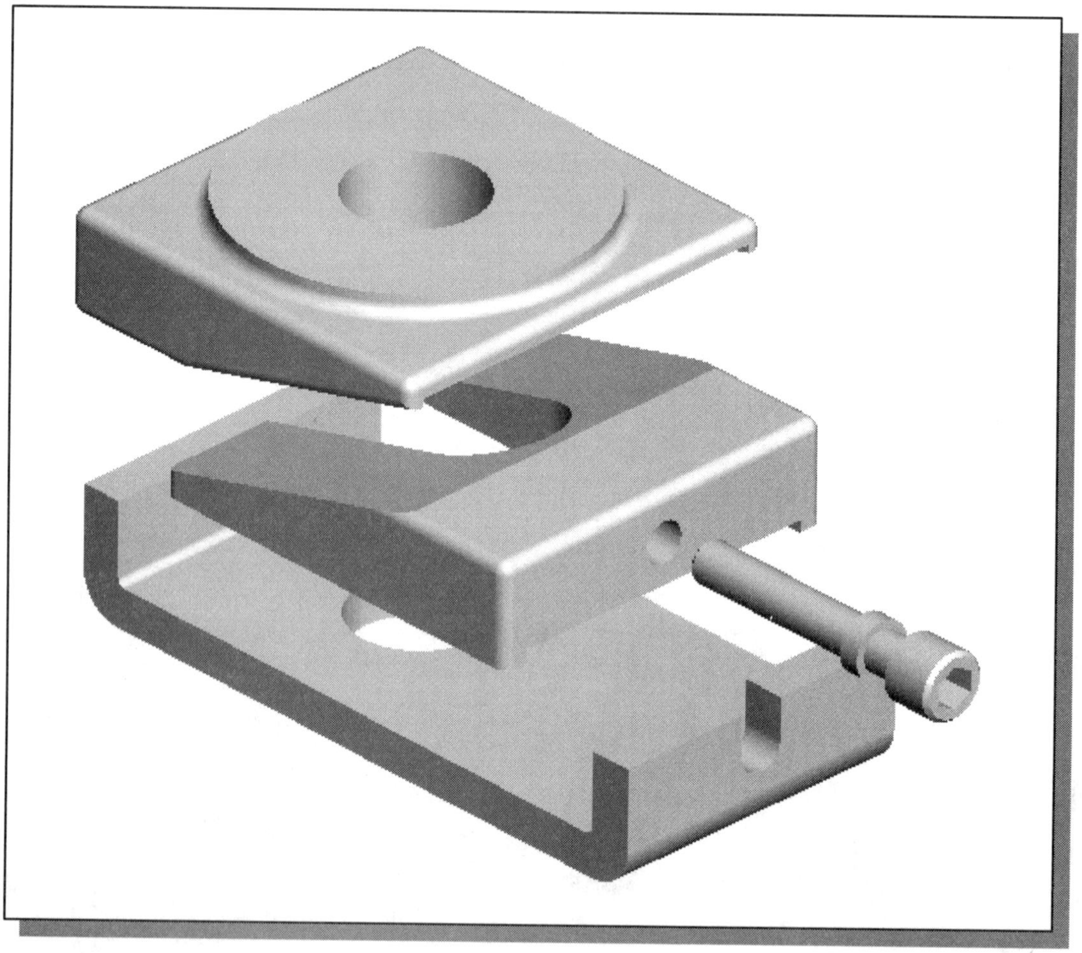

(a) Base Plate

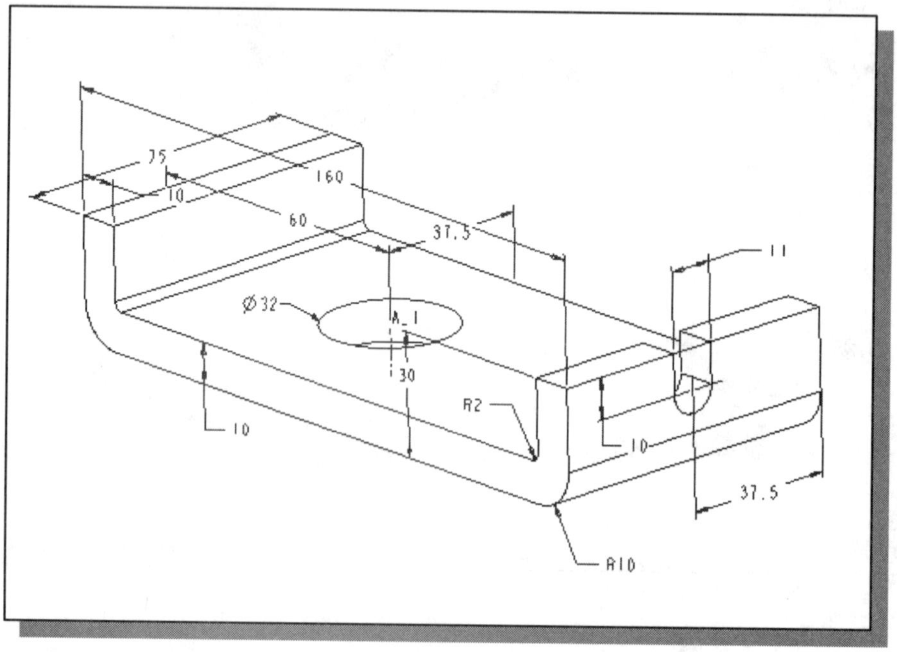

(b) **Sliding Block** (Rounds and Fillets: R3)

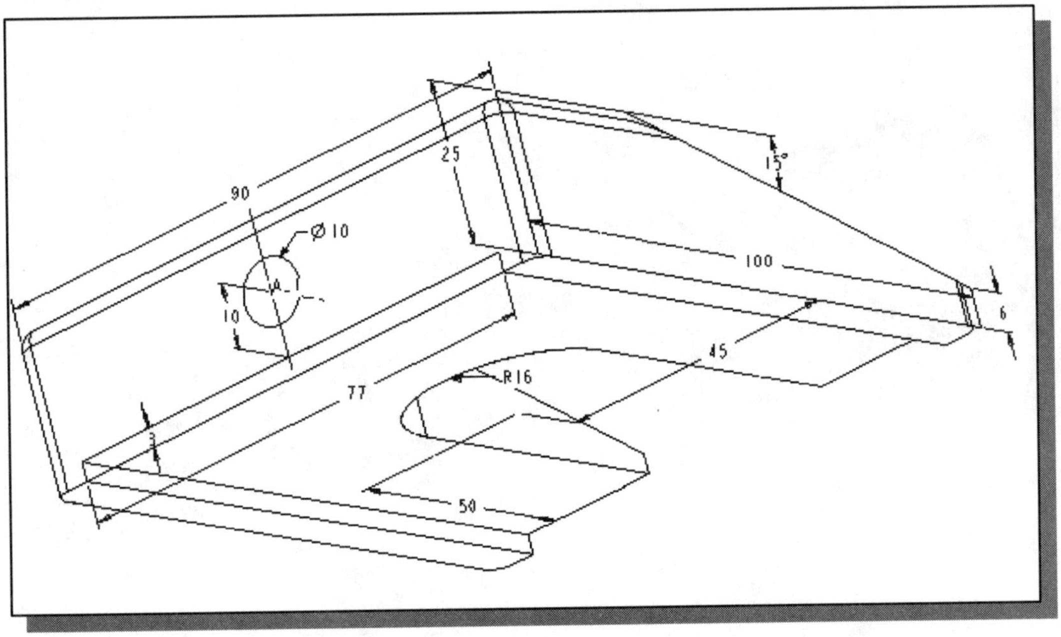

(c) **Lifting Block** (Rounds and Fillets: R3)

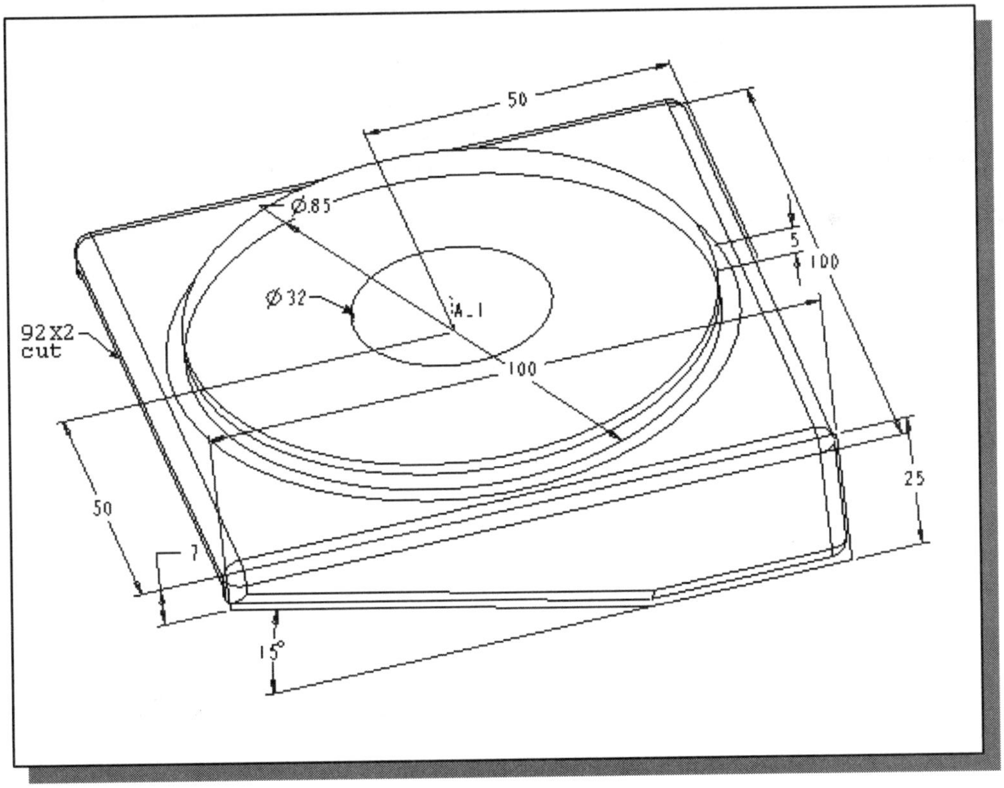

(d) **Adjusting Screw** (M10 × 1.5)

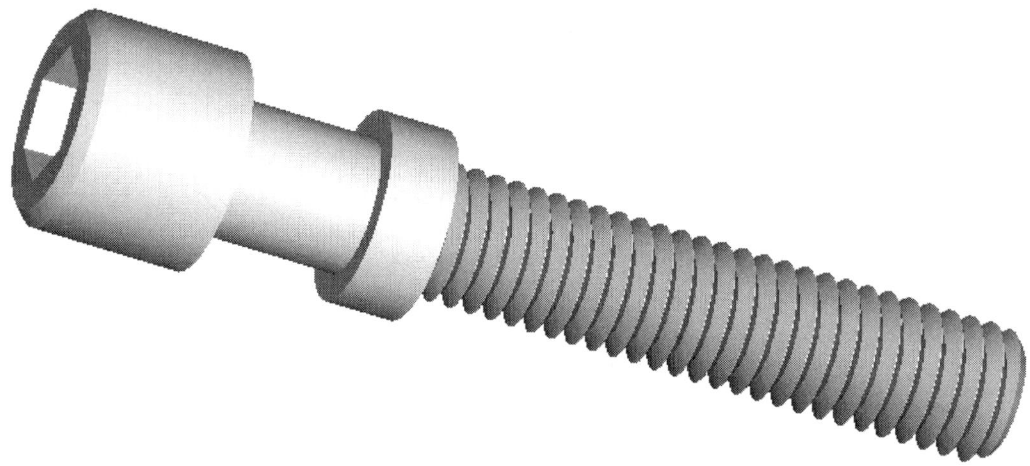

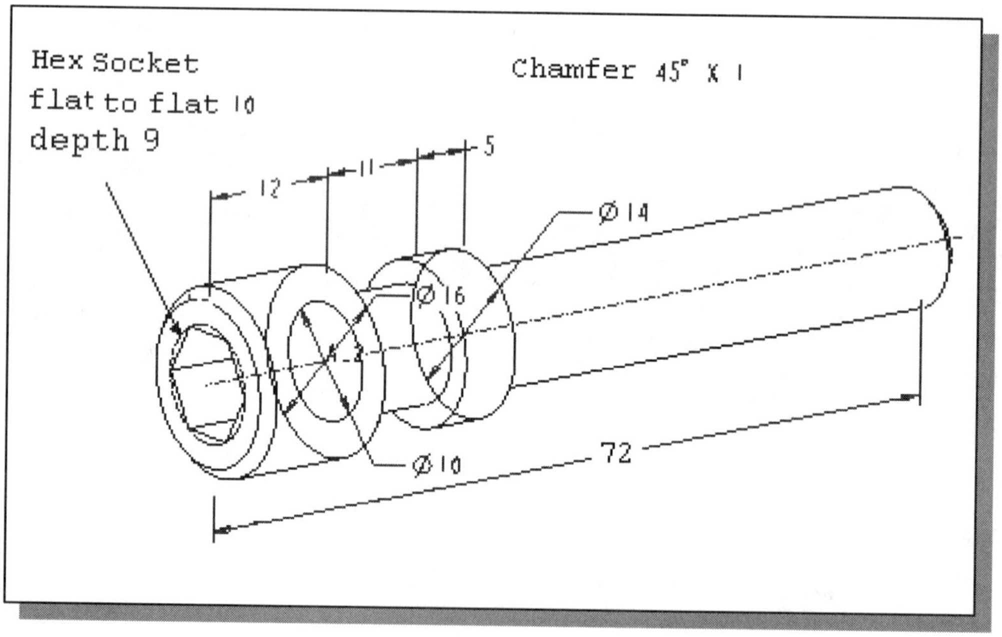

INDEX

E

F

G

H

I

L

M

N

O

P

R

S

T

Notes:

Notes:

Notes:

Notes:

Notes: